Progress in Inorganic Chemistry

Volume 58

Advisory Board

PROGRESS IN INORGANIC CHEMISTRY

Edited by

Kenneth D. Karlin

DEPARTMENT OF CHEMISTRY
JOHNS HOPKINS UNIVERSITY
BALTIMORE, MARYLAND

VOLUME 58

WILEY

Library of Congress Catalog Number: 59-13035

ISBN 978-1-118-79282-7

Printed in the United States of America

10 9 8 7 6 5 4 3 2 1

Contents

Tris(dithiolene) Chemistry: A Golden Jubilee

STEPHEN SPROULES

West CHEM, School of Chemistry, University of Glasgow, Glasgow, G12 8QQ, United Kingdom

CONTENTS

Progress in Inorganic Chemistry, Volume 58, First Edition. Edited by Kenneth D. Karlin.
© 2014 John Wiley & Sons, Inc. Published 2014 by John Wiley & Sons, Inc.

I. INTRODUCTION

The search for organometallic compounds with sulfur-donor ligands gave inorganic chemistry its first tris(dithiolene) coordination compound in 1963 (1). Anticipating a combination of CO and sulfur-donor ligands, King (1) apathetically described the product of the reaction of bis(trifluoromethyl)dithiete with molybdenum hexacarbonyl as a hexavalent metal coordinated to three as yet unidentified dithiols. The first bis(dithiolene) homologues with late transition metals appeared in the literature the previous year (2, 3). Seduced by the remarkable properties exhibited by these compounds, three research groups led the investigation in the early 1960s: Gray and his cohort at Columbia and then at Caltech; Schrauzer and co-workers in Munich, the Shell Development Company, and University of California at San Diego; and finally the Harvard quartet of Davison, Edelstein, Holm, and Maki. The competitive environment that ensued significantly advanced this emerging field into what we now know as transition metal dithiolene chemistry. Bis(dithiolene) compounds elicited greater interest than their tris(dithiolene) analogues despite both being strongly chromophoric, exhibiting multiple reversible electron-transfer processes, and possessing unprecedented molecular geometries. Bis(dithiolenes) were found to be persistently square planar (3–5), an outcome that could only arise from ligand participation in the frontier orbitals. Therefore, this sulfur–donor ligand with an unsaturated carbon backbone is regarded as the first noninnocent chelating ligand as it can exist in one of three forms: a dianionic dithiolate, a monoanionic dithienyl radical, and a neutral dithione (Scheme 1). Gray and co-workers (6, 7) worked on the premise that these were metal stabilized radical-ligand systems. Schrauzer and co-workers (8, 9) could never escape calling these dithioketones, whereas Holm, Maki, and co-workers (4, 10) avoided applying such definitive terms. An innovative compromise was brokered by McCleverty (11, 12) when he introduced the term dithiolene, obviating the need to specifiy discrete oxidation levels (13, 14). Only the metal was assigned, and for the archetypal bis(dithiolene) complexes of group 10 (VIII B) metals, it was unanimously agreed that a low-spin d^8 ion lay at the center of the neutral, monoanionic, and dianionic species, with each differing in the occupation of ligand-based valence orbitals.

The prosperity enjoyed by transition metal dithiolene complexes abruptly faded at the end of the 1960s. Interest in bis(dithiolene) compounds continued given their structural resemblance to tetrathiafulvalene (TTF), which pushed this field into the new areas of photonics and conductitvity (15). In contrast, tris(dithiolenes) chemistry languished until an unexpected revival in the mid-1990s. The spark came from the discovery of dithiolene ligands in biological systems (16, 17). The protein structure of an aldehyde:ferridoxin oxidoreductase

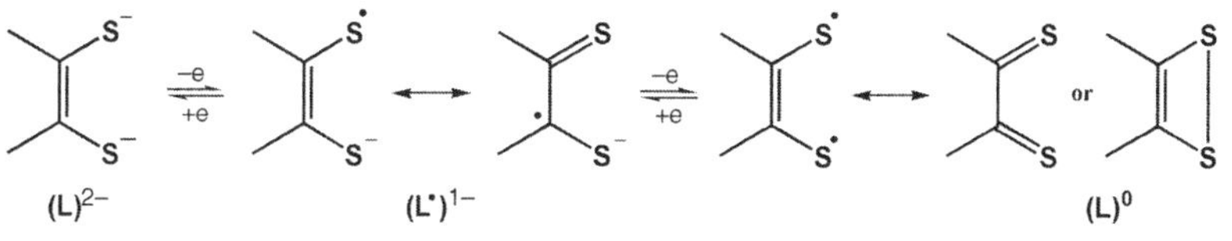

Scheme 1.   The three oxidation levels for a dithiolene ligand (L = dithiolene).

consisting of a tungsten ion coordinated by two dithiolate ligands did for dithiolene chemistry what the Cambrian Explosion did for life on Earth (18). Oxo-molybdenum and -tungsten bis(dithiolene) synthetic dead ends were now in vogue as small molecule analogues for the active sites of oxotransferase enzymes (19–23). The fortunes of tris(dithiolene) compounds were similarly transformed. The last decade has been the most insightful in the 50-year history of tris(dithiolene) chemistry as these old compounds became the subject of scrutiny by new techniques, modern instrumentation, and advanced computational methodology (24–26). The focus was on their molecular and electronic structures that had never been completely resolved. The following account tracks the history and evolution of tris(dithiolene) chemistry in this its golden jubilee year.

II. LIGANDS

Dithiolene ligands can be categorized into three groups based on the nature of the C–C bond in the elementary $\{S_2C_2\}$ unit: (1) arene dithiolates where the double bond is part of an aromatic system (Fig. 1a); (2) alkene dithiolates with an olefinic double bond (Fig. 1b); and (3) neutral dithiones with a C–C single bond and unsaturated S–C bonds (Fig. 1c). The synthetic route to dithiolene ligands depends largely on the metal ion it binds. Arene dithiolates are traditionally isolated as dithiols. The analogous alkene-1,2-dithiols are unstable (27–29), and preferably handled in situ as pro-ligands or alkali salts before combining with an appropriate metal reagent. The recurrent step in all ligand synthesis is protection of the 1,2-dithiolate unit, which takes on a multitude of forms from simple protonation, to alkyl, ketyl, thione, and silyl entities that prevent exposed sulfur atoms from partaking in counterproductive side reactions. The protecting groups also allow the carbon backbone to be functionalized so that the physical properties of the final dithiolene ligand and complex are tailored to suit the desired application.

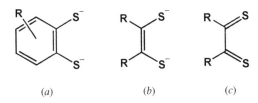

(a)   (b)   (c)

Figure 1.   General classes of dithiolene ligand.

A.   Arene Dithiolates

The archetypal member of this group of ligands is the ubiquitous benzene-1,2-dithiolate, $(bdt)^{2-}$. It first appeared in the literature in 1966 in the synthesis of square-planar bis(dithiolene) complexes with Co, Ni, Cu (30), and was generated by treating o-dibromobenzene with cuprous butyl mercaptan (31, 32), followed by cleavage of the thioether with sodium in liquid ammonia (Scheme 2a) (33). A high-yielding synthesis involving the addition of sodium 2-propanethiolate to 1,2-dichlorobenzene in dimethylacetamide at 100 °C produces 1,2-bis(isopropylthio)benzene, which is readily deprotected to form benzene-1,2-dithiol, bdtH$_2$ (Scheme 2b) (34, 35). The procedure was updated some 30 years later with the ortholithiation of thiophenol, and subsequent reaction with elemental sulfur followed by acidification, which gave large quantities of dithiol (Scheme 2c) (36, 37). In both procedures, specific functional groups can be introduced in the first stage, for example, the preparation of veratrole-4,5-dithiol (vdtH$_2$) from 1,2-dimethoxybenzene following Scheme 2a (38), and 3,4,5,6-tetrafluorobenzene-1,2-dithiol (bdtF$_4$H$_2$) from 1,2,3,4-tetrafluorobenzene following Scheme 2b (39).

All known arene dithiolate ligands utilized in the formation of tris(dithiolene) complexes are presented in Fig. 2. Dithiol versions of toluene-3,4-dithiolate, $(tdt)^{2-}$, xylene-4,5-dithiolate, $(xdt)^{2-}$, and the crown ether homologues can be prepared from their corresponding o-dibromo precursors following Scheme 2a (30, 38, 40). Alkyl protection of the thiolate groups followed by ortholithiation leads to 3,6-bis(trimethylsilyl)ethene-1,2-dithiolate, $(tms)^{2-}$ (41). This methodology has been exploited by Kreickman and Hahn (42) to generate an

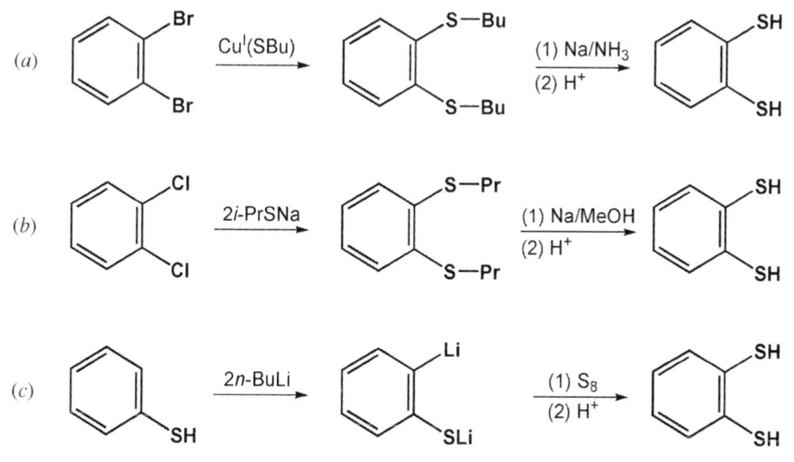

Scheme 2.   Preparation of benzene-1,2-dithiol.

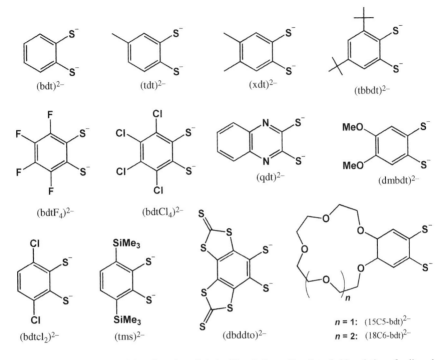

Figure 2.   Arene-1,2-dithiolate ligands and their abbreviations. (See list of abbreviations for ligand identification.)

inventory of $(bdt)^{2-}$ moieties linked to each other or catecholate, $(cat)^{2-}$, via an amide bridge. The ligands pertinent to this topic are displayed in Fig. 3.

Tetrachlorobenzene-1,2-dithiol ($bdtCl_4H_2$) is prepared by boiling hexachlorobenzene with sodium sulfide and iron powder in N,N'-dimethylformamide (DMF) (30, 43). Addition of base precipitates the iron compound, $[Fe(bdtCl_4)_2]_n$, and treatment with ZnO in boiling MeOH liberates $bdtCl_4H_2$. The four-step synthesis of 3,5-di-*tert*-butylbenzene-1,2-dithiol, tbbdtH₂, starts with commercially available 3,5-di-*tert*-butyl-2-aminobenzoic acid as outlined in Scheme 3 (44). Sulfur-rich 2,5-dithioxobenzo[1,2-*d*:3,4-*d'*]bis[1,3]dithiolene-7,8-dithiolate, $(dbddto)^{2-}$, begins with thiolation of hexachlorobenzene with benzylmercaptan to form hexakis(benzylthio)benzene (45). Treatment with sodium in liquid ammonia followed by protonation affords benzenehexathiol. Only the hexathiol reacts with carbon disulfide in pyridine to give nearly quantitative yields of the pyridinium salt of 7-mercapto-2,5-dithioxobenzo[1,2-*d*:3,4-*d'*]bis[1,3]dithiole-8-thiolate, the precursor to $(dbddto)^{2-}$ (46). Quinoxaline-2,3-dithiol, qdtH₂, is conveniently formed in the reaction of 2,3-dichloroquinoxaline with excess thiourea in refluxing ethanol (47).

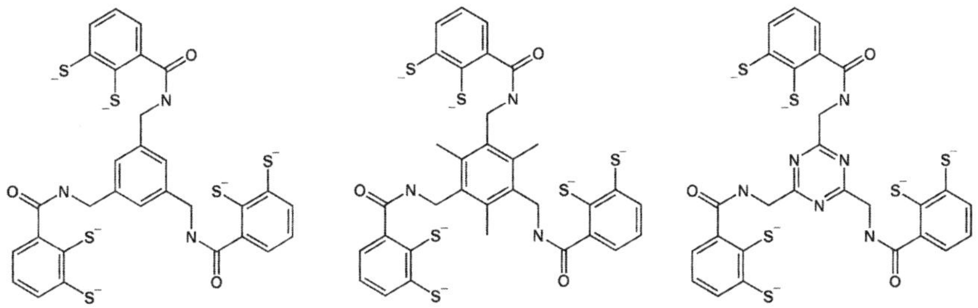

$(bn\text{-}bdt_3)^{6-}$ $(ms\text{-}bdt_3)^{6-}$ $(tr\text{-}bdt_3)^{6-}$

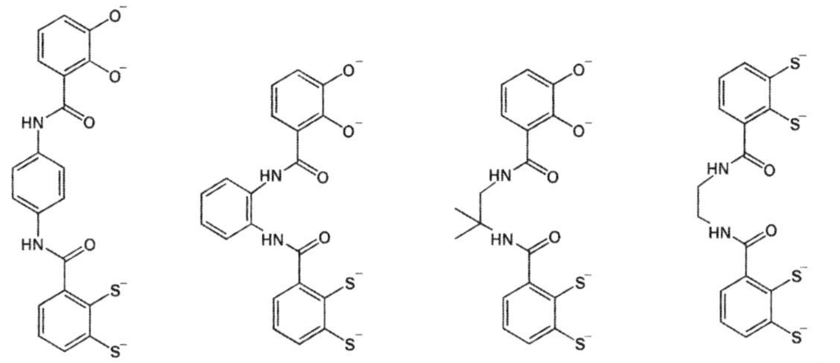

(bdt-**1**-cat)$^{4-}$ (bdt-**2**-cat)$^{4-}$ (bdt-**3**-cat)$^{4-}$ (bdt-**4**-bdt)$^{4-}$

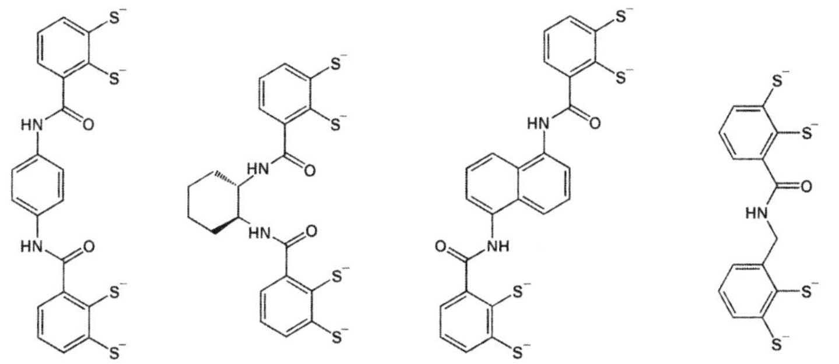

(bdt-**5**-bdt)$^{4-}$ (bdt-**6**-bdt)$^{4-}$ (bdt-**7**-bdt)$^{4-}$ (bdt-**8**-bdt)$^{4-}$

Figure 3. Benzene-1,2-dithiolate-based polydentate ligands. (See list of abbreviations for ligand identification.)

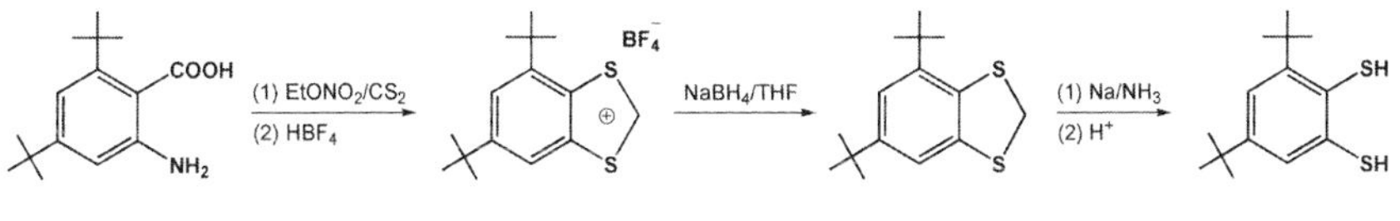

Scheme 3. Preparation of $tbbdtH_2$.

B. Alkene Dithiolates

The tremendous variety of dithiolene ligands with an alkene backbone presented in Fig. 4 highlight the desire to adapt the basic motif in order to equip the complex with specific properties. They have been grouped here based on the sulfur source used in preparing the ligand, principally elemental sulfur, phosphus pentasulfide, and carbon disulfide.

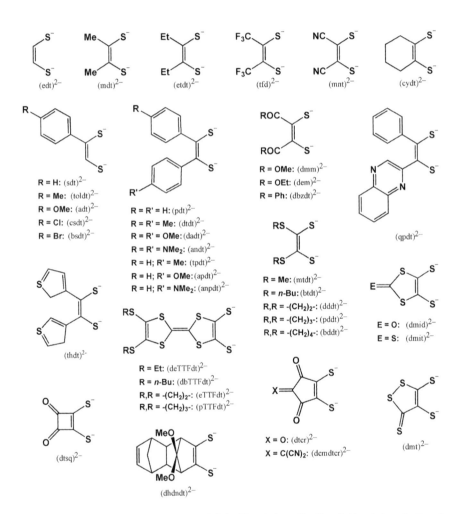

Figure 4.   Alkene-1,2-dithiolate ligands and their abbreviations. (See list of abbreviations for ligand identification.)

1. Sulfur

With the notable proficiency by which sulfur atoms readily bond to each other, it is rather surprising to find only one genuine example of a dithiolene ligand that is introduced to a metal as a dithiete (Fig. 5a) (1, 10). The synthesis of bis(trifluoromethyl)dithiete is conducted under conditions that would violate modern Health and Safety protocols: hexafluoro-2-butyne is bubbled through molten sulfur, after which the malodorous and poisonous dithiete is obtained as a liquid via fractional distillation of the reaction mixture (48). Its structure as that of a dithiete rather than a dithione was confirmed by vapor-phase X-ray diffraction revealing an S−S distance of 2.05 Å (49). Bis(trifluoromethyl)dithiete is an oxidizing agent and it is readily reduced by sodium to form the dithiolate, Na_2tfd (50).

The paucity of dithiete entities stems from the propensity of sulfur atoms to target neighboring molecules to form oligomeric and polymeric mixtures (51). The reported preparation of benzene-1,2-dithiete (52) was revised as a mixture of sulfur-bridged species, the smallest being bis(o-phenylene)tetrasulfide (53). Sulfuryl chloride oxidation of $[Cp_2Ti(dmm)]$ (54), where $(dmm)^{2-}$ is dimethylmaleate-2,3-dithiolate and Cp is cyclopentadienyl, releases dimethy-1,2-dithiete dicarboxylate (Fig. 5b) (55). This dithiete is sufficiently stable for X-ray diffraction studies (Fig. 6), and has been characterized with an S−S bond length of 2.07 Å.

Both dithiobenzil and 4,4′-bis(dimethylamino)dithiobenzil have been generated by irradiation of the corresponding dithiocarbonate releasing CO (56). It is speculated that the former exists exclusively as diphenyl-1,2-dithiete, whereas the latter is a dithione (Fig. 5c). In the presence of $[Mo(CO)_6]$, dark green, neutral tris(dithiolene) compounds are isolated from the reaction (56). Similarly, photolysis of 3-(methylthio)-5,6-tetramethylene-1,4,2-dithiazine yields tetramethylene-dithiete/cyclobutanedithione, which is scavenged by $[Mo(CO)_6]$ to form $[Mo(cydt)_3]$ $(cydt^{2-} = cyclohexene-1,2-dithiolate)$ (57).

Schrauzer and Mayweg (3, 58) stumbled into dithiolene chemistry via the esoteric reaction of nickel sulfide and "tolan", more commonly known as diphenylacetylene, which produced a black crystalline solid formulated

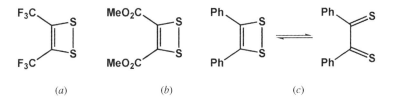

      (a)   (b)   (c)

Figure 5. Representative structures of (a) bis(trifluoromethyl)dithiete, (b) dimethyl-1,2-dithiete dicarboxylate, and (c) the equilibrium between diphenyl-1,2-dithiete and dithiobenzil.

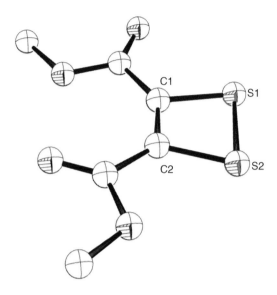

Figure 6.   Molecular structure of dimethyldithiete dicarboxylate.

$NiS_4C_4Ph_4$. The one-pot reaction produced the first neutral bis(dithiolene)nickel complex that the authors described as square planar and diamagnetic (3). Although Schrauzer's laboratory would divert to an alternative method of complex synthesis (8, 9), their approach was used by others investigating the reaction of unsaturated organic molecules with sulfur in the presence of metal ions. Dimethyl- and diethyl-acetylene dicarboxylate are considered activated because the highly electron-withdrawing ester substituents weaken the triple bond. Therefore, they are primed to react with metal–sulfur units (e.g., $\{MS_2\}$ and $\{M(S_4)\}$) to form a five-membered ring: A metallodithiolene. This procedure has successfully produced V, Mo, and W complexes with three $(dmm)^{2-}$, diethylmaleate-2,3-dithiolate, $(dem)^{2-}$, dibenzoylethene-1,2-dithiolate $(dbzdt)^{2-}$, bis(trifluoromethyl)ethene-1,2-dithiolate $(tfd)^{2-}$, and 1-quinoxalyl-2-phenylethene-1,2-dithiolate $(qpdt)^{2-}$ ligands in mediocre yields (50, 59–64).

2.   Carbon Disulfide

The most widely encountered ligand in dithiolene chemistry is 1,2-dicyanoethene-1,2-dithiolate abbreviated $(mnt)^{2-}$, and so-named from the cis-orientation of the cyanide substituents found in maleonitrile. The disodium salt of maleonitriledithiolate is easily prepared from the combination of sodium cyanide and carbon disulfide in DMF to form $[S_2CCN]^{1-}$ (Eq. 1). Two equivalents of this adduct decompose to give

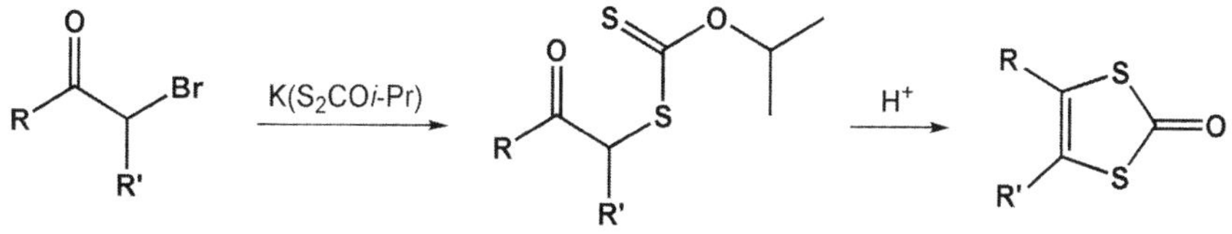

Scheme 4.　Synthetic route to 1,3-dithiole-2-one.

Na$_2$mnt and elemental sulfur (65).

$$NaCN + CS_2 + 3HC(O)NMe_2 \rightarrow NCCS_2Na \cdot 3HC(O)NMe_2$$
$$2NCCS_2Na \cdot 3HC(O)NMe_2 \rightarrow Na_2S_2C_2(CN)_2 + 6HC(O)NMe_2 + 2S \qquad (1)$$

Despite the noted virtues of the thiophosphorester synthetic approach, it is restricted to just a handful of commercially obtainable acyloins. Modifying the ligand appendages to alter solubility, electronics, or sterics, relies on a different tactic, and the methods providing the most variety are syntheses of functionalized 1,3-dithiole-2-ones or vinylene dithiocarbamates (Scheme 4).

These are produced by combining an α-bromoketone with the sulfiding agent potassium o-isopropyl xanthate, K(S$_2$COi-Pr) (66), forged in the reaction of isopropyl alcohol, potassium hydroxide, and carbon disulfide (67). This procedure leads to phenylethene-1,2-dithiolate, (sdt)$^{2-}$, and its analogues tolylethene-1,2-dithiolate, (toldt)$^{2-}$, anisylethene-1,2-dithiolate, (adt)$^{2-}$, p-chlorophenylethene-1,2-dithiolate, (csdt)$^{2-}$, and p-bromophenylethene-1,2-dithiolate, (bsdt)$^{2-}$ (68); bis(3-thienyl)ethene-1,2-dithiolate, (thdt)$^{2-}$ (69); several substituted 1,2-diphenyl-1,2-dithiolates: 1,2-ditolylethene-1,2-dithiolate, (dtdt)$^{2-}$, 1,2-dianylethene-1,2-dithiolate, (dadt)$^{2-}$, 1-tolyl-2-phenylethene-1,2-dithiolate, (tpdt)$^{2-}$, 1-anisyl-2-phenylethene-1,2-dithiolate, (apdt)$^{2-}$, and 1-anilyl-2-phenylethene-1,2-dithiolate, (anpdt)$^{2-}$, (68, 70); and a more tractable form of 1,2-dimethylethene-1,2-dithiolate, (mdt)$^{2-}$ (71). The latter was crystallographically characterized with average S$-$C and C$-$C distances within the {S$_2$C$_2$} unit of 1.754(1) Å and 1.340 (2) Å, respectively (72), offering baseline bond lengths in the "free" ligand (Fig. 7). Base hydrolysis cleaves the ketyl protecting group affording the dianionic dithiolate ligand poised for complexation.

Dithiolene ligands saturated with sulfur atoms find favor in electrically conducting salts and charge-transfer complexes (15, 73, 74). The motivation for ligands of this type followed the discovery of heterocyclic TTF (C$_6$H$_4$S$_4$) described as an "organic metal" (75), and bearing a striking resemblance to bis(dithiolene) complexes (Fig. 8), and therein the anionic coordination complexes provide an ideal complement to the TTF radical cation (15, 73).

The analogy motivated a new synthetic direction in the preparation of hetero-cyclic dithiolene ligands to design coordination complexes with enhanced

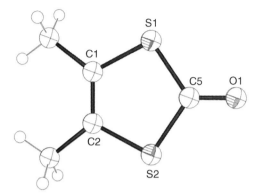

Figure 7.   Molecular structure of 4,5-dimethyl-1,3-dithiole-2-one.

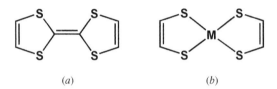

(a)   (b)

Figure 8.   Structure of (a) TTF and (b) a generic bis(dithiolene) complex.

electronic, photonic, and magnetic properties (15, 77). The progenitor of this sulfur-rich collection of ligands is 1,3-dithiole-2-thione-4,5-dithiolate abbreviated (dmit)$^{2-}$ from its original name dimercaptoisotrithione (78). The preparation is exceedingly simple: Carbon disulfide is reacted with an alkali metal (Na or K) in DMF (79, 80). Importantly, the formation of a carbon disulfide adduct (a thioxanthate) has provided a simple route to multi-gram amounts of the ligand. The ligand is stabilized when ZnSO$_4$ is added in the final step to generate [Zn(dmit)$_2$]$^{2-}$ salts (Scheme 5) (81), an efficient ligand delivery reagent (78, 82). The ligand has also been structurally characterized as an air-sensitive NMe$_4$$^+$ salt where again the S−C and C−C distances of 1.724(6) and 1.371(8) Å, respectively (83), offer baseline intraligand bond lengths (Fig. 9).

A variety of ligands are generated from (dmit)$^{2-}$, such as 2-oxo-1,3-dithiole-4,5-dithiolate, (dmid)$^{2-}$ (84), and 1,2-dithiole-3-thione-4,5-dithiolate, (dmt)$^{2-}$ (85, 86), by protecting the thiolates with benzoyl groups (81, 86, 87), or alkyl-substituted 1,4-dithiin-2,3-dithiolates by alkylating the dithiolate, followed by converting and cleaving the thione (Scheme 5) (79, 88, 89). The TTF based dithiolates have also been prepared from P(OEt)$_3$ promoted coupling of protected (dmit)$^{2-}$ and (dmid)$^{2-}$ moieties (Scheme 5) (90, 91).

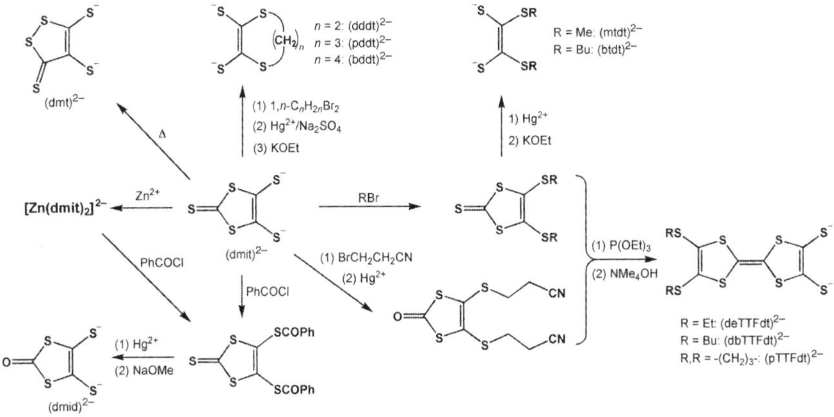

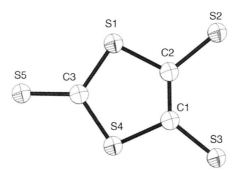

Scheme 5.   Derivatives of (dmit)$^{2-}$.

Figure 9.   Molecular structure of (NMe$_4$)$_2$dmit.

3.   Phosphorus Pentasulfide

Schrauzer and Finck (92) discovered in their one-pot synthesis of [Ni(pdt)$_2$] that dithiobenzil cannot be isolated; rather it is stabilized by forming covalent bonds with itself or a transition metal ion. The scale of the reaction was improved by heating the α-hydroxybenzoin with phosphorus pentasulfide in xylene or dioxane (Scheme 6) (9, 93, 94). The dithiolene is stabilized as a thiophosphorester that readily relinquishes the ligand to a waiting metal center (94). The poor yields are offset by the multi-gram scale of the reaction and inexpensive reagents (95). A variety of substituted α-hydroxyketones (acyloins) can be used to generate a range of dithiolene substitution patterns, for example acetoin to form complexes with (mdt)$^{2-}$ (96).

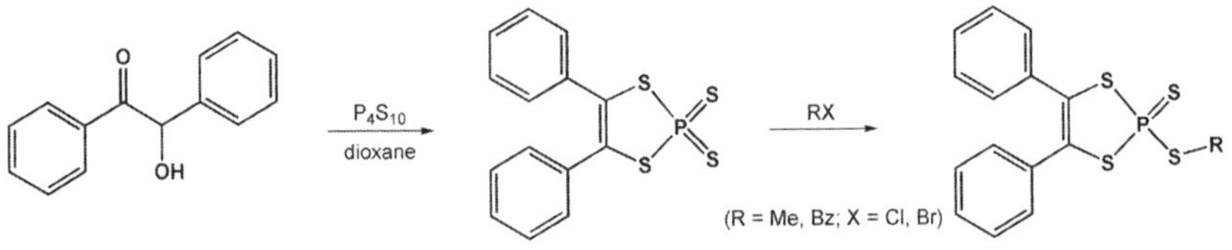

Scheme 6.   Preparation of thiophosphoresters and their alkyl stabilized derivatives.

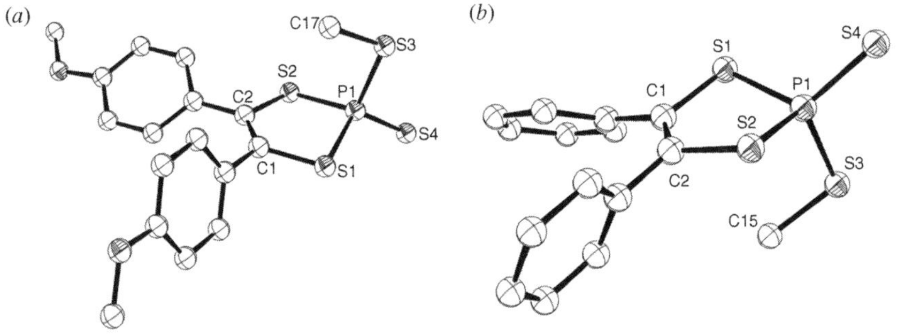

(a)   (b)

Figure 10.   Crystallographic structures of (a) (dadt)P(S)(SMe) and (b) (pdt)P(S)(SMe).

Recently, Donahue and co-workers (95) confirmed the constitution of these thiophosphoresters by adding an alkylating agent to the amber reaction mixture (Scheme 6). This reaction affords more tractable thiophosphoryl thiolates, $(R_2C_2S_2)P(S)(SR')$ [R' = Me, Bz (benzyl)], as precursors to 1,2-diphenylethene-1,2-dithiolate, $(pdt)^{2-}$, and $(dadt)^{2-}$. Several have been structurally characterized and the average S−C (1.773 Å) and C−C (1.342 Å) bond lengths serve as useful benchmarks of the intraligand distances of alkene-1,2-dithiolates in the absence of a metal ion (Fig. 10). The dithiolene ligand can be liberated from the thiophosphoryl thiolate by straightforward base hydrolysis. The benefit is that well-defined stoichiometric amounts of ligand can be added to metals circumventing bis- and tris (dithiolene) thermodynamic dead ends (22, 23, 95).

4.   Other Sulfur Sources

The simplest of all dithiolenes, ethene-1,2-dithiolate, $(edt)^{2-}$, is derived from the combination of cis-1,2-dichloroethylene, benzoyl chloride, and thiourea (Scheme 7) (29). The benzoyl-protected thioether is cleaved via base hydrolysis to give multi-gram quantities of $(Li/Na)_2edt$ (97). The diethyl-substituted analogue is formed via a Pd catalyzed cross-coupling of bis(triisopropylsilyl)disulfide and hex-3-yne to give 3,4-bis(triisopropylsilanylsulfanyl)hex-3-ene (98). The silyl groups are jettisoned during the reaction with the metal precursor. This method

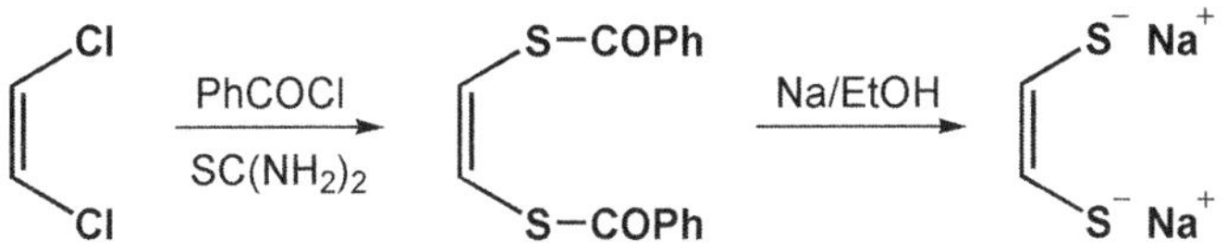

Scheme 7. Synthesis of Na$_2$edt.

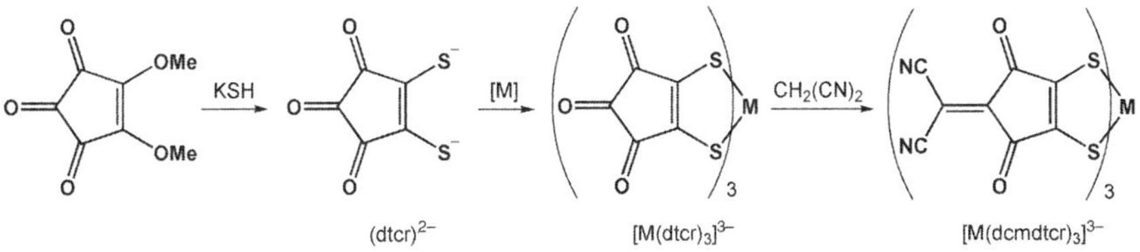

Scheme 8. Preparation of (dtcr)$^{2-}$ and (dcmdtcr)$^{2-}$ complexes with M = Cr(III), Fe(III), and Co(III).

has been used to assemble mono(dithiolene) analogues of the active sites of pyranopterin-containing Mo and W enzymes (99, 100). To date, only 1,2-diethylethene-1,2-dithiolate, (etdt)$^{2-}$, has been complexed with transition metal ions, but the versatility of this approach has been demonstrated with numerous substituted alkynes, and the products can be further trapped by including methyl iodide to form 1,2-alkylthioolefins as precursors to alkene-1,2-dithiolates (101). The ligand has also been prepared as a thiophosphoryl dithiolene (95).

Dipotassium salts of 1,2-dithiocroconate, K$_2$dtcr (102, 103), and 1,2-dithiosquarate, K$_2$dtsq (104), are prepared by treating the corresponding dimethylcroconate and diethylsquarate molecules with potassium hydrogen sulfide (Scheme 8). Modification of the five-membered croconate ring is accomplished postcomplexation. For instance, malononitrile displaces a ketyl group to give tris(dithiolene) complexes with 4-dicyanomethyl-1,2-dithiocroconate, (dcmdtcr)$^{2-}$, ligands (Scheme 8) (103).

C. Dithiones

Collectively known as dithioxamides, these dithiones are the only known ligands of this type found in dithiolene chemistry (Fig. 11). The entry level compound is rubeanic acid, (SCNH$_2$)$_2$, first identified two centuries ago (105). It is prepared by bubbling H$_2$S through an aqueous solution of KCN and [Cu(NH$_3$)$_4$]$^{2+}$ (106), though today it is readily acquired from chemical suppliers. Infrared (IR) spectral data confirmed the dithione structure for the molecule (107). Dthiooxamide has found wide ranging use as a metal deactivator in petroleum products, inhibitor of certain bacteria and dehydrogenases, accelerator of

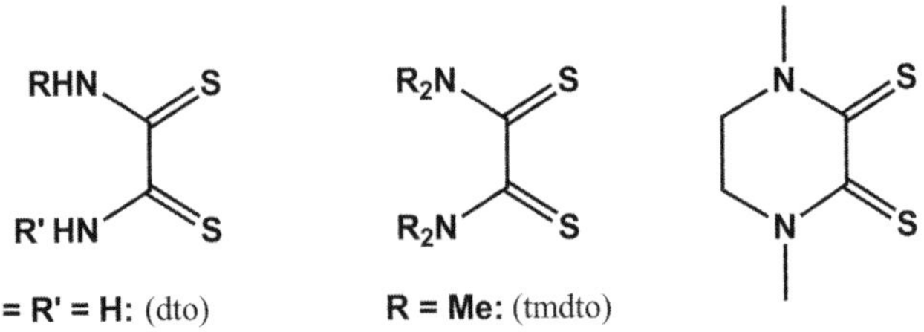

R = R' = H: (dto)
R = R' = Me: (mdto)
R = H; R' = Me: (dmdto)

R = Me: (tmdto)
R = Et: (tedto)

(Me_2pipdt)

Figure 11. 1,2-Dithione ligands and their abbreviations. Dithiooxamide = dto, methyldithiooxamide = mtdo, dimethyldithiooxamide = dmdto, tetramethyldithiooxamide = tmdto, tetraethyldithiooxamide = tedto, 1,4-Dimethylpiperazine-2,3-dithione = Me_2pipdt.

vulcanization, dichroic stain in light polarizing films, and to detect the presence of cuprous ions (108, 109). These molecules have a long history in coordination chemistry as they can either bind through the sulfur or nitrogen atoms, or both, depending on the preference of the targeted metal ion (108–113). In alkaline media, deprotonation of the amine groups generates the dianionic form, $[S_2C_2(NH)_2]^{2-}$, which has a propensity to form insoluble polymeric substances with metals (110). *N*-Alkyl-substituted variants are prepared by reacting the parent dithiooxamide with a primary amine, or treating the corresponding *N,N'*-dialkyloxamide with phosphorus pentasulfide (114, 115), depending on the desired substitution pattern. Tetra-alkyl substituted dithiooxamides lack amine protons (Fig. 11), and therefore bulky groups favor sulfur coordination of metal ions, a conclusion based on electronic and IR spectroscopy (116–118). However, the existence of $\{MS_6\}$ polyhedra is entirely speculative in the absence of structural evidence to prove three chelating dithiones.

The preparation of Me_2pipdt (Fig. 11), commences with the cyclocondensation of *N,N'*-dimethyl-1,2-diaminoethane with dimethyl oxalate in refluxing toluene to form the *N,N'*-dimethyloxamide (119). This compound is converted to the corresponding dithiooxamide with *p*-methoxyphenylthioxophosphine. Several variants are known with different alkyl groups and these have been structurally characterized (120, 121). The short S–C length of 1.668(2) Å in Me_2pipdt is synonymous with a double bond and the long C–C distance of 1.523(2) Å is consistent with a single bond (Fig. 12). The short C–N distance of 1.352(2) Å and the planar nature of the amide-like units point to a degree of electron delocalization over the SCN atoms, which stabilizes the dithione form and dissuades dimerization with a neighboring molecule.

Neutral square-planar compounds of the type $[M(L^{2-})(L^0)]$, containing a dianionic dithiolate and a neutral dithione, have found application as near-infrared (NIR) dyes and nonlinear optical (NLO) materials (122, 123). These "push–pull"

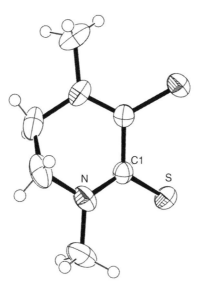

Figure 12.   Crystal structure of Me$_2$pipdt.

complexes are so-named because the electron-donating dithione pushes and the electron-withdrawing dithiolate pulls. Complexes with two or three dithione ligands are the only known cationic species in transition metal dithiolene chemistry. A single tris(dithiolene) complex is known, this being [Fe(Me$_2$pipdt)$_3$]$^{2+}$ (124). Dithiones are considerably weaker donor ligands than their dithiolate counterparts and struggle to chelate early transition metal ions.

III.   COMPLEXES

Attaching the aforementioned dithiolene ligands to metal ions is rather trivial in comparison to the synthesis of the ligands themselves. The vast majority of tris(dithiolene) complexes are prepared from combining the correct stoichiometric ratio of free dithiol or dialkali dithiolate with an appropriate metal reagent, principally metal chlorides. Alternative methods include transmetalation and alkyne reduction by a metal sulfide.

A.   Metathesis

The metathetical approach works for most arene-1,2-dithiolates [bdt, tdt, xdt, qdt, tbbdt (3,5-di-*tert*-butylbenzene-1,2-dithiolate), bdtF$_4$ (3,4,5,6-tetrafluorobenzene-1,2-dithiolate), bdtCl$_4$ (3,4,5,6-tetrachlorobenzene-1,2-dithiolate), bdtCl$_2$, tms, bdt-crown ethers, bn-bdt$_3$ [(1,3,5-tris(amidomethylbenzenedithiolate)benzene], dbddto], Na$_2$mnt, deprotected 1,3-dithiole-2-ones, (Na/K)$_2$dmit, Na$_2$dmt, dipotassium 5,6-dihydro-1,4-dithiin-2,3-dithiolate, K$_2$dddt, and K$_2$dtcr. The assembled tris(dithiolene) complexes are typically anionic and charge balanced by alkali metals that are replaced by larger ammonium, phosphonium, and arsonium cations to facilitate crystallization. Neutral complexes are formed by exposing the reaction mixture to the atmosphere. Group 4 (IV B) and 5 (V B) complexes with (bdt)$^{2-}$ and (tdt)$^{2-}$ ligands are innovatively prepared from metal amide rather than chloride precursors (Eq. 2) (125, 126).

$$[M(NR_2)_x] + (6-x)NaHS_2C_6H_4 + (x-3)(HS)_2C_6H_4 \rightarrow Na_{6-x}[M(S_2C_6H_4)_3] + (6-x)HNR_2$$

$$(M = Ti, Zr, Hf, x = 4; M = Nb, Ta, x = 5; R = Me, Et)$$

$$(2)$$

The procedure requires both the free dithiol and its singly deprotonated form, (bdtH)$^{1-}$, achieved by including a base (e.g., sodium cyclopentadienide or n-butyllithium). The alkali metal is replaced with a bulkier countercation. The dianionic [Sn(bdt)$_3$]$^{2-}$ and [Sn(tdt)$_3$]$^{2-}$ analogues were produced from the reaction of [Sn(NMe$_2$)$_2$] with 3 equiv of dithiol (127). Switching to the alkoxide, [Ti(i-PrO)$_4$], affords the same result as for the amido precursor (128), and remains the reagent of choice for the elegant array of supramolecular constructs by Hahn and co-workers (129–133).

The tris(dithiolene) complexes [Fe(mnt)$_3$]$^{3-}$ (134), [Fe(mnt)$_3$]$^{2-}$ (134, 135), [Co (mnt)$_3$]$^{3-}$ (135–137), and [Fe(bdtCl$_4$)$_3$]$^{2-}$ (43), are formed during the preparation of the corresponding bis(dithiolene) complexes: [Fe(mnt)$_2$]$_2$$^{2-}$ [Co(mnt)$_2$]$^{2-}$, and [Fe(bdtCl$_4$)$_2$]$_n$, respectively. These two metals mark the point along the first-row, where homoleptic complexes transition from tris(dithiolene) species to bis(dithiolene) ones (135). An extra dose of ligand (Na$_2$mnt or bdtCl$_4$H$_2$) generates the tris(dithiolene) complex. A plethora of related Lewis base adducts has been documented (138).

Thiophosphoresters formed in the reaction of an acyloin and a three- to sixfold excess of P$_4$S$_{10}$ are added to metal chlorides or high-valent oxides of V, Mo, W, and Re (94, 96). The simultaneous addition of dilute HCl cleaves the ester and generates the dithiol form of the ligand, which is in turn stabilized by complexation. Neutral complexes, [M(pdt)$_3$], prevail for Mo, W, and Re, as the reaction is stirred in air. The analogous complexes [M(mdt)$_3$] (M = Mo, W, Re) (96, 139, 140), [W(dtdt)$_3$] (96, 139), [W(dadt)$_3$] (96, 139), and [W(andt)$_3$], where (andt)$^{2-}$ is 1,2-dianilylethene-1,2-dithiolate (141), have been prepared by this method. Hydrazine added to the thiophosphorester solution prior to the addition of [VO(acac)$_2$], where (acac)$^{1-}$ is acetylacetonate, and NEt$_4$Br results in the monoanionic complex, NEt$_4$[V(pdt)$_3$] (96). Neutral [V(dtdt)$_3$] and [V(dadt)$_3$] stem from oxidation of the corresponding monoanions (139). Davison et al. (142)

used anhydrous VCl_3 in combination with dilute NaOH to isolate this compound. Hydrazine or hydroxide can reduce the neutral species to its monoanionic complex. As mentioned above, the thiophosphorester can be alkylated by introducing an appropriate reagent after removal of excess P_4S_{10}. These isolable and structurally characterized thiophoshoryl dithiolenes can be activated for metal chelation following standard base hydrolysis conditions (95).

High-valent oxides of V ([VO(acac)$_2$], [VO]SO$_4\cdot x$H$_2$O), Mo (Na$_2$[MoO$_4$]· 2H$_2$O, (NH$_4$)$_6$[Mo$_7$O$_{24}$]·4H$_2$O, [MoO$_2$(acac)$_2$]), Tc(K[TcO$_4$]), W (Na$_2$[WO$_4$]· 2H$_2$O), and Re([Re$_2$O$_7$]), are alternative reagents for the preparation of tris(dithiolene) complexes with (tdt)$^{2-}$ (31, 143, 144), (bn-bdt$_3$)$^{6-}$, where bn = benzene (145), (tr–bdt$_3$)$^{6-}$, where tr = triazole (145), (edt)$^{2-}$ (96), (sdt)$^{2-}$ (146), and (pdt)$^{2-}$ (94, 96). The reaction uses HCl to labilize the oxo ligands.

Heteroleptic complexes are sparingly encountered in tris(dithiolene) chemistry. All known examples are prepared by metathetical procedures, where the dithiol or dithiolate were added to a metallo-bis(dithiolene) unit with one or two labile ligands. The earliest reported complexes are [Mo(pdt)$_2$(mnt)] and [Mo(pdt)$_2$(edt)] formed by acidification of a mixture of [Mo(pdt)$_2$(CO)$_2$] and Na$_2$mnt or Na$_2$edt, respectively (147). Katakis and co-workers (148) combined equimolar amounts of 4-(4-methoxyphenyl)-1,3-dithiole-2-one and 4-(4-bromophenyl)-1,3-dithiole-2-one with [WBr$_4$(MeCN)$_2$] to form four different complexes. Each was separated by column chromatography: [W(adt)$_3$] with 1:9 benzene/cyclohexane as eluent; [W(adt)$_2$(bsdt)] with 3:7 benzene/cyclohexane; [W(adt)(bsdt)$_2$] with 3:2 benzene/cyclohexane; and [W(bsdt)$_3$] with neat benzene.

A truly elegant series of tris(dithiolene) complexes, [Mo(tfd)$_x$(bdt)$_{3-x}$] ($x =$ 0–3), where tfd = bis(trifluoromethyl)ethene-1,2-dithiolate, was recently prepared by Fekl and co-workers (149). The heteroleptic combinations are described in Scheme 9, where an oxo ligand is replaced by (bdt)$^{2-}$ to give [Mo(tfd)$_2$(bdt)]$^{1-}$ prior to oxidation by [Mo(tfd)$_3$], converting it to the neutral form. Labile phosphine ligands are displaced by bis(trifluoromethyl)dithiete to give neutral

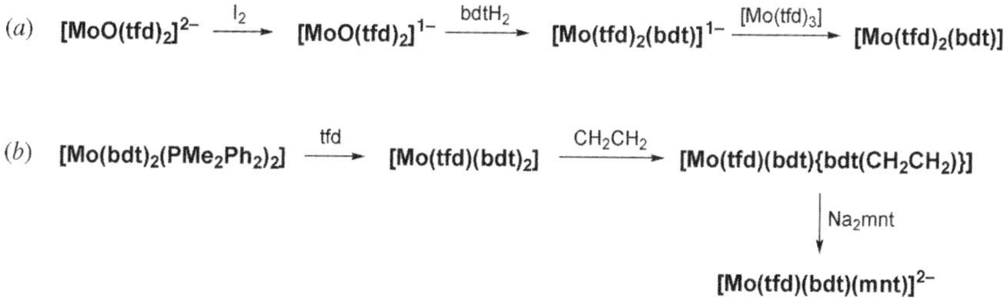

(a) [MoO(tfd)$_2$]$^{2-}$ $\xrightarrow{\text{I}_2}$ [MoO(tfd)$_2$]$^{1-}$ $\xrightarrow{\text{bdtH}_2}$ [Mo(tfd)$_2$(bdt)]$^{1-}$ $\xrightarrow{\text{[Mo(tfd)}_3]}$ [Mo(tfd)$_2$(bdt)]

(b) [Mo(bdt)$_2$(PMe$_2$Ph$_2$)$_2$] $\xrightarrow{\text{tfd}}$ [Mo(tfd)(bdt)$_2$] $\xrightarrow{\text{CH}_2\text{CH}_2}$ [Mo(tfd)(bdt){bdt(CH$_2$CH$_2$)}]

$\downarrow$ Na$_2$mnt

[Mo(tfd)(bdt)(mnt)]$^{2-}$

Scheme 9.   Reaction sequences for preparation of heteroleptic complexes, (a) [Mo(tfd)$_2$(bdt)], and (b) [Mo(tfd)(bdt)$_2$] and [Mo(tfd)(bdt)(mnt)]$^{2-}$.

[Mo(tfd)(bdt)$_2$]. Addition of ethylene to this species generates a complex adduct [Mo(tfd)(bdt){bdt(CH$_2$CH$_2$)}], where the alkene adds across the S $\cdots$ S unit of (bdt)$^{2-}$ to form 2,3-dihydro-1,4-benzodithiin, bdt(CH$_2$CH$_2$). Such nucleophilic addition reactions have been performed previously (150), and the weakly bound thioether ligand was displaced by (mnt)$^{2-}$ to give [Mo(tfd)(bdt)(mnt)]$^{2-}$, the only known complex with three different dithiolene ligands (149).

B. Redox

The first compound of this type, [Mo(tfd)$_3$], was formed from the reaction of bis(trifluoromethyl)dithiete and molybdenum hexacarbonyl (1). The tungsten analogue was also prepared, but necessitated longer reaction times (72 h) due to the inherent reluctance of [W(CO)$_6$] to surrender its ligands. The reaction with [Cr(CO)$_6$] proceeds to completion in <5 h to form [Cr(tfd)$_3$] (10, 65). Davison et al. (134) claimed to have formed [Fe(tfd)$_3$] via the same procedure, however, this has been revised as dimeric [Fe$_2$(tfd)$_4$] (152, 153), following structural characterization of [Co$_2$(tfd)$_4$] (154). The conversion to the tris(dithiolene) complex is carried out in the high boiling point solvents methyl or ethyl cyclohexane at 100–130 °C wherein the zero-valent metal is oxidized to formally a +VI ion by three dithietes. The corresponding vanadium complex, [V(tfd)$_3$]$^{1-}$, is oxidized to a formally +V ion (142). The complex dianions [M(tfd)$_3$]$^{2-}$ (M = Mo, W) have been isolated from the reaction of [MS$_4$]$^{2-}$ and 3 equiv of hexafluorobut-2-ene (50). Here, the high-valent tetrathiometalate reduces the alkyne to an alkene via induced internal electron transfer. Similar reactions with activated alkynes, principally the aforementioned dialkylacetylene dicarboxylates, led to tris(dithiolene) dianions of V, Mo, and W from the corresponding tetrathiometalates or [MoS(S$_4$)$_2$]$^{2-}$ (53–58). Following the successful production of [Ni(pdt)$_2$] (3, 58), Schrauzer et al. (93) isolated the [M(pdt)$_3$] (M = Cr, Mo, W) from the combination of zero-valent hexacarbonyl, sulfur, and diphenylacetylene. Rauchfuss and co-workers (155) utilized the same reagents in the reaction with 3 equiv of tetrathiapentalenedione, which gave [M(dmid)$_3$]$^{2-}$ (M = Mo, W). This esoteric reaction generates 3 equiv of COS; neither CO$_2$ nor CS$_2$ were detected. The peripheral ketones on each of the (dmid)$^{2-}$ ligands can then be hydrolyzed and alkylated to give three 1,4-dithiin-2,3-dithiolates bound to the metal ion; an example exists with Mo (155).

Photolysis of a metal hexacarbonyl in the presence of a dithiete or a dithione provides an alternative method to forming tris(dithiolene) complexes, following the same procedure for quinones (156, 157). Carbon monoxide can be liberated from 1,3-dithiole-2-ones using ultraviolet (UV) light, and the ensuing dithione is captured by a transition metal ion. The dark green neutral complexes [Mo(pdt)$_3$] (56), [Mo(andt)$_3$] (56), [Mo(thdt)$_3$] (69) [Mo(cydt)$_3$] (57), and [Mo(mtdt)$_3$] (mtdt^{2-} = 1,2-bis(methylthio)ethene-1,2-dithiolate) (158), have all been prepared in this manner.

Access to different members of each tris(dithiolene) electron-transfer series is accomplished by using an appropriate oxidizing/reducing agent. Many complexes are synthesized in reactions open to the atmosphere and therein the most air-stable form prevails; relevant examples being neutral complexes of Mo, W, and Re following Schrauzer's thiophosphorester synthetic approach (8, 96). These complexes are readily transformed by mild reducing agents (e.g., hydroxide) to the corresponding monoanions. More potent reagents (e.g., hydrazine, n-butyllithium, elemental sodium, and cobaltocene) have been used to generate more reduced species. Selecting the appropriate reducing agent is dependent on both the reduction potential for the parent compound and the solvent in which to perform the conversion in order to eliminate counterproductive side reactions. Alternatively, reduced tris(dithiolene) complexes isolated from anaerobic reactions, such as $Li_2[Mo(bdt)_3]$ (159), can be sequentially oxidized. Many oxidizing agents have been used from simple ferrocenium salts and halogens, to neutral complexes [Ni (tfd)$_2$] and [Mo(tfd)$_3$] (160). Complete or partial oxidation with radical cation salts of TTF leads to paramagnetic materials with attractive magnetic and conductive properties (15, 73, 77, 161).

C. Transmetalation

An efficient and high-yielding approach to various tris(dmit) complexes involves combining the metal chloride with $[Zn(dmit)_2]^{2-}$ (78, 81). This complex is highly soluble in a wide range of solvents and the ligand stability is greatly enhanced when coordinated to Zn(II) (162). The combination of (NBu$_4$)$_2$[Zn(dmit)$_2$] with anhydrous VCl$_3$ and ReCl$_5$ led to the clean isolation of (NBu$_4$)$_2$[V(dmit)$_3$] and (NBu$_4$)$_2$[Re(dmit)$_3$], respectively (161, 163). Using Na$_2$dmit mainly gave $[VO(dmit)_2]^{2-}$ and $[ReO(dmit)_2]^{1-}$ (161). Analogues with TTF based dithiolene ligands have been prepared from the reaction of their Zn(II) salt and VCl$_3$ to create a series, $[V(R_2TTFdt)_3]^{2-}$ (R = Et, n-Bu; $R_2 = -CH_2CH_2CH_2-$) (91). Dithiolate salts of Zn obtained from a commercial source were used to prepare [W(tdt)$_3$] and [Re(tdt)$_3$] from acidified aqueous solutions of Na$_2$[WO$_4$] and NH$_4$[ReO$_4$], respectively (144).

Transmetalation with dithiolenes dates to experiments in the mid-1960s by Schrauzer et al. (147), who monitored the transference of ligands from [Ni(pdt)$_2$] to [M(CO)$_6$] (M = Cr, Mo, W). The major product was the thermodynamically favored [M(pdt)$_3$] (M = Cr, Mo, W), although rather forcing conditions were employed. The desired heteroleptic carbonyl–dithiolene complexes were preferably isolated via UV irradiation of mixtures of [Ni(pdt)$_2$] and [M(CO)$_6$] (M = Cr, Mo, W) forming products that retained CO ligands. Holm and co-workers (22, 23) used transmetalation from bis(dithiolene)nickel complexes to generate Mo and W small molecule analogues of oxotransferase enzyme active sites. They frequently encountered the facile formation of tris(dithiolene) species of Mo and W. These

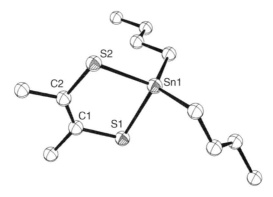

Figure 13.   Molecular structure of [Sn(mdt)(n-Bu)₂].

reaction sinks are conveniently separated from the product mixture by column chromatography.

The isolation of mixed dithiolene–carbonyl complexes of Mo and W has been elegantly performed using Sn protected dithiolenes (72). The process involves base hydrolysis of 4,5-dimethyl-1,3-dithiole-2-one to form $(mdt)^{2-}$, which is readily scavenged by $[SnCl_2(n\text{-}Bu)_2]$ giving a colorless precipitate. Purification by column chromatography successfully led to isolation of $[Sn(mdt)(n\text{-}Bu)_2]$ in modest yields, but on a multi-gram scale. The Sn complex was crystallographically characterized exhibiting S−C distances 1.778(6) and 1.776(3) Å, and a C−C bond length of 1.338(8) Å (Fig. 13).

These distances are synonymous with a dianionic dithiolate ligand, and the 0.3 Å increase reflects the loss of a complete set of π bonds around the 1,3,2-dithiastannisole ring system compared to the 1,3-dithiole-2-one (71). Tin-dithiolene complexes have been structurally characterized with $(mnt)^{2-}$ and $(dmit)^{2-}$ ligands (164). This ligand chaperone substantially improved the yield of $[Ni(mdt)_2]$ in comparison to the original Schrauzer method (166) negating the need for further purification. The technique was applied to the synthesis of $[W(mdt)_3]$ in 80% yield (Scheme 10), with $[SnCl_2(n\text{-}Bu)_2]$ re-formed in the process. This product was the first tris(dithiolene) complex synthesized using a Sn based dithiolene-transfer reagent despite being successful in the preparation of other dithiolene complexes (167–169).

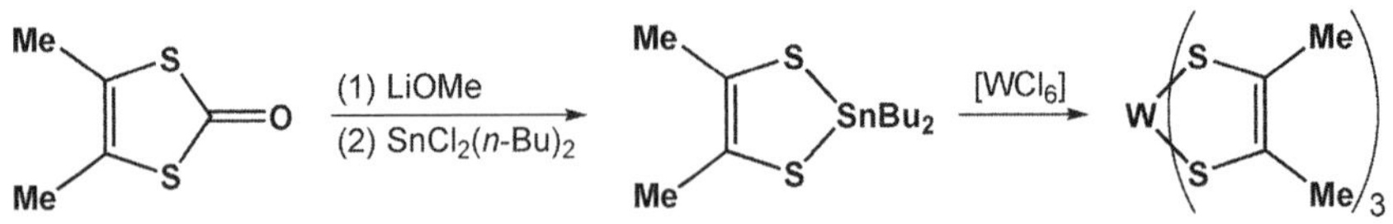

Scheme 10.   Synthesis of [W(mdt)₃].

Although not a metal, silyl-protected (etdt)$^{2-}$ cleanly transfers to a variety of metals without a separate deprotection step (98). Akin to the aforementioned Sn chemistry, the reaction is driven by the formation of strong M$-$S in the tris(dithiolene) complexes and Si$-$F/Cl/O bonds in the byproducts.

IV. STRUCTURES

Since the most recent structural update in this Forum (170), the number of reported tris(dithiolene) crystal structures found in the Cambridge Structural Data Centre has trebled (24). Figure 14 displays the distribution of structurally

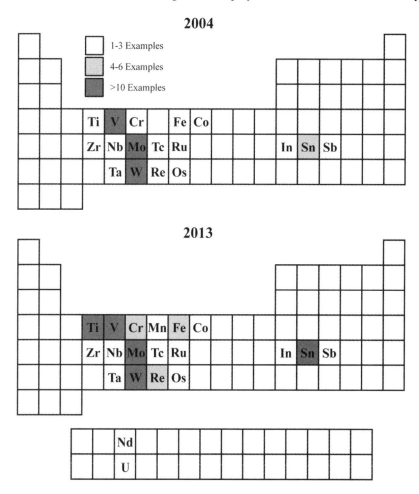

Figure 14.   Distribution and frequency of tris(dithiolene) complexes in 2004 and 2013.

characterized complexes across the periodic table along with a comparison to the 2004 compilation published by Stiefel and co-workers (172). For the most part, the increase in population number stems from repeats of existing structures particularly in the case of Ti and Sn. On the other hand, the first structures containing Mn and f block elements Nd and U are welcomed into the family.

Intrigue in transition metal dithiolene compounds stemmed from their vibrant colors and rich redox chemistry. It is perhaps not surprising that the dawn of dithiolene chemistry and related coordination complexes coincided with the rise of single-crystal X-ray diffractometry. Although an established technique, it was not until the 1950s that automated diffractometers become available and accessible to chemists. The impact on the field was profound, especially in regard to the unusual non-octahedral geometry adopted by tris(dithiolene) compounds (24, 25). Prior to X-ray crystallography, structures and formulas were derived from spectroscopic data [electronic absorption and electron paramagnetic resonance (EPR)] and accurate elemental analysis. At the same time, the concept of ligand field theory (LFT) was developed representing an amalgamation of crystal field and molecular orbital (MO) theory. It seemed appropriate to apply the benefits of LFT to these new dithiolene complexes, and this required their molecular structure to be defined.

A. Beginnings

1. Neutral Complexes

Eisenberg, armed with crystals from Schrauzer, undertook the structure determination of [Re(pdt)$_3$] at the Brookhaven National Laboratory with Ibers (24, 25). This research followed from the prior year's successful characterization of square-planar (NMe$_4$)$_2$[Ni(mnt)$_2$] (5, 171). After laboriously sifting through diffraction data and manually estimating their intensities, the structure was defined as a trigonal prismatic (TP) array of sulfur atoms about a central Re atom (Fig. 15a) (172, 173). This structure was the first example of a coordination complex bearing TP geometry that was considered implausible for six-coordination complexes as an octahedral arrangement minimizes interligand repulsion. Each {ReS$_2$C$_2$} metallodithiolene ring was planar and the polyhedron adopted D_{3h} point symmetry remarkably similar to the Mo site in molybdenite (174). Almost simultaneously, a second neutral tris(dithiolene) complex was crystallographically characterized (Fig. 15b), namely, [Mo(edt)$_3$], with the same TP polyhedron (175). This compound exhibited a previously unseen structural distortion that lowered the molecule to C_{3h} point symmetry by virtue of a pronounced bend along the S $\cdots$ S vector of each ligand.

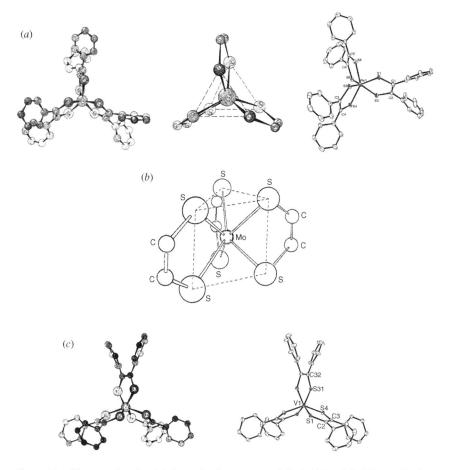

Figure 15.   Diagrams showing (*a*) the molecular structure of [Re(pdt)$_3$] and its TP coordination geometry as originally determined in 1964 compared with the 2006 redetermination shown as a thermal ellipsoid plot. [Adapted from (170, 171, 174)]; (*b*) The molecular structure of [Mo(edt)$_3$] showing TP geometry and folded dithiolene ligands [adapted from (173)]. (*c*) Comparison of the original and modern molecular structure of [V(pdt)$_3$] from the 1966 and 2010 determinations. [Adapted from (177–179)].

The periodic diagonal was completed in 1966 with the structural characterization of [V(pdt)$_3$] by Eisenberg et al. (177, 178) exhibiting TP geometry (Fig. 15*c*). The maxim that six-coordinate equated to octahedral was shattered by these structure determinations, and tris(dithiolene) complexes distinguished themselves with this unique geometry. Moreover, it was clear that these were not anomalous results imposed by lattice forces, as electronic spectra revealed the geometry persisted in solution (177, 180, 181). In the absence of diffraction quality single crystals, X-ray powder diffraction was sought to ascertain how widespread this motif was across tris(dithiolene) complexes. Using [Re(pdt)$_3$] as a calibrant, and adhering to the general idea that isomorphous materials are isostructural, powder patterns were recorded for a vast number of compounds with conflicting results. Neutral complexes [W(pdt)$_3$], [W(bdt)$_3$], and [W(tdt)$_3$] showed similar patterns to the corresponding Re species, and were diagnosed as TP (172, 180). Similar powder patterns were obtained for [V(pdt)$_3$]$^{1-}$, [V(pdt)$_3$], [Cr(pdt)$_3$], and [Mo(pdt)$_3$], and these were deemed TP despite differing from their third-row analogues (172, 173). The subsequent structural report of [V(pdt)$_3$] consolidated this conclusion (177, 178). Schrauzer and co-workers (175) reported the powder patterns of [M(edt)$_3$] (M = V, Mo, W, Re), concluding that the Mo and W species were definitely different from the Re compound, and the V species was decidedly different from the others. This finding disagreed with the conclusions drawn by the team at Columbia University stating that [W(pdt)$_3$] is isomorphous with isoelectronic [V(pdt)$_3$]$^{1-}$, [Cr(pdt)$_3$], and [Mo(pdt)$_3$], but not [Re(pdt)$_3$]. Powder diffraction studies of [Ru(pdt)$_3$] and [Os(pdt)$_3$] gave distinctly different patterns to their group 6 (VI B) and 7 (VII B) analogues, and were even distinct from each other (175). The veracity of these speculations mandated single-crystal diffraction studies, which in most cases, were not forthcoming for several decades.

   The prism dimensions in these three structures are strikingly similar (177). The M−S distances are invariant across the series despite the increase in the ionic radius descending the d block (Table I) (182, 183). The original text quoted radii

TABLE I

Comparison of Metric Parameters for Three Tris(dithiolene) Complexes and Molybdenite, MoS$_2$

Complex	Ionic Radius[a]	Θ^b	M−S[c]	$S\cdots S_{intra}{}^d$	$S\cdots S_{inter}{}^e$	S−C[c]	C−C[c]
[V(pdt)$_3$]	68	4.3	2.34	3.06	3.06	1.69	1.41
[Mo(edt)$_3$]	75	0	2.33	3.10	3.11	1.70	1.34
[Re(pdt)$_3$]	72	3.4	2.32	3.03	3.05	1.69	1.34
MoS$_2{}^f$	79	0	2.41	3.17	3.13		

[a] Ionic radii in picometers (pm) were taken from (182) with the metal ion in the +V oxidation state in each complex and +IV oxidation state for MoS$_2$.
[b] Mean twist angle, $\Theta = (\Theta_1 + \Theta_2 + \Theta_3)/3$.
[c] Average distance in angstroms (Å).
[d] Average distance between S atoms within the dithiolene ligand, in angstroms (Å).
[e] Average distance between adjacent S atoms related by the threefold axis, in angstroms (Å).
[f] The $S\cdots S$ values define the dimensions of the TP array of sulfur atoms in the structure presented in (174).

from Paulings seminal work *The Nature of the Chemical Bond*, however, the values have been refined and reinterpreted over the years. This text utilizes corrected ionic radii by Shannon (182), which for all purposes are identical to those catalogued by Pauling. A recent update of covalent radii has been reported and lists the values for V, Mo, and Re as 153, 154, and 151 pm, respectively, which indicates that the M−S bond length and size of the prism should be identical in all three structures. However, this was not known at the time and has been omitted from the main text so as not to undermine the argument made by proponents of interligand bonding. The size of the prism, where the length is given by the intraligand distance ($S \cdots S_{intra}$) and width by the interligand distance ($S \cdots S_{inter}$) of the trigonal SSS face, were only marginally smaller than the same motif characterized for the infinitely extended molybdenite lattice with the empirical formula MoS_2 (Table I) (174). This observation led researchers to hypothesize the unusual geometry is stabilized by an $S \cdots S$ bonding interaction as the interligand distance of ~3 Å lies well within the sum of the van der Waals radii (Fig. 16) (172, 173, 177, 178, 184). The same interligand distance was observed in bis(dithiolene) complexes (171, 184–187). The bonding interaction was suggested to arise from partial oxidation of the ligand system. Neutral [V(pdt)$_3$] exemplifies this point, as the formal oxidation state of the metal is +VI if the ligands are regarded as dianionic dithiolates (181). Interestingly, *ab initio* theoretical studies on the sulfur analogue of superoxide, $(S_2^{\bullet})^{1-}$, produced S−S bond

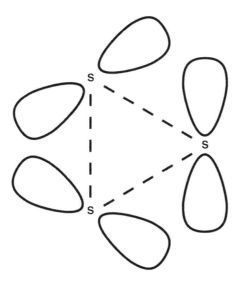

Figure 16.   Pictorial representation of the sulfur bonding interaction between trigonal prismatically arranged dithiolene ligands.

distances ranging from 2.8–2.9 Å that are marginally shorter than the interligand contacts in these TP structures (188, 189). The envelope distortion of the
dithiolene ligands in [Mo(edt)$_3$] attenuates the interligand interaction resulting
in a slightly larger prism (Fig. 15b). The structure of a diselenolene analogue,
[Mo{Se$_2$C$_2$(CF$_3$)$_2$}$_3$], also exhibited this distortion to C_{3h} symmetry (190).
The likelihood of oxidized ligands was further evidenced by the intraligand
bond distances in these neutral complexes. Removal of electrons from a
dithiolate leads to a shortening of the S—C bonds and lengthening of
the C—C bond as they attain more double- and single-bond character, respectively (Scheme 1). The average S—C and C—C bond distances listed in Table I
express a degree of ligand oxidation as these values are distinct from
the dithiolate dianions characterized in the structures of [Ni(mnt)$_2$]$^{2-}$ (171),
[Co(mnt)$_2$]$^{2-}$ (186), and [Cu(mnt)$_2$]$^{2-}$ (187) (1.73 and 1.32 Å, respectively).
However, the low quality of the data undermined the usefulness of these ligand
metrics.

Several conventions have been developed to describe the topology in these
TP complexes. The trigonal twist angle (Θ) is defined as the dihedral angle
between the S atoms in the two-dimensional projection along the threefold axis
(Fig. 17) (191–195). In six-coordinate complexes, Θ ranges from 0° in a perfect
TP to 60° in a regular octahedron (or trigonal antiprism). However, chelate
ligands by their very composition possess a maximum twist angle as a function
of the M—L bond length and ligand bite (L—M—L) angle (194–197). For
example, Θ is limited to ~48° in (bdt)$^{2-}$ complexes of early transition metals (198). Therefore, the trigonal twist represents a $D_{3h} \rightarrow D_3$ symmetry reduction. The trigonal twist angles for [V(pdt)$_3$], [Mo(edt)$_3$], and [Re(pdt)$_3$] are listed

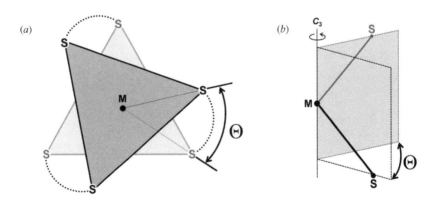

Figure 17.   Depiction of trigonal twist angle, Θ, as viewed along (a) the C_3 axis and (b) the C_2' axis in
D_3 point symmetry.

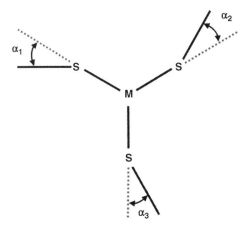

Figure 18. Depiction of the dithiolene folding angles, α_1, α_2, α_3 viewed along the threefold axis.

in Table I, and show each to be a (nearly) perfect prism. Other parameters used to gauge the complex geometry include the trans SMS angle ($136°$ for TP; $180°$ for octahedral) and the dihedral angle between the MS_2 and trigonal SSS planes ($90°$ for TP; $\sim 55°$ for octahedral) (170). The folding of the dithiolene ligands in [Mo(edt)$_3$] represents a $D_{3h} \rightarrow C_{3h}$ symmetry reduction, and is defined by the dihedral angle, α (Fig. 18). Together these two parameters (Θ and α) describe the symmetry of any tris(dithiolene) complex.

2. Reduced Complexes

A largely antitrigonal prismatic geometry was reported for the crystal structure of (NMe$_4$)$_2$[V(mnt)$_3$] reported in 1967 (Fig. 19) (199). With $\Theta = 38°$ and the $S \cdots S_{inter}$ stretched to 3.15 Å, it was postulated that the twist away from TP alleviated the increased ligand repulsion generated by two extra valence electrons compared to [V(pdt)$_3$]. The ligands were considered dianionic dithiolates, and the intraligand S–C distance averaging 1.72 Å and C–C bond length of 1.34 Å correlate with the reduced dithiolene level. Therefore, the twisted structure was regarded as a compromise between residual $S \cdots S$ bonding and interligand repulsion.

X-ray powder diffraction patterns were collected for first-row [M(mnt)$_3$]$^{2-}$ (M = Ti, V, Cr, Mn, Fe) complexes (199, 200). All were described as isostructural with distorted octahedral structures and +IV metal ions coordination by three (mnt)$^{2-}$ ligands. The crystal structure of (AsPh$_4$)$_2$[Fe(mnt)$_3$] solved in 1973 cemented this conclusion, with $\Theta = 49°$ (201). The average Fe–S distance at 2.266 Å is appreciably shorter than the corresponding distance of 2.36 Å in

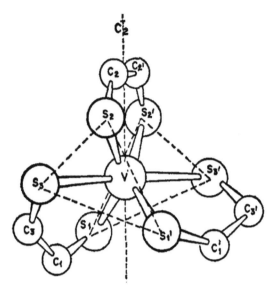

Figure 19.   Depiction of the structure of (NMe$_4$)$_2$[V(mnt)$_3$]. [Adapted from (199).]

[V(mnt)$_3$]$^{2-}$ (199), which can be ascribed to the difference in the covalent radius of the metal ion. The consequence were shorter interligand contacts at 3.10 Å, similar in magnitude to the three neutral complexes described above. The increased Fe−S covalency overrides ligand repulsion in this reduced complex, as potential S···S bonding is greatly perturbed by the twist to octahedral and subsequent misalignment of adjacent S p orbitals (Fig. 16). A similar outcome was observed for (AsPh$_4$)$_2$[M(mnt)$_3$] (M = Mo, W) with pronounced twist angles ~28° and flat dithiolate ligands (202). The average M−S distances of 2.373 and 2.371 Å for Mo and W species, respectively, are only slightly longer than for V, which is the reason why the lighter analogue was cocrystallized with the Mo compound for single-crystal EPR studies (203, 204).

The structural characterization of [In(mnt)$_3$]$^{3-}$ was the first complex without a transition metal ion (205). The long In−S distance of 2.604 Å is typical of main group dithiolate complexes, and the intraligand (mnt)$^{2-}$ bond lengths match the other structures (171, 186, 187, 199). The molecule possesses a twist angle of 47.3° and irregular buckling of the dithiolene ligands whose fold angles average 30.3°. The latter is the result of crystal packing and is a feature of similar complexes with In, Sn, and Sb (Table II).

TABLE II
Listing of Structurally Characterized Tris(dithiolene) Complexes[a]

Complex[b]	Counterion[c]	M–S[d]	S–C[d]	C–C[d]	Θ[e]	α[f]	Reference
$[Ti(bdt)_3]^{2-\ g}$	$NH_2Me_2^+$	2.428	1.761	1.402		22.5	206
	$AsPh_4^+$	2.413	1.758	1.413	37.9	12.8	130
$[Ti(ms\text{-}bdt)_3)]^{2-}$	$AsPh_4^+$	2.413	1.758	1.413	37.9	12.8	130
$[Ti_2(tr\text{-}bdt_3)_2]^{4-\ h}$	NMe_3Ph^+	2.416	1.755	1.414	37.0	9.3	132
$[Ti_4(bn\text{-}bdt_3)_4]^{8-\ g,h}$	Li^+, K^+, NEt_4^+	2.417	1.745	1.411			207
$[Ti(bdt\text{-}\mathbf{1}\text{-}cat)_3Ti]^{4-\ i}$	Na^+, PNP^+	2.398	1.724	1.410	39.2	4.0	208
$[Ti(bdt\text{-}\mathbf{3}\text{-}cat)_3Ti]^{4-\ i}$	Na^+, PNP^+	2.407	1.755	1.408	34.4	6.1	208, 209
$[Ti(bdt\text{-}\mathbf{4}\text{-}bdt)_3Ti]^{4-\ h,i}$	Li^+, PNP^+	2.410	1.759	1.416	37.0	6.3	133, 210
$[Ti(bdt\text{-}\mathbf{5}\text{-}bdt)_3Ti]^{4-\ h,i}$	PNP^+	2.409	1.760	1.409	37.9	0.9	133
$[Ti(bdt\text{-}\mathbf{7}\text{-}bdt)_3Ti]^{4-\ h,i}$	NEt_4^+	2.418	1.766	1.399	30.8	7.6	129, 131
	$AsPh_4^+$	2.417	1.758	1.415	30.5	17.2	129, 131
	NBu_4^+	2.410	1.748	1.412	35.4	7.8	129
$[Ti(dddt)_3]^{2-}$	$AsPh_4^+$	2.417	1.736	1.368	36.3	14.7	211
$[Zr(bdt)_3]^{2-}$	NMe_4^+	2.543	1.766	1.388	38.7	3.0	126, 198
$[V(bdt)_3]^{2-}$	NEt_4^+	2.362	1.742	1.382	37.3	12.3	212
$[V(bdt)_3]^{1-}$	PPh_4^+	2.350	1.739	1.411	1.3	22.5	179
$[V(tdt)_3]^{2-}$	NEt_4^+	2.363	1.740	1.367	38.9	8.6	212
$[V(dbddto)_3]^{2-}$	PPh_4^+	2.369	1.739	1.393	40.6	11.1	213
$[V(edt)_3]$		2.347	1.665	1.356	1.6	0.9	212
$[V(mnt)_3]^{2-\ g}$	NMe_4^+	2.36	1.72	1.34			199
	TTF^+	2.358	1.707	1.362	39.7	12.8	214
	PPh_4^+	2.368	1.732	1.362	38.0	4.3	179
$[V(pdt)_3]$		2.338	1.690	1.407	4.3	2.2	177, 178
		2.344	1.702	1.397	4.9	2.5	179
$[V(pdt)_3]^{1-}$	NEt_4^+	2.329	1.731	1.375	0.8	23.7	179
$[V(dmm)_3]^{2-}$	NEt_4^+	2.367	1.726	1.361	34.3	3.3	61
$[V(dmit)_3]^{2-\ g}$	NBu_4^+	2.375	1.724	1.375			85
	TTF^+	2.359	1.699	1.381	4.6	24.8	215
	NMP^+	2.378	1.727	1.343	38.2	12.0	216
$[V(dddt)_3]$		2.349	1.696	1.381	15.4	0.5	217
$[V(dddt)_3]^{1-}$	NBu_4^+	2.343	1.717	1.353	1.8	22.5	218
	TTF^+	2.342	1.717	1.362	5.0	22.9	217
$[V(bddt)_3]^{1-}$	NEt_4^+	2.350	1.722	1.314	7.2	22.9	89
$[V(deTTFdt)_3]^{1-\ g}$	NEt_4^+				~0	~20	91
$[Nb(bdt)_3]^{1-}$	$AsPh_4^+$	2.441	1.745	1.395	1.4	24.9	126, 219
$[Ta(bdt)_3]^{1-}$	$AsPh_4^+$	2.430	1.748	1.382	30.8	17.4	220
$[Cr(bdtCl_2)_3]^{2-}$	NBu_4^+	2.332	1.742	1.406	45.6	10.7	221
$[Cr(tbbdt)_3]^{1-}$	NBu_4^+	2.299	1.750	1.409	39.1	12.8	222
$[Cr(mnt)_3]^{2-}$	PPh_4^+	2.344	1.734	1.330	45.4	4.8	223
$[Cr(mnt)_3]^{3-}$	PPh_4^+	2.394	1.734	1.354	51.5	9.2	223
$[Mo(bdt)_3]$		2.367	1.727	1.411	0.0	23.0	126, 224
$[Mo(bdt)_3]^{1-}$	NBu_4^+	2.382	1.724	1.377	33.5	7.9	225
	PNP^+	2.386	1.749	1.392	34.8	9.2	159
$[Mo(bdt)_3]^{2-}$	PNP^+	2.396	1.759	1.396	24.9	0.6	159
$[Mo(bdtCl_2)_3]^{2-}$	$NHEt_3^+$	2.391	1.737	1.437	2.8	6.8	226
$[Mo(tbbdt)_3]$		2.364	1.737	1.412	1.5	23.7	227

(*continued*)

TABLE II
(*Continued*)

Complex[b]	Counterion[c]	M−S[d]	S−C[d]	C−C[d]	Θ[e]	α[f]	Reference
[Mo(tbbdt)₃]¹⁻	NBu₄⁺	2.378	1.757	1.407	31.7	8.5	227
[Mo(bdt-**1**-cat)₃Ti]⁴⁻ⁱ	Li⁺,Na⁺, PNP⁺	2.373	1.749	1.414	22.5	2.8	223
[Mo(qdt)₃]¹⁻	PPh₄⁺	2.393	1.745	1.434	14.6	12.2	229
[Mo(qdt)₃]²⁻	PPh₄⁺	2.393	1.737	1.442	4.5	6.0	230
[Mo(bdt)₂(tfd)]ʲ		2.364	1.721	1.420	2.9	23.8	149
		2.367	1.711	1.371	1.2	17.9	
[Mo(bdt)(tfd)(mnt)]²⁻ʲ	NEt₄⁺	2.381	1.742	1.400	3.2	2.7	149
		2.378	1.746	1.323	2.8	0.4	
		2.396	1.717	1.346	2.3	0.7	
[Mo(edt)₃]ᵍ		2.33	1.70	1.37	0	18	96, 175
[Mo(etdt)₃]		2.363	1.713	1.369	0.5	15.5	98
[Mo(mdt)₃]		2.367	1.714	1.358	2.4	15.9	23
[Mo(mdt)₃]¹⁻	NEt₄⁺	2.374	1.726	1.354	1.6	4.0	23
[Mo(mdt)₃]²⁻	NEt₄⁺	2.396	1.755	1.335	2.6	1.1	231
[Mo(mnt)₃]²⁻	AsPh₄⁺	2.373	1.735	1.324	28.2	2.0	202
	PPh₄⁺	2.382	1.734	1.355	28.7	2.2	232
	[Ag(μ-dppa)]⁺	2.384	1.729	1.345	8.2	0.6	233
	[Cu(μ-dppm)]⁺	2.385	1.726	1.355	20.7	3.3	234
	[Cu(μ-dppm)]⁺	2.392	1.734	1.361	15.1	2.3	235
	[Ag(μ-dppm)]⁺	2.364	1.721	1.338	19.5	6.8	236
[Mo(tfd)₃]		2.355	1.696	1.367	0	17.7	50
[Mo(tfd)₃]¹⁻	NEt₄⁺	2.372	1.742	1.339	15.9	1.2	50
[Mo(tfd)₃]²⁻	NEt₄⁺	2.375	1.737	1.336	16.1	1.5	50
[Mo(pdt)₃]		2.366	1.719	1.383	3.0	17.4	232
[Mo(pdt)₃]¹⁻	PPh₄⁺	2.373	1.739	1.355	18.0	11.0	232
[Mo(dmm)₃]²⁻	PPh₄⁺	2.394	1.742	1.338	10.6	4.3	59, 60
[Mo(dmit)₃]²⁻	NBu₄⁺	2.404	1.734	1.350	16.6	3.7	237
	NBu₄⁺	2.406	1.733	1.349	16.6	3.7	238
[W(bdt)₃]		2.380	1.744	1.420	0	25.2	239
[W(bdt)₃]¹⁻	NMe₄⁺	2.387	1.779	1.423	29.6	5.4	240
	PHMe₂Ph⁺	2.366	1.713	1.395	32.3	2.8	241
[W(bdt)₃]²⁻	AsPh₄⁺	2.386	1.717	1.391	23.0	2.7	242
	NEt₄⁺	2.391	1.751	1.397	1.9	2.0	21
[W(bdtCl₂)₃]¹⁻	NEt₄⁺	2.375	1.750	1.388	7.4	2.3	226
	PPh₄⁺	2.384	1.748	1.409	23.7	1.4	226
[W(bdtCl₂)₃]²⁻	NEt₄⁺	2.381	1.752	1.399	33.0	3.9	226
	PPh₄⁺	2.380	1.750	1.400	34.6	9.0	226
[W(tbbdt)₃]		2.361	1.747	1.405	1.0	26.2	227
[W(tbbdt)₃]¹⁻	NBu₄⁺	2.375	1.762	1.390	31.5	10.1	227
[W(mdt)₃]		2.362	1.735	1.367	0.7	13.8	72
[W(mdt)₃]¹⁻	NEt₄⁺	2.375	1.730	1.354	2.9	7.2	231
[W(mdt)₃]²⁻	NEt₄⁺	2.388	1.763	1.317	2.4	1.6	231
	NEt₄⁺	2.394	1.763	1.338	2.5	1.6	232
[W(mnt)₃]²⁻	AsPh₄⁺	2.371	1.732	1.322	27.8	2.0	202
	PPh₄⁺	2.384	1.739	1.363	28.8	2.0	232
	[Cu(μ-dppm)]⁺	2.387	1.735	1.362	15.1	1.9	243
	[Ag(μ-dppm)]⁺	2.391	1.751	1.347	19.4	6.3	236

TABLE II
(*Continued*)

Complex[b]	Counterion[c]	M−S[d]	S−C[d]	C−C[d]	Θ[e]	α[f]	Reference
[W(pdt)$_3$]		2.363	1.743	1.380	3.2	12.4	22
		2.385	1.715	1.451	1.0	19.5	232
[W(pdt)$_3$]$^{1-}$	NMe$_3$Bz$^+$	2.371	1.742	1.360	14.3	13.7	22
	PPh$_4^+$	2.380	1.749	1.359	22.8	11.0	232
[W(dmit)$_3$]$^{1-}$	Fc$^+$	2.379	1.712	1.340	0.8	10.8	238
[W(dmit)$_3$]$^{2-}$	NBu$_4^+$	2.394	1.730	1.340	15.5	4.2	238
[W(dmid)$_3$]$^{2-}$	PPh$_4^+$	2.394	1.731	1.346	24.9	1.7	155
[Mn(bdt)$_3$]$^{2-}$	PNP$^+$	2.331	1.760	1.399	44.7	2.4	244
[Mn(bdtCl$_2$)$_3$]$^{2-}$	NEt$_4^+$	2.347	1.760	1.415	47.9	8.3	221
[Mn(mnt)$_3$]$^{2-}$	PPh$_4^+$	2.328	1.747	1.359	49.9	8.0	221
[Tc(bdt)$_3$]$^{1-}$	AsPh$_4^+$	2.339	1.728	1.389	8.0	2.4	245
[Tc(mnt)$_3$]$^{2-}$	AsPh$_4^+$	2.362	1.722	1.367	38.9	2.9	246
[Re(bdtCl$_2$)$_3$]$^{1-}$	[C$_8$H$_{16}$N]$^+$	2.340	1.739	1.400	24.8	1.7	247
[Re(tms)$_3$]$^{1-}$	[C$_8$H$_{16}$N]$^+$	2.342	1.745	1.397	0.5	4.0	247
[Re(mnt)$_3$]$^{2-}$	PPh$_4^+$	2.364	1.734	1.359	38.3	3.0	248
[Re(pdt)$_3$]		2.324	1.685	1.338	3.4	2.5	172, 173
		2.332	1.725	1.363	3.8	1.3	176
[Re(pdt)$_3$]$^{1-}$	PPh$_4^+$	2.342	1.747	1.349	26.3	4.4	248
[Fe(mnt)$_3$]$^{2-}$	AsPh$_4^+$	2.263	1.730	1.356	49.0	7.2	201
	PPh$_4^+$	2.263	1.724	1.349	49.2	6.9	223
	PPh$_4^+$	2.269	1.727	1.345	47.5	5.5	223
	PPh$_4^+$	2.274	1.733	1.363	46.6	6.3	249
[Fe(mnt)$_3$]$^{3-}$	PPh$_4^+$	2.287	1.735	1.368	53.4	12.0	249
[Ru(bdt)$_3$]$^{2-}$	NBu$_4^+$	2.356	1.754	1.409	44.5	5.4	250
[Ru(mnt)$_3$]$^{2-}$	AsPh$_4^+$	2.343	1.722	1.362	47.1	5.4	251
[Ru(mnt)$_3$]$^{3-}$	NEt$_4^+$	2.347	1.709	1.390	50.2	0	251
[Os(bdt)$_3$]$^{2-}$	PHCy$_3^+$	2.356	1.756	1.398	43.0	10.7	252
[Co(dtcr)$_3$]$^{3-}$	PPh$_4^+$	2.265	1.675	1.402	58.7	10.1	101
[In(mnt)$_3$]$^{3-}$	NEt$_4^+$	2.604	1.722	1.352	47.3	30.3	205
[In(dtsq)$_3$]$^{3-}$	NBu$_4^+$	2.600	1.664	1.384	48.4	3.1	253
	NMe$_3$Bz$^+$	2.642	1.693	1.401	45.5	14.3	253
[Sn(bdt)$_3$]$^{2-}$	NH$_2$Me$_2^+$	2.526	1.769	1.404	48.3	26.3	127
[Sn(tdt)$_3$]$^{2-}$	NH$_2$Me$_2^+$	2.537	1.800	1.450	47.2	34.0	127
[Sn(mnt)$_3$]$^{2-}$	Na$^+$, NEt$_4^+$	2.536	1.734	1.360	45.7	24.3	254
[Sn(dmit)$_3$]$^{2-}$	NMe$_4^+$	2.558	1.702	1.392	50.6	44.6	255
	NEt$_4^+$	2.545	1.742	1.353	52.3	17.0	256
	Me$_2$py$^+$	2.547	1.744	1.357	44.3	20.7	255
	Me$_2$py$^+$	2.540	1.737	1.356	46.3	20.2	255
	NEt$_4^+$	2.543	1.737	1.345	52.2	22.0	255
	PPh$_4^+$	2.551	1.747	1.361	48.6	25.0	255
	NBu$_4^+$	2.537	1.742	1.343	49.6	19.3	257
	NMe$_4^+$	2.558	1.738	1.358	35.3	45.4	257
	AsPh$_4^+$	2.546	1.739	1.361	48.8	25.1	258
	PPh$_4^+$	2.559	1.738	1.351	50.8	27.9	259
[Sn(dmid)$_3$]$^{2-}$	NEt$_4^+$	2.556	1.742	1.373	51.5	39.4	255
	NBu$_4^+$	2.541	1.741	1.354	50.3	19.4	260

(*continued*)

TABLE II
(*Continued*)

Complex[b]	Counterion[c]	M–S[d]	S–C[d]	C–C[d]	Θ[e]	α[f]	Reference
	PPh_4^+	2.553	1.746	1.350	47.2	33.3	260
$[Sb(bdt)_3]^{1-}$	NEt_4^+	2.478	1.758	1.404	49.2	17.4	261
$[Sb(dmit)_3]^{1-}$	Me_2py^+	2.531	1.736	1.355	52.0	47.3	262
$[Nd(dddt)_3]^{3-}$	Na^+	2.857	1.766	1.361	18.2	86.7	263
$[U(dddt)_3]^{2-}$	$[Na(18C6)]^+$	2.738	1.759	1.370	6.7	81.3	264

[a] Solvents of crystallization have been omitted. Values are given for only one structural unit in cases where multiple structures are contained within the unit cell.
[b] Ligand abbreviations are given in Figs. 2–4.
[c] Abbreviations: PNP^+ = Bis(triphenylphosphoranylidene)ammonium, TTF^+ = tetrathiafulvalenium, NMP^+ = *N*-methylphenazinium, dppa = bis(diphenylphosphino)amine, dppm = bis(diphenylphosphino)methane, NMe_3Bz^+ = trimethylbenzylammonium, Fc^+ = ferrocenium, $C_8H_{16}N^+$ = 5-azonia-spiro-[4,4]nonane, $PHCy_3^+$ = tricyclohexylphosphonium, Me_2py^+ = 1,4-dimethylpyridinium, 18C6 = 18-crown-6.
[d] Average values.
[e] Mean twist angle, $\Theta = (\Theta_1 + \Theta_2 + \Theta_3)/3$. Value may differ from cited literature depending on the calculation method.
[f] Mean fold angle, $\alpha = (\alpha_1 + \alpha_2 + \alpha_3)/3$.
[g] Data from original reference.
[h] Metrics reported for only one TiS_6 polyhedron as there is little difference between the two sites within each molecule.
[i] Number specifies amide link connecting $(cat)^{2-}$ and $(bdt)^{2-}$, or two $(bdt)^{2-}$ units as shown in Fig. 3.
[j] Data listed for each ligand type in the order in which they appear in the formula.

3. Isoelectronic Series

The first decade of single-crystal diffraction studies was completed by a comparison of the geometry and metric parameters in the isoelectronic series, $[Zr(bdt)_3]^{2-}$, $[Nb(bdt)_3]^{1-}$, $[Mo(bdt)_3]$ (Fig. 20) (126). Formally, all are d^0 metal

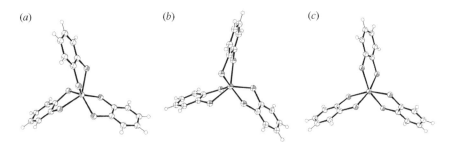

(a)   *(b)*   *(c)*

Figure 20. Ellipsoid representations of the crystal structures of (*a*) $[Mo(bdt)_3]$, (*b*) $[Nb(bdt)_3]^{1-}$, and (*c*) $[Zr(bdt)_3]^{2-}$.

TABLE III

Salient Metric Parameters for Isoelectronic Series of Complexes

Complex	$M-S^a$	$S \cdots S_{intra}^{\ b}$	$S \cdots S_{inter}^{\ c}$	$S-C^a$	$C-C^a$	Θ^d	α^e
$[Zr(bdt)_3]^{2-}$	2.544	3.264	3.586	1.762	1.388	38.7	3.0
$[Nb(bdt)_3]^{1-}$	2.441	3.150	3.232	1.743	1.395	1.4	24.9
$[Mo(bdt)_3]$	2.367	3.110	3.091	1.726	1.411	0	23.0

^{a} Average distance in angstroms (Å).
^{b} Average distance between S atoms within the dithiolene ligand, in angstroms (Å).
^{c} Average distance between adjacent S atoms related by the threefold axis, in angstroms (Å).
^{d} Mean twist angle, $\Theta = (\Theta_1 + \Theta_2 + \Theta_3)/3$.
^{e} Mean fold angle, $\alpha = (\alpha_1 + \alpha_2 + \alpha_3)/3$.

ions coordinated by fully reduced ligands. The salient parameters are presented in Table III and show that the size of the $\{MS_6\}$ coordination sphere shrinks as the series is traversed. Concomitantly, the intraligand bond distances indicate the ligand system in the neutral Mo species is clearly oxidized compared to the Zr complex.

The Nb and Mo structures possess C_{3h} point symmetry with a TP array of sulfur atoms and pronounced dithiolene fold; the Zr species is distorted octahedral with D_3 symmetry and planar ligands. Interestingly, the Nb compound was the first to have TP geometry without the apparent stabilizing influence of interligand interactions. The sulfur atoms are a further 0.14 Å apart than in the $\{MoS_6\}$ prism, the largest known at the time. Cowie and Bennett (198, 219, 224) published each structure back-to-back some years later and quashed the notion of interligand S $\cdots$ S bonding stabilizing this unique geometry. Instead, they invoked an electronic argument, specifically matching the energies of the metal d and dithiolene π orbitals that generate strong bonding interactions for prismatic $\{MS_6\}$ coordination spheres (181).

B.   Redux

Active research in tris(dithiolene) chemistry faded dramatically at the end of the 1960s with just a handful of publications in the following two decades. There was a lengthy hiatus until the structural determination of the active site of a tungsten-containing aldehyde:ferridoxin oxidoreductase (18) infused transition metal dithiolene chemistry with new vigor. During the lull, a range of new techniques had been developed that would enable us to explore these systems in greater detail. X-ray diffractometry remained a cornerstone of these studies and the use of modern instrumentation provided high-resolution structures and greater confidence in the metric parameters.

1.   Trigonal Twist

It was not until the structural characterization of [V(mnt)$_3$]$^{2-}$ that we learned TP geometry was only associated with the oxidized members of any given tris(dithiolene) electron-transfer series. The neutral complexes [Re(pdt)$_3$], [Mo(edt)$_3$], and [V(pdt)$_3$] exhibited twist angles $\sim0°$ (D_{3h}); the reduced complex did not, owing to the two additional electrons that were said to populate ligand-based orbitals, promoting a distorted octahedral structure (D_3) in response to interligand repulsion (Fig. 21).

X-ray powder patterns and electronic absorption spectra alluded to similar geometries for the other isolated [M(mnt)$_3$]$^{2-/3-}$ salts, where M = Ti, Cr, Mn, Fe, Co, Mo, W, Re. With the exception of (NEt$_4$)$_3$[Ru(mnt)$_3$], all have been crystallized with the same counterion, PPh$_4$$^+$. The claim was proven correct with the successive structural characterization of these compounds over the last 40 years. Considering the bite angle of (mnt)$^{2-}$, the first-row complexes are defined with twist angles close to the octahedral limit, $\Theta = 38.0–53.4°$ (Table II). The variation depends on the M−S bond distance, which is a function of periodicity with the smaller, more electropositive ions possessing shorter bonds. This finding is a general trend for all tris(dithiolene) complexes where the M−S distance varies with the size of the metal ion and its effective nuclear charge or oxidation state (Fig. 22). Interestingly, second-row [Ru(mnt)$_3$]$^{2-/3-}$ complexes have essentially the same twist angles as their Fe counterparts despite ~0.05-Å longer M−S bonds (251).

The opposing trend is apparent within an electron-transfer series, where the trianionic species, [M(mnt)$_3$]$^{3-}$ (M = Cr, Fe, Ru) have longer bonds, but larger twist angles than the corresponding dianions (Table II). Therefore, the higher metal oxidation state in the complex dianions draws the ligands closer, and together with the smaller ionic radius, reduces the twist angle. The Mo and W complexes follow this pattern, with neutral complexes bearing shorter distances than their

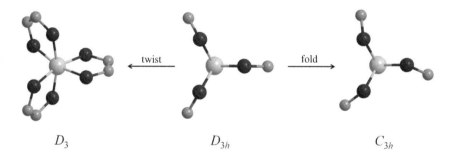

D_3 D_{3h} C_{3h}

Figure 21.   Representations of the distortion of TP (D_{3h}) geometry to distorted octahedral (D_3) by twisting and the envelope-type conformation (C_{3h}) from dithiolene folding.

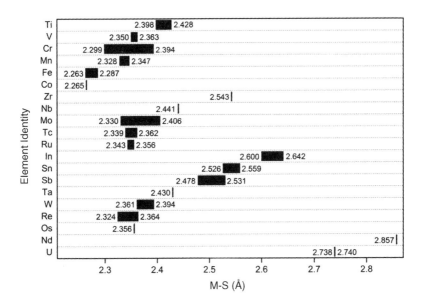

Figure 22.   Ranges of M—S bond distanced in crystallographically characterized complexes.

corresponding dianions (Table II). This result also accounts for $\Theta = 38°$ in $[V(mnt)_3]^{2-}$, which formally and physically possesses a V(IV) central ion (179, 204). This complex has two fewer electrons than $[Mn(mnt)_3]^{2-}$, which are removed from π^* orbitals that are antibonding with respect to a TP. Therefore, the observed twist angle is purely dependent on the interligand repulsion between $(mnt)^{2-}$ and the formation of strong π bonding orbitals with the metal. The same logic applies to $[Mo(mnt)_3]^{2-}$ and $[W(mnt)_3]^{2-}$, with low-spin d^2 central ions and no electrons in the π^* level (232). Consequently, the M—S bond distances at ~ 2.38 Å are the longest of any $(mnt)^{2-}$ complex with a transition metal ion, yet both have small twist angles $\sim 29°$. The stark increase in the twist angle to $38.3°$ in $[Re^{IV}(mnt)_3]^{2-}$ with its low-spin d^3 ion clearly demonstrates the impact of adding an electron to the π^* level (248).

The twist angle is a delicate balance between two opposing forces, namely, the metal–ligand π overlap (covalency) that is maximized in TP, though dependent on the energy difference between metal and ligand fragments and the number of valence electrons, and interligand repulsion that is minimized at the octahedral limit. However, a study by Sugimoto et al. (226) clarified the input of counter-cations on the resultant complex geometry, with $[W(bdtCl_2)_3]^{1-}$ $\{(bdtCl_2)^{2-} = $ 3,6-dichlorobenzene-1,2-dithiolate$\}$ structurally determined with NEt_4^+ and PPh_4^+ cations. The former produced a twist angle of $7.4°$ whereas the latter was characterized with $\Theta = 23.7°$ (Fig. 23). A similar result was discovered for

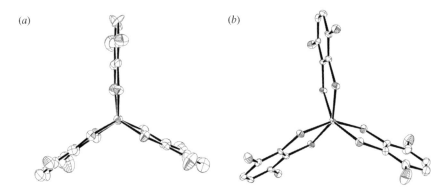

Figure 23. Pictorial representations of the anion $[W(bdtCl_2)_3]^{1-}$ in single crystals with cation (a) NEt_4^+ ($\Theta = 7.4°$) and (b) PPh_4^+ ($\Theta = 23.7°$), viewed down the threefold axis.

$(NEt_4)_2[W(bdt)_3]$, $\Theta = 1.9°$ (21), and $(AsPh_4)_2[W(bdt)_3]$, $\Theta = 23.0°$ (242). It is not clear if this geometric disparity stems from the size of the countercation or its ability to instigate nonbonding interactions with the periphery of the complex anion.

Several triple-stranded helicates have been structurally characterized by Hahn and co-workers (42, 129, 131, 133, 209, 210, 228, 265) using bdt-cat and bis(bdt) ligands (Fig. 3) that clasp metal ions at each end. The best results are offered by titanium, though there is a heterometallic Ti/Mo helicate that has the {TiO_6} center connected to a {MoS_6} center (228). The latter provides an opportunity to assess the loss of helicity by oxidation of the terminal $\{MoS_6\}^{2-}$ unit prompting a change to a more TP geometry. The group has also constructed tris(bdt) ligands, $(bn-bdt_3)^{6-}$, $(ms-bdt_3)^{6-}$ (ms = mesitylene), and $(tr-bdt_3)^{6-}$, which either encapsulate a single metal ion (130, 145), or bridge two centers to form a double-stranded, banana-shaped helicate (132). Balancing the stoichiometry and reaction conditions led to the formation of an exquisite tetramer, $LiK(NMe_4)_6[Ti_4(bn-bdt_3)_4]\cdot6DMF$ (207). The crystal structure showed a Ti(IV) ion at the vertex of a regular tetrahedron, a ligand on each triangular face, and a K^+ ion cornered inside.

2. Dithiolene Fold

The phenomenon of dithiolene bending first appeared in the crystal structure of $[Mo(edt)_3]$ and has since been encountered in a range of other tris(dithiolene) compounds (23, 50, 89, 98, 126, 149, 175, 179, 217, 218, 224, 227, 232), and many mono(dithiolene) species (266–268).

To date, five monoanionic vanadium complexes have been characterized with C_{3h} symmetry and a "paddle wheel" motif when viewed along their threefold axis (Fig. 24); the fold angles lie in the very narrow range of 22.5°–23.7° (89, 179, (*a*)   (*b*)

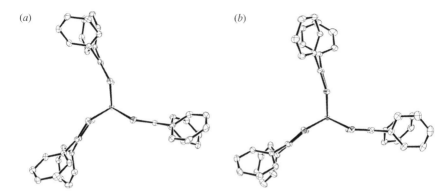

Figure 24.   Pictorial representations of the structures of (*a*) [V(pdt)$_3$]$^{1-}$ ($\alpha = 23.7°$) and (*b*) [Mo(pdt)$_3$] ($\alpha = 17.4°$), as viewed along their threefold axis.

217, 218). Likewise, eight neutral molybdenum complexes, which are categorized into two groups with the arene-1,2-dithiolates giving $\alpha = 21.8°–23.6°$ (126, 149, 224, 227), and alkene-1,2-dithiolates with $\alpha = 15.5°–18°$ (23, 50, 98, 149, 175, 232). The difference in dithiolene type is more noticeable with the five neutral tungsten complexes: $\alpha = 25.7°$ for arene-1,2-dithiolates (227, 239) compared with $\alpha = 15.2°$ for alkene-1,2-dithiolates (22, 72, 232). The structure of [Nb(bdt)$_3$]$^{1-}$ has a large fold angle of 24.9° much like isoelectronic [Mo(bdt)$_3$] (219). On the basis of theoretical calculations, Campbell and Harris (269) suggested a second-order Jahn–Teller distortion as the cause, where the highest occupied (HOMO) and lowest unoccupied molecular orbitals (LUMO) are ligand and metal based, respectively, and can only mix with a reduction to lower symmetry.

Different fold angles identified for the two types contrast the π donating abilities of arene- and alkene-1,2-dithiolates, where the former produce large distortions in order to generate a more stable complex. This result is elegantly seen in the structure of [Mo(tfd)(bdt)$_2$] with both ligand types, where the folding of the two (bdt)$^{2-}$ ligands at 23.8° is significantly larger than in the (tfd)$^{2-}$ ligand at 17.9° (Fig. 25*a*) (149). It is not certain if the folding aids in formation of the ethene adduct with (bdt)$^{2-}$ instead of (tfd)$^{2-}$ to generate [Mo(tfd)(bdt){bdt(CH$_2$CH$_2$)}], or whether it exclusively stems from the different redox potentials of each ligand. The 2,3-dihydro-1,4-benzodithiin is displaced by (mnt)$^{2-}$ to generate the only known tris(dithiolene) complex with different ligands (Fig. 25*b*). This dianionic species is highly TP with no dithiolene fold (149).

Dithiolene folding is taken to an extreme in the two crystallized complexes with f block metals (Fig. 26) (263, 264). The fold approaches right angles, which facilitates an interaction between the alkene π electron density of the ligand and the metal ion (270). This brings the {S$_2$C$_2$} unit almost side-on to the metal with M$\cdots$C=C contacts of ~3.1 Å for Nd(III) and ~2.8 Å for U(IV). The central

(a)   (b)

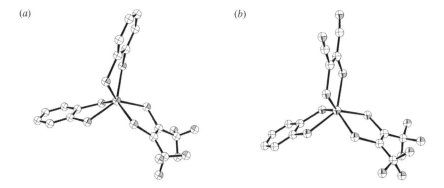

Figure 25.   Crystal structures of (a) [Mo(tfd)(bdt)$_2$] and (b) [Mo(tfd)(bdt)(mnt)]$^{2-}$.

{MS$_6$} core for [Nd(dddt)$_3$]$^{3-}$ and [U(dddt)$_3$]$^{2-}$ are essentially identical with intermediate twist angles. These are presumably governed by the assortment of alkali metal ions, crown ethers, and solvent molecules that generate a number of interactions with the ligand periphery (Fig. 26a). The oxidized version, [U (dddt)$_3$]$^{2-}$ is more TP than the trianionic Nd complex. The intrinsic electronic effects at play are succinctly detailed in a DFT study (270), which defines the importance of back-donation from U into the alkene π bond in [U(dddt)$_3$]$^{3-/2-}$ (the trianion was generated *in silico*). In contrast, steric crowding in the polymeric structure of [Nd(dddt)$_3$]$^{3-}$ is most likely responsible for the folding of the metallocycle. The results underscore a potential application of dithiolenes in lanthanide/actinide differentiation.

(a)   (b)

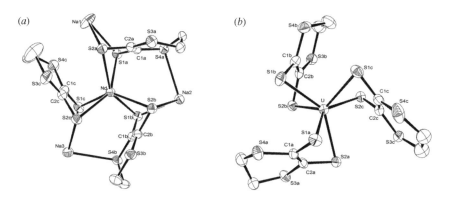

Figure 26.   Pictorial representations of the complex anions (a) [Nd(dddt)$_3$]$^{3-}$ and (b) [U(dddt)$_3$]$^{2-}$.

3. Oxidized Ligands

Oxidation of coordinated arene-1,2-dithiolates leads to a "quinoidal" distortion of the aromatic ring so-named for the archetypal dioxolene, catecholate, a molecule that is readily converted to the benzosemiquinone and benzoquinone forms (Scheme 11) (156, 271, 272). X-ray crystallographic studies have established ligand metrics to distinguish each redox level within a given transition metal complex. Both catechol and benzoquinone themselves have been structurally characterized (273, 274).

The fully reduced form has six equidistant $C-C$ bonds in the aromatic ring and $C-O$ single bonds. The average aromatic $C-C$ distance increases when the unit is oxidized with a short–long–short alternating sequence of $C-C$ and $C-O$ double bonds (Scheme 11). Like their dioxolene counterparts, dithiolene $S-C$ and $C-C$ bonds are sensitive to the ligand oxidation level (275–279). This situation occurs because electrons are removed from the π_3 MO of the rudimentary $\{S_2C_2\}$ unit, which has $S-C$ antibonding and $C-C$ bonding character, shortening the former and lengthening the latter (Fig. 27). High-resolution X-ray crystallography at cryogenic temperatures (100 K) conducted on a suite of bis(dithiolene) complexes had shown that the ligand oxidation levels could be determined, or at the very least approximated when corroborated with spectroscopic and theoretical data (277).

X-Ray crystallographic diagnosis of the ligand oxidation levels in tris(dithiolene) complexes was only first investigated when a complete electron-transfer series was structurally characterized. This event occurred in 2002 when Hahn and co-workers (239) added the structure of neutral [W(bdt)$_3$] to the previously obtained ones for the dianion and monoanion (21, 241). Unfortunately, the intraligand bond distances did not feature in the analysis, and ligands were assigned as dithiolates. The opportunity was not wasted 3 years later when Hahn (159) added the molecular structure of [Mo(bdt)$_3$]$^{2-}$ to the corresponding Mo series (126, 224, 225).

The intraligand bond distances for both series are displayed in Fig. 28, and clearly show a shortening of the $S-C$ bonds and lengthening of the average aromatic $C-C$ bond as the series is transcended from dianion to neutral species. The dianionic members of each series have long $S-C$ bonds of ~ 1.75 Å commensurate with their interpretation as dithiolate ligands. The bond becomes mildly shorter upon one-electron oxidation to the monoanionic species, and

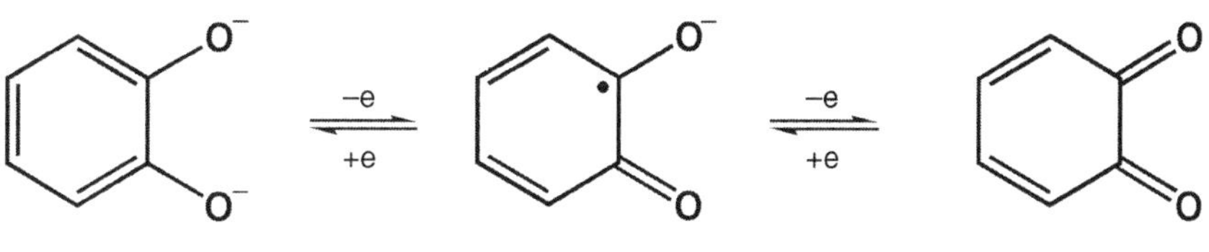

Scheme 11. Dioxolene redox levels.

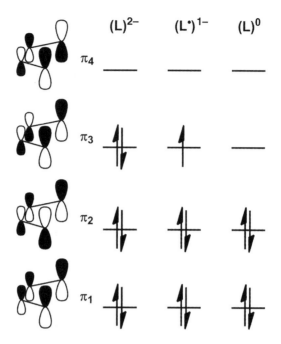

Figure 27. Population of dithiolene π orbitals for the dithiolate, $(L)^{2-}$, dithienyl radical, $(L^\bullet)^{1-}$, dithione, and $(L)^0$ oxidation levels.

noticeably shorter to 1.727(6) Å for [Mo(bdt)$_3$] where the quinoidal distortion is clearly evident (224). Although the W series is undermined by low-quality crystals, the 1.744(6) Å S–C bond length in [W(bdt)$_3$] suggests a somewhat higher metal oxidation state consistent with its position in the bottom row of the d

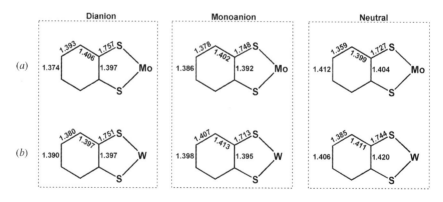

Figure 28. Average intraligand bond distances for crystallographically characterized series (a) [Mo(bdt)$_3$]^z and (b) [W(bdt)$_3$]^z, ($z = 2-$, $1-$, 0).

(a)

(b)

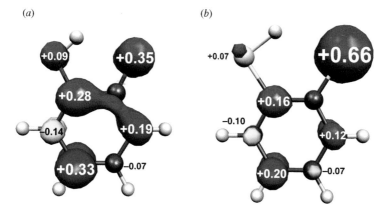

Chapter 1, Figure 29. Mulliken spin distribution of the radicals in (a) benzosemiquinone and (b) dithiobenzosemiquinone.

Progress in Inorganic Chemistry, Volume 58, First Edition. Edited by Kenneth D. Karlin.
© 2014 John Wiley & Sons, Inc. Published 2014 by John Wiley & Sons, Inc.

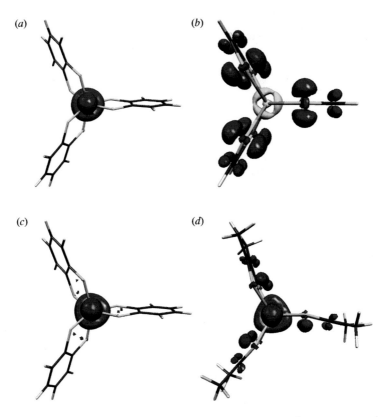

Chapter 1, Figure 49. Mulliken spin density plots for (a) [Mo(bdt)$_3$]$^{1-}$, (b) [Mo(edt)$_3$]$^{1-}$, (c) [W(bdt)$_3$]$^{1-}$, and (d) [W(mdt)$_3$]$^{1-}$, derived from DFT calculations as viewed down the C_3 axis.

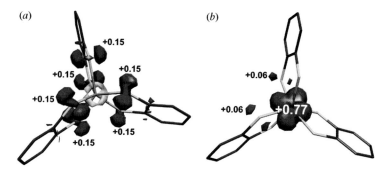

Chapter 1, Figure 51. Mulliken spin density plots and populations for (*a*) [Re(bdt)$_3$] and (*b*) [Re(bdt)$_3$]$^{2-}$. [Adapted from (247).]

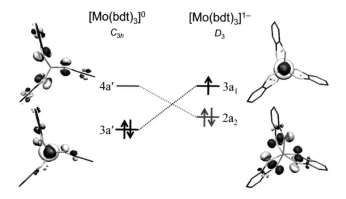

Chapter 1, Figure 58. Depiction of the destabilization of the d$_{z^2}$ orbital in [Mo(bdt)$_3$]$^{1-}$ upon reduction of [Mo(bdt)$_3$] and shift in symmetry from C_{3h} to D_3.

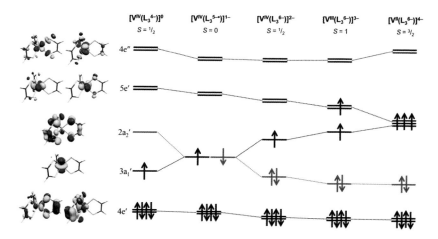

Chapter 1, Figure 62. Qualitative MO scheme depicting the ordering of the frontier orbitals for the [V(L$_3$)]^z ($z = 0$, 1−, 2−, 3−, 4−) electron transfer series. The MOs shown left are annotated with D_{3h} symmetry labels; the redox-active 2a$_{2'}$ is highlighted in gray.

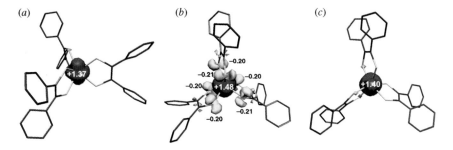

Chapter 1, Figure 63.   Mulliken spin density plots and populations for (*a*) [V(pdt$_3$)], (*b*) [V(pdt$_3$)]$^{1-}$, and (*c*) [V(pdt$_3$)]$^{2-}$ obtained from DFT calculations. [Adapted from (26).]

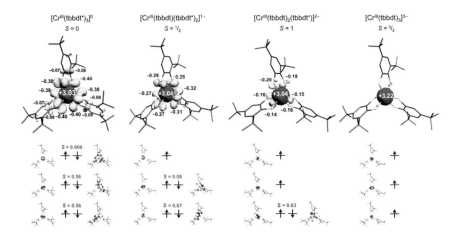

Chapter 1, Figure 64.   Spin density plots of the [Cr(tbbdt)$_3$]^z (*z* = 0, 1−, 2−, 3−) electron-transfer series, as derived from DFT calculations together with values of the spin density of the Mulliken analyses, and qualitative MO diagrams of the corresponding pairs of magnetic orbitals.

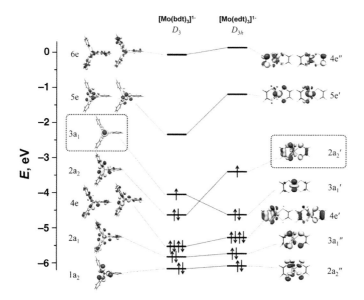

Chapter 1, Figure 65. Molecular orbital scheme for $[\mathrm{Mo(bdt)_3}]^{1-}$ (D_3 symmetry labels) and $[\mathrm{Mo(edt)_3}]^{1-}$ (D_{3h} symmetry labels) with the SOMO identified in each case. Neutral complexes resemble the manifold for $[\mathrm{Mo(edt)_3}]^{1-}$, but with C_{3h} point symmetry and one less electron in the $2a_2'$ MO. [Adapted from (232).]

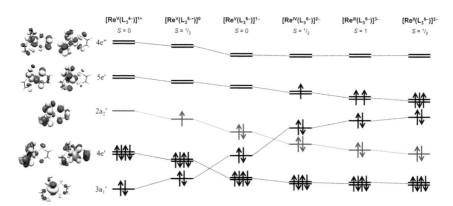

Chapter 1, Figure 66. Qualitative MO scheme depicting the ordering of the frontier orbitals for the $[\mathrm{Re(L_3)}]^z$ ($z = 1+, 0, 1-, 2-, 3-, 4-$) electron-transfer series. The MOs are annotated with D_{3h} symmetry labels and the redox-active $2a_{2'}$ is highlighted in gray.

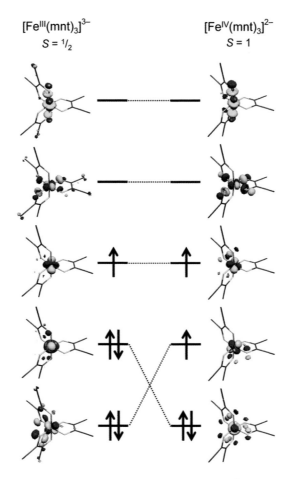

Chapter 1, Figure 67. Qualitative MO scheme for [Fe(mnt)$_3$]$^{3-}$ and [Fe(mnt)$_3$]$^{2-}$.

(a)   (b)

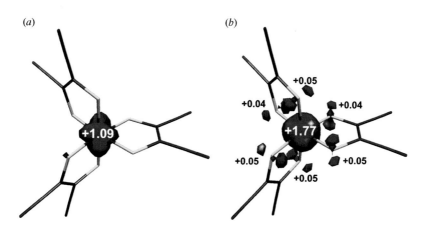

Chapter 1, Figure 68.   Mulliken spin density plots and populations for (a) [Fe^{III}(mnt)$_3$]$^{3-}$, and (b) [Fe^{IV}(mnt)$_3$]$^{2-}$.

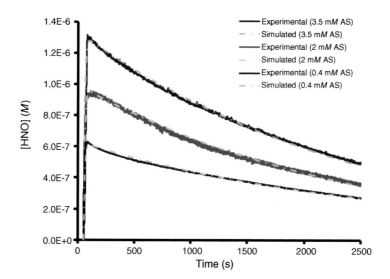

Chapter 2, Figure 2.   Current intensity vs time measured for three different concentrations of AS. Dotted lines show the corresponding simulations for the current intensity obtained from Co^{III}PNO$^-$ oxidation, by using the model shown in Scheme 14.

Chapter 2, Compound II.   Structure of COT1.

(a)   (b)

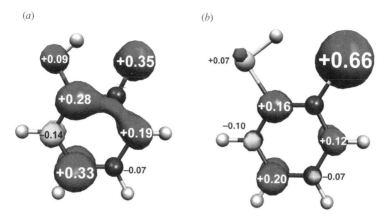

Figure 29.   Mulliken spin distribution of the radicals in (a) benzosemiquinone and (b) dithiobenzo-semiquinone. (See color version of this figure in the color plates section.)

block (239). Unfortunately, these changes are rarely outside the 3σ confidence level even across the electron transfer series, and therefore cannot be used exclusively to determine the electronic structure. This stems from two inherent traits of sulfur, namely, the ligands are symmetrically equivalent such that the bond distance changes are distributed over all three dithiolenes, and the propensity of spin density to remain on the sulfur atoms rather than dispersing through the π network (Fig. 29). This trait is contrasted by dioxolenes (e.g., catecholate), where >60% of the spin density is found on the aromatic ring in the benzosemiquinone radical. This generates large structural changes with each redox step well within the resolution of X-ray crystallography even at room temperature. In addition, there are examples of ligand mixed valency in tris(dioxolene) compounds, particular with Cr, where one ligand can be differentiated from the other two (280). Tris(dithiolene) ligands, on the other hand, are collectively oxidized or reduced, and within their threefold symmetry, exist as three dianionic dithio-lates, $(L_3)^{6-}$ $(S = 0)$, the one-electron oxidized unit, $(L_3)^{5-\bullet}(S = 1/2)$ or two-electron oxidized form, $(L_3)^{4-}$ $(S = 0)$ (179, 281). It is preferable to use this formalism to represent the electronic structure (oxidation level) of the ligands in these complexes. Only the neutral tris(dithiolene) chromium complex is touted as having a central Cr(III) ion and three monoanionic dithiolene radical ligands, $(L^\bullet)^{1-}$ (221), though no crystal structure exists to validate this theoretical hypothesis.

After 42 years, the crystals entrusted to Eisenberg by Schrauzer were mounted on a Bruker SMART APEX-II CCD Platform diffractometer at the University of Rochester (176). The result was a vindication of the original determination (Fig. 15a) (172, 173). The bond distances and angles were refined with greater accuracy, and the calculated twist angle nudged slightly to $\Theta = 3.8°$ (Table II). The corresponding vanadium complex was also redetermined with a similar outcome (Fig. 15c) (179). The structural characterization of neutral and monoanionic $(pdt)^{2-}$ compounds was recently completed for V, Mo, W, and Re. For the first time, a comparison of the intraligand S–C and C–C bonds for the same ligand could be compared across a variety of D_{3h}, D_3, and C_{3h} molecules with different metal ions. These bond distances are charted in the histogram shown in Fig. 30. Both EPR and X-ray absorption spectroscopy (XAS) confirm a V(IV) central ion in [V(pdt)$_3$] (178, 179, 181), which implies an $(L_3)^{4-}$ ligand set. Therefore, the two-electron oxidized tris(dithiolene) unit is characterized with an average S–C bond of 1.70 Å and C–C bond of 1.40 Å. Likewise, [Re(pdt)$_3$] is spectroscopically characterized with a one-electron oxidized tris(dithiolene) unit, $(L_3)^{5-•}$, and the average S–C and C–C bonds of 1.73 and 1.36 Å, respectively, represent this redox level. The fully reduced ligand system has characteristic dithiolate bonds: S–C ~1.76 Å and C–C ~1.33 Å. Using these values as a guide, and the overall structural changes contrasting members of the same electron-transfer series, a $(L_3)^{5-•}$ ligand set can be assigned to [V(pdt)$_3$]$^{1-}$ and [Mo(pdt)$_3$]$^{1-}$. The structural change between neutral and monoanionic vanadium complexes is more

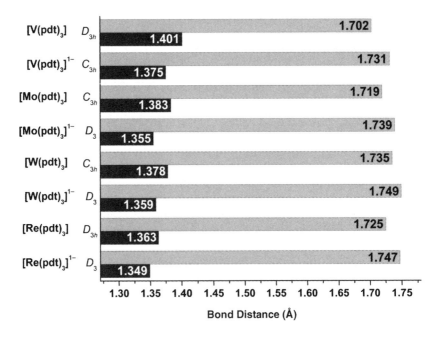

Figure 30.   Comparison of the intraligand S–C (light gray) and C–C (charcoal) bond distances in crystallographically characterized [M(pdt)$_3$]^z (M = V, Mo, W, Re; $z = 0$, 1–).

apparent than the Mo and W analogues, which in light of recent EPR and X-ray absorption spectroscopic studies (179), highlight the different electronic effect governing the C_{3h} distortion. An $(L_3)^{6-}$ unit is preferably associated with [W (pdt)$_3$]$^{1-}$ and [Re(pdt)$_3$]$^{1-}$, while [Mo(pdt)$_3$] exhibits more $(L_3)^{4-}$ character than [W(pdt)$_3$]; the latter is closer to an $(L_3)^{5-\bullet}$ unit. These data do not wholly permit an explicit assignment in Mo and W species (compared with V and Re complexes) because of the pervasive covalency that blurs the boundaries for integer assignment of oxidation levels (232). Nevertheless, it provides a meaningful trend especially when correlated with spectroscopic and theoretical data.

V. THEORY

A. Hückel

Theoretical calculation of the molecular orbitals began in earnest as the first crystallographic results were announced. With knowledge of the three-dimensional structure, and therein the point symmetry, a bonding picture of early transition metal tris(dithiolene) complexes arose from semiempirical Hückel MO theory. Gray and co-workers (181) based their calculations on [Re(pdt)$_3$], which had been crystallographically confirmed as D_{3h} symmetric. Schrauzer and Mayweg (96) working with [Mo(edt)$_3$] utilized the same point group, citing the distortion to C_{3h} symmetry as relatively unimportant. Therefore, D_{3h} symmetry labels were applied and the coordinate system defined with the z-axis aligned with the threefold axis and the x-axis bisecting one dithiolene ligand (Fig. 31). Neither involved contributions beyond the {MS$_6$C$_6$} core.

The ligand field (LF) splitting of any metal ion in threefold symmetry sees the traditional t$_{2g}$ ($d_{xy,xz,yz}$) and e$_g$ ($d_{x^2-y^2,z^2}$) states of an octahedron transformed into three levels: a$_1'$ (d_{z^2}), e' ($d_{x^2-y^2,xy}$), e'' ($d_{xz,yz}$), in ascending order (Fig. 32). The designation of each specific d orbital reflects the change in coordinate system from TP to octahedral. Any twist away from a perfect TP, which represents a $D_{3h} \rightarrow D_3$ symmetry reduction, only alters the energy separation and not the LF ordering (Fig. 32). The change in Mulliken labels to a$_1$, e and e, are ignored for convenience, and D_{3h} labels are maintained throughout.

The original semiempirical calculations were based on simple LFT assuming the tris(dithiolene) unit exhibits some degree of oxidation and therein dithioketone character. This assessment is supported by the intraligand bond distances (172, 173, 175). Both sets of calculations combined eight dithiolene orbitals, four of which are π-type (nominally denoted π_{1-4}, Fig. 27) from the p$_z$ atomic orbitals of the {S$_2$C$_2$} unit (96, 181). Two sets of σ-type orbitals from the sp^2 hybridized sulfur atoms (Fig. 33), one symmetric (S$_{\sigma'}$) and antisymmetric (S$_{\sigma''}$) pair

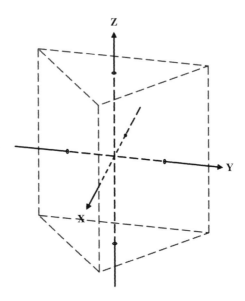

Figure 31.   The coordinate system defined for TP complexes.

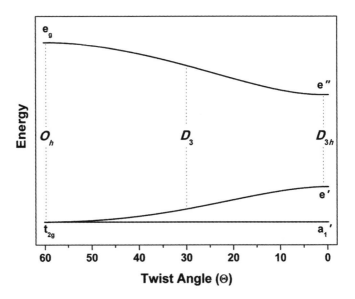

Figure 32.   Plot depicting the correlation between LF splitting and trigonal twist in tris(dithiolene) complexes.

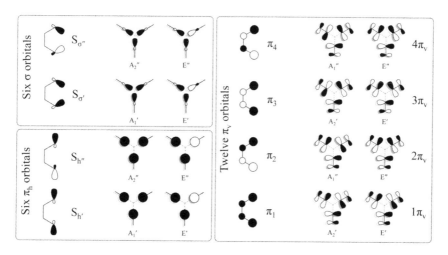

Figure 33.   Combination of the dithiolene ligand orbitals into three sets of σ, π_h, and π_v MOs for the tris(dithiolene) unit, as viewed along the threefold axis.

so-named because they are projected toward the metal ion (Fig. 34). The remaining sp^2 hybrid rotated 120° in the dithiolene plane was labeled as nonbonding by Schrauzer in contrast to Gray's designation as forming a mix of σ- and π-overlap with the d orbitals (termed "h" for horizontal, labeled $S_{h'}$ and $S_{h''}$). These dithiolene orbitals multiply to 24 orbitals when a tris(dithiolene) unit is assembled as shown in Fig. 33, with 6 σ, 6 π_h, and 12 π_v orbitals, where "v" (vertical) describes the {S_2C_2} π system orthogonal to the ligand plane. Gray labeled these $1\pi_v \rightarrow 4\pi_v$, while Schrauzer represented the degenerate and non-degenerate orbitals separately.

Combining 24 ligand orbitals with 9 from the metal ion (five d, one s, three p orbitals) generates the 33 outputted MOs from the computation, and explains the

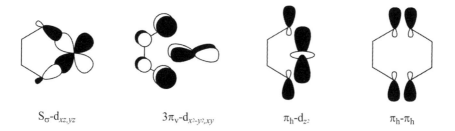

Figure 34.   Key orbital interactions in D_{3h} tris(dithiolene) complexes.

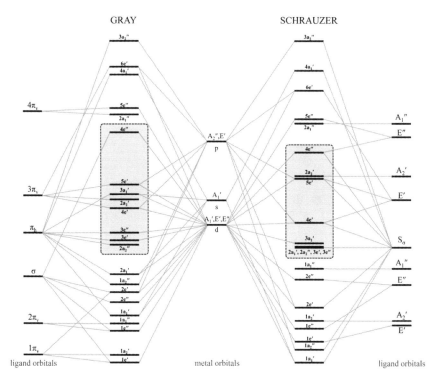

Figure 35.   Molecular orbital schemes developed independently by Gray and co-workers (181), and Schrauzer and Mayweg (96), for TP complex (D_{3h}) of the type [M(S$_2$C$_2$R$_2$)$_3$]. The key frontier levels are highlighted.

origin of the orbital numbering still in use today. The two schemes are contrasted in Fig. 35. The most conspicuous difference is the energetic ordering. Gray and co-workers (178) claimed the value of $-82,000\,\text{cm}^{-1}$ for the energy of the sp^2 hybrid orbital used by Schrauzer and Mayweg was unrealistic compared to his value of $-110,000\,\text{cm}^{-1}$. Furthermore, Schrauzer neglects contributions from the non-bonding S sp^2 orbitals (lone pair), whereas Gray mixes these with the d$_{z^2}$ orbital (Fig. 34). The proficiency of this interaction must greatly depend on the dithiolene ligand oxidation level, as the sp^2 hybrid orbitals are transformed into rudimentary p orbitals in the dithiolate form leading to increased overlap with the d orbitals. Schrauzer and co-workers (139) retorted that Gray used unreasonable Coulombic terms for the Re d orbitals that led to an incorrect bonding picture with an electron surplus on the ligands.

On the whole, the compositions of the key frontier MOs were largely consistent. The metal contribution dominates the 3a$_1$' MO, though Gray added that this is the antibonding component from a combination of ligand π_h and metal d_{z^2} orbitals. Its $2a_1'$ bonding counterpart is largely ligand based and stabilizes the trigonal prism, however, it was not regarded as an interligand bonding interaction (Fig. 16). The metal d orbitals are energetically matched to the $3\pi_v$ orbital. The $2a_2'$ MO is purely ligand based and defines the oxidation level of the tris(dithiolene) unit: filled it is $(L_3)^{6-}$; empty it is $(L_3)^{4-}$. The key bonding interaction is the $3\pi_v$ overlap with the metal $d_{x^2-y^2,xy}$ orbitals. While both the filled $4e'$ (bonding) and empty $5e'$ (antibonding) MOs were described as extensively delocalized over the complex framework, there is more metal character in the antibonding level. The overlap is maximized in a TP conformation and outweighs the interligand repulsion that manifests in the $2a_2'$ MO that is antibonding with respect to this geometry. It is easy to appreciate how $(L_3)^{6-}$ units favor a distorted octahedron.

The assignment of a metal oxidation state derives from the composition of the $4e'$ MOs as either metal or ligand centered, and the ordering of the four key orbitals: $3a_1'$, $2a_2'$, $4e'$, $5e'$. Schrauzer and Mayweg (96) skirted the issue by evoking conventional resonance theory, for example, the ground state of neutral [Mo(edt)$_3$] (and its group 6 (VI B) congeners) was represented by the three canonical forms shown in Fig. 36. The same approach was applied to describe the ground state in neutral bis(dithiolene) complexes with group 10 (VIII B) metals (282–284). For two dithiolates and one dithioketone, a +IV oxidation state is implied. The electron configuration for these neutral group 6 (VI B) complexes is $(3a_1')^2(4e')^4$. The paramagnetic neutral vanadium complex with one less electron has a $(3a_1')^2(4e')^3$ electron configuration while neutral rhenium is identified as $(3a_1')^2(4e')^4(5e')^1$. The authors did not indicate any metal oxidation states. The ordering of these frontier MOs was found to be independent of the metal, so Schrauzer could describe the reduced complexes by simply adding more electrons to the manifold, leading to a $(3a_1')^2(4e')^4(5e')^1$ electron configuration for the group 6 (VI B) monoanions and vanadium dianion, and a $(3a_1')^2(4e')^4(5e')^2$ configuration for the group 6 (VI B) dianions. These conclusions were assembled in light of known magnetic moments and appearance of metal hyperfine coupling in EPR spectra for the $S = 1/2$ species (96).

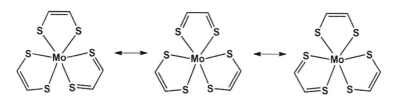

Figure 36.   Set of resonance structures describing the ground state of [Mo(edt)$_3$]. [Adapted from (96).]

It comes as no surprise that Gray (178, 181) presented a different set of ground-state electron configurations based on the same experimental evidence, principally EPR. The neutral rhenium and isoelectronic vanadium dianion were defined with a $(4e')^4(2a_2')^2(3a_1')^1$ electron configuration, whereas neutral vanadium has a $(4e')^4(3a_1')^1(2a_2')^0$ configuration with the 5e' level unoccupied. The key ingredient was the identical vanadium hyperfine coupling in the two $S = 1/2$ complexes that led to the decision to re-order the nonbonding $3a_1'$ and $2a_2'$ orbitals. Both are described with a $^2A_1$ ground term (177, 178). This result was at odds with conclusions drawn by the team at Harvard who preferred a $^2A_2$ ground term (10, 142).

Translating these electron configurations into oxidation states was treated very carefully in light of the mixed composition of the 4e' orbitals. Gray noted (181):

"If we consider 4e' as an orbital derived from $(d_{x^2-y^2,xy})$, and 5e' (empty) as essentially a $3\pi_v$ level, we then assign five electrons to the Re (four in 4e', one in $3a_1'$) and this would have a d^5 Re(II) configuration. The d orbital ligand field splitting then appears to be $xz,yz > z^2 > xy$, x^2-y^2. In this scheme, the two electrons in the $2a_2'$ symmetry orbital of the $3\pi_v$ set give the L_3 ligand unit a charge of -2. Thus in this limiting formulation of $[Re^{II}(L_3{}^{2-})]$, the ligand possesses considerable radical character. The other limiting formulation assigns 5e' as a $(d_{x^2-y^2,xy})$ level and considers 4e' as being derived from $3\pi_v$. Thus the ground-state configuration is $[4e' \ (3\pi_v)]^4[2a_2' \ (3\pi_v)]^2[3a_1' \ (d_{z^2})]^1$. This is a d^1 Re(VI) configuration and the apparent d orbital splitting is $xz, yz > xy$, $x^2-y^2 > z^2$. In accordance with this, the levels derived from $3\pi_v$ ($2a_2'$ and 4e') are filled, and the ligand unit assumes the configuration $(L_3)^{6-}$. In other words, in this limiting formulation the ligands are in classical dianionic form."

A similar opinion was adopted for $[V(pdt)_3]$ following its structural characterization (178), with a ground state somewhere between the $[V^V(L_3{}^{5-•})]$ and $[V^0(L_3{}^0)]$ extremes, bypassing the nonsensical notion of $[V^{VI}(L_3{}^{6-•})]$ that demands removal of an electron from the vanadium 3p level. Ultimately, the team from Columbia University subscribed to the traditional view of the 4e' MOs as predominately ligand based, and the corresponding 5e' as metal based.

Recent theoretical studies have confirmed Gray's manifold as the more accurate bonding representation in tris(dithiolene) complexes. However, despite strategically covering all possible metal oxidation states and complex formulations, $[Re(pdt)_3]$ was still cast as a metal-centered paramagnet along similar lines to $[V(pdt)_3]$ and $[V(pdt)_3]^{2-}$ with a $^2A_1$ ground term (177, 178, 181). It was 10 years before the ground state differences proposed by Gray ($^2A_1$) and Schrauzer ($^2E'$) were mediated in an EPR study of $[Re(pdt)_3]$ and $[Re(tdt)_3]$ by Al-Mowali and Porte (285). The trick involved averaging the Coulombic energies used by Gray and Schrauzer that gave a different ordering of the frontier orbitals in better agreement with the EPR data (Fig. 37). The ground state was reassigned as $^2A_2$

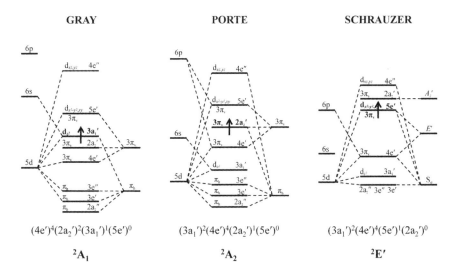

Figure 37.   Qualitative MO schemes for [Re(pdt)₃] as proposed by Gray and co-workers (181), Al-Mowali and Porte (285), and Schrauzer and Mayweg (96). The HOMO, electron configuration and ground term are indicated in each case.

with the unpaired electron circulating in the purely ligand-based $2a'_2$ MO, an idea originally rejected by Schrauzer and Mayweg. However, despite identifying the correct electronic structure for neutral Re tris(dithiolenes), the spectra were erroneously interpreted (247).

By building on these early computations, extended Hückel calculations were able to be applied to the group 5 (V B) complexes $[Nb(edt)_3]^{1-}$ and $[Ta(edt)_3]^{1-}$ (97), and $[V(bddt)_3]^{0/1-}$ (bddt^{2-} = 1,4-butanediyldithioethene-1,2-dithiolate) (89). Tatsumi et al. (97) defined the Nb and Ta complexes as possessing a $(4e')^4(2a_2')^2(3a_1')^0(5e')^0$ electron configuration with a +V metal ion and closed-shell ligands. Furthermore, they charted the relative energies of the frontier orbitals as a function of Θ, and showed a dramatic stabilization of the $2a_2'$ MO expected with the increased interligand separation and a concomitant destabilization of the $4e'$ level because of attenuated metal–ligand overlap. Isoelectronic $[V(dddt)_3]^{1-}$ and $[V(bddt)_3]^{1-}$ were characterized with the same electronic structure despite the geometric distortion favoring a C_{3h} polyhedron in contrast to the distorted octahedron for $[Nb(edt)_3]^{1-}$ (89). The $2a_2'$ HOMO and $3a_1'$ LUMO are separated by 0.6 and 0.3 eV for $[V(dddt)_3]^{1-}$ and $[V(bddt)_3]^{1-}$, respectively. The marginally larger covalency in the former led the authors to propose a slightly lower oxidation state for vanadium and some oxidized character to the (dddt)$^{2-}$ ligands. This conclusion is supported by the shorter S–C and longer C–C intraligand bond distances found for $[V(bddt)_3]^{1-}$ (218), though the resolution is not sufficient to make a clear assertion.

B. Fenske–Hall

In the three decades that followed the crystallographic characterization of [Mo (edt)$_3$], three more compounds were shown to possess the distortion to C_{3h} geometry such that it was no longer considered an anomaly of crystal packing. Campbell and Harris (269) used Fenske–Hall calculations in an effort to identify the cause and discovered it stemmed from a second-order Jahn–Teller distortion driven by a small HOMO–LUMO energy gap in neutral tris(dithiolene)molybdenum complexes. The HOMO is identified as the ligand-centered 2a$_2'$ MO; the LUMO is the metal-based 3a$_1'$ (d$_{z^2}$), and the energy gap is smallest for a regular trigonal prism ($\alpha = 0°$). This near degeneracy of orthogonal orbitals is alleviated by a geometric distortion that lowers the total molecular symmetry such that the HOMO and LUMO transform to the same symmetric representation, a$'$. This explanation is the definition of a second-order Jahn–Teller distortion, and the $D_{3h} \rightarrow C_{3h}$ dithiolene fold stabilizes the HOMO and destabilizes the LUMO, resulting in a more stable system (Fig. 38).

The energy gains from the distortion are dependent on the occupancy of the 2a$_2'$ and 3a$_1'$ orbitals, with the effect centered on neutral molybdenum and tungsten complexes, and their isoelectronic vanadium and niobium monoanions, as these diamagnetic species have two electrons in the 2a$_2'$ orbital and none in the 3a$_1'$ MO. In complexes with fewer electrons (e.g., neutral vanadium) or extra electrons (e.g., neutral rhenium), the benefits of folding are greatly diminished. The premise is that the mixing of formally metal-based LUMO and ligand-based HOMO produces stronger M−S bonds as d$_{z^2}$ character is introduced into the ground state with a concomitant increase in the dithiolene contribution to the LUMO (Fig. 39).

Large fold angles generate greater covalency as reflected in the metal–ligand overlap populations. Interligand repulsion is attenuated by decreased overlap between adjacent sulfur atoms in the envelope structure contributing to its stabilization. Despite the pronounced increase in metal contribution to the HOMO, Campbell and Harris continued to classify these complexes as d^0 systems [i.e., a Mo(VI) central ion]. They also deemed the input of the filled orbitals as inconsequential to the distortion (Fig. 38). In the C_{3h} structure, the 3π_v orbitals that constitute the 4e$'$ and 5e$'$ levels are no longer orthogonal to the d$_{z^2}$, and a second bonding interaction ensues. However, this notion was dismissed because it weakened the existing bond between 3π_v and d$_{x^2-y^2,xy}$ orbitals.

Although identifying the correct orbital interaction that drives the dithiolene fold, Campbell and Harris erroneously assigned the HOMO and LUMO, which are in fact metal and ligand centered, respectively (281). Furthermore, for the polarizable V−S bonds in the isoelectronic vanadium tris(dithiolene) monoanions,

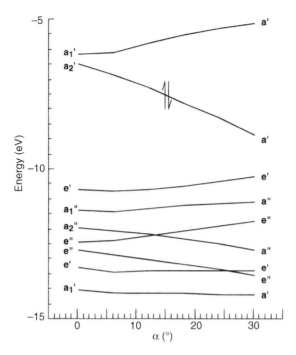

Figure 38. Variations in the energies of the LUMO and 12 highest occupied orbitals of [Mo(edt)$_3$] as a function of the fold angle, α. [Adapted from (269).]

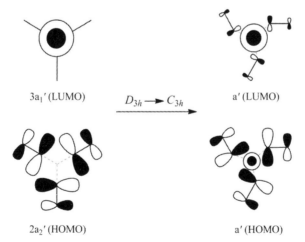

Figure 39. Orthogonal $2a_2'$ and $3a_1'$ orbitals in D_{3h} symmetry overlap upon folding the dithiolene ligands along the S $\cdots$ S vector to produce a' orbitals of mixed metal and ligand character in C_{3h} symmetry. [NB: the position of the S p orbitals is such that they are not overlapping with the torus of the d_{z^2} orbital in the a' MOs].

the fold aids in maximizing the antiferromagnetic coupling between metal- and ligand-centered magnetic orbitals (179, 248). These revelations coincided with the advent of DFT, the salient difference being that these calculations targeted spectroscopic observables rather than exclusively focusing on geometric phenomena.

VI.   ELECTROCHEMISTRY

The team at Harvard of Davison, Edelstein, Holm and Maki were the first to recognize that $[Mo(tfd)_3]$ and $[Cr(mnt)_3]^{3-}$ differed by three valence electrons (10). Reversible redox behavior was one of the key properties exhibited by all transition metal dithiolene complexes and, in the case of tris(dithiolene) species, a change in geometry. As such, a wide variety of electrochemical techniques have been applied to these systems. The eclectic array of working electrodes, solvents, electrolytes, and reference electrodes used in different setups renders reduction potentials difficult to compare since they are referenced to different internal and external standards. Redox potentials listed in Table IV are reported versus the internal ferrocenium/ferrocene $(Fc^{+/0})$ couple, as endorsed by the electrochemistry commission (286). Those potentials not referenced in this manner have been converted as noted in Table IV by using known redox potentials for $Fc^{+/0}$ for different solvents, electrolytes, and reference electrodes (160); in some cases reasonable guesses have been made. Note that redox events are notoriously sensitive to a host of factors (287), and these converted values can vary dramatically in different systems.

Most tris(dithiolene) complexes form multi-membered electron-transfer series with complex charges ranging $1+ \rightarrow 4-$ according to Eq. 3.

$$[M(L_3)]^{1+} \underset{-e^-}{\overset{+e^-}{\rightleftharpoons}} [M(L_3)]^{0} \underset{-e^-}{\overset{+e^-}{\rightleftharpoons}} [M(L_3)]^{1-} \underset{-e^-}{\overset{+e^-}{\rightleftharpoons}} [M(L_3)]^{2-} \underset{-e^-}{\overset{+e^-}{\rightleftharpoons}} [M(L_3)]^{3-} \underset{-e^-}{\overset{+e^-}{\rightleftharpoons}} [M(L_3)]^{4-}$$

$$(3)$$

Two series have five members: $[Re(bdt)_3]^z$ and $[Re(tdt)_3]^z$ $(z = 1+, 0, 1-, 2-, 3-)$; the cyclic voltammogram of $[Re(bdt)_3]$ is shown in Fig. 40 (247). Complexes with Mn, Fe, Co, Ru, and Rh typically display one reversible process; the accessibility of different charge states is more dependent on the metal rather than the properties of the ligand. Lewis acidic Nb and Ta also have one reversible $1-/2-$ redox couple, in contrast with V, which is found in a range of states, akin to the group 6 (VI B) metals where stable 0, $1-$, and $2-$ species are common. One obvious purpose of these measurements is to identify which members of the species can be potentially isolated, and the appropriate chemical reagent to effect their oxidation/ reduction. These data ultimately led to the very recent isolation and characterization of $[Re(pdt)_3](SbCl_6)$ (248), after it was forecast to be accessible given the mild potential of the $1+/0$ couple (287).

TABLE IV
Reduction Potentials for Tris(dithiolene) Complexes[a]

Complex	Solvent	Correction	1+/0	0/1−	1−/2−	2−/3−	3−/4−	References
[Ti(bdt)$_3$]	MeCN	b			−0.24 ir	−2.06 r		125
[Ti(bdtCl$_4$)$_3$]	CH$_2$Cl$_2$	c			+0.18 ir	−1.43 r		288
[Ti(dddt)$_3$]	DMF	d			−0.39 ir	−1.99 r		211
[Zr(bdt)$_3$]	MeCN	b				−2.25 r		125
[V(bdt)$_3$]	CH$_2$Cl$_2$			+0.12 ir	−0.62 r	−1.92 r		179
[V(tdt)$_3$]	CH$_2$Cl$_2$			+0.09 ir	−0.66 r	−1.93 r		179
[V(bdtCl$_4$)$_3$]	CH$_2$Cl$_2$	c		+0.62 ir	−0.21 r	−1.35 r		288
[V(etdt)$_3$]	MeCN	e		−0.36 r	−1.32 r			98
[V(mnt)$_3$]	CH$_2$Cl$_2$		+1.12 ir	+0.13 r	−0.92 r	−2.09 r		179
[V(tfd)$_3$]	CH$_2$Cl$_2$	c		+0.82 ir	−0.39 r	−1.44 r		288
[V(sdt)$_3$]	CH$_2$Cl$_2$	f		−0.30 r	−1.23 r			146
[V(pdt)$_3$]	CH$_2$Cl$_2$		+0.68 ir	−0.17 r	−1.19 r	−2.21 r		179
[V(dtdt)$_3$]	DMF	g		−0.19 r	−1.12 r			139
[V(dadt)$_3$]	DMF	g		−0.26 r	−1.16 r			139
[V(dmit)$_3$]	CH$_2$Cl$_2$	h		+0.17 r	−0.33 r			161
[V(dddt)$_3$]	DMF	d		−0.08 r	−0.98 r	−1.91 r		218
[V(bddt)$_3$]	DMF	d		+0.18 r	−0.82 r	−1.86 r		89
[Nb(bdt)$_3$]	CH$_2$Cl$_2$	i			−0.92 r			125
[Nb(edt)$_3$]	DMSO	j			−1.32 r			97
[Nb(etdt)$_3$]	MeCN	e			−1.52 r			98
[Nb(dmit)$_3$]	CH$_2$Cl$_2$	h		−0.24 ir	−1.28 r			161
[Nb(dddt)$_3$]	DMF	d		+0.06 ir	−1.10 r	−2.13 r		211
[Ta(bdt)$_3$]	CH$_2$Cl$_2$	i			−1.25 r			125
[Ta(edt)$_3$]	DMSO	j			−1.49 r			97
[Cr(bdtCl$_4$)$_3$]	CH$_2$Cl$_2$	c		+0.47 r	−0.27 r	−0.74 r		288
[Cr(bdtCl$_2$)$_3$]	CH$_2$Cl$_2$			+0.19 ir	−0.36 r	−0.85 r		221
[Cr(tbbdt)$_3$]	CH$_2$Cl$_2$			−0.27 r	−0.94 r	−1.57 r		222
[Cr(mnt)$_3$]	CH$_2$Cl$_2$			+0.85 ir	+0.29 r	−0.34 r		221
[Cr(tfd)$_3$]	MeCN	k		+0.65 r	−0.09 r			16
[Cr(pdt)$_3$]	CH$_2$Cl$_2$	l		−1.45 r				289
[Cr(dtcr)$_3$]	MeCN	m				+0.01 r		103
[Cr(dcmdtcr)$_3$]	MeCN	m				+0.23 r		103
[Mo(bdt)$_3$]	CH$_2$Cl$_2$		+0.93 ir	−0.20 r	−0.75 r			232
[Mo(tdt)$_3$]	CH$_2$Cl$_2$	f		−0.27 r	−0.93 r			38
[Mo(bdtCl$_2$)$_3$]	MeCN	n		+0.14 r	−0.45 r			226
[Mo(tbbdt)$_3$]	CH$_2$Cl$_2$		+0.91 ir	−0.48 r	−0.99 r			227
[Mo(vdt)$_3$]	CH$_2$Cl$_2$	f		−0.53 r	−1.00 r			38
[Mo(15C5-bdt)$_3$]	CH$_2$Cl$_2$	f		−0.53 r	−1.00 r			38
[Mo(qdt)$_3$]	CH$_2$Cl$_2$			+0.58 ir	−0.24 r	−1.87 r		232
[Mo(edt)$_3$]	CH$_2$Cl$_2$		+0.94 ir	−0.61 r	−1.12 r			232
[Mo(mdt)$_3$]	MeCN	e		−0.74 r	−1.26 r			231
[Mo(etdt)$_3$]	MeCN/THF	e		−0.76 r	−1.30 r			98
[Mo(mnt)$_3$]	CH$_2$Cl$_2$		+0.70 qr	−0.15 r	−1.51 r	−2.55 ir		232
[Mo(tfd)$_3$]	MeCN			+0.28 r	−0.28 r			50
[Mo(sdt)$_3$]	CH$_2$Cl$_2$		+0.85 ir	−0.61 r	−0.99 r			232
[Mo(toldt)$_3$]	CH$_2$Cl$_2$	o		−0.63 r	−1.00 r			68
[Mo(adt)$_3$]	CH$_2$Cl$_2$	o		−0.68 r	−1.04 r			68
[Mo(pdt)$_3$]	CH$_2$Cl$_2$		+0.54 ir	−0.60 r	−1.06 r			232
[Mo(apdt)$_3$]	CH$_2$Cl$_2$	o		−0.62 r	−1.07 r			68
[Mo(qpdt)$_3$]	MeCN	p		−0.24 r	−0.68 r			64

(*continued*)

TABLE IV
(Continued)

Complex	Solvent	Correction	1+/0	0/1−	1−/2−	2−/3−	3−/4−	References
[Mo(dmm)$_3$]	CH$_2$Cl$_2$	q		+0.00 r	−0.46 r			60
[Mo(dmid)$_3$]	CH$_2$Cl$_2$	r		−0.05 r	−0.40 r	−1.82		155
[Mo(dmit)$_3$]	CH$_2$Cl$_2$	h		−0.10 r	−0.30 r			238
[Mo(dhdndt)$_3$]	CH$_2$Cl$_2$	m		−0.16 r	−0.63 r			290
[W(bdt)$_3$]	CH$_2$Cl$_2$		+1.00 ir	−0.23 r	−0.92 r			232
[W(tdt)$_3$]	DMF	s		−0.21 r	−1.04 r			181
[W(bdtCl$_4$)$_3$]	CH$_2$Cl$_2$	c		+0.16 r	−0.43 r			288
[W(bdtCl$_2$)$_3$]	MeCN	n		−0.01 r	−0.67 r			226
[W(tbbdt)$_3$]	CH$_2$Cl$_2$		+0.95 ir	−0.47 r	−1.11 r			227
[W(vdt)$_3$]	CH$_2$Cl$_2$	f		−0.52 r	−1.12 r			38
[W(15C5-bdt)$_3$]	CH$_2$Cl$_2$	f		−0.51 r	−1.12 r			38
[W(edt)$_3$]	DMF	g		−0.65 r	−1.22 r			139
[W(mdt)$_3$]	CH$_2$Cl$_2$		+0.07 ir	−0.89 r	−1.38 r			232
[W(etdt)$_3$]	MeCN/THF	e		−0.77 r	−1.37 r			98
[W(mnt)$_3$]	CH$_2$Cl$_2$			+0.67 r	+0.07 r	−1.80 r	−2.32 ir	232
[W(tfd)$_3$]	MeCN			+0.22 r	−0.11 r			50
[W(sdt)$_3$]	CH$_2$Cl$_2$	o		−0.62 r	−1.07 r			68
[W(toldt)$_3$]	CH$_2$Cl$_2$	o		−0.69 r	−1.16 r			68
[W(adt)$_3$]	CH$_2$Cl$_2$	o		−0.82 r	−1.23 r			68
[W(csdt)$_3$]	CH$_2$Cl$_2$	o		−0.56 r	−0.93 r			68
[W(bsdt)$_3$]	CH$_2$Cl$_2$	o		−0.59 r	−1.01 r			68
[W(pdt)$_3$]	CH$_2$Cl$_2$		+0.84 ir	−0.61 r	−1.09 r			232
[W(dtdt)$_3$]	DMF	g		−0.65 r	−1.22 r			139
[W(dadt)$_3$]	DMF	g		−0.61 r	−1.13 r			139
[W(andt)$_3$]	DMF	g		−0.66 r	−1.13 r			139
[W(apdt)$_3$]	CH$_2$Cl$_2$	o		−0.65 r	−1.12 r			68
[W(anpdt)$_3$]	CH$_2$Cl$_2$	o		−0.76 r	−1.21 r			68
[W(dem)$_3$]	MeCN	t		+0.49 r	+0.04 r			62
[W(dmit)$_3$]	CH$_2$Cl$_2$	h		−0.16 r	−0.34 r			238
[W(dmid)$_3$]	CH$_2$Cl$_2$	r		−0.11 r	−0.45 r			155
[W(btdt)$_3$]	CH$_2$Cl$_2$	r		−0.78 r	−1.01 r			155
[Mn(bdtCl$_4$)$_3$]	CH$_2$Cl$_2$	c		+0.23 ir	+0.02 r	−1.30 r	−1.50 ir	288
[Mn(bdtCl$_2$)$_3$]	CH$_2$Cl$_2$				−0.12 ir	−1.53 r		221
[Mn(mnt)$_3$]	CH$_2$Cl$_2$					−0.70 r		221
[Mn(dddt)$_3$]	DMF	d			−0.34 ir	−1.02 qr		211
[Tc(tdt)$_3$]	CH$_2$Cl$_2$	u		+0.16 r	−0.25 r	−1.34 r		144
[Re(bdt)$_3$]	CH$_2$Cl$_2$			+0.43 r	−0.01 r	−1.43 r	−2.25 r	247
[Re(tdt)$_3$]	DMF	s		+0.43 r	−0.02 r	−1.54 r	−2.34 r	180
[Re(bdtCl$_4$)$_3$]	CH$_2$Cl$_2$	c	+0.96 ir	+0.33 r	−1.06 r			288
[Re(bdtCl$_2$)$_3$]	CH$_2$Cl$_2$			+0.45 r	−1.14 r			247
[Re(tms)$_3$]	CH$_2$Cl$_2$			+0.27 r	−0.30 r	−1.89 r		247
[Re(mnt)$_3$]	CH$_2$Cl$_2$	c		+0.90 r	−0.45 r	−1.13 r	−2.22 qr	291
[Re(pdt)$_3$]	DMF	s	+0.20 r	−0.30 r	−1.77 r	−2.55 ir		180
[Re(dmit)$_3$]	CH$_2$Cl$_2$	h		+0.26 ir	−1.11 r	−1.76 r		163
[Fe(bdtCl$_4$)$_3$]	CH$_2$Cl$_2$	c			−0.11 r	−1.15 r	−1.89 r	288
[Fe(qdt)$_3$]	MeCN	v				−0.07 r		292
[Fe(mnt)$_3$]	CH$_2$Cl$_2$				+0.25 ir	−0.75 r		249
[Fe(dtcr)$_3$]	MeCN	m				−0.31 r		103
[Fe(dcmdtcr)$_3$]	MeCN	m				−0.16 r		103
[Ru(mnt)$_3$]	CH$_2$Cl$_2$					−0.70 r		251
[Os(bdt)$_3$]	CH$_2$Cl$_2$	e		−0.22 r	−0.48 r	−0.98 ir		252

TABLE IV
(*Continued*)

Complex	Solvent	Correction	1+/0	0/1−	1−/2−	2−/3−	3−/4−	References
[Os(tdt)$_3$]	DMF	w		−0.20 r	−1.05 r	−2.04 r		293
[Co(qdt)$_3$]	DMSO	x				+0.11 r		292
[Co(mnt)$_3$]	CH$_2$Cl$_2$	c				−0.35 r		135
[Co(dtcr)$_3$]	MeCN	m				+0.02 r		103
[Co(dcmdtcr)$_3$]	MeCN	m				+0.20 r		103
[Rh(mnt)$_3$]	CH$_2$Cl$_2$	c		+0.21 ir	−0.15 qr			294
[Sn(dmit)$_3$]	CH$_2$Cl$_2$	y		+0.48 ir	+0.24 ir			257
[Sn(eTTFdt)$_3$]	CH$_2$Cl$_2$	y	+0.60 ir	+0.11 ir	−0.20 ir			257

[a] In V versus Fc$^{+/0}$. Ligand abbreviations are presented in Figs. 2 and 4. DMSO = dimethyl sulfoxide, MeCN = acetonitrile, DMF = N,N'-dimethylformamide, THF = tetrahydrofuran, r = reversible, qr = quasireversible, ir = irreversible.

[b] Referenced to Ag/AgCl electrode with [N(n-Pr)$_4$]ClO$_4$ as electrolyte. Potentials are shifted −0.54 V.

[c] Referenced to saturated calomel electrode (SCE) with [NEt$_4$]ClO$_4$ as electrolyte. Potentials are shifted −0.38 V.

[d] Referenced to Ag/AgCl with [N(n-Bu)$_4$]ClO$_4$ as electrolyte. Potentials are shifted −0.47 V.

[e] Experimental Fc$^{+/0}$ potential at +0.40 V.

[f] Experimental Fc$^{+/0}$ potential at +0.56 V.

[g] Referenced to Ag/AgCl electrode with LiClO$_4$ as electrolyte. Potentials are shifted −0.52 V.

[h] Referenced to SCE with [N(n-Bu)$_4$]ClO$_4$ as electrolyte. Potentials are shifted −0.48 V.

[i] Referenced to Ag/AgCl electrode with [N(n-Bu)$_4$]PF$_6$ as electrolyte.

[j] Referenced to SCE with [N(n-Bu)$_4$]ClO$_4$ as electrolyte. Potentials are shifted −0.45 V.

[k] Referenced to SCE with [N(n-Pr)$_4$]ClO$_4$ as electrolyte. Potentials are shifted −0.49 V.

[l] Referenced to SCE electrode with [N(n-Bu)$_4$]PF$_6$ as electrolyte. Potentials are shifted −0.42 V.

[m] Referenced to SCE with [N(n-Bu)$_4$]BF$_4$ as electrolyte. Potentials are shifted −0.46 V.

[n] Experimental Fc$^{+/0}$ potential at +0.48 V.

[o] Referenced to Ag/AgCl electrode with [N(n-Bu)$_4$]ClO$_4$ as electrolyte. Potentials are shifted −0.75 V.

[p] Experimental Fc$^{+/0}$ potential at +0.45 V.

[q] Referenced to SCE with [N(n-Bu)$_4$]ClO$_4$ as electrolyte. Potentials are shifted −0.48 V.

[r] Referenced to Ag/AgCl electrode with [N(n-Bu)$_4$]PF$_6$ as electrolyte. Potentials are shifted −0.38 V.

[s] Referenced to Ag/AgClO$_4$ with [N(n-Pr)$_4$]ClO$_4$ as electrolyte. Potentials are shifted +0.04 V.

[t] Experimental Fc$^{+/0}$ potential at +0.44 V.

[u] Referenced to SCE with [N(n-Bu)$_4$]ClO$_4$ as electrolyte. Potentials are shifted −0.43 V.

[v] Referenced to Ag/AgCl electrode with [N(n-Bu)$_4$]ClO$_4$ as electrolyte. Potentials are shifted −0.38 V.

[w] Referenced to SCE with [N(n-Bu)$_4$]ClO$_4$ as electrolyte. Potentials are shifted −0.47 V. Data were markedly different to [Os(bdt)$_3$], and the values have been moved to correspond with more reasonable redox couples. This erroneous assignment may stem from an incorrect resting potential.

[x] Referenced to Ag/AgCl electrode with [N(n-Bu)$_4$]ClO$_4$ as electrolyte. Potentials are shifted −0.45 V.

[y] Referenced to Ag/AgCl electrode with [N(n-Bu)$_4$]ClO$_4$ as electrolyte. Data are uncorrected.

The redox potentials for any given complex are dependent on the dithiolene substituents (13, 139). Those with electron-withdrawing groups, such as (mnt)$^{2−}$, (tfd)$^{2−}$, and (qdt)$^{2−}$, tend to favor reduced complexes (dianions and trianions) and stabilize tris(dithiolene) species with Fe, Co, and Ru. This finding presumably stems from the difficulty of oxidizing these dithiolene ligands, which are also strong π-donors. Oxidized compounds are promoted by dithiolene ligands with electron-donating

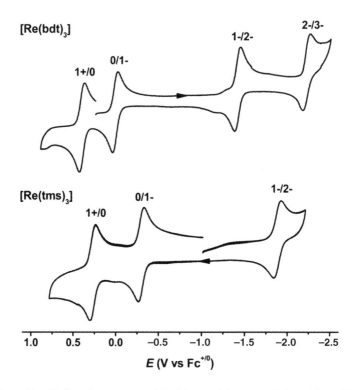

Figure 40.   Cyclic voltammograms of [Re(bdt)$_3$] and [Re(tms)$_3$]. [Adapted from (247).]

groups (e.g., alkyl-substituted arene-1,2-dithiolates and dialkyl alkene-1,2-dithio-lates). The latter are more oxidizable because they lack the stabilizing aromatic group, and the redox potential varies with the ligand substituent. The was neatly conveyed in a seminal study of a series of W complexes with varying para substituents on the extremity of the dithiolene ligand (68). For the series [W(R-sdt)$_3$], where R = H, Me, OMe, Cl, Br (Fig. 4), the ease of reduction followed the trend Cl > Br > H > Me > OMe, spanning a range of 263 mV. A linear relationship with the Hammett constant was demonstrated (148). The same trend is operative for aryl dithiolates, with [W(bdt)$_3$] more difficult to reduce than [W(vdt)$_3$], but markedly easier than [W(bdtCl$_2$)$_3$], following the trend Cl > H > OMe.

Diagnoses of the redox event as (predominantly) metal or ligand centered can be determined by the difference in the peak potentials for successive redox events. For example, the cyclic voltammogram for [Re(bdt)$_3$] shown in Fig. 40 has four reversible waves with the first two separated by ~0.3 V compared with ~1.5 and ~0.9 V for the next two (247). The first two reductions can be regarded as ligand centered with minimal shift of the redox potential corresponding to the absence of any substantial structural rearrangement of the complex. Further reduction to the dianion is metal-centered [Re(V) → Re(IV), as determined by spectroscopy and theory] with the formation of a Jahn–Teller distorted, low-spin d^3 ion (247, 248). The shift in the potential can be related to the adjustment of the complex structure to accommodate an additional electron on the metal. The spacings are similar for [Re (tms)$_3$] (Fig. 40) and other Re containing complexes such that this trend holds in these systems. The redox chemistry in the Cr complexes is regarded as completely ligand based (221, 222), and, although the redox potentials reflect the nature of the dithiolene ligand, the difference between successive events is very similar irrespective of the ligand type. The first two reduction potentials for Mo and W homologues have very similar values that reflect their ligand-centered nature (38, 50, 68, 226, 232). On the whole, Mo complexes are only slightly easier to reduce than their W counterparts, which indicate a minor dependency of the reduction potential on the metal ion. The functionalization of (bdt)$^{2-}$ with 15-crown-5 (15C5) provides a binding site for alkali metal cations (295). The redox potentials for [M(15C5-bdt)$_3$] (M = Mo, W and 15C5-bdt = 2,3,5,6,8,9,11,12,14,15-decahydro-1,4,7,10,13,16-benzohexaoxacyclooctadecine-18,19-dithiolate) become more positive (up to 110 mV) upon inclusion of Li$^+$ cations while retaining their reversibility. More complicated voltammograms are observed for Na$^+$ binding, which arise from changes in the complex stability upon reduction of the neutral species to the monoanionic and dianionic forms, in addition to overlapping processes for the bound and unbound ligands in the system.

A similar positive shift of the reduction potential of [Mo(qdt)$_3$]$^{2-}$ was observvsed when the complex was treated with acid (296). The process became irreversible while the oxidation wave remained unchanged. One nitrogen atom of the (qdt)$^{2-}$ ligand is protonated and the shift consistent with a decrease in negative charge. Further addition of acid leads to protonation of each dithiolene ligand and isolation of turquoise [Mo(qdtH)$_3$]$^{1+}$ (230). This species retains the one-electron oxidation wave and also exhibits a three-electron quasireversible reduction that the authors speculated involved simultaneous reduction of all three (qdtH)$^{1-}$ ligands and a loss of aromaticity of the pyrazine ring. Interestingly, cyclic voltammetry of quinoxaline-2,3-dithiol in the presence of acid does not exhibit this electrochemical behavior and it was inferred that the formation of quinoxalinium radicals requires a metal ion to be present.

VII. MAGNETOMETRY

Magnetic moments of assorted tris(dithiolene) complexes were principally determined at room temperature using a Gouy balance. This efficient technique determines the spin ground state for all members of any given electron-transfer series. Generally, the ground states vary between diamagnetic ($S = 0$) and spin-doublets ($S = 1/2$) for the vast majority of compounds of this type given the

TABLE V
Spin Ground States for $[M(\text{dithiolene})_3]^z$ ($z = 1+, 0, 1-, 2-, 3-, 4-$) Complexes[a]

Metal	1+	0	1−	2−	3−	4−
Ti			$S = 1/2$	$S = 0$	$S = 1/2$	
Zr				$S = 0$	$S = 1/2$	
Hf				$S = 0$		
V	$S = 0$	$S = 1/2$	$S = 0$	$S = 1/2$	$S = 1$	$S = 3/2$
Nb		$S = 1/2$	$S = 0$	$S = 1/2$	$S = 0$	
Ta			$S = 0$	$S = 1/2$		
Cr		$S = 0$	$S = 1/2$	$S = 1$	$S = 3/2$	
Mo	$S = 1/2$	$S = 0$	$S = 1/2$	$S = 0$	$S = 1/2$	$S = 1(?)$
W	$S = 1/2$	$S = 0$	$S = 1/2$	$S = 0$	$S = 1/2$	
Mn			$S = 1$	$S = 3/2$	$S = 2(?)$	
Tc	$S = 0$	$S = 1/2$	$S = 0$	$S = 1/2$		
Re	$S = 0$	$S = 1/2$	$S = 0$	$S = 1/2$	$S = 1(?)$	$S = 1/2(?)$
Fe				$S = 1$	$S = 1/2$	$S = 0(?)$[b]
Ru				$S = 1$	$S = 1/2$	
Os		$S = 0$	$S = 1/2$	$S = 1$	$S = 1/2$	
Co				$S = 1/2$	$S = 0$	
Rh				$S = 1/2$	$S = 0$	
Pt				$S = 0$		

[a] List of redox states generated based on available experimental evidence; ? indicates tentative assignment.
[b] The complex $[\text{Fe}^{\text{II}}(\text{Me}_2\text{pipdt})_3]^{2+}$ is diamagnetic (124).

low-spin preference for heavy metals (Table V). The diamagnetic species gave sharp NMR signals and the $S = 1/2$ compounds were routinely examined by EPR spectroscopy. As such, detailed magnetic susceptibility measurements are restricted to the handful of compounds with $S = 1$ ground states. Spin-quartet species, of which there are only four examples, are analyzed by both magnetic susceptibility and EPR.

The magnetic moments for $[\text{Cr}(\text{bdtCl}_2)_3]^z$ ($z = 1-, 2-, 3-$) elegantly display the utility of such measurements across an electron-transfer series (Fig. 41) (221). Similar Cr based series with different redox-active ligands trend in this way (222, 280, 297–306). The spin ground state is instantly identified as $S = 1/2$ for the monoanion, $S = 1$ for the dianion, and $S = 3/2$ for the trianion, as the room temperature magnetic moment is equivalent to the spin-only value for 1 (1.73 μ_B), 2 (2.82 μ_B), and 3 (3.87 μ_B) unpaired electrons, respectively, where μ_B is the Bohr magneton. The drop in the effective magnetic moment (μ_{eff}) below 20 K stems from field saturation and intermolecular interactions in $[\text{Cr}(\text{bdtCl}_2)_3]^{1-}$, and the presence of zero-field splitting (ZFS, parameterized by D and E) for the $S = 1$ and $S = 3/2$ species. It is therefore difficult to accurately determine D from a

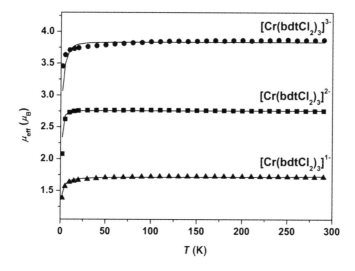

Figure 41.   Temperature dependence of the magnetic moment, μ_{eff}, μ_B of powder samples of $[Cr(bdtCl_2)_3]^z$ ($z = 1-, 2-, 3-$) at 1 T external field. Symbols represent experimental data and solid lines the fit. [Adapted from (219).]

straightforward magnetic measurement, but its magnitude can be estimated. For $[Cr(bdtCl_2)_3]^{3-}$, having a noninteger spin state (Kramer's doublet), EPR spectroscopy is more adept at determining ZFS (see Section VIII.D.2); this value can be then fixed when fitting magnetic data. For non-Kramer's systems, EPR is not practical at the frequencies available in most research laboratories, therefore magnetic data alone are used to estimate ZFS. Importantly, susceptibility measurements only provide the total spin ground state for a molecule, so these Cr dianions ($S = 1$) were classified with a low-spin Cr(II) d^4 ion rather than the true configuration of a Cr(III) $d^3(S = 3/2)$ ion antiferromagnetically coupled to one dithiolene radical ($S = 1/2$). This determination relies on an element specific probe (e.g., XAS).

The spin-triplet ground state of $[Fe(mnt)_3]^{2-}$ is clearly seen in the magnetic susceptibility plot with an effective magnetic moment of 2.86 μ_B between 50–290 K (Fig. 42) (249). The decrease below 50 K is due to magnetization saturation at 1 T and the influence of ZFS. Isofield magnetization experiments at 1, 4, and 7 T were fit for $g = 2.02$ and $D = +10\,cm^{-1}$, identical to previous studies (307), and corroborated by applied-field Mössbauer data (308). The one-electron reduced species $[Fe(mnt)_3]^{3-}$ has a temperature-independent magnetic moment of 1.82 μ_B (10–200 K) indicative of an $S = 1/2$ ground state and $g = 2.09$ (Fig. 42). Interestingly, the progressive rise in μ_{eff} at 200 K is characteristic of a thermal spin-crossover from low-spin Fe(III) $S = 1/2$ to high-spin Fe(III) $S = 5/2$ in the solid state. For homologues with heavier metal ions at their core, orbital contributions

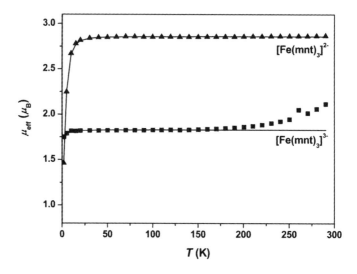

Figure 42.   Temperature dependence of the magnetic moment, μ_{eff}, μ_B of powder samples of [Fe(mnt)$_3$]^z ($z = 2-$, 3$-$) at 1 T external field. Symbols represent experimental data and solid lines the fit. [Adapted from (249).]

attenuate the magnetic moment, from a room temperature value of 1.69 μ_B for [Ru(mnt)$_3$]$^{3-}$ (251), down to a meager 1.1 μ_B for both [Rh(mnt)$_3$]$^{2-}$ (294) and [Os(tdt)$_3$]$^{1-}$ (293).

VIII.   SPECTROSCOPY

A vast array of spectroscopic instrumentation have been employed to test, poke, and prod tris(dithiolenes) over the decades in order to unlock the intricacies of their molecular and electronic structure. The techniques range from vibrational and electronic absorption, to magnetic resonance and element specific probes (e.g., XAS and Mössbauer spectroscopy). A summary of the insight that each provided is detailed in this section.

A.   Vibrational

To say that vibrational spectroscopy has been sparingly applied to tris(dithiolene) complexes is to flirt recklessly with understatement. Instead they live in the shadow cast by their bis(dithiolene) counterparts, with no all-encompassing assignment of IR and Raman bands for any members of this class. The exception are main group tris(dithiolenes)—principally tin—who have been recipients of a thorough examination of their vibrational spectra to validate band assignments and test the accuracy of theoretical modeling (309–311). However, these investigations are only distantly relevant to the electronic structure. For transition metals where dithiolene redox activity is operative, an initial foray by Schrauzer and Mayweg (96) identified five IR bands (denoted ω_{1-5}), which represent the perturbed C=C, C=S, RC(=S)C, and M−S stretching vibrations of the metallodithiolene unit. In D_{3h} symmetry, the C=C stretch is ω_1 and of the type A_2'', the C−S modes ω_2 and ω_3 are E' and A_2'', respectively, as are the M−S stretching modes labeled ω_4 and ω_5. The A_1' and E' modes are only Raman active; E'' works for both. Despite the limited set of compounds, Schrauzer and Mayweg (166) identified a similar trend to that displayed by bis(dithiolene) compounds, namely, the increase in ν(C−C) with simultaneous decrease in the ν(C−S) and ν(M−S) as the complex becomes more electron deficient. Parallel findings have been reported for [M(dmit)$_3$] (M = V, Mo, Re) complexes (163, 216, 237, 238). Subsequent studies have played down this assignment since the C−S and M−S modes exist in concert with in-plane deformation of the metallodithiolene unit. Moreover, the ν(C−S) was reassigned in the preceding overview of this subject (312) in line with an in-depth analysis performed by Schläpfer and Nakamoto (313) on bis(dithiolene) nickel complexes with (edt)$^{2-}$, (tfd)$^{2-}$, (pdt)$^{2-}$, and (mnt)$^{2-}$. The band ranging from 800–900 cm^{-1} was designated as predominantly ν(C−S) in nature with the higher energy feature at $\sim$1100 cm^{-1} arising from the combined input of ν(C−S) and ν(C−R), where R is the dithiolene substituent. This logical reevaluation ties in with the dithiolene radical fingerprint identified in the resonance Raman spectra of [M (tbbdt)$_2$]^z (M = Ni, Pd, Pt, Cu, Au; $z = 0$, 1−, 2−) complexes (314). Enhancement of a band at $\sim$1100 cm^{-1} (B_{3u} vibration in D_{2h} symmetry—IR and Raman active) with significant ν(C−S) character was seen in spectra obtained in resonance with the intervalence charge transfer (IVCT) electronic transition, dithiolate → dithienyl radical. This finding is evidence for the presence of a coordinated dithiolene radical. Complementary IR spectra also display a prominent band at $\sim$1100 cm^{-1} for complexes bearing dithiolene radicals, which is noticeably absent in those species with closed-shell ligands (276, 314). However, the intensity of this feature is dependent on the arene-1,2-dithiolate substituents: Pronounced for (tbbdt)$^{2-}$ complexes and nonexistent in (bdt)$^{2-}$ homologues. Despite its varying reliability, the IR band was identified in solid-state (KBr) and dichloromethane solution spectra for [Cr (tbbdt)$_3$]$^{1-}$ (222), highlighting the existence of dithiolene radicals in this compound (Fig. 43). In concert with other spectroscopic markers, its electronic structure was formulated [Cr^{III}(tbbdt)(tbbdt$^{\bullet}$)$_2$]$^{1-}$ ($S = 1/2$). More interestingly was the discovery of the band for the neutral and dianionic members of the electron-transfer series, which shifted $\sim$150 cm^{-1} to lower energy in the spectrum of the trianion (Fig. 43). It remains the only experimental evidence for a [Cr^{III}(tbbdt$^{\bullet}$)$_3$]^0 electronic structure description for the neutral complex in line with neutral tris(dioxolene) and tris (diimine) species (222, 280, 305, 306, 315, 316); in fact, octahedral Cr complexes not

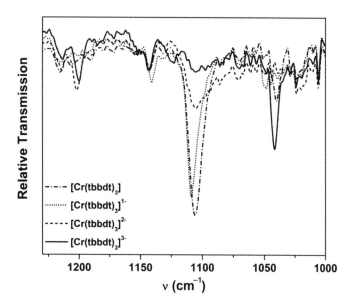

Figure 43. Overlay of the IR spectra for the four-membered electron transfer series $[\mathrm{Cr}^{III}(\mathrm{tbbdt})_3]^z$ ($z = 0$, $1-$, $2-$, $3-$) in $\mathrm{CH_2Cl_2}$ solution at $-25\,^{\circ}\mathrm{C}$. [Adapted from (222).]

in the +III oxidation state are a rare commodity in coordination chemistry and underpin the inherent stability of the d^3 electron configuration (304). Additionally, the IR band is present in $[\mathrm{M}(\mathrm{tbbdt})_3]$ ($\mathrm{M} = \mathrm{Mo}$, W), but not the one-electron reduced $[\mathrm{M}(\mathrm{tbbdt})_3]^{1-}$ species (227). The former is described as having an oxidized tris(dithiolene) unit, whereas the monoanions are characterized with three closed-shell dithiolates (232, 281).

Tatsumi et al. (97) endeavored to chart the magnitude of the trigonal twist angle with the energy difference between the $\nu(\mathrm{M-S})$ A_2 and E (in D_3 symmetry; formally A_2'' and E' in D_{3h} in perfect TP geometry) stretching vibrations in the far-IR region of the spectrum. By plotting the calculated A_1, A_2, $E(1)$, and $E(2)$ modes as a function of Θ, they demonstrated that the distortion to octahedral geometry diminishes the energetic separation between these two IR active bands, A_2 and E (1) (Fig. 44). For example, distorted octahedral $[\mathrm{V}(\mathrm{mnt})_3]^{2-}$ and $[\mathrm{Re}(\mathrm{mnt})_3]^{2-}$ have separations of 20 and $13\,\mathrm{cm}^{-1}$, respectively (317), whereas $[\mathrm{Nb}(\mathrm{edt})_3]^{1-}$ and $[\mathrm{Ta}(\mathrm{edt})_3]^{1-}$ investigated by Tatsumi et al. (97) showed larger separations of 49 and $47\,\mathrm{cm}^{-1}$, respectively. Their result is difficult to substantiate with the absence of structural data for the two group 5 (V B) monoanions, although it was argued they would resemble isoelectronic $[\mathrm{Mo}(\mathrm{edt})_3]$ (175); certainly this is true for $[\mathrm{Mo}(\mathrm{bdt})_3]$ and $[\mathrm{Nb}(\mathrm{bdt})_3]^{1-}$ (126)—both are C_{3h}, which strictly speaking, lies outside the scope of the point symmetry examined. The Ta analogue is distorted octahedral,

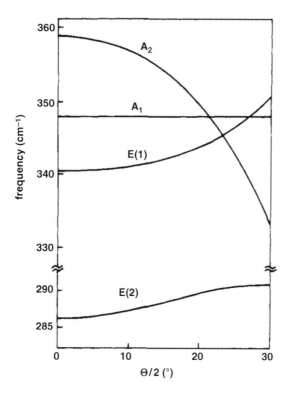

Figure 44.   Calculated $\nu(M-S)$ frequencies for threefold symmetric MS_6 unit as a function of Θ. [Adapted from (97).]

$\Theta = 30.8°$ (220). Tatsumi et al. (97) never considered the correlation between the $\nu(M-S)$ bands of these $(bdt)^{2-}$ complexes in light of their structural characterization (125). The key parameters collated in Table VI demonstrate a difficulty of gaining even a vague notion of the complex geometry. This result stems from an assumption that the identified vibrational stretches arise from the $\nu(M-S)$ modes alone. In fact, these vibrations are coupled to the breathing modes of the metallodithiolene ring and ligand deformations that were omitted from this analysis (310–312, 318). Naturally, this is of no consequence for the compact $(edt)^{2-}$ ligand, but clearly the notion is challenged by larger dithiolene ligands.

Having tracked reversible redox events in a suite of $[M(mnt)_3]^z$ (M = V, Cr, Mn, Fe, Mo, W, Re) complexes by following the $\nu(C\equiv N)$ band and, in some cases, the $\nu(M-S)$ stretching vibrations (319), Best et al. (317) developed a spectrochemical cell that enabled IR spectra to be recorded for highly reduced tetra-anions of V, Mo and Re. Infrared spectroscopy is one of the few techniques that records data faster than the lifetime of the complex. These results are the only experimental data

TABLE VI

Comparison of $\nu(M-S)$ Stretching Frequencies with Crystallographic Twist Angle Θ

Complex	$\nu(M-S)^a$		Δ^b	Θ^c	Reference
	E(1)	A_2			
$[Ti(bdt)_3]^{1-}$	388	343	45	37.9	125
$[Ti(dddt)_3]^{2-}$	420	345	75	36.3	211
$[Zr(bdt)_3]^{2-}$	370	319	51	38.7	125
$[V(edt)_3]$	385	361	24	1.6	96
$[V(pdt)_3]$	406	346	60	4.3	96
$[V(pdt)_3]^{1-\,d}$	398	349	49	0.8	96
$[V(dddt)_3]^{1-\,d}$	430	385	45	1.8	217
$[V(bddt)_3]^{1-}$	450	395	55	7.2	89
$[V(dddt)_3]^{2-}$	345	325	20	e	217
$[V(mnt)_3]^{2-}$	354	334	20	38.0	282
$[Nb(tdt)_3]^{1-}$	364	336	28	1.4^f	125
$[Nb(edt)_3]^{1-}$			49		97
$[Ta(tdt)_3]^{1-}$	392	339	53	30.8^f	125
$[Ta(edt)_3]^{1-}$			47		97
$[Cr(pdt)_3]$	421	356	65		96
$[Mo(edt)_3]^{\,d}$	380	354	26	0.0	96
$[Mo(pdt)_3]^{\,d}$	403	356	47	3.0	96
$[W(pdt)_3]^{\,d}$	403	359	44	1.0	96
$[Re(pdt)_3]$	373	359	14	3.8	96
$[Re(pdt)_3]^{1-}$	361	350	11	26.3	96
$[Re(mnt)_3]^{2-}$	321	308	13	38.3	282

^{a} Infrared active modes under D_3 symmetry in reciprocal centimeters (cm^{-1}).

^{b} Here $\Delta = E(1) - A_2$, in reciprocal centimeters (cm^{-1}).

^{c} Data taken from Table II in degrees.

^{d} Has C_{3h} point symmetry.

^{e} Anticipated to be distorted octahedral.

^{f} Value is that of the $(bdt)^{2-}$ structure.

known for these species. Typically, the $\nu(C\equiv N)$ band is independent of the metal ion and complex charge (Table VII), however makes a decidedly abrupt shift when the metal is pushed to an extremely low oxidation state, M(II). It has been cited as unfortunate that the $\nu(C=C)$ and $\nu(C-S)$ modes are drowned out by absorption of solvents and supporting electrolyte, and so cannot provide an indication of the ligand redox changes as the system is oxidized. However, given the resistance of $(mnt)^{2-}$ to surrender an electron, and the absence of significant structural changes upon ligand oxidation, that is, $[Cr(mnt)_3]^{2-}$ and $[Cr(mnt)_3]^{3-}$, it is unlikely to yield any meaningful response in the dithiolene stretching frequencies. But the data neatly demonstrate a particular efficacy of $(mnt)^{2-}$ to stabilize a range of metal oxidation states, from high-valent Fe(IV) and Ru(IV) (249, 251), to low-valent M(II) in tetra-anions. As well as being a strong π-donor, the low-lying π^* orbital that

TABLE VII
Infrared $\nu(C\equiv N)$ Frequencies in $[M(mnt)_3]^z$ Complexes[a]

Metal	$z = 1-$	$z = 2-$	$z = 3-$	$z = 4-$	Reference
V	2224(sh), 2216	2215(sh), 2203	2188	2164	317, 319
Cr	2214	2203	2191		319
Mo	2224(sh), 2216	2213(sh), 2203	2178	2134	317, 319
W	2225(sh), 2217	2213(sh), 2202	2193(sh), 2170		319
Mn		2211(sh), 2201	2205(sh), 2192		319
Tc		2200			246
Re	2226(sh), 2216	2218(sh), 2218	2200(sh), 2179	2133	317, 319
Fe		2211(sh), 2200	2185		319
Ru		2198	2183		251

[a] Combined solid-state (KBr) and solution data, in reciprocal centimeters (cm^{-1}); sh = shoulder.

is stabilized by conjugation of the $\{S_2C_2\}$ kernel with the cyanide substituents provides electron-rich metals with the opportunity to offload charge onto the ligands. Unique spectral features associated with this π^* orbital surface in pre-edge X-ray absorption spectra of both metal and sulfur K-edges (320).

B. Electronic

The deep colors of dithiolene complexes underscored their use as analytical reagents in the sequestration of metal ions (321). Their rich spectral profile across the entire visible region renders most of the compounds dark blue, purple, and green, with some of the more oxidized species nearly black. Strong absorptions in the low-energy part of the visible region, 600–800 nm, are synonymous with tris(dithiolene) species (Fig. 45).

Schrauzer & Mayweg (96) and Gray and co-workers (177, 181) simultaneously began addressing key features in the electronic spectra of neutral V, Mo, W, and Re complexes to further understand and validate their electronic structure. These compounds are characterized by two dominant absorption bands with the lower energy one possessing the greater intensity (Table VIII). Their position is influenced by the metal, type of dithiolene ligand, and remote substituents. For example, $[Mo(bdt)_3]$ has the two bands at 667 and 426 nm (232), which are slightly red-shifted in the presence of electron-donating groups in $[Mo(xdt)_3]$ to 694 and 450 nm, respectively (40). They also exhibit a weak shoulder feature >850 nm. A red shift of 110 nm is effected by OMe and crown ether substituted $(bdt)^{2-}$ that submerge the weak shoulder (38). Compounds with alkene-1,2-dithiolates trend in the same manner and provide a wider range of transition energies, from 602 and 406 nm in $[Mo(edt)_3]$ (232) to 732 and 488 nm in $[Mo(dmid)_3]$ (155), but lack the weak shoulder transition found in arene-1,2-dithiolates. The $(dmid)^{2-}$ ligand appears to be a strong electron donor in stark contrast to the related $(mtdt)^{2-}$

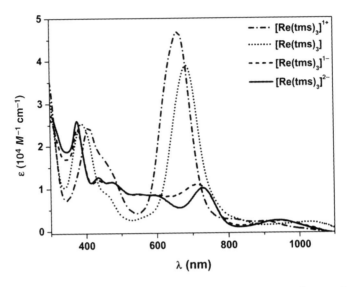

Figure 45.   Overlay of the electronic absorption spectra for the series [Re(tms)$_3$]^z (z = 1+, 0, 1−, 2−).
[Adapted from (247).]

and 1,2-bis(butylthio)ethene-1,2-dithiolate, (btdt)$^{2-}$, ligands that effect a dramatic
blue shift to ~530 and ~350 nm, respectively (155). However, it is ~60 mV more
difficult to reduce than [Mo(edt)$_3$], which suggests the lowest energy band in
[Mo(dmid)$_3$] has a completely different electronic origin. Therefore, electron-
withdrawing ligands can be readily identified based on the significant red shift of
these two prominent bands, which trends across all tris(dithiolene) compounds.

Isoelectronic tungsten species have essentially identical spectra except that the
heavier metal effects a 30–50-nm red shift and a slight drop in the intensity. The
arene-1,2-dithiolates also display a low-energy shoulder not present for alkene-
1,2-dithiolates, and the substituent effects trend in the same way as for Mo with
Katakis and co-workers (68) identifying a linear relationship of the peak maxima
with the Hammett parameter across the series [W(R-sdt)$_3$] (R = H, Me, OMe, Cl,
Br). The same spectral profile is evident for the tris(dithiolene)vanadium monoanions with the second dominant band shifted to ~550 nm. The neutral V species,
of which there are only two characterized by electronic spectroscopy, show a blue
shift and noticeable increase in the intensity of the lowest energy peak, and a new
band (900–1000 nm) in the NIR (179). Tris(dithiolene)rhenium monocations have
the same valence electron count as neutral Mo and W species, and possess the
same spectral profile, although the peaks are noticeably less intense (247, 322).
The neutral species exhibit a minute red shift of these peaks and a further decrease
in intensity of the ~700-nm band (Fig. 45). These complexes also display a
modestly intense peak in the NIR, measured at 1215 nm in [Re(pdt)$_3$] (322), which

TABLE VIII

Electronic Spectral Data for [M(dithiolene)$_3$]^z Complexes[a]

Complex	Solvent	z = 1+	z = 0	z = 1–	z = 2–	z = 3–	Reference
[Ti(bdt)$_3$]	DMF				568	434	206
[Ti(tdt)$_3$]	MeCN				437	414	125
[Ti(bdtCl$_4$)$_3$]	CH$_2$Cl$_2$				813 (0.03) 559 (0.71) 441 (1.60)		43
[Ti(bdt-**6**-bdt)$_3$Ti][b]	DMF				806 (0.03) 556 (0.76) 429 (1.32) 546 (1.55) 428 (2.35)		133
[Ti(mnt)$_3$]	Nujol				640 464		323
[Ti(dddt)$_3$]	DMF				625 (0.66) 541 (0.68) 481 (0.54)		211
[Zr(bdt)$_3$]	MeCN				500 (sh, 0.20) 279 (sh, 2.52)		125
[Zr(tdt)$_3$]	MeCN				360 (1.14) 310 (sh, 2.10)		125
[Hf(tdt)$_3$]	MeCN				391 (0.54) 302 (1.84)		125
[V(bdt)$_3$]	CH$_2$Cl$_2$			874 (sh, 0.27) 670 (sh, 0.88) 605 (0.98) 519 (0.90) 460 (sh, 0.64)	878 (0.42) 681 (sh, 0.46) 553 (0.97) 426 (0.85)	891 (0.51) 552 (0.90) 458 (sh, 0.56) 423 (1.05)	179

(*continued*)

TABLE VIII
(Continued)

Complex	Solvent	$z = 1+$	$z = 0$	$z = 1-$	$z = 2-$	$z = 3-$	Reference
[V(tdt)₃]	CH₂Cl₂			871 (sh, 0.31) 700 (sh, 0.89) 606 (0.94) 528 (0.93) 457 (sh, 0.62)	900 (0.46) 559 (0.82) 428 (0.94) 332 (1.66)		179
[V(bdtCl₄)₃]	CH₂Cl₂			939 (0.16) 662 (0.85) 515 (0.70)	862 (0.65) 540 (1.18) 437 (1.15)		43
[V(edt)₃]	CH₂Cl₂		911 (0.05) 656 (0.40) 513 (0.49) 405 (0.14)	639 (sh, 0.21) 576 (0.40) 526 (0.29) 363 (sh, 0.21)	1010 (0.09) 587 (0.21) 422 (sh, 0.12) 381 (0.21)		179
[V(etdt)₃]	MeCN			678 (0.64) 597 (1.01) 532 (0.71)			98
[V(mnt)₃]	CH₂Cl₂			639 (0.82) 566 (1.19) 508 (sh, 0.71) 386 (sh, 0.38) 341 (0.93)	954 (0.31) 658 (sh, 0.20) 578 (0.50) 527 (0.38) 430 (0.67) 372 (sh, 0.98)	672 (0.24) 530 (sh, 0.27) 443 (sh, 0.78) 421 (0.92) 339 (sh, 1.48)	179
[V(sdt)₃]	CH₂Cl₂			695 (sh, 0.82) 605 (1.04) 550 (sh, 0.85)			146
[V(pdt)₃]	CH₂Cl₂		1014 (0.46) 783 (3.13) 558 (1.61) 421 (0.47)	700 (1.32) 597 (1.40) 545 (sh, 1.04) 421 (sh, 0.36)	955 (0.38) 637 (sh, 0.47) 590 (0.78) 420 (sh, 0.85)		179

70

Compound	Solvent			Ref.
[V(dmit)₃]	Acetone	699 (0.09), 515 (0.18), 441 (0.16)		85
[V(dmt)₃]	Acetone	690 (0.07), 465 (0.23), 417 (0.21)		85
[V(dddt)₃]	DMF		816 (1.18), 631 (1.17), 602 (1.23), 403 (0.53)	218
[V(bddt)₃]	DMF	660 (0.61), 409 (0.56)	565 (1.07), 368 (1.38)	89
[Nb(bdt)₃]	MeCN		581 (1.07), 383 (1.19)	125
[Nb(tdt)₃]	MeCN		556 (0.18), 460 (0.17), 411 (0.20)	125
[Nb(edt)₃]	MeCN		554 (0.22), 493 (0.31), 411 (0.36)	97
[Nb(etdt)₃]	MeCN		719 (1.10), 465 (1.00)	98
[Nb(dddt)₃]	DMF		448 (0.49), 348 (1.45)	211
[Ta(bdt)₃]	MeCN		493 (0.45), 419 (0.42), 371 (0.61)	125
[Ta(edt)₃]	MeCN			97
[Cr(bdtCl₄)₃]	CH₂Cl₂	787 (0.30), 612 (0.63), 457 (0.94)	787^c, 599, 543, 465	43

(continued)

71

TABLE VIII
(Continued)

Complex	Solvent	z = 1+	z = 0	z = 1–	z = 2–	z = 3–	Reference
[Cr(bdtCl₂)₃]	CH₂Cl₂			1960 (0.07) 770 (0.14) 620 (sh, 0.23) 530 (0.35)	1695 (0.06) 780 (0.08) 590 (0.17) 454 (0.26)	630 (0.05) 400 (0.39)	221
[Cr(tbbdt)₃]	CH₂Cl₂		1000 (0.22) 700 (0.88) 590 (2.03)	1750 (0.28) 910 sh 780 (0.52) 520 (0.81)	1720 (0.25) 910 (0.30) 700 (0.59) 590 (0.72)	790 (0.15) 590 (0.24) 460 (sh, 0.41)	222
[Cr(mnt)₃]	CH₂Cl₂				2071 (0.18) 1005 (0.03) 686 (0.47) 571 (sh, 0.39) 481 (0.77)	668 (0.11) 560 (0.22) 470 (0.66) 425 (0.98)	221
[Cr(pdt)₃]	CHCl₃		687 (0.30) 581 (2.10)				177
[Cr(dtcr)₃]	MeCN					526 (4.47) 498 (4.27) 416 (sh, 2.09)	101
[Cr(dcmdtcr)₃]	MeCN					660 (6.17) 608 (sh, 4.79)	101
[Mo(bdt)₃]	CH₂Cl₂		855 (0.30) 667 (2.34) 426 (2.01)	836 (sh, 0.30) 679 (0.72) 621 (0.77)	585 (1.69) 355 (2.81)		232
[Mo(tdt)₃]	CH₂Cl₂		870 sh 683 (2.34) 435 (2.20)	930 sh 730 (0.75) 622 (0.78) 491 (1.36)	675 sh 596 (1.34) 418 sh 352 (2.39)		144

Complex	Solvent					Ref
[Mo(xdt)₃]	CHCl₃	694 (1.72) 450 (1.53)				40
[Mo(bdtCl₂)₃]	MeCN			574 (1.10) 352 (2.05)		226
[Mo(tbbdt)₃]	CH₂Cl₂	881 (0.32) 693 (2.48) 430 (2.14)	862 (sh, 0.23) 701 (0.80) 638 (0.88) 479 (1.16)	628 (0.64) 575 (1.71) 413 (0.33) 353 (2.50)		227
[Mo(vdt)₃]	CH₂Cl₂	773 (1.45) 477 (0.80)				38
[Mo(15C5-bdt)₃]	CH₂Cl₂	777 (1.34) 478 (0.76)				38
[Mo(18C6-bdt)₃]	CH₂Cl₂	778 (1.21) 475 (0.69)				38
[Mo(qdt)₃]	CH₂Cl₂		782 (sh, 0.60) 697 (1.19) 613 (1.54)	563 (5.22) 397 (4.32)	760 (0.57) 563 (1.10)	232
[Mo(tfd)(bdt)₂]	CH₂Cl₂	666 (1.10) 594 (sh, 0.80) 418 (1.10)				149
[Mo(tfd)₂(bdt)]	CH₂Cl₂	630 (1.10) 580 (sh, 0.70) 406 (0.90)				149
[Mo(edt)₃]	CH₂Cl₂	602 (1.15) 412 (1.03)	999 (0.06) 687 (0.59) 733 (0.52) 527 (0.07) 413 (0.30)	673 (0.52) 378 (0.59)		232
[Mo(mdt)₃]	MeCN	641 (1.68) 512 (0.38) 436 (1.40)		735 (0.89) 537 (0.05) 415 (0.51)		231
[Mo(etdt)₃]	Toluene	642 (1.22) 434 (1.00)				98
[Mo(mnt)₃]	CH₂Cl₂		964 (0.10) 692 (1.03)	669 (0.66) 390 (1.05)	961 (0.18) 563 (1.23)	232

(*continued*)

TABLE VIII
(Continued)

Complex	Solvent	z = 1+	z = 0	z = 1-	z = 2-	z = 3-	Reference
[Mo(tfd)$_3$]	MeCN		581^d 387		622 328		50
[Mo(sdt)$_3$]	CH$_2$Cl$_2$		668 (1.54) 443 (1.11)	1008 (0.07) 747 (0.88)	686 (0.61) 400 (1.15)		232
[Mo(toldt)$_3$]	CHCl$_3$		677 (1.20) 449 (0.96)				324
[Mo(adt)$_3$]	CHCl$_3$		705 (1.28) 459 (0.85)				324
[Mo(pdt)$_3$]	CH$_2$Cl$_2$		691 (5.97) 447 (3.55)	990 (sh, 0.24) 756 (3.73)	682 (1.29) 338 (2.42)		232
[Mo(apdt)$_3$]	CHCl$_3$		720 (1.06) 455 (0.72)				324
[Mo(dmm)$_3$]	MeCN				650 450 sh 356		59
[Mo(dmit)$_3$]	MeCN		855 580 sh 510 sh 405	845 595 480 425	755 490		237
[Mo(dmid)$_3$]	CH$_2$Cl$_2$		732^e 488		766 sh 642 430 sh 346		155
[Mo(mtdt)$_3$]	CHCl$_3$		526 350				155
[Mo(btdt)$_3$]	Benzene		528 342				155
[Mo(dhdndt)$_3$]	CH$_2$Cl$_2$		641 (1.82)				290

Complex	Solvent	λ / nm (ε)			Ref
[W(bdt)₃]	CH₂Cl₂	451 (1.80) 762 (0.20) 623 (1.72) 381 (1.26)		548 (0.57) 475 (1.46)	232
[W(tdt)₃]	Acetone	780 sh 640 (1.28) 430 sh 383 (1.36)	920 sh	665 shᶠ 502 (0.79) 485 (1.66) 330 (2.58)	144
[W(xdt)₃]	CHCl₃	649 (2.69) 390 (1.54)			40
[W(bdtCl₄)₃]	CH₂Cl₂		870 (0.04) 645 (0.39) 556 (0.65) 353 (1.60)	680 (0.09) 546 (0.88) 478 (1.95) 340 (2.25)	43
[W(bdtCl₂)₃]	MeCN		650 sh 540 (0.52) 360 (0.94)	535 (0.47) 470 (0.83)	226
[W(tbbdt)₃]	CH₂Cl₂	790 (0.35) 649 (2.53) 390 (1.90) 353 (1.25)	632 (0.61) 537 (1.17) 360 (1.91)	539 (0.90) 468 (2.45) 348 (0.40) 344 (1.84)	227
[W(vdt)₃]	CH₂Cl₂	740 (2.85) 434 (1.32)			38
[W(15C5-bdt)₃]	CH₂Cl₂	745 (1.56) 434 (1.05)			38
[W(edt)₃]	CHCl₃	576 (0.67) 404 (0.41) 368 (0.40)			96
[W(mdt)₃]	MeCN	606 (1.77) 508 (0.39) 401 (0.99)	667 (0.59) 595 (sh, 0.29) 466 (0.31) 370 (0.52)	705 (0.08) 621 (0.43) 529 (0.88) 358 (sh, 0.52)	231

(continued)

TABLE VIII
(*Continued*)

Complex	Solvent	z = 1+	z = 0	z = 1–	z = 2–	z = 3–	Reference
[W(etdt)$_3$]	Toluene		608 (0.64) 505 (0.28) 390 (0.43)				98
[W(mnt)$_3$]	CH$_2$Cl$_2$			900 (0.03) 618 (0.63)	571 (0.43) 491 (0.33)	1033 (0.25) 528 (0.96)	232
[W(trd)$_3$]	MeCN				566 sh 505 433 sh		50
[W(sdt)$_3$]	CH$_2$Cl$_2$		631 (2.16) 503 (0.47) 407 (1.40)				146
[W(toldt)$_3$]	CHCl$_3$		644 (1.78) 410 (1.08)				324
[W(adt)$_3$]	CHCl$_3$		670 (1.81) 413 (1.12)				324
[W(csdt)$_3$]	CHCl$_3$		634 (1.94) 407 (1.17)				324
[W(bsdt)$_3$]	CHCl$_3$		634 (2.01) 406 (1.21)				324
[W(pdt)$_3$]	CH$_2$Cl$_2$		657 (3.12) 414 (1.39)	901 (0.10) 687 (1.54)	616 (0.53) 525 (0.88)		232
[W(dadt)$_3$]	CHCl$_3$		682 (1.63) 407 (0.72)				324
[W(tpdt)$_3$]	CHCl$_3$		668 (1.57) 412 (0.83)				324
[W(apdt)$_3$]	CHCl$_3$		679 (1.91) 415 (1.05)				324

76

Complex	Solvent					Ref
[W(anpdt)₃]	CHCl₃	793 (2.30) 441 (1.51)				324
[W(dem)₃]	MeCN			602 sh 548 sh 474		62
[W(dmit)₃]	MeCN	810 570 410	780 490 430	686 574 478		238
[W(dmid)₃]	CH₂Cl₂			704 590 356		155
[Mn(bdt)₃]	MeCN			727 (0.07) 575 (0.10) 502 (0.13)		244
[Mn(tdt)₃]	MeCN			741 (0.13) 582 (0.15) 508 (0.19)		244
[Mn(bdtCl₄)₃]	CH₂Cl₂		714^g 568 521	714 (0.48) 571 (0.52) 521 (0.72)		43
[Mn(bdtCl₂)₃]	CH₂Cl₂			714 (0.09) 571 (0.09) 511 (0.12)		221
[Mn(mnt)₃]	CH₂Cl₂			833 (0.24) 634 (0.26) 388 (3.60)	388 (5.10)	319
[Mn(dddt)₃]	DMF			558 (0.18) 397 (0.90)		212
[Tc(tdt)₃]	Benzene	710 (1.66) 526 sh 454 (2.01)	730 sh^h 540 (1.36) 462 (1.37)			144

(*continued*)

TABLE VIII
(*Continued*)

Complex	Solvent	$z = 1+$	$z = 0$	$z = 1-$	$z = 2-$	$z = 3-$	Reference
[Re(bdt)₃]	CH₂Cl₂		1062 (0.22)	942 (0.57)	956 (0.12)		247
			674 (3.89)	742 (1.49)	765 (0.44)		
			442 (sh, 1.54)	502 (0.86)	670 (0.90)		
			390 (2.50)	429 (2.36)	605 (0.83)		
					429 (sh, 2.36)		
[Re(tdt)₃]	CH₂Cl₂	661	1262 (0.02)	960			144
		460 sh	1085 (0.05)	745 (0.93)			
		415	692 (1.60)	510 sh			
			401 (1.10)	442 (1.96)			
[Re(bdtCl₄)₃]	CH₂Cl₂			934 (0.39)			43
				741 (1.16)			
				500 (0.41)			
				424 (1.40)			
[Re(bdtCl₂)₃]	CH₂Cl₂			921 (0.57)	737 (0.38)		247
				730 (1.81)	658 (0.84)		
				507 (0.56)	603 (0.73)		
				411 (2.22)	429 (1.72)		
[Re(tms)₃]	CH₂Cl₂	905 (0.22)	1026 (0.19)	941 (0.24)	941 (0.21)		247
		811 (0.27)	841 (0.23)	712 (1.09)	729 (1.01)		
		661 (4.67)	684 (3.89)	599 (0.85)	599 (0.85)		
		454 (1.67)	459 (sh, 0.87)	548 (0.87)	548 (0.87)		
		409 (2.45)	388 (2.54)	437 (1.15)	473 (1.15)		
[Re(bdt-**4**-bdt)₃Re]ⁱ	DMF			730 (1.42)			265
				507 (sh, 0.53)			
				430 (sh, 2.00)			
[Re(bdt-**5**-bdt)₃Re]ⁱ	DMF			735 (1.00)			265
				510 (sh, 0.43)			
				440 (sh, 1.43)			

Compound	Solvent						Ref
[Re(mnt)₃]	CH₂Cl₂			1030 (0.14) 806 (0.48) 345 (1.73)	943 (sh, 0.11) 676 (0.38) 385 (1.42) 350 (sh, 1.23)	847 (sh, 0.13) 641 (0.34) 513 (0.63) 380 (1.14)	319
[Re(pdt)₃]	CH₂Cl₂	703 (2.06) 499 (0.47) 433 (0.62)	1215 (0.07) 711 (1.45) 491 (0.23) 426 (0.59)	1080 (0.14) 835 (0.57) 464 (sh, 0.34) 426 (0.59)	747 (0.22) 684 (0.25) 410 (0.86)		322
[Re(apdt)₃]	CHCl₃		732 (1.90) 432 (0.85)				324
[Re(dmit)₃]	CH₂Cl₂		866 (1.27) 553 (0.98) 378 (2.33)	628 (sh, 0.45) 538 (sh, 0.76) 478 (1.18) 378 (2.49)			163
[Fe(bdtCl₄)₃]	CH₂Cl₂			1315^g 565	1333 (0.27) 769 (0.53) 545 (0.38)		43
[Fe(mnt)₃]	CH₂Cl₂				806 (0.33) 610 (0.11) 405 (0.80)	990 (0.07) 714 (0.11) 602 (sh, 0.12) 363 (3.70)	319
[Fe(dtcr)₃]	MeCN					680 sh 484 (2.88) 458 (sh, 2.69)	103
[Fe(dcmdtcr)₃]	MeCN					680 sh 604 (3.89) 522 (sh, 2.95)	103
[Ru(mnt)₃]	CH₂Cl₂				681 (0.34) 649 (0.34) 376 (1.13)	500 sh 405 (1.49) 327 (1.91)	251

(continued)

79

TABLE VIII
(Continued)

Complex	Solvent	z = 1+	z = 0	z = 1−	z = 2−	z = 3−	Reference
[Os(tdt)₃]	CH₂Cl₂		1709 (0.09) 1235 sh 926 (0.30) 649 sh 481 (0.35)	1709 (0.05) 1049 (0.31) 649 sh 578 (0.46) 481 (0.49)			292
[Co(mnt)₃]	MeCN					665 (0.04) 461 (0.40)	323
[Co(dtcr)₃]	MeCN					560 (4.57) 388 (2.40) 308 (4.79)	103
[Co(dcmdtcr)₃]	MeCN					708 (10.72) 658 (4.47) 542 (1.51)	103
[In(mnt)₃]	Acetone					367 (0.92) 348 (1.02)	309
[Tl(mnt)₃]	Acetone					358 (1.05) 249 (2.23)	309
[Ge(mnt)₃]	CH₂Cl₂				350 (sh, 2.06) 323 (2.06)		325
[Sn(mnt)₃]	CH₂Cl₂				366 (sh, 1.67) 333 (1.94)		325
[Sn(dmit)₃]	MeCN				458		255
[Sn(eTTFdt)₃]	DMF		1111 (0.04) 833 (0.22) 444 (0.44)		488 (0.65)		255

80

[Sb(bdt)$_3$]	MeCN	490 (0.25) 395 (0.22)	125
[Sb(edt)$_3$]	DMSO	543 (0.15) 333 (0.20)	27

[a] Data presented as absorption maxima in nanometers (nm) and intensity $10^4\ M^{-1}\ cm^{-1}$ in parentheses. Abbreviations: DMSO = dimethyl sulfoxide, MeCN = acetonitrile, DMF = N,N'-dimethylformamide, sh = shoulder.

[b] Each end of the helicate is [Ti(bdt)$_3$]$^{2-}$. Ligand abbreviation given in Fig. 3.

[c] Tentative formulation, however, a purple color is consistent with other monoanions.

[d] Spectrum recorded in pentane.

[e] Spectrum recorded in benzene.

[f] Spectrum recorded in water.

[g] Sample prepared by chemical oxidation of the dianion with Br$_2$; formulation is speculative.

[h] Spectrum recorded in CH$_2$Cl$_2$.

[i] Each end of the helicate is [Re(bdt)$_3$]$^{1-}$; intensities per molecule. Ligand abbreviations given in Fig. 3.

is blue-shifted ~ 150 nm for aryl dithiolates (247). Furthermore, the ~ 400-nm peak bares a shoulder to lower energy for both the monocation and neutral species. Using these spectra, Schrauzer and Mayweg (96) described the ~ 700-nm peak as a $\pi \rightarrow \pi^*$ transition ($4e' \rightarrow 5e'$), and the other as either the second $\pi \rightarrow \pi^*$ transition ($2a_2' \rightarrow 5e'$) or $M \rightarrow \pi^*$ transitions ($3a_1' \rightarrow 5e'$, $2a_2'$), or a combination of both. Gray's (181) alternative MO manifold led to the assignment of the more intense transition as $2a_2' \rightarrow 5e'$, which was a combined ligand-to-metal (LMCT) and ligand-to-ligand charge transfer (LLCT) band given the mixed nature of the acceptor level. The higher energy band was assigned as a combination of the $2a_2'' \rightarrow 3a_1'$ and $3e' \rightarrow 3a_1'$ LMCT transitions, where the intensity stems from the ligand (π_h) character in the final state. They added that TP geometry is likely for complexes with the same spectral structure, namely, neutral complexes of V, Cr, Mo, W, Re, and the monoanion of vanadium (177, 180, 181). The claim was supported with the observation of identical spectra in the solid state. Schrauzer's assignment has been disregarded in light of the broad acceptance of the Gray MO scheme (Fig. 35). These assignments have been substantiated and re-evaluated by the application of resonance Raman spectroscopy. Although a vibrational spectroscopic technique (see Section VIII.A), the enhancement of molecular bonds with a single frequency of light that corresponds to a specific electronic transition can be used to diagnose the origin of the excitation. Clark and Turtle (326) investigated the three-membered electron-transfer series for vanadium: [V(pdt)$_3$], [V(pdt)$_3$]$^{1-}$, and [V(mnt)$_3$]$^{2-}$. At this time, the terminal members had been structurally characterized as TP (178) and distorted octahedral (199), respectively, and the monoanion was presumed to be TP based on the likeness of its electronic spectrum to [V(pdt)$_3$] (177). Their data agreed with Gray's earlier assignment of the lesser of the two dominant bands as the $3e' \rightarrow 3a_1'$ LMCT transition, but they described the more intense one as the $4e' \rightarrow 3a_1'$ LMCT transition. The resonance enhancement was centered on the symmetric stretching mode of the VS$_6$ core (A$_1'$), and therefore this level is logically the recipient of these one-electron excitations. The low-energy feature at 1015 nm also displayed enhancement of the A$_1'$ mode and was assigned as the $3a_1' \rightarrow 2a_2'$ MLCT transition given its intensity. The weak feature at 1215 nm in the spectrum of [Re(pdt)$_3$] was also given this assignment, rather than the $3a_1' \rightarrow 5e'$ LF excitation proposed by Gray (181). The dianion [V(mnt)$_3$]$^{2-}$ also exhibited a similar feature assigned as this LF transition (here the $2a_2'$ MO is filled). In contrast, [V(pdt)$_3$]$^{1-}$ does not, which supported the conclusion it contains a d^0 central ion.

The consequence of confirming the reversibility of multiple redox processes by spectroelectrochemistry was an enormous influx of new spectral data. In concert with other spectroscopic and computational studies, we now have a more or less concrete idea of the electronic structure for tris(dithiolene) systems. All complexes with oxidized ligands possess the intense low-energy band ~ 700 nm, and this covers monocationic Re, monoanionic V, and neutral species of V, Mo, W, and Re. This hallmark band, directly analogous to the feature seen in bis(dithiolene) compounds (275, 277, 279), is more likely the $4e' \rightarrow 2a_2'$ transition that is predominantly LLCT in nature with a hint of MLCT character based on the metal content of the $4e'$ level, and thus found at slightly lower energy for Mo compared to W. The $2a_2'$ MO is empty for $[V(pdt)_3]$ and $[Re(pdt)_3]^{1+}$, so the band is significantly more intense than the corresponding feature in the spectra of $[V(pdt)_3]^{1-}$ and $[Re(pdt)_3]$, where the acceptor orbital is half occupied (179, 247). The band is lost upon further reduction to $[V(pdt)_3]^{2-}$ and $[Re(pdt)_3]^{1-}$, respectively, indicating the orbital is filled. This assignment is nicely corroborated by a comparison of $[Mo(bdt)_3]^{0/1-}$ and $[Mo(pdt)_3]^{0/1-}$ (232): both neutral complexes have this intense band, but it is noticeably absent in the spectrum of $[Mo(bdt)_3]^{1-}$, which is characterized with a fully reduced tris (dithiolene) unit by EPR. Its alkene-1,2-dithiolate homologues are regarded as having a one-electron oxidized ligand moiety, and hence exhibit the signature band (Fig. 46).

Electronic spectra are the clearest indicator of the dithiolene ligand dependency on the electronic structure. Moreover, this band is not present in the spectra of monoanionic Nb and Ta species, which, unlike their congener V, are described as fully reduced ligands bound to a +V central ion. Their spectra can be listed with the group 4 (IV B) metals that are undoubtedly +IV ions and closed-shell dithiolates. The single spectrum recorded for $[Tc(tdt)_3]$ is sufficient to characterize its electronic structure the same as Re and distinctly different from Mn (144); the 3d metal is clearly unable to stabilize a dithiolene ligand radical. The species

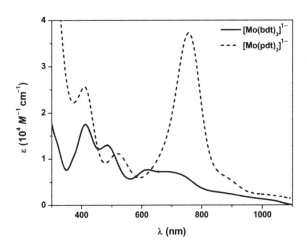

Figure 46.   Overlay of the electronic spectra of $[Mo(bdt)_3]^{1-}$ and $[Mo(pdt)_3]^{1-}$. [Adapted from (232).]

identified as $[Mn(bdtCl_4)_3]^{1-}$ (288), prepared by *in situ* oxidation, does not show the hallmark dithiolene radical band; in fact, it is almost identical to the dianion and perhaps incorrectly formulated.

The NIR features of neutral tris(dithiolene)rhenium compounds could plausibly arise from the $2a_2' \rightarrow 5e'$ excitation unavailable to the monocationic species. Its intensity is a measure of the ligand content in the $5e'$ orbitals (and consequently the $4e'$ orbitals) that contributes electric dipole character to the transition. The transition should also be seen for the monoanion, as here the $2a_2'$ MO is doubly occupied and red shifts the transition, though this is countered by a blue shift from the larger twist angle stabilizing the $5e'$ orbitals. It could be construed that the peaks between 930–960 nm for the arene-1,2-dithiolates, and 1030–1080 nm for alkyl-1,2-dithiolates, correspond to this particular excitation. Certainly the LF $3a_1' \rightarrow 5e'$ transition is expected to appear in the NIR region for reduced complexes with distorted octahedral geometries, as it is no longer forbidden by symmetry.

The appearance of a weak band in the NIR region of neutral Mo and W complexes with aromatic-1,2-dithiolates could logically stem from the $3a_1' \rightarrow 2a_2'$ excitation. It is important to recall the paddle wheel structure of these neutral compounds is instigated by a second-order Jahn–Teller distortion to alleviate the near degeneracy of the HOMO and LUMO (Fig. 38). The switch to C_{3h} symmetry and mixing of these orbitals confers electric-dipole intensity to what is presumably the lowest energy transition for these compounds. Aryl dithiolates have larger fold angles than their alkyl counterparts, which attest to their weaker π donating ability, and therefore a smaller HOMO–LUMO gap. The corresponding band for the alkene-1,2-dithiolates is expected to reside at lower energy (and intensity) corresponding to a smaller structural distortion. Based on the plot developed by Campbell and Harris (269), the $4e' \rightarrow 2a_2'$ transition remains unperturbed by dithiolene folding, thus the trend seen between different types of dithiolenes matches with the Re compounds that are persistently D_{3h} symmetric (247, 248).

Chromium complexes have a decidedly unique electronic arrangement as evidenced by their electronic spectra; they can be regarded as the only system with localized ligand radicals. As such they exhibit the quintessential NIR dithiolate $\rightarrow$ dithienyl radical IVCT band seen in most Cr complexes with three redox-active ligands (302, 306). This band ranges from 1695–2070 nm and is present in all monoanionic and dianionic species, clearly indicating at least one dithienyl radical in each (221, 222); the band is absent from the neutral and trianionic members of the electron-transfer series (Fig. 47). Coupled with IR, EPR, and XAS data, the electronic structures of this series are readily formulated: $[Cr^{III}(L^•)_3]$ ($S = 0$); $[Cr^{III}(L)(L^•)_2]^{1-}$ ($S = 1/2$); $[Cr^{III}(L)_2(L^•)]^{2-}$ ($S = 1$); $[Cr^{III}(L)_3]^{3-}$ ($S = 3/2$), where L is any dithiolene ligand.

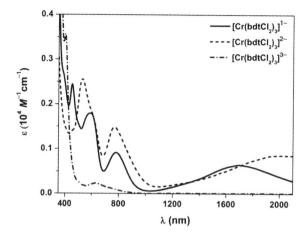

Figure 47.   Overlay of the electronic absorption spectra for the series $[Cr(bdtCl_2)_3]^z$ ($z = 1-, 2-, 3-$). [Adapted from (219).]

Prominent low-energy bands are seen in the electronic spectra of $[Os(tdt)_3]^{0/1-}$ (293), $[Fe(bdtCl_4)_3]^{2-}$ (288), and $[Sn(eTTFdt)_3]$ ($eTTFdt^{2-}$ = dimethylenetetrathiafulvalenedithiolate) (255). For the latter, where the TTF based ligand is oxidized, the band at 1111 nm could be the IVCT transition between dithiolate and radial forms of TTF (255). For group 8 (VIII B) compounds, the transition does not signify radical character, rather a LF transition that is infused with electric-dipole character via ligand contribution to the "metal-based" orbitals. All other compounds listed in Table VIII are representative of metal ions coordinated by three dithiolate dianions, with LMCT bands expected to dominate the 300–500-nm region. A band at ~400 nm in $(mnt)^{2-}$ systems is categorized as the $\pi \rightarrow \pi^*$ band of the ligand. It is found in the spectrum Na_2mnt (327), and clearly evident in main group complexes that provide baseline data free from the intrusion of d orbitals (289). Similar is the strong absorption at ~480 nm in $(dmit)^{2-}$ complexes (328).

C.   Nuclear Magnetic Resonance

Nuclear magnetic resonance (NMR) spectroscopy is a standard analytical method for most chemical synthesis. Synthesized dithiolene ligands and complexes are examined by ^1H NMR (proton NMR) to check for purity and occasionally paramagnetism. In special cases, other nuclei were targeted, such as the ^{19}F signal in $[Mo(tfd)_3]$ (1). It has rarely been used beyond reporting rudimentary chemical shifts of different proton environments. One such study highlighted aromaticity in the $\{MS_2C_2\}$ metallocycle by tracking the chemical shift of a proton directly bound to the unit. The idea first surfaced with the simplest dithiolene ligand $(edt)^{2-}$ from early work of Schrauzer and Mayweg (96, 282). Though diamagnetic neutral complexes $[Mo(edt)_3]$ and $[W(edt)_3]$ proved too insoluble (96), a sufficiently concentrated DMSO solution of $[V(edt)_3]^{1-}$ exhibited a sharp ^1H NMR signal at the high end of the aromatic region (~9 ppm), similar to $[Ni(edt)_2]$ (282). The corresponding resonances for $[Mo(edt)_3]^{2-}$ and $[Re(edt)_3]^{1-}$ were less deshielded, but still far removed from Na_2edt. Tatsumi et al. (97) measured dithiolenic chemical shifts of 7.79 and 7.96 ppm for $[Nb(edt)_3]^{1-}$ and $[Ta(edt)_3]^{1-}$, respectively. The extent of the downfield shift corresponds to the degree of aromaticity in the metallodithiolene unit, and the values here suggest the ligands are more oxidized in $[V(edt)_3]^{1-}$ than $[Mo(edt)_3]^{2-}$ and $[Re(edt)_3]^{1-}$, which is certainly true in light of other spectroscopic data (179, 247, 248). Interestingly, the metallodithiolene rings in $[V(edt)_3]^{1-}$ are presumably nonplanar, as seen with other dithiolene ligands in a C_{3h} symmetric molecule. The Mo and Re compounds are D_3 symmetric, and the influence of geometry on the chemical shift was not ascertained. The presence of an induced diamagnetic current in the metallodithiolene ring was first detected in the isoelectronic series $[M(sdt)_3]$ (M = Mo, W) and $[V(sdt)_3]^{1-}$ (146). The dithiolenic proton shifted from 6.43 ppm in the alkyl-protected ligand, Me_2sdt, to 9.52, 9.77, and 8.83 ppm for the Mo, W, and V complexes, respectively. The ^1H resonances for the phenyl ring protons were essentially invariant across these four compounds. The effect commutes with the interspin distance, and is therefore only applicable to ligands with dithiolenic protons, specifically $(edt)^{2-}$ and variants of $(sdt)^{2-}$. For example, the ^{19}F NMR singlet for $[Mo(tfd)_3]$ has a chemical shift δ −56.4 ppm; the two-electron reduced $[Mo(tfd)_3]^{2-}$, δ −53 ppm (50). Crystallographic and spectroscopic analysis has revealed the tris(dithiolene) redox level corresponds to the magnitude of the downfield shift in this series; the monoanionic complex, $[V(sdt)_3]^{1-}$, was diagnosed with an $(L_3)^{5-\cdot}$ unit compared with an almost fully oxidized $(L_3)^{4-}$ unit for the two neutral species (179, 248). Ring current is absent from dianionic dithiolate ligands, giving a resonance typical of an olefinic proton. The input of remote dithiolene ligand substituents was probed by ^1H NMR spectroscopy on a series of $[W(R-sdt)_3]$ complexes developed by Katakis and co-workers (68), where R represents the para substituents H, Me, OMe, Cl and Br. The chemical shift followed the trend MeO < Me < Cl ~ Br ~ H with a linear correlation to the Hammett parameter (148). A similar conclusion was shown in a series of square-planar gold complexes with functionalized $(sdt)^{2-}$ ligands (329). Furthermore, investigation of the dithiolenic proton chemical shift across the mixed ligand series, $[W(adt)_x(bsdt)_{3-x}]$ ($x = 0$–3), revealed the para substituent on the phenyl group led to a greater downfield shifting of the dithiolenic proton on the other ligands rather than its adjacent proton (148). The phenomenon arises from the extensive delocalization throughout the entire $\{MS_6C_6\}$ moiety. The ^{95}Mo and ^{183}W chemical shifts in these neutral tris(dithiolene) complexes showed these metal ions to be more electron deficient than the corresponding ions in $[MoO_4]^{2-}$ and $[WCl_6]$, respectively. Therefore, they were assigned a +VI oxidation state that clearly conflicts with a putative +IV assignment given earlier (153, 175), which has only very recently has been experimentally ratified (232, 248, 281). It is important to recognize that a formal +VI oxidation state may not be an accurate description for the calibrants $[MoO_4]^{2-}$ and $[WCl_6]$ because in larger metal ions the strong covalent bonds are difficult to characterize with conventional (integer) oxidation states.

D.   Electron Paramagnetic Resonance

1.   Spin Doublet

The ease with which tris(dithiolene) complexes can take up or release electrons means every second step along any electron-transfer series is a paramagnetic species. With few exceptions, these have one unpaired electron giving a total spin ground state of $S = 1/2$ (Table V). Magnetic susceptibility measurements of spin-doublet species can almost be considered redundant. The electronic structure is preferably examined by EPR spectroscopy, which involves flipping the electron spin with microwave radiation (X-band frequency is ~9.5 GHz) in a magnetic field. The field position of the resonance is described by a dimensionless parameter known as the Landé g-factor given by the relationship $g = h\nu/\mu_B B$—essentially the ratio between the frequency (ν) of the transition in a magnetic field (B). For a free electron in a vacuum $g_e = 2.0023$, and this value shifts when the electron is located in an orbital of an atom or molecule (i.e., the medium impacts its magnetic moment). Therefore, g-values are diagnostic of the chemical environment of the free electron, and can vary dramatically with the element that acts as host, either the metal or dithiolene ligand. Furthermore, the electron spin (S) can mingle with any vocal nuclear spins ($I > 0$) nearby, providing what is termed hyperfine coupling (parameterized by A). Various metals have different isotopes with nuclear spins and the magnitude of the interaction aids in defining the location of the unpaired spin. Electron paramagnetic resonance becomes very complicated beyond this elementary description of the basic experiment, but for early investigators, simply determining the g-value in a room temperature fluid solution (isotropic) spectrum could give an immediate indication of the electronic structure of the compound, that is, is the unpaired spin metal or ligand based, an approach not available for the diamagnetic ($S = 0$) complexes that are naturally probed by NMR.

The EPR studies of tris(dithiolene) compounds proved to be rather complicated compared with their bis(dithiolene) analogues, as many of the reported conclusions were re-evaluated by successive generations. This reappraisal is not a condemnation of the pioneers of this field, rather recognition that there was so little material to compare and contrast the EPR parameters to gain a clear assignment of the electronic structure. For example, in the first EPR study, $[V(mnt)_3]^{2-}$ was simulated with $g_{iso} = 1.980$ and $A_{iso} = 63.3$ G from a room temperature spectrum of a 1 : 1 $CHCl_3/DMF$ solution of the complex (10). The authors astutely pointed out that the hyperfine coupling to ^{51}V ($I = 7/2$, 100% abundant) was little more than one-half of that for the aquo ion of vanadyl, $[V^{IV}O]^{2+}$, and preferred to position the unpaired electron away from the metal and onto the ligands, formulated as $[V^{III}(mnt_3^{5-\bullet})]^{2-}$ with a $(4e')^4(3a_1')^2(2a_2')^1(5e')^0$ electron configuration. Both the group 6 (VI B) (Cr, Mo, W) monoanions and the aforementioned vanadium dianion were more logically cast as ligand-based paramagnets with a $(3a_1)^2(2a_2)^1$ electron configuration assuming they retained their D_3 symmetry in solution (10).

The neutral vanadium compound, $[V(pdt)_3]$, was identified as having the same ground state term as the dianion, and an orbital reversal occurs when the dianion is oxidized via stepwise depopulation of the metal-based $3a_1$ orbital (142). Single-crystal EPR studies unequivocally revealed the metal ion as V(IV) d^1 with a $^2A_1$ ground term (203, 204). The smaller hyperfine coupling is now considered standard for non-oxo vanadium(IV). The single-crystal data were simulated using the recognizable spin-Hamiltonian shown in Eq. 4, where $\mathbf{g}$ and $\mathbf{A}$ are the 3×3 electron Zeeman and magnetic hyperfine interaction matrices, respectively.

$$\hat{H} = \mu_B \cdot \mathbf{B} \cdot \mathbf{g} \cdot \mathbf{S} + \mathbf{S} \cdot \mathbf{A} \cdot \mathbf{I} \qquad (4)$$

The benefit of a magnetically dilute sample (1% in diamagnetic $[Mo(mnt)_3]^{2-}$) is that the spectral linewidths are at their narrowest, so Kwik and Stiefel (204) could improve upon the previous interpretation derived from a frozen solution sample (10, 142). Spectra for D_{3h} and D_3 molecules are characterized with nearly isotropic g-values and highly anisotropic, axial A-values with the now trademark $g_\parallel > g_\perp$ and $A_\perp > A_\parallel$ pattern; $[V(mnt)_3]^{2-}$ is the benchmark member of this class with simulation yielding $g = (1.978, 1.980, 1.989)$ and $A = (-88, -84, -3) \times 10^{-4}$ cm^{-1} (179). These parameters are slightly different from those published by Kwik and Stiefel (204) who elected to have the smallest A-value as positive, but this was changed to conserve the motional averaging $\langle A \rangle = (A_1 + A_2 + A_3)/3 \approx A_{iso}$; the miniscule size of this value means that A_3 can be either positive of negative with no impact on the overall interpretation. The sign of all three values as negative assumes a dominant Fermi contact contribution (s orbital character) to the hyperfine, and therefore the sign follows that of the nuclear g-value for the element. Kwik and Stiefel (204) also described $\mathbf{g}$ and $\mathbf{A}$ as coaxial, with the z components aligned along the molecular C_3 axis. Atherton and Winscom (203) arrived at the same conclusions about the electronic construct of $[V(mnt)_3]^{2-}$, but they assumed C_{2v} symmetry seemingly oblivious of the actual structure of the molecule. All tris(dithiolene)vanadium neutral and dianionic complexes have very similar parameters irrespective of the dithiolene ligand type (Tabel IX) (179). The same description applies for $[Nb(mnt)_3]^{2-}$ ($S = 1/2$) as Nb(IV) (329).

TABLE IX
Spin Hamiltonian Parameters for $S = 1/2$ Species Derived from Simulation of EPR Spectra

Complex	g_{iso}[a]	g_1	g_2	g_3	$\langle g \rangle$[b]	Δg[c]	A_{iso}[d]	A_1	A_2	A_3	$\langle A \rangle$[e]	Reference
[Ti(bdt)$_3$]$^{3-}$		2.18	2.04	2.00	2.073	0.18						206
[V(edt)$_3$]	1.990	1.991	1.989	1.988	1.989	0.003	−57.5	−8	−81	−84	−57.7	179
[V(pdt)$_3$]	1.991	1.993	1.991	1.989	1.991	0.004	−57.2	−5	−80.5	−83	−56.2	179
[V(dmit)$_3$]					1.966[f]							216
[V(dddt)$_3$]					1.9990[f]			−5	−95	−95	−60.5	91
[V(deTTFdt)$_3$]					2.0094[f]							91
[V(dbTTFdt)$_3$]					2.0088[f]							91
[V(pTTFdt)$_3$]					2.0088[f]							91
[V(bdt)$_3$]$^{2-}$	1.980	1.993	1.977	1.974	1.981	0.019	−60.2	−2	−86	−89	−59.0	179
[V(tdt)$_3$]$^{2-}$	1.980	1.994	1.976	1.973	1.981	0.021	−60.2	−3	−87	−88	−59.3	179
[V(bdtCl$_4$)$_3$]$^{2-}$	1.9795						−58.1					43
[V(dbdcto)$_3$]$^{2-}$					1.97[f]							213
[V(edt)$_3$]$^{2-}$		1.993	1.977	1.976	1.982	0.017	−58.9	−2	−88	−89	−59.7	179
[V(mnt)$_3$]$^{2-}$	1.982	1.989	1.980	1.978	1.982	0.011	−57.4	−3	−84	−88	−58.3	179
[V(tfd)$_3$]$^{2-}$	1.9829											142
[V(pdt)$_3$]$^{2-}$	1.983	1.992	1.980	1.976	1.983	0.016	−58.3	−4	−85	−87	−58.7	179
[V(dmit)$_3$]$^{2-}$	1.979	2.000	1.971	1.964	1.978	0.036	−63.9	3.3	−94	−113	−67.9	85
[V(dmt)$_3$]$^{2-}$	1.978	2.000	1.965	1.960	1.973	0.040	−61.9	4	−96.7	−113.3	−68.7	85
[V(dddt)$_3$]$^{2-}$	1.9917				2.007[f]		−61.2					218
[Nb(mnt)$_3$]$^{2-}$	1.988				1.988[g]		94.4	127.2	127.2	−28.8	94.4	330
[Nb(dmit)$_3$]					2.011[f]							161
[Cr(bdtCl$_4$)$_3$]$^{1-}$	1.9964						14.0					43
[Cr(bdtCl$_2$)$_3$]$^{1-}$		2.010	2.006	2.006	2.007	0.004		7	18	18	14.3	221
[Cr(tbbdt)$_3$]$^{1-}$		2.0074	1.9947	1.9947	1.999	0.0127						222
[Cr(mnt)$_3$]$^{1-}$		2.001	1.995	1.993	1.999	0.008		8	20	20	16.0	221
[Cr(tfd)$_3$]$^{1-}$	1.9941						15.2					10
[Cr(pdt)$_3$]$^{1-}$	1.996						17.6					96

(continued)

TABLE IX
(Continued)

Complex	$g_{iso}{}^a$	g_1	g_2	g_3	$\langle g \rangle^b$	Δg^c	$A_{iso}{}^d$	A_1	A_2	A_3	$\langle A \rangle^e$	Reference
[Mo(bdt)₃]¹⁻	2.0061	2.0207	2.0076	1.9944	2.0076	0.0263	27.1	39	29	13	27.0	232
[Mo(tdt)₃]¹⁻	2.003						26.2					181
[Mo(tbbdt)₃]¹⁻	2.0050	2.0203	2.0081	1.9955	2.0080	0.0248	24.6	39	24	13	25.3	232
[Mo(qdt)₃]¹⁻	2.0004	2.0123	2.0059	1.9930	2.0037	0.0193	28.7	39	40	7	28.7	232
[Mo(tfd)(bdt)₂]¹⁻	2.0041	1.992	1.992	2.005	1.996	0.013	20.8					331
[Mo(tfd)₂(bdt)]¹⁻	2.0051	2.006	2.006	2.006	2.006	0	14.9					331
[Mo(edt)₃]¹⁻	2.0106	2.0095	2.0099	2.0124	2.0106	0.0029	12.6	9	17	14.5	13.5	232
[Mo(mnt)₃]¹⁻	2.0108	2.0107	2.0109	2.0132	2.0116	0.0025	11.8	8	16	13	12.3	232
[Mo(tfd)₃]¹⁻	2.0092	2.016	2.010	2.004	2.010	0.012	10.9					331
[Mo(sdt)₃]¹⁻	2.0101	2.0110	2.0119	2.0121	2.0117	0.0011	11.6	12	15	11	12.7	232
[Mo(pdt)₃]¹⁻	2.0125	2.0120	2.0121	2.0156	2.0132	0.0036	9.4	6	12.5	10	9.5	232
[Mo(dmit)₃]¹⁻	2.008				1.988^f							238
[W(bdt)₃]¹⁻	1.9825	2.0237	2.0069	1.9448	1.9918	0.0789	-48.0	-59	-70	-26	-51.7	232
[W(tdt)₃]¹⁻	1.974						-46.5					181
[W(bdtCl₄)₃]¹⁻	1.985											43
[W(mdt)₃]¹⁻	1.9941	2.0094	2.0012	1.9878	1.9995	0.0216	-27.3	-32	-34	-11	-25.7	232
[W(mnt)₃]¹⁻	1.9946	1.9864	1.9896	2.0006	1.9922	0.0142	-29.4	-18	-34	-32	-28.0	232
[W(tfd)₃]¹⁻	1.9910											10
[W(pdt)₃]¹⁻	1.9933	2.0075	2.0018	1.9907	2.0000	0.0168	-27.0	-30	-31	-12	-24.3	232
[W(dmit)₃]¹⁻	1.988				1.987^f							238
[Tc(bdt)₃]		2.014	2.014	2.018	2.015	0.004		6.5	6.5	7.5	6.8	332
[Tc(tdt)₃]	2.001						12.3					144
[Re(bdt)₃]	2.010	2.012	2.012	2.014	2.013	0.002		-4.5	-30	-29	-21.2	238
[Re(tdt)₃]	2.000						-12.7					144
[Re(bdtCl₂)₃]		2.015	2.015	2.015	2.015	0		-6	-25	-32	-21.0	232
[Re(tms)₃]		2.014	2.015	2.016	2.015	0.002		-4	-27	-31	-20.7	232
[Re(edt)₃]					2.010^f							96

Complex	$\langle g \rangle$[a]	g_1	g_2	g_3	$\langle g \rangle$[b]	Δg[c]	A_{iso}[d]	A_1	A_2	A_3	Ref
[Re(pdt)$_3$]	2.015	2.012	2.016	2.016	2.015	0.004	−8.6	−20.3	−20.1	−16.3	322
[Re(dmit)$_3$]					2.013[f]						161
[Re(bdt)$_3$]$^{2-}$		1.701	1.680	1.515	1.632	0.186	−310	−690	−100	−367	232
[Re(mnt)$_3$]$^{2-}$		1.704	1.252	1.847	1.601	0.595	−200	−170	−570	−313	322
[Fe(mnt)$_3$]$^{3-}$		2.27	2.09	1.97	2.09	0.30					249
[Fe(dtcr)$_3$]$^{3-}$		2.47	2.07	1.85	2.13	0.62					103
[Fe(dcmdtcr)$_3$]$^{3-}$		2.37	2.16	1.87	2.13	0.50					103
[Ru(mnt)$_3$]$^{3-}$		2.120	2.026	1.968	2.038	0.152					333
[Os(mnt)$_3$]$^{3-}$		2.19	2.01	1.82	2.01	0.37					333
[Sn(eTTFdt)$_3$]$^{1-}$					2.008[f]						257

[a] Isotropic g-value from fluid solution measurements.

[b] Here $\langle g \rangle = (g_1 + g_2 + g_3)/3$, from powder or frozen solution measurements.

[c] Here $\Delta g = g_1 - g_3$.

[d] The A_{iso} from fluid solution measurement in units of $10^{-4}\,\mathrm{cm}^{-1}$. Metal hyperfine splitting: ^{53}Cr, $I = 3/2$; ^{95,97}Mo, $I = 5/2$; ^{93}Nb, $I = 9/2$; ^{185,187}Re, $I = 5/2$; ^{99}Tc, $I = 9/2$; ^{51}V, $I = 7/2$; ^{183}W, $I = 1/2$.

[e] Here $\langle A \rangle = (A_1 + A_2 + A_3)/3$, from powder or frozen solution measurements. The sign derives from the sign of the gyromagnetic ratio of the metal assuming a dominant Fermi contact contribution to the interaction.

[f] Spectrum recorded on powder sample.

[g] $A_3 = 3A_{iso} - (A_1 + A_2)$. Individual g-values could not be resolved.

Chromium monoanionic and trianionic members of the electron-transfer series possess noninteger ground states of $S = 1/2$ and $S = 3/2$, respectively. The EPR spectra of the spin-doublet species was originally described as most likely ligand centered along with its group 6 (VI B) homologues based on the small ^{53}Cr ($I = 3/2$, 9.53% abundant) hyperfine coupling (10). It was subsequently revealed that the tris(dithiolene) series parallels the tris(dioxolene) and tris(diimine) ones, where all members are invariantly Cr(III) and the $S = 1/2$ ground state derived from antiferromagnetic coupling of the ligand- and metal-based SOMOs (222, 280, 297–306). As with magnetochemical measurements, EPR does not reveal whether the Cr based spin arises from a +I, +III, or +V metal ion. However, with the aid of other sources, the monoanion is formulated $[Cr^{III}(L)(L^{•})_2]^{1-}$, which gives a net unpaired electron on the metal generating the familiar $g_{\parallel} > g_{\perp}$ and $A_{\perp} > A_{\parallel}$ pattern (Table IX). Moreover, this profile is similar to other monoanions of Cr(III) with three chelating ligands (222, 306, 316).

Fluid solution EPR spectra of monoanionic Mo and W complexes revealed a dependency of the hyperfine coupling on the type of dithiolene ligand, aryl or alkyl based (232). This finding is most clearly expressed comparing the ^{95,97}Mo ($I = 5/2$, 25.47% abundant) hyperfine coupling constant for $[Mo(bdt)_3]^{1-}$ and $[Mo(edt)_3]^{1-}$, which differ by a factor of 2 (Fig. 48). There is also a noticeable difference in g_{iso} for the two groups that is beautifully presented in the series $[Mo(bdt)_x(tfd)_{3-x}]^{1-}$ ($x = 0$–3), where g_{iso} sequentially increases with a concomitant reduction in the hyperfine coupling as the series is traversed (331). The former has greater metal character in the ground state wavefunction than the latter, and was described as $[Mo^V(bdt)_3]^{1-}$ (232). The DFT derived composition of the SOMO and Mulliken spin population analysis (Fig. 49) reflect the different electronic arrangements (Table X). Curiously, the fluid solution EPR spectrum of $[Mo^V(abt)_3]^{1-}$, where $(abt)^{2-}$ = aminobenzenethiolate, has been reported with a hyperfine coupling constant of $\sim 40 \times 10^{-4}\,cm^{-1}$ (334).

Tris(dithiolene)tungsten monoanions display a broad, featureless resonance at X-band, where the hyperfine satellites from the ^{183}W ($I = 1/2$, 14.31% abundant) isotope are buried in the spectral linewidth. Greater resolution is obtained at S-band ($\sim 3.7\,GHz$) that allows for these hyperfine lines to be observed (232). The isotropic g-value is lower than the corresponding Mo compounds because of the very large spin–orbit coupling constant for the third-row metal (335). Additionally, the W complexes yield much larger isotropic hyperfine coupling constants with values ranging 27.0–29.4 × $10^{-4}\,cm^{-1}$ for the alkene-1,2-dithiolates, which are dwarfed by $A_{iso} = 48 \times 10^{-4}\,cm^{-1}$ for $[W(bdt)_3]^{1-}$ (Table IX).

Frozen solution spectra exhibit the opposing pattern to that seen for the first-row metals with $g_{\perp} > g_e > g_{\parallel}$ for all monoanions with aryl-1,2-dithiolate ligands. This highlights greater covalency in the Mo/W−S bonds, distributing more spin density on the ligands. The Mulliken spin population in $[Mo(bdt)_3]^{1-}$ and $[W(bdt)_3]^{1-}$ sees ≤ 1 spin at the metal ion, consistent with a smaller hyperfine interaction

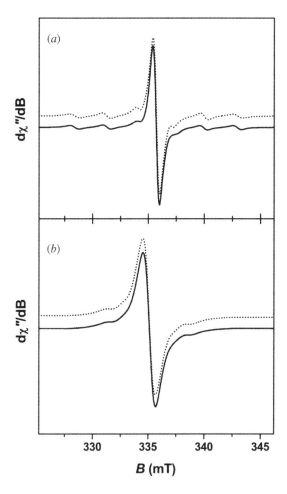

Figure 48.   Fluid solution X-band EPR spectra of (a) [Mo(bdt)$_3$]$^{1-}$ and (b) [Mo(edt)$_3$]$^{1-}$ in THF solution at 200 K. Experimental data are shown in black and simulations represented by a dotted trace.

compared with vanadium (Table X). The g-anisotropy (Δg) is three times larger for W than Mo, in line with the aforementioned spin–orbit coupling constants for these third- and second-row transition metals, respectively. Monoanionic molybdenum tris(dithiolene) complexes, where the dithiolene is olefinic exhibit nearly isotropic frozen solution spectra. Such a manifestation is a clear indication that the unpaired electron is no longer residing on the metal, but rather delocalized across the ligand framework (Fig. 49b). The g-anisotropy is miniscule, an order of magnitude lower than their aryl-1,2-dithiolate counterparts. The axial symmetry is maintained with the maximum value corresponding to $g_\parallel$ with only a very minor

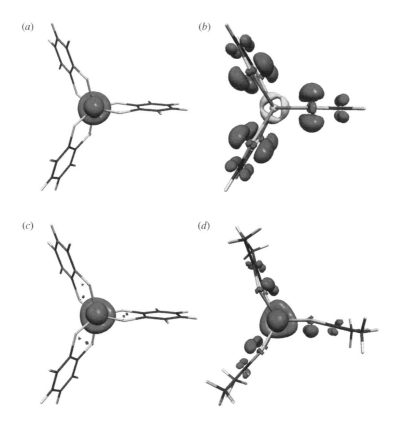

Figure 49.   Mulliken spin density plots for (a) [Mo(bdt)$_3$]$^{1-}$, (b) [Mo(edt)$_3$]$^{1-}$, (c) [W(bdt)$_3$]$^{1-}$, and (d) [W(mdt)$_3$]$^{1-}$, derived from DFT calculations as viewed down the C_3 axis. (See color version of this figure in the color plates section.)

TABLE X
Singly occupied molecular orbital (SOMO) Composition and Mulliken Spin Population Analysis for Selected Complexes[a]

			%Composition			Mulliken Population		
Complex	Symmetry	SOMO	M d	S p	C p[b]	M	S	C[b]
[Mo(bdt)$_3$]$^{1-}$	D_3	3a$_1$	68.5	12.2	3.6	+1.01	−0.09	+0.05
[Mo(edt)$_3$]$^{1-}$	D_{3h}	2a$_2'$	0.1	66.4	22.8	−0.18	+0.94	+0.32
[W(bdt)$_3$]$^{1-}$	D_3	3a$_1$	63.9	12.1	4.4	+0.87	+0.01	+0.08
[W(mdt)$_3$]$^{1-}$	C_{3h}	4a$'$	36.0	33.4	13.7	+0.45	+0.38	+0.16

[a] Data taken from (232).
[b] Metallodithiolene carbon atoms only.

splitting of $g_\perp$; the opposite pattern to that seen with arene-1,2-dithiolates. The g-values are all greater than g_e as is the case for isoelectronic neutral Re analogues; the positive g-shift most likely stems from the reasonably large spin–orbit coupling constant for sulfur (336). Molybdenum superhyperfine is coupling present in the spectra of each complex, and resolved at low frequency. The hyperfine interaction is weakly axial, and the approximate $A_x < A_y \sim A_z$ pattern suggests the **A** is rotated away from **g** such that the larger A-values are aligned with the smaller g-values.

In contrast to the molybdenum complexes, the spectra for tris(dithiolene) tungsten monoanions exhibit a pronounced g-splitting such that the anisotropy is only one-third of the value determined for $[W(bdt)_3]^{1-}$ (Table IX). This arises from a larger splitting of $g_\perp$ leaving $g_x < g_e$, highlighting a different ground state composition than for molybdenum. It was reported that these species adopt a C_{3h} symmetric structure that destabilizes the ligand-centered $2a_2'$ MO with respect to the metal-centered $3a_1'$ MO, but also facilitates mixing of these orbitals that transform as a' in C_{3h} point symmetry (232). Therefore, dithiolene folding installs W d_{z^2} character into the ground state, estimated at 36% (Table X), and accounts for the g-splitting fuelled by the large spin–orbit coupling constant of the metal. Moreover, the tungsten hyperfine interaction is also much larger than for the Mo homologues. These data support the notion of a C_{3h} solution structure for W monoanions with alkene-1,2-dithiolate ligands in contrast to the D_{3h} or D_3 symmetric geometries found by X-ray crystallography (Table II). It appears tungsten's preference for higher oxidation states places the true electronic structure somewhere between the $[W^{IV}(L_3^{5-\bullet})]^{1-}$ and $[W^V(L_3^{6-})]^{1-}$ limiting forms.

Neutral tris(dithiolene)rhenium complexes having an $S = 1/2$ ground state have been analyzed by EPR spectroscopy and proved to be the key evidence that motivated Al-Mowali and Porte (285) to revise the published orbital manifolds of Gray and co-workers (181) and Schrauzer & Mayweg (96). Gray's and Schrauzer's electronic structures both position the unpaired electron on the metal (Fig. 37). Porte and co-workers (285) keenly observed that the frozen solution spectra of [Re(pdt)_3] and [Re(tdt)_3] were drastically different from that of [ReOCl_4]. The EPR spectrum of the latter is characterized by large g-anisotropy driven by spin–orbit coupling. Additionally, there is an even larger magnetic hyperfine interaction from the ^{185,187}Re ($I = 5/2$, 100% abundant) nuclei affording a spectrum that spans 500 mT. In contrast, [Re(tdt)_3] has a narrow signal spanning 15 mT (Fig. 50), and a $(3a_1')^2(4e')^4(2a_2')^1$ electronic configuration was proposed with the unpaired electron located in a pure ligand orbital (Fig. 51). Frozen glass spectra of all neutral Re tris(dithiolenes) are indistinguishable with a central signal at $g \sim 2.013$ flanked on either side by two features that Porte incorrectly assigned as g components. In fact, these signals stem from Re superhyperfine coupling of $\sim 20 \times 10^{-4}\,cm^{-1}$ for arene-1,2-dithiolates (247), and slightly smaller for [Re(pdt)_3] (322). The entire spectral profile is dominated by a colossal quadrupole

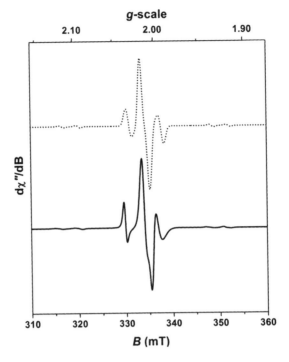

Figure 50.   X-band EPR spectrum of [Re(bdt)$_3$] in CH$_2$Cl$_2$ at 10 K. The experimental spectrum is shown in black and simulation represented by the dotted trace. [Adapted from (247).]

interaction, and this term is included in the spin-Hamiltonian (Eq. 5), where **P** is the 3×3 tensor matrix that expresses the nuclear spin–nuclear spin coupling.

$$\hat{H} = \mu_B \cdot B \cdot \mathbf{g} \cdot S + S \cdot \mathbf{A} \cdot I + I \cdot \mathbf{P} \cdot I \tag{5}$$

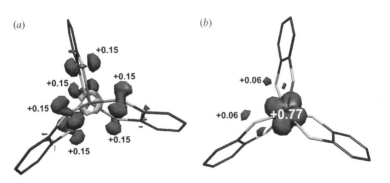

Figure 51.   Mulliken spin density plots and populations for (a) [Re(bdt)$_3$] and (b) [Re(bdt)$_3$]$^{2-}$. [Adapted from (247).] (See color version of this figure in the color plates section.)

Nuclear quadrupole interactions can result in perturbations of the intensity and spacing of the $\Delta m_I = 0$ (hyperfine) transitions (337–339). Such an effect is operative here, with the expected "six-line" magnetic hyperfine interaction pattern (given isotropic g and the Re nuclear spin) compressed into two flanking features of the central resonance. Additionally, we find a small pair of satellite signals 15 mT away from the central resonance (Fig. 50) that are quadrupole-allowed transitions rarely encountered in EPR spectra. These are assigned as the $\pm 1/2 \rightarrow \pm 3/2$ ($\Delta m_I = 1$) transitions since these have the smallest energy separation, and their position gives a clear indication of the quadrupolar interaction at $24.2 \times 10^{-4}\,\mathrm{cm}^{-1}$ for [Re(bdt)$_3$] (247) and $21.9 \times 10^{-4}\,\mathrm{cm}^{-1}$ for [Re(pdt)$_3$] (322). This very unique situation, where the quadrupole interaction is larger than the magnetic hyperfine interaction, can only arise if the spin is located on the ligand and the diamagnetic Re ion has a $(d_{z^2})^2$ electronic configuration.

This arrangement generates the very large valence contribution to the electric field gradient producing the observed quadrupole-allowed transitions in the spectrum. Large quadrupole couplings that exceed the magnetic hyperfine inter-action have been observed for square-planar bis(dithiolene)gold(III) (275). There-fore, a $^2A_2'$ ground term (D_{3h} point symmetry) can be unambiguously assigned to these neutral species with the formula [Re^V(L$_3^{5-\bullet}$)].

The doubly reduced dianionic species also have an $S = 1/2$ ground state, though there was no recorded EPR spectrum for such a complex; McCleverty and co-workers (291) commented that (PPh$_4$)$_2$[Re(mnt)$_3$] did not yield a signal at room temperature. The EPR spectrum of [Re(bdt)$_3$]$^{2-}$ shown in Fig. 52 is distinctly different from the spectrum of the neutral species above (247). The spectrum spans 450 mT, akin to the Re(VI) d^1 signals recorded by Abrams et al. (240–242), and clearly depicts a metal-centered paramagnet. The spectrum is dominated by six broad hyperfine lines. Quadrupole effects produce the characteristic uneven splitting of the six hyperfine lines, and quadrupole-allowed transitions are seen in the low-field region. The simulation parameters for [Re(bdt)$_3$]$^{2-}$ listed in Table IX show anisotropic g- and A-values, and a quadrupole coupling of 20×10^{-4} cm^{-1} (247). It is clear from this spectrum that the unpaired electron is on the metal ion and the pronounced anisotropy is indicative of an almost orbitally degenerate ^2E ground term for the [Re^{IV}(L$_3^{6-}$)]$^{2-}$ electronic structure (Fig. 51).

Group 8 (VIII B) trianionic complexes possess low-spin d^5 metal ions and exhibit highly rhombic EPR spectra. The g-anisotropy is driven by two phenomena, these being M$-$S covalency and a subtle structural distortion that lifts the $^2T_2$ ground term of a t$_{2g}$ hole to a ^2A ground term (333). For [Fe^{III}(mnt)$_3$]$^{3-}$, structural data confirm a Jahn–Teller distortion along the C_2 axis orthogonal to the main axis of the molecule (249). The threefold symmetry splits the t$_{2g}$ orbitals into a$_1$ and e levels and the Jahn–Teller destroys orbital degeneracy in the latter. The distortion is greater for [Fe^{III}(mnt)$_3$]$^{3-}$ than its dithiocarbamate, (dtc)$^{1-}$, counterpart [Fe^{III}(dtc)$_3$] because of the larger bite angle supplied by the dithiolene ligand. The nearly degenerate ground state in

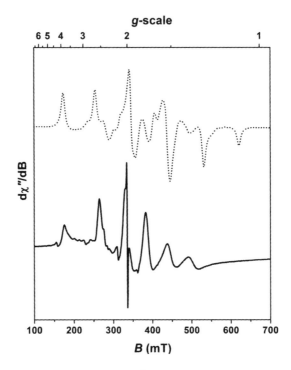

Figure 52. X-band EPR spectrum of $[Re(bdt)_3]^{2-}$ in CH_2Cl_2 at 20 K. The experimental spectrum is shown in black and simulation represented by the dotted trace. [Adapted from (247).]

$[Fe^{III}(dtc)_3]$ produces a highly anisotropic low-spin (HALS) spectrum (249). The g-anisotropy in $[Ru^{III}(mnt)_3]^{3-}$ is less than that in the Fe homologue (333). Its C_3 symmetry structure shows no evidence for a Jahn–Teller distortion, presumably modulated by the increased covalency (251). Anisotropy increases in the Os variant as the sizeable spin–orbit coupling constant of the 5d metal imposes itself on the LF splitting, shifting g_z below 2.

2. Spin Quartet

Isoelectronic Cr trianions and Mn dianions have an $S = 3/2$ spin ground state (103, 221, 222). With ZFS larger than the microwave quantum ($D \gg h\nu$), their EPR spectra bear the hallmark $g_{eff} = (4, 4, 2)$ for the allowed transitions between of $m_s = \pm 1/2$ levels—a g-value for each canonical direction x, y, and z, with $g_x = g_y$ and the pattern known as axial (Fig. 53). The ZFS is found to be small, and contributes to the EPR spectrum as a forbidden transition at $g_{eff} = 6$ from the $m_s = \pm 3/2$ Kramer's doublet. Its intensity is proportional to the magnitude of the ZFS, parameterized by D. The sign of D is known by the temperature dependence of this signal; D is negative because this

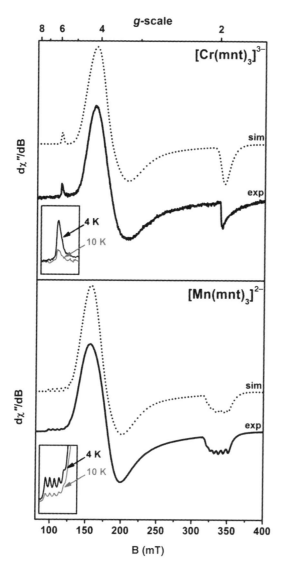

Figure 53.   X-band EPR spectra of $[Cr(mnt)_3]^{3-}$ and $[Mn(mnt)_3]^{2-}$ recorded at 10 K. Experimental spectrum displayed as a solid black line; simulation as the dotted trace. Insets show the response in intensity of the $g_{eff} = 6$ signal from the excited $m_s = \pm 3/2$ Kramer's doublet upon cooling to 4 K. [Adapted from (221).]

TABLE XI
Spin Hamiltonian Parameters for $S = 3/2$ Species Derived from Simulation of EPR Spectra

Complex	g_1	g_2	g_3	D^a	E/D	Reference
$[Cr(bdtCl_2)_3]^{3-}$	1.980	1.980	1.960	−3.3	0.080	221
$[Cr(mnt)_3]^{3-}$	1.970	1.970	1.980	−1.6	0.073	221
$[Cr(dtcr)_3]^{3-b}$	5.75	3.55	1.98			103
$[Cr(dcmdtcr)_3]^{3-b}$	5.73	4.68	1.96			103
$[Mn(bdtCl_2)_3]^{2-}$	2.020	2.020	1.970	−1.6	0	221
$[Mn(mnt)_3]^{2-}$	2.039	2.039	2.010	−1.3	0.071	221

^{a} In units of reciprocal centimeters (cm^{-1}).
^{b} Only effective g-values were reported.

feature diminishes in intensity with respect to the $g_\perp \sim 4$ of the $m_s = \pm 1/2$ levels as the temperature increases (Fig. 53 inset). The spectra were simulated using the familiar spin-Hamiltonian for high-spin systems (Eq. 6).

$$\hat{H} = \mu_B \cdot \mathbf{B} \cdot \mathbf{g} \cdot \mathbf{S} + \mathbf{S} \cdot \mathbf{D} \cdot \mathbf{S} = \mu_B \cdot \mathbf{B} \cdot \mathbf{g} \cdot \mathbf{S} + D[\hat{S}_z^2 - 1/3S(S+1) + E/D(\hat{S}_x^2 + \hat{S}_y^2)]$$
(6)

The simulation parameters collated in Table XI are characteristic of near-octahedral d^3 ions with miniscule rhombicity (E/D). These ZFS parameters are used to fit the magnetic susceptibility data (see Section VII).

E. X-Ray Absorption Spectroscopy

When the previous compilation appeared in print almost 10 years ago there were no X-ray absorption spectra (XAS) recorded on tris(dithiolene) compounds (343). The situation has improved markedly in light of a seminal study by Solomon, Holm, and co-workers (344) on the $[Ni(mdt)_2]^{0/1-/2-}$ series. This work provided the first direct measure of the sulfur content of the ground-state orbital via sulfur K-edge X-ray absorption spectroscopy (XAS) and refined the electronic structure description of this well-known series. The subsequent investigation of a broader set of compounds further clarified the covalent bonding and radical character in bis(dithiolene) systems in general by the combination of S K-edge and time-dependent density functional theory, TD–DFT, calculations (345). Since 2008, tris(dithiolenes) have received long awaited scrutiny by XAS at both metal and ligand edges.

1. Metal Edges

K-edge X-ray absorption spectra arise from promotion of a core 1s electron to a vacant orbital on an absorbing atom (Fig. 54). This part of the spectrum is named

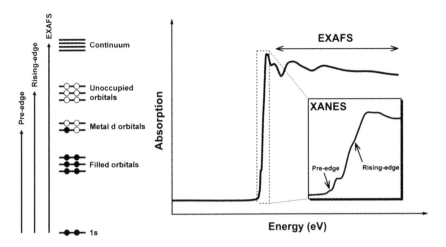

Figure 54.   Simplified depiction of the components of a metal K-edge XAS experiment: Excitation of a core 1s electron to vacant d orbitals generates the pre-edge features, whereas the rising-edge is dominated by dipole-allowed 1s → np transitions (XANES). Beyond the edge, the input energy from the X-ray beam is sufficient to eject the electron from the absorbing atom which emanates as a photoelectron that interacts with electrons of the surrounding atoms giving rise to oscillations in the spectrum (EXAFS).

X-ray absorption near-edge structure (XANES) and provides electronic structure information about the material under examination. As more energy is applied (beyond the edge) the core electron is ejected from the absorbing atom and emanates outward as a photoelectron interacting with the immediate neighbors generating an oscillating pattern in the spectrum. This post-edge region of the spectrum is termed extended X-ray absorption fine structure (EXAFS) since the oscillations arising from the interference of the photoelectron with electrons in the surrounding atoms yield geometric information about the system, mainly distances.

The XANES part of the experiment is subdivided into pre-edge and rising-edge regions. The rising-edge is dominated by electric dipole-allowed 1s → np transitions (Fig. 54). The pre-edge region in metal K-edge spectra arises from electric dipole-forbidden, but quadrupole-allowed, 1s → nd transition that can gain intensity through mixing of metal p character via departure from centrosymmetry (e.g., a twist from square planar to tetrahedral). The geometric modification also affects the ligand field and hence the transition energy. The K-edge is a direct measure of the effective nuclear charge (Z_{eff}) of an element, (i.e., its oxidation state), with the core 1s orbital shifting to deeper binding energy with increasing Z_{eff}. This gives rise to the generally accepted rule that a 1 eV shift in the edge to higher energy corresponds to an increase in the oxidation state by one unit.

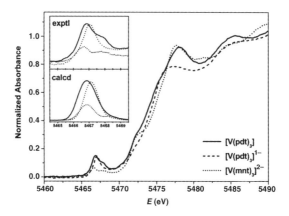

Figure 55.   Overlay of the normalized V K-edge XAS spectra of [V(pdt)₃], [V(pdt)₃]¹⁻, and [V(mnt)₃]²⁻. Inset shows expansion of the experimental and calculated pre-edge region. [Adapted from (179).]

The tris(dithiolene)vanadium monoanion is diamagnetic and impervious to scrutiny by EPR, so it remained ill-defined in the literature. Clearly evident in the V K-edge spectra for [V(pdt)₃], [V(pdt)₃]¹⁻, and [V(mnt)₃]²⁻ shown in Fig. 55 is that both the pre- and rising-edge energies are very similar across this three-membered series (179). The pre-edge energies differ maximally by 0.5 eV for all measured V compounds indicating a common oxidation state for the metal (Table XII). With EPR data confirming a V(IV) central ion in the neutral and dianionic $S = 1/2$ species, it can be concluded that the monoanion is $[V^{IV}(L_3{}^{5-•})]^{1-}$. This result is validated by the calculated (TD–DFT) pre-edge spectra that closely match the experiment (Fig. 55 inset). Moreover, the first pre-edge peak in the spectrum of a genuine V(III) compound, [V(dtc)₃], occurs at 5465.7 eV, which is 1 eV lower in energy (179). The pre-edge is noticeably varied across the series, with [V(mnt)₃]²⁻ less intense than $[V^{IV}(pdt)_3]$ and $[V^{IV}(pdt)_3]^{1-}$ (Fig. 55 inset) The difference lies in the trigonal twist angle, with the larger twist away from centrosymmetric octahedral facilitating greater V 4p mixing into the 5e′ MOs. Thus, the pre-edge intensities of $[V^{IV}(pdt)_3]$ (Θ = 4.6°) and $[V^{IV}(pdt)_3]^{1-}$ (Θ = 0.8°) are three times larger than for [V(mnt)₃]²⁻ (Θ = 38.0°), with up to 4% V 4p character in the 5e′ level. Moreover, the intensity remains unchanged in solution spectra, thus proving the TP geometry is retained in the absence of lattice packing. This useful geometry marker was utilized for electrochemically generated [V(mnt)₃]¹⁻ (S = 0). While the pre-edge peak energy supported a +IV oxidation state for the metal ion, the intensity is consistent with a TP structure for this monoanion, as found for all crystallographically characterized tris(dithiolene)vanadium monoanions (89, 91, 179, 215, 217, 218).

TABLE XII
Pre-Edge and Rising-Edge Transition Energies

Complex	Edge[a]	Pre-Edge[b]	Rising-Edge[c]	Reference
$[V(edt)_3]$	K	5466.8	5474.2	179
$[V(pdt)_3]$	K	5466.9	5473.9	179
$[V(pdt)_3]^{1-}$	K	5467.0	5473.7	179
$[V(mnt)_3]^{1-}$	K	5467.2	5474.8	179
$[V(mnt)_3]^{2-}$	K	5466.7	5475.0	179
$[V(tdt)_3]^{2-}$	K	5466.7	5473.9	179
$[Cr(tbbdt)_3]^{1-}$	K	5989.7		222
$[Cr(bdtCl_2)_3]^{1-}$	K	5990.0	5997.2	221
$[Cr(bdtCl_2)_3]^{2-}$	K	5990.3	5997.0	221
$[Cr(mnt)_3]^{2-}$	K	5990.4	5998.0	221
$[Cr(bdtCl_2)_3]^{3-}$	K	5990.1	5996.7	221
$[Cr(mnt)_3]^{3-}$	K	5990.3	5997.7	221
$[Mn(bdtCl_2)_3]^{2-}$	K	6540.3	6547.7	221
$[Mn(mnt)_3]^{2-}$	K	6540.3	6549.3	221
$[Fe(mnt)_3]^{2-}$	K	7112.3	7116.9	249
$[Fe(mnt)_3]^{3-}$	K	7112.0	7116.1	249
$[Re(bdt)_3]$	L_1		12533.6	247
$[Re(bdt)_3]^{1-}$	L_1		12533.9	247
$[Re(bdt)_3]^{2-}$	L_1		12532.6	247

[a] Reference edge energy: V, 5465 eV; Cr, 5989 eV; Mn, 6539 eV; Fe, 7112 eV; Re, 12527 eV.
[b] Lowest energy peak maxima.
[c] Determined at the first inflection point of the rising-edge.

Six-coordinate tris(dithiolene)chromium complexes have been known to form a four-membered electron-transfer series where each member is related by reversible one-electron transfer waves (10, 135, 221, 222, 288, 323). For dioxolene (catechol) (347) and diimine (bipyridine) (306) analogues, the electrochemistry (cyclic voltammetry or polarography) encompasses all seven members within the solvent window, $[Cr(L)_3]^z$ ($z = 3+, 2+, 1+, 0, 1-, 2-, 3-$). The redox noninnocence of dithiolene ligands bound to Cr is evident from electronic absorption and IR spectroscopy, which indicated that the tris(dithiolene) series mirrors the tris (dioxolene) one (221, 222). As such, the K-edge profile for all Cr compounds is essentially identical (Table XII). For the three-membered series, $[Cr(bdtCl_2)_3]^z$ ($z = 1-, 2-, 3-$), the pre-edge energy is found within experimental error at 5990.1 ± 0.2 eV, consistent with no change of Z_{eff} across this series (221). The pre-edge peak for $[Cr(tbbdt)_3]^{1-}$ was found at 5989.7 eV (222), the subtle difference reflecting the varying electronic properties of a t-Bu versus a Cl substituent. Two pre-edge peaks were found in the Cr K-edges of $[Cr(mnt)_3]^{2-}$ and $[Cr(mnt)_3]^{3-}$, with a rather noticeable low-energy shoulder in the former at 5989.4 eV. For the first time, these pre-edge peaks were assigned with the aid of TD–DFT calculations. For the Werner-type $[Cr(mnt)_3]^{3-}$, one peak was assigned as predominately due to the Cr 1s $\rightarrow$ t$_{2g}$ transition (i.e., excitation to the three half-empty Cr t$_{2g}$ orbitals). The spectrum of [Cr(mnt)$_3$]$^{2-}$ also has this feature as well as a low-energy shoulder ascribed as the Cr 1s $\rightarrow$ (mnt$^\bullet$)$^{1-}$ transition, where the absorbing orbital is computed to comprise 36% Cr 3d, and most importantly, 1.4% Cr 4p character, which delivers electric-dipole character (221).

The history of tris(dithiolene) chemistry of Mn and Fe is rather thin owing to the paucity of available compounds. With only five isolated tris(dithiolene)manganese complexes (135, 221, 244, 288, 323), all of which are $S = 3/2$ dianions, the electronic structure is readily formulated as [Mn^{IV}(L)$_3$]$^{2-}$. Not surprisingly, their pre-edge energies are identical, while the rising-edge energies reflect the different dithiolene ligand types (221). Prior to 2009, only [Fe(mnt)$_3$]$^{2-}$ had been structurally characterized (201, 223, 249), and described as a low-spin Fe(IV) d^4 ion coordinated by three closed-shell dithiolate ligands. This assignment was supported by a change in the Mössbauer isomer shift of 0.14 mm s^{-1} (see Section VIII.F). Comparison of the Fe K-edge spectra of [Fe(mnt)$_3$]$^{3-}$ and [Fe(mnt)$_3$]$^{2-}$ reveal a shift of 0.8 eV in the rising-edge energy consistent with an increase in Z_{eff} upon oxidation of the trianion (249). In contrast, the pre-edge energy decreases by 0.3 eV. The expected ~1 eV shift is counteracted by a decrease in the trigonal twist angle ([Fe(mnt)$_3$]$^{3-}$, $\Theta = 53.4°$; [Fe(mnt)$_3$]$^{2-}$, $\Theta = 46.6°$) that lowers the pre-edge transition energy for the dianion (249).

Excitation of a 2s electron produces an L$_1$-edge, and such spectra were recorded for the [Re(bdt)$_3$]^z ($z = 0$, 1$-$, 2$-$) series because of the prohibitively high energy required to access the K shell (247). Edge resolution deteriorates with increasing energy due to shortening of the core-hole lifetime (the missing core electron is rapidly replaced by any means available to the absorbing atom). The short lifetime leads to substantial broadening of the rising-edge obscuring all pre-edge structure. The Re L$_1$-edge spectra are featureless, however, they do clearly identify a lower oxidation state in [Re(bdt)$_3$]$^{2-}$ and an identical one for [Re(bdt)$_3$]$^{0/1-}$ (Table XII), corroborating conclusions drawn from EPR.

2. Sulfur K-Edge

The perfect compliment to a metal K-edge is the sulfur K-edge. It follows the same principles described above except that its pre-edge features result from electric dipole-allowed S 1s $\rightarrow$ 3p transitions (Fig. 56). There are two ways to create a vacant site in the S 3p orbitals, namely, removing an electron to generate a sulfur-centered radical, or a partial hole from covalent M$-$S bonding that shifts electron density from filled S 3p orbitals to vacant metal d orbitals. Thus, S K-edge probes all singly occupied and unoccupied orbitals of a transition metal complex that have S 3p character, and is a direct measure of the covalency of a metal–sulfur bond with the peak intensity reflecting the S 3p content in the final state (347–349). The oxidation level of the sulfur atoms will be reflected in the pre- and rising-edge

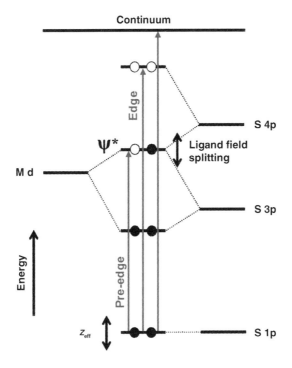

Figure 56.   Simplified scheme identifying the origins of pre- and rising-edge transitions in sulfur K-edge XAS.

energies, but with more than one sulfur atom present, as is the case for tris (dithiolene) compounds, a 1 eV shift is rarely encountered. Furthermore, changes in complex geometry can counteract this shift since the LF is modified.

The enormous impact of sulfur K-edge XAS studies lies in its ability to fingerprint the existence of a monoanionic dithiolene radical ligand, $(L^{\bullet})^{1-}$, as a low-energy pre-edge peak (typically $<2470\,eV$) defined as the S $1s \rightarrow (L^{\bullet})^{1-}$ transition. In many cases, particularly when S K-edge spectra of two or more members of an electron-transfer series are overlaid, a straightforward qualitative examination determines the metal and ligand oxidation levels.

The S K-edge spectra for the $[Cr(bdtCl_2)_3]^z$ ($z = 1-$, $2-$, $3-$) series are exceedingly straightforward (221). Overlooking the small amount of oxidized impurity, the spectrum of $[Cr(bdtCl_2)_3]^{3-}$ has no prominent pre-edge features. The spectra of the monoanion and dianion have two pre-edge peaks: The first is indicative of a dithiolene ligand radical at 2469.7 eV, where the intensity of this peak reflects the number of radicals (or vacancies in the ligand 3p shell). Complexes with $(mnt)^{2-}$ (221) and $(tbbdt)^{2-}$ (222) match this more complete series (Table XIII).

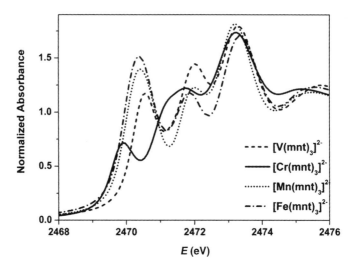

Figure 57.   Overlay of the normalized S K-edge XAS spectra of [M(mnt)$_3$]$^{2-}$ (M = V, Cr, Mn, Fe). [Adapted from (26).]

The pre-edge region for [Fe(mnt)$_3$]$^{3-/2-}$ can be deconvoluted to reveal three overlapping peaks assigned as transitions to the half-filled t$_{2g}$ orbital(s), empty Fe$-$S antibonding e$_g$ orbitals, and the π^* of the (mnt)$^{2-}$ ligands. The salient feature of this comparison is the significant increase in the intensity of the S 1s $\rightarrow$ t$_{2g}$ peak in [Fe(mnt)$_3$]$^{2-}$, and its shift to lower energy due to the higher Z$_{eff}$ of the Fe(IV) ion, rather than the creation of a ligand hole. This change in metal Z$_{eff}$ is elegantly demonstrated in Fig. 57, which shows an overlay of the S K-edge spectra of [M (mnt)$_3$]$^{2-}$ (M = V, Cr, Mn, Fe) (179, 221, 249). Only the chromium complex has a genuine ligand radical pre-edge peak at 2469.9 eV. The remaining complexes have three (mnt)$^{2-}$ ligands coordinating a +IV central ion. Upon moving from left to right across the first-row of the d block, the S 1s $\rightarrow$ t$_{2g}$ transition shifts to lower energy due to the increasing Z$_{eff}$ of the metal ion.

In the S K-edge spectra of the tris(dithiolene)vanadium series, a ligand hole peak is clearly evident at 2470.2 eV for [V(pdt)$_3$] and [V(pdt)$_3$]$^{1-}$, and absent in the spectrum of [V(pdt)$_3$]$^{2-}$. It is described as the S 1s $\rightarrow$ 2a$_2'$ excitation, with the acceptor orbital singly occupied in the monoanionic complex, (L$_3$)$^{5-\bullet}$, and empty for the neutral complex, (L$_3$)$^{4-}$. The first pre-edge peak [V(pdt)$_3$]$^{2-}$ at 2470.4 eV, albeit close in energy to the dithiolene radical peak, is in fact a S 1s $\rightarrow$ 5e$'(\alpha)$ transition (α representing a "spin up" electron) that corresponds to the shoulder at 2471.0 eV for [V(pdt)$_3$]$^{1-}$ and 2471.1 eV for [V(pdt)$_3$]; the shift to lower energy is consistent with a fully reduced tris(dithiolene) ligand set in the dianion. Additionally, the change in geometry also contributes to the transition energy and intensity by altering the S 3p content in the frontier orbitals. The transition to the 5e′(β) (β representing a "spin down" electron) MOs is at higher energy buried under transitions to the dithiolene S–C π* orbital and the rising-edge itself due to significant spin polarization of the V–S bonds.

Two separate studies of tris(dithiolene)molybdenum complexes have been published. The first comprised $[Mo(tbbdt)_3]^{0/1-}$, where the neutral species has two pre-edge peaks; the lowest energy assigned as the S $1s \rightarrow (L^·)^{1-}$ transition (227). The loss of this feature upon reduction to the monoanion led the authors to define these species as $[Mo^V(tbbdt)_2(tbbdt^·)]$ and $[Mo^V(tbbdt)_3]^{1-}$, respectively. This conclusion was modified by Solomon and co-workers (281) who extended their interest from $[Ni(mdt)_2]^z$ to $[Mo(mdt)_3]^z$ ($z = 0$, 1−, 2−). The S K-edge spectra for the neutral and monoanionic complexes clearly display a low-energy peak at 2470.4 and 2470.1 eV, respectively, assigned as the S $1s \rightarrow 2a_2′$ transition. This peak was absent in the spectrum of the dianion. Solomon was the first to identify a varying electronic structure among monoanionic complexes of Mo, one that is dependent on the type of dithiolene ligand. Arene-1,2-dithiolates bind to a more oxidized metal ion and consequently are fully reduced. In contrast, complexes with a tris(alkene-1,2-dithiolate) moiety are one-electron oxidized and bound to a Mo(IV) ion. The outcome results from configurational mixing between the Mo d_{z^2} ($3a_1′$) orbital and a filled ligand orbital at deeper binding energy that destabilizes the d orbital above the redox-active ligand-based $2a_2′$ MO (Fig. 58). This interaction is favored by a distorted octahedral (D_3) structure and forbidden in D_{3h} symmetric $[Mo(mdt)_3]^{1-}$ ($\Theta = 1.6°$). Solomon and co-workers (281) further added that the neutral complexes, $[Mo(tbbdt)_3]$ and $[Mo(mdt)_3]$, are closer to Mo(IV) with two holes in the ligands because the intensity of the radical pre-edge peak is similar in both cases. There are two well-resolved pre-edge features in the spectra of

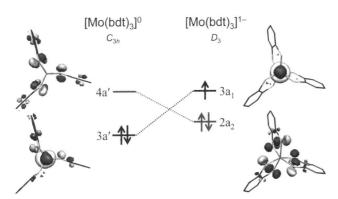

Figure 58. Depiction of the destabilization of the d_{z^2} orbital in $[Mo(bdt)_3]^{1-}$ upon reduction of $[Mo(bdt)_3]$ and shift in symmetry from C_{3h} to D_3. (See color version of this figure in the color plates section.)

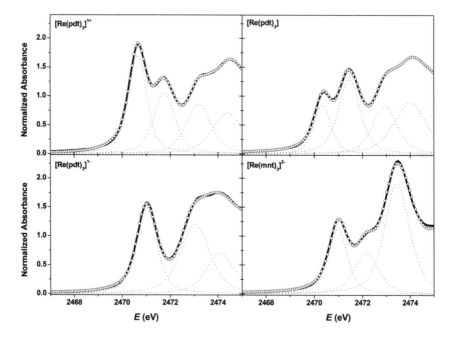

Figure 59.   Pseudo-Voight deconvolution of the S K-edge XAS spectra of $[Re(pdt)_3]^{1+/0/1-}$ and $[Re(mnt)_3]^{2-}$. Circles represent the experimental data; dotted lines the pseudo-Voigt peaks; and the solid line the sum of the fit. [Adapted from (248).]

$[Re(pdt)_3]^{1+/0}$ (Fig. 59), with the lowest energy peak occurring at 2470.65 and 2470.42 eV, respectively (248). This peak is 1 eV lower in energy than the second pre-edge peak (Table XIII). The low-energy transition is the S $1s \rightarrow 2a_2'$ excitation where the acceptor orbital is distributed over the tris(dithiolene) unit.

As unambiguously established by EPR (247), $[Re^V(L_3^{5-\bullet})]$ has one oxidative hole, and the incremental increase in intensity of the first pre-edge peak shown by pseudo-Voigt deconvolution of the spectrum demonstrates a ligand-based oxidation to form the monocation, $[Re^V(L_3^{4-})]^{1+}$ (Fig. 59). The anions $[Re(pdt)_3]^{1-}$ and $[Re(mnt)_3]^{2-}$ exhibit one pre-edge peak each, characterized as the S $1s \rightarrow 5e'$ ($d_{x^2-y^2,xy}$) transition; the absence of the lower energy feature is diagnostic of a fully reduced ligand, $(L_3)^{6-}$. For $[Re(pdt)_3]^{1+/0/1-}$, the progression to lower energy of this peak (2471.69 to 2471.42 to 2471.03 eV) across the series demonstrates the ligand-centered nature of this reduction with retention of a Re(V) central ion. The variation in transition intensity is commensurate with the a drift away from TP. The decrease in intensity of the pre-edge peak in $[Re(mnt)_3]^{2-}$ compared to $[Re(pdt)_3]^{1-}$ is in keeping with reduction of the metal ion to Re(IV), with an electron added to the $5e'$ level (Fig. 59).

TABLE XIII
Experimental and Calculated S K-Pre-Edge Transition Energies

| Complex | Pre-Edge | | |
	Experimental	Calculated[a]	Reference
$[V(bdt)_3]^{2-}$	2470.36	2470.37	179
	2471.77	2471.07	
$[V(tdt)_3]^{2-}$	2470.37	2470.37	179
	2471.70	2471.09	
$[V(pdt)_3]$	2470.21	2470.35	179
	2471.08	2471.66	
	2472.04	2472.46	
$[V(pdt)_3]^{1-}$	2470.24	2470.02	179
	2471.00	2471.29	
	2471.98	2472.07	
$[V(pdt)_3]^{2-}$	2470.44	2470.61	179
	2471.92	2471.37	
$[V(mnt)_3]^{2-}$	2470.54	2470.37	179
	2471.47	2471.55	
$[Cr(bdtCl_2)_3]^{1-}$	2469.75	2469.40	221
	2471.20	2470.67	
		2471.34	
$[Cr(bdtCl_2)_3]^{2-}$	2469.74	2469.37	221
	2471.31	2471.18	
$[Cr(bdtCl_2)_3]^{3-}$	2471.59	2471.35	221
$[Cr(tbbdt)_3]^{1-}$	2469.79	2469.51	222
	2471.19	2470.70	
		2471.29	
$[Cr(mnt)_3]^{2-}$	2469.89	2469.64	221
	2471.14	2471.33	
	2471.85		
$[Cr(mnt)_3]^{3-}$	2470.30	2470.94	221
	2471.52		
$[Mn(bdtCl_2)_3]^{2-}$	2470.94	2470.40	221
		2471.78	
$[Mn(mnt)_3]^{2-}$	2470.40	2470.65	221
	2471.78	2471.37	
$[Fe(mnt)_3]^{2-}$	2470.34	2470.54	249
	2471.78	2471.72	
$[Fe(mnt)_3]^{3-}$	2470.70	2471.32	249
	2471.42		
$[Mo(tbbdt)_3]$	2470.18		227
	2471.26		
$[Mo(tbbdt)_3]^{1-}$	2470.02		227
	2470.85		
$[Mo(mdt)_3]$	2470.43		281
	2471.54		
$[Mo(mdt)_3]^{1-}$	2470.13		281
	2471.31		

(*continued*)

TABLE XIII
(*Continued*)

Complex	Pre-Edge		Reference
	Experimental	Calculated[a]	
$[Mo(mdt)_3]^{2-}$	2471.31		281
$[Re(tms)_3]$	2470.05	2470.09	247
	2471.20	2471.32	
$[Re(tms)_3]^{1-}$	2470.94	2471.05	247
$[Re(pdt)_3]^{1+}$	2470.65	2470.63	248
	2471.69	2471.71	
$[Re(pdt)_3]$	2470.42	2470.38	248
	2471.42	2471.45	
$[Re(pdt)_3]^{1-}$	2471.03	2471.02	248
$[Re(mnt)_3]^{2-}$	2470.98	2470.84	248

[a] Derived from TD–DFT calculations.

Sulfur K-edge XAS has been successfully utilized to directly probe metal–ligand bonding in coordination complexes and enzymes (349, 350). In this context, covalency is defined as the ligand contribution to bonding, which is experimentally determined from the intensity (D_0) of the S 1s → 3p excitations to acceptor orbitals with some sulfur character, given by Eq. 7.

$$D_0(S\ 1s \to \Psi^*) = \beta^2 h I_s/3n \tag{7}$$

Here, β^2 is the sulfur 3p content (i.e., covalency), h is the number of holes in the acceptor orbitals, I_s is the radial transition dipole integral of the electric dipole-allowed S 1s → 3p transition, $|\langle S_{1s}|r|S_{3p}\rangle|^2$, and n is the number of absorbing sulfur atoms (281, 320, 344). It has been shown through a combination of experiment and theory that the transition dipole integral, I_s, varies linearly with the energy of the S 1s → 4p transition, and the I_s values for a variety of S-donor ligands have been estimated by this method (320). From these experimental values, the covalencies were determined from the peak areas (D_0). The I_s value for the $[Mo(mdt)_3]^{0/1-/2-}$ compounds was estimated at 14.22 based on a experimental S 1s → 4p peak identified at ~2476.7 eV and visible in the second derivative of the spectrum (281). For $[Re(pdt)_3]^{1+/0/1-}$, there is a decrease in the S 1s → 4p transition concomitant with successive reduction of the tris(dithiolene) unit across this series from monocation (2477.6 eV) to neutral (2477.2 eV) to monoanion (2476.6 eV) (248); I_s values were determined using the calibration plot developed by Solomon and co-workers (320). Experimental covalencies (β^2) for the two pre-edge transitions in $[Re(pdt)_3]^{1+/0}$, and one in $[Re(pdt)_3]^{1-}$ and $[Re(mnt)_3]^{2-}$ were estimated using Eq. 7 (Table XIV). For the S 1s → $2a_2'$ excitation, the estimated S 3p content is clearly too high (>80%), a distinct overestimation compared to previous experimental and calculated composition of the dithiolene π_3 orbital

TABLE XIV
Calculated and Experimental Covalencies for the Tris(dithiolene) Complexes[a]

Complex	Symmetry	I_{s^b}	Transition	h^c	Experimental Covalency[d] A	Experimental Covalency[d] B	Calculated Covalency[e]
$[Re(pdt)_3]^{1+}$	D_{3h}	15.78	$1s \rightarrow 2a_2'$	2	87.3	62.8	51.5
			$1s \rightarrow 5e'$	4	31.7	32.1	40.5
$[Re(pdt)_3]$	D_{3h}	15.12	$1s \rightarrow 2a_2'$ (β)	1	84.5	58.3	58.3
			$1s \rightarrow 5e'$	4	41.4	40.2	40.2
$[Re(pdt)_3]^{1-}$	D_3	14.10	$1s \rightarrow 5e$	4	47.2	42.7	39.4
$[Re(mnt)_3]^{2-}$	D_3	12.92	$1s \rightarrow 5e$	3	45.0	37.4	33.8
$[V(pdt)_3]$	D_{3h}	14.57	$1s \rightarrow 2a_2'$	2	78.6	52.2	62.1
			$1s \rightarrow 3a'(\beta)$	1			4.1
			$1s \rightarrow 5e'(\alpha)$	2	33.9	31.7	28.2
$[V(pdt)_3]^{1-}$	C_{3h}	13.45	$1s \rightarrow 4a'(\alpha)$	1	117.4	72.0	64.0
			$1s \rightarrow 3a'(\beta)$	1			9.7
			$1s \rightarrow 5e'(\alpha)$	2	35.3	30.5	29.9
$[Mo(mdt)_3]$	C_{3h}	14.22	$1s \rightarrow 4a'$	2	69	44.8	52.5
			$1s \rightarrow 5e'$	4	27	24.6	36.9
$[Mo(mdt)_3]^{1-}$	D_{3h}	14.22	$1s \rightarrow 2a_2'$ (β)	1	75	48.4	68.9
			$1s \rightarrow 5e'$	4	37	33.5	34.3
$[Mo(mdt)_3]^{2-}$	D_{3h}	14.22	$1s \rightarrow 5e'$	4	36	32.6	32.5

[a] Data taken from (248).
[b] Radial transition dipole integral.
[c] Number of holes in the orbital.
[d] Presented as S 3p% per hole, h. Method A: experimental I_s value. Method B: $[Re(pdt)_3]$ calibrated.
[e] Derived from DFT calculations.

(179, 221, 231, 232, 247, 279, 281, 344, 351). The values obtained for the S 3p contribution to the 5e' level were more reasonable, matching DFT derived values for $[Re(bdt)_3]^z$ ($z = 1+$, 0, 1−, 2−, 3−) (247).

The difficulty in identifying an obscure transition beyond the edge is circumvented by employing the TD–DFT approach successfully used to quantify covalency in bis(dithiolene) complexes (345). This entails calculating the oscillator strength of the S $1s \rightarrow 3p$ transition since neither I_s nor β^2 are experimentally observed. One compound is chosen as the standard, in this instance $[Re(pdt)_3]$ because its electronic structure is so unambiguously defined by EPR (247). Overall, the agreement between experiment and theory across the Re series is excellent (Fig. 60). This improvement is due to the inherent assumption that the charge on the S atoms is the dominant contributor to I_s. Orbitals involved in π bonding are more contracted than their σ counterparts, and as such, their π^* equivalents are less antibonding and have smaller I_s values. This undermines the position that all S $1s \rightarrow 3p$ transitions are the same, exposing an intrinsic flaw in this covalency estimation.

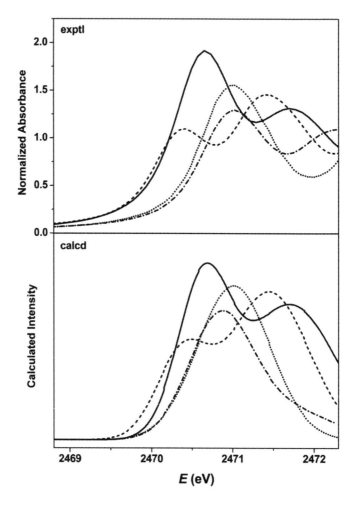

Figure 60. Comparison of the experimental (exptl) and calculated (calcd) S K-pre-edge spectra of [Re(pdt)$_3$]$^{1+}$ (—), [Re(pdt)$_3$] (-----), [Re(pdt)$_3$]$^{1-}$ (·····), and [Re(mnt)$_3$]$^{2-}$ (·–·–·). [Adapted from (248).]

This contrast was extended to [V(pdt)$_3$]$^{0/1-}$, which have the same two pre-edge features (179). Again the [Re(pdt)$_3$] calibrated covalency estimates are in better agreement with experiment than the I_s derived values (Table XIV). Moreover, the data highlight the large polarization of the V–S bonds in this series, with the second pre-edge peak corresponding to excitations to the 5e′(α) orbitals only. For [V(pdt)$_3$], the addition of the 1s → 3a′(β) transition to this second pre-edge feature brings the total intensity to 32.3%, in excellent agreement with the experimental estimate of 31.7%.

The intensity of the first-edge peak in [Mo(mdt)$_3$] is considerably less intense than for [Re(pdt)$_3$]$^{1+}$ (26), despite both complexes being characterized with an (L$_3$)$^{4-}$

ligand unit attached to a d^2 ion. The covalency analysis was similarly applied to $[Mo(mdt)_3]^z$ ($z = 0$, $1-$, $2-$) with the results posted in Table XIV (248). For $[Mo(mdt)_3]$, there is a significant difference between all three values, with the $[Re(pdt)_3]^0$ calibrated value being closer to the theoretical one. All neutral molybdenum and tungsten tris(dithiolene) complexes exhibit a geometric distortion that represents a lowering in symmetry $D_{3h} \rightarrow C_{3h}$, mixing the HOMO and LUMO. On the other hand, only a D_{3h} structure was ever computed for isoelectronic $[Re^V(L_3^{4-})]^{1+}$ indicating a much larger energy gap exists between the $3a_1'$ (HOMO-1) and $2a_2'$ (LUMO), which is not unexpected for this heavy metal. Therefore, the first pre-edge transition is to the nonbonding $2a_2'$ LUMO, whereas it is to the $4a'$ MO in $[Mo(mdt)_3]$—a mix of metal and ligand character governed by the fold angle. The electronic structure is best represented by the resonance structures $\{[Mo^{IV}(L_3^{4-})]^0 \leftrightarrow [Mo^V(L_3^{5-\bullet})]^0\}$, where $[Mo(mdt)_3]$ ($\alpha = 16°$) (23) is closer to the $[Mo^{IV}(L_3^{4-})]$ formulation, while $[Mo(tbbdt)_3]$ ($\alpha = 23.8°$) (227) has a greater contribution from the Mo(V) form. In the isoelectronic vanadium tris(dithiolene) monoanions, the highly polarizable V−S bonds ensue a $[V^{IV}(L_3^{5-\bullet})]^{1-}$ formulation, where strong antiferromagnetic coupling between metal and ligand radicals drives the distortion rather than HOMO–LUMO mixing (179).

F. Mössbauer

High-energy gamma rays of appropriate energy promote nuclear excitations in the target element during the Mössbauer experiment (352). Similar to XAS, the technique is element specific and can conveniently define the oxidation state, coordination geometry, and coordination number from the isomer (or chemical) shift and quadrupole splitting. Performed in the presence of an external magnetic field, additional spin-Hamiltonian parameters (e.g., the g-value, zero-field, and magnetic hyperfine interactions) can be determined. It is therefore quite discouraging to learn that so few elements have suitable sources to enable these experiments to be performed away from a synchrotron. In practice, ^{57}Fe is by far the most common element studied because its source, the decay of ^{57}Co by electron capture to an excited state of ^{57}Fe, is abundant, sufficiently active, and long-lived (half-life = 272 days). The gamma ray emitted during the decay is at the correct energy (14.4 keV) to be absorbed by ^{57}Fe nuclei in the sample. The high energy and exceedingly narrow linewidths endows Mössbauer with immense sensitivity and the ability to detect very low concentrations. The natural abundance of ^{57}Fe (2.15%) is sufficient for solid samples of Fe compounds; dilute solutions necessitate isotopic enrichment.

Only two tris(dithiolene) complexes have been the subject of a Mössbauer study, $[Fe(mnt)_3]^{2-/3-}$ (249, 307, 308, 353). The results corroborate the XAS and crystallographic data in that the trianion contains a low-spin Fe(III) $S = 1/2$ ion and oxidation of this species is predominantly metal based to a low-spin Fe(IV) $S = 1$ center (249). The zero-field spectra are presented in Fig. 61 and show a drop in the isomer shift (referenced to α-Fe) upon oxidation of the trianion ($\delta = 0.36$ mm s^{-1}) to the dianion ($\delta = 0.24$ mm s^{-1}), commensurate with an increase in the iron oxidation state. Applied-field measurements yielded a sign for the quadrupole splitting: [Fe(mnt)$_3$]$^{3-}$, $\Delta E_Q = -1.68$ mm s^{-1}; [Fe(mnt)$_3$]$^{2-}$, $\Delta E_Q = -1.56$ mm s^{-1}. Again, DFT calculations can reliably compute these Mössbauer parameters (Table XV). The sign of the quadrupole splitting for [Fe(mnt)$_3$]$^{3-}$ is reversed in the calculation. Interestingly, the asymmetry parameter, η, was found to vary

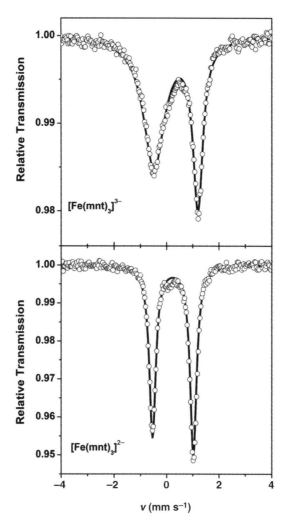

Figure 61. Zero-field Mössbauer spectra for powder samples of [Fe(mnt)$_3$]$^{3-}$ and [Fe(mnt)$_3$]$^{2-}$ at 80 K. Solid lines represent best fits of the data shown as open circles. [Adapted from (249).]

TABLE XV
Comparison of the Calculated and Experimental (in parentheses) Mössbauer Parameters[a]

Complex	δ (mm s^{-1})		ΔE_Q (mm s^{-1})		η [b]	
[Fe(mnt)$_3$]$^{2-}$	0.22	(0.24)	−2.03	(−1.56)	0.05	(0.14)
[Fe(mnt)$_3$]$^{3-}$	0.37	(0.36)	+1.90	(−1.68)	0.98	(0.85)

[a] Data taken from (249).
[b] Asymmetry parameter of the electric field gradient.

TABLE XVI
The ^{119}Sn Mössbauer Data for Tris(dithiolene) Complexes[a]

Complex	δ (mm s^{-1})	Γ (mm s^{-1})
(NBu$_4$)$_2$[Sn(tdt)$_3$]	1.10	1.04
(NMe$_4$)$_2$[Sn(mnt)$_3$]	1.04	1.84
(NEt$_4$)$_2$[Sn(mnt)$_3$]	1.12	1.14
(NBu$_4$)$_2$[Sn(mnt)$_3$]	1.23	1.02
(AsPh$_4$)$_2$[Sn(mnt)$_3$]	0.97	0.96

[a] Data are taken from (355) and (356).

considerably between two similar studies, with Niarchos and Petridis' (308) value of 0.7 surprisingly larger than Milsmann et al. (224) at 0.14; the latter is more similar to other {FeS$_6$} systems. The ZFS parameters were determined for [Fe(mnt)$_3$]$^{2-}$, with $D = 10.5$ cm^{-1} and $E/D = 0.2$, in agreement with magnetic susceptibility data (249, 307). The ^{57}Fe hyperfine matrix was found to be temperature independent; $A = (0.85, 0.78, 0.60)$ mm s^{-1}.

Tin is also a Mössbauer amenable element and two tris(dithiolene) tin complexes have been included in studies of the isomer shifts and quadrupole splitting across a range of organotin compounds (355, 356). The data posted in Table XVI reveal a significant dependence of the isomer shift (referenced to BaSnO$_3$) on the counter-cation. There is a complete absence of quadrupole splitting ($\Delta E_Q \sim 0$ mm s^{-1}) given the symmetric distribution of ligands, and therein the electron density about Sn(IV). The linewidth (Γ) of the Lorentzian used to fit the data is twice as large as those for ^{57}Fe and envelopes the modest quadrupole splitting anticipated in these complexes.

IX. SUMMARY

The last decade has seen prodigious surge in the number of studies targeting the molecular and electronic structure of tris(dithiolene) compounds. This growth arose from a general redux in metallodithiolene chemistry and a reacquaintance with the term "dithiolene radical". In this section, the data are compiled to give an overview of the electronic structures of tris(dithiolene) compounds, as outlined in Chart 1.

CHART 1
Summary of Electronic Structures for each Electron-Transfer Series, $[M(L)_3]^z$ (L=dithiolene; $z = 1+, 0, 1-, 2-, 3-, 4-$)

Metal	1+	0	1−	2−	3−	4−
Ti			$[Ti^{IV}(L_3^{5\bullet})]^{1-}$	$[Ti^{IV}(L_3)]^{2-}$	$[Ti^{III}(L)_3]^{3-}$	
Zr				$[Zr^{IV}(L_3)]^{2-}$	$[Zr^{III}(L)_3]^{3-}$	
Hf				$[Hf^{IV}(L_3)]^{2-}$		
V	$[V^V(L_3^{4-})]^{1+}$	$[V^{IV}(L_3^{4-})]^0$	$[V^{IV}(L_3^{5\bullet})]^{1-}$	$[V^{IV}(L_3^{6-})]^{2-}$	$[V^{III}(L_3)]^{3-}$	$[V^{II}(L_3)]^{4-}$
Nb		$[Nb^V(L_3^{5-})]^0$	$[Nb^V(L_3^{6-})]^{1-}$	$[Nb^{IV}(L_3^{6-})]^{2-}$	$[Nb^{III}(L_3)]^{3-}$	
Ta			$[Ta^V(L_3^{6-})]^{1-}$	$[Ta^{IV}(L_3^{6-})]^{2-}$		
Cr		$[Cr^{III}(L\bullet)_3]^0$	$[Cr^{III}(L)(L\bullet_2)]^{1-}$	$[Cr^{III}(L_2)(L\bullet)^{2-}]^{2-}$	$[Cr^{III}(L)_3]^{3-}$	
Mo	$[Mo^V(L_3^{4-})]^{1+}$	$[Mo^V(L_3^{4-})]^0 \leftrightarrow [Mo^{IV}(L_3^{4-})]^0$	$[Mo^V(L_3^{6-})]^{1-}$ or $[Mo^{IV}(L_3^{5\bullet})]^{1-}$	$[Mo^{IV}(L_3^{6-})]^{2-}$	$[Mo^{III}(L_3^{6-})]^{3-}$	$[Mo^{II}(L_3^{6-})]^{4-}$
W	$[W^V(L_3^{4-})]^{1+}$	$[W^V(L_3^{5-})]^0 \leftrightarrow [W^{IV}(L_3^{4-})]^0$	$[W^V(L_3^{6-})]^{1-} \leftrightarrow [W^{IV}(L_3^{5\bullet})]^{1-}$	$[W^{IV}(L_3^{6-})]^{2-}$	$[W^{III}(L_3^{6-})]^{3-}$	
Mn			$[Mn^V(L_3)]^{1-}$	$[Mn^{IV}(L_3)]^{2-}$	$[Mn^{III}(L_3)]^{3-}$	
Tc	$[Tc^V(L_3^{4-})]^{1+}$	$[Tc^V(L_3^{5-})]^0$	$[Tc^V(L_3^{6-})]^{1-}$	$[Tc^{IV}(L_3^{6-})]^{2-}$	$[Tc^{III}(L_3^{6-})]^{3-}$	
Re	$[Re^V(L_3^{4-})]^{1+}$	$[Re^V(L_3^{5-})]^0$	$[Re^V(L_3^{6-})]^{1-}$	$[Re^{IV}(L_3^{6-})]^{2-}$	$[Re^{III}(L_3^{6-})]^{3-}$	$[Re^{II}(L_3^{6-})]^{4-}$
Fe				$[Fe^{IV}(L_3)]^{2-}$	$[Fe^{III}(L_3)]^{3-}$	$[Fe^{II}(L_3)]^{4-\,a}$
Ru				$[Ru^{IV}(L_3)]^{2-}$	$[Ru^{III}(L_3)]^{3-}$	
Os		$[Os^{VI}(L_3)]^0$	$[Os^V(L_3)]^{1-}$	$[Os^{IV}(L_3)]^{2-}$	$[Os^{III}(L_3)]^{3-}$	
Co				$[Co^{IV}(L_3)]^{2-}$	$[Co^{III}(L_3)]^{3-}$	
Rh			$[Rh^V(L_3)]^{1-}$	$[Rh^{IV}(L_3)]^{2-}$	$[Rh^{III}(L_3)]^{3-}$	
Pt				$[Pt^{IV}(L_3)]^{2-}$		

^{a}The complex $[Fe^{II}(Me_2pipdt)_3]^{2+}$ is diamagnetic (124).

A. Group 4 (IV B)

All tris(dithiolene) compounds with group 4 (IV B) metals Ti, Zr, Hf are dianions and their electronic structure is exceedingly straightforward: a +IV ion bound by three closed-shell dithiolate ligands. Aside from $[Zr(bdt)_3]^{2-}$, all structures come from Ti compounds that are consistently octahedral ($\Theta > 35°$). No structure yet exists for Hf, but given the similarity in electronic spectra, replete with LMCT bands, it will be distorted octahedral. One-electron reduction to a trianionic species is reversible for Ti and Zr, but the one-electron oxidation is not. This result indicates a d^0 metal ion will not stabilize a dithiolene radical.

B. Group 5 (V B)

Vanadium tris(dithiolene) complexes are spectroscopically rich because of their ability to form multi-membered electron-transfer series. A total of five charge states are known for V, $0 \rightarrow 4-$. A sixth (the monocation) has only been seen as an irreversible event by cyclic voltammetry. A synopsis of the electronic structures for this electron-transfer series is shown in Fig. 62. The di-, tri-, and tetra-anionic members are +IV, +III, and +II central ions bound by three closed-shell dithiolates. Their structures, crystallographic and geometry optimized, are distorted octahedral. Vanadium K-edge spectra prove that the neutral and monoanionic complexes have the same metal oxidation state as the dianion, namely, +IV, and S K-edge, electronic and IR spectroscopy bear the signature of oxidized ligands.

It is important to realize that there are no unpaired electrons on the tris(dithiolene) moiety in the neutral species. The ligand unit is two-electron oxidized, but these electrons come out of the same orbital, the redox-active $2a_2'$ MO. This observation is neatly shown by the Mulliken spin population analysis (Fig. 63). With the tris(dithiolene) ligand at its most electron deficient, the geometry is TP because of the prevailing strength of the V$-$S bonds ($4e'$). There is no additional distortion to C_{3h} because no energy is gained from HOMO–LUMO mixing with just a single electron in the highest occupied state, $3a_1'$. Reduction of the neutral compound to the monoanion deposits one electron into this orbital, as marked by the radical fingerprint in both S K-edge and electronic absorption spectra. The crystal structures of all known tris(dithiolene)vanadium monoanions exhibit a significant dithiolene chelate fold, $\alpha > 20°$ (Table II), which represents a lowering of the symmetry from D_{3h} to C_{3h}. This fold allows the metal- ($3a_1'$) and ligand-centered ($2a_2'$) SOMOs to mix, since both transform as a'. The two SOMOs strongly antiferromagnetically couple giving rise to a diamagnetic ground state ($S=0$) driving the structural distortion. The Mulliken spin density plot shown in Fig. 63b underpins the polarizability of the V$-$S bonds, where >1 electron is found on the metal and dithiolene ligands. The formation of a monocation is unfavorable

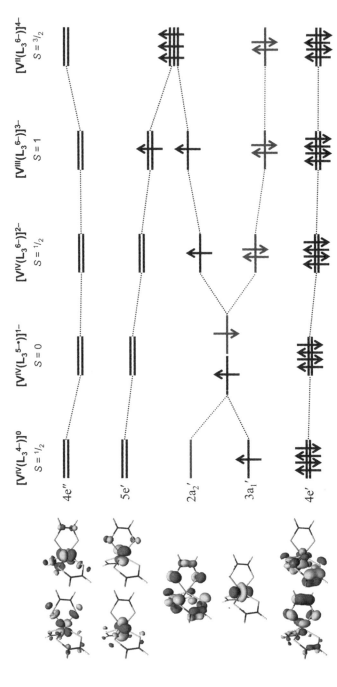

Figure 62. Qualitative MO scheme depicting the ordering of the frontier orbitals for the $[V(L_3)]^z$ ($z = 0, 1-, 2-, 3-, 4-$) electron-transfer series. The MOs shown left are annotated with D_{3h} symmetry labels; the redox-active $2a_2'$ is highlighted in gray. (See color version of this figure in the color plates section.)

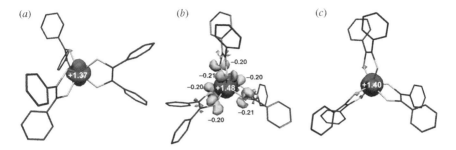

Figure 63.   Mulliken spin density plots and populations for (a) [V(pdt$_3$)], (b) [V(pdt$_3$)]$^{1-}$, and (c) [V (pdt$_3$)]$^{2-}$ obtained from DFT calculations. [Adapted from (26).] (See color version of this figure in the color plates section.)

because the event is metal centered and a d^0 ion cannot stabilize oxidized dithiolene ligands.

There are fewer Nb and Ta compounds in the literature, but both are distinctly different from their lighter congener. These are more akin to the group 4 (IV B) metals, preferring a high oxidation state. The monoanions possess +V ions and closed-shell dithiolates; the dianions +IV ions. Their electronic spectra lack low-energy bands typical of oxidized ligands. Interestingly, [Ta(bdt)$_3$]$^{1-}$ is distorted octahedral where the Nb homologue is not (Table II). This result arises from the HOMO–LUMO energy gap. Both have the correct electron configuration to induce a dithiolene folded structure (HOMO doubly occupied), but for Ta the energy gap is too large for the distortion to be of any value. For Nb, its C_{3h} molecular structure indicates a much smaller HOMO–LUMO energy gap, and is the only complex known where the HOMO is ligand based and the LUMO is predominantly the d$_{z^2}$ orbital (the reverse is found for Mo). The magnitude and even existence of the dithiolene fold will depend on the type of ligand coordinated. At this stage only one structure is known (126,219), but it cannot be presumed that a C_{3h} structure would be present for all Nb monoanions. For a stronger π-donor, such as (pdt)$^{2-}$, the order of the orbitals may switch leading to a more reduced Nb ion and oxidized tris(dithiolene) unit.

C.   Group 6 (VI B)

The idiosyncrasies of first-row metals compared with their second- and third-row ones are most apparent with Cr, Mo, and W. Tris(dithiolenes) of chromium likely possess localized ligand radicals by analogy to dioxolenes and diimines. Of course, this is not strictly the case as each is D_3 symmetric, which splits the three ligand orbitals into an a$_1$ and e set, and spin density is deposited uniformly as the complex is oxidized (Fig. 64). Nevertheless, the ligands are treated individually in

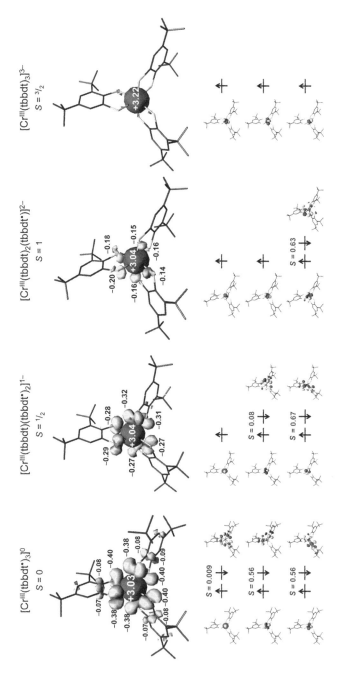

Figure 64. Spin density plots of the $[Cr(tbbdt)_3]^z$ ($z = 0$, $1-$, $2-$, $3-$) electron-transfer series, as derived from DFT calculations together with values of the spin density of the Mulliken analyses, and qualitative MO diagrams of the corresponding pairs of magnetic orbitals. (See color version of this figure in the color plates section.)

120

Chart 1. The calculated electronic structures for the $[Cr(tbbdt)_3]^z$ ($z = 0$, $1-$, $2-$, $3-$) series are representative of all compounds of this type (Fig. 64). The trianion is a rudimentary Werner-type compound with three closed-shell dithiolates surrounding a Cr(III) ion. The series is stepped by successive one-electron oxidation of each dithiolene ligand with the addition of a ligand-based SOMO to the orbital manifolds for the dianionic, monoanionic, and neutral species. The spectroscopic data are replete with dithiolene radical fingerprints: electronic spectra with the low-energy IVCT band; the $\nu(C-S^{\bullet})$ stretch at $\sim 1100 \, cm^{-1}$ for all but the trianion; identical pre- and rising-edge energies in Cr K-edge spectra; low-energy pre-edge peak at $\sim 2469.7 \, eV$ in the S K-edge spectra. The invariance of the $+III$ oxidation state at Cr stems from intrinsic stability of a d^3 ion in an octahedral LF, hence near-to-octahedral molecular structures. This stability runs through to the missing member of the series, $[Cr(L)_3]$. Only two molecules have ever been isolated, these being $[Cr(pdt)_3]$ by Schrauzer and Mayweg (96), and $[Cr(tfd)_3]$ by Holm and co-workers (10). Based on X-ray powder diffraction and electronic absorption spectroscopy, Eisenberg et al. (177) suggested that $[Cr(pdt)_3]$ is TP given its similarity to $[V(pdt)_3]$ (known at the time) and $[V(pdt)_3]^{1-}$, which was recently shown to be "more TP" than the neutral compound (179). The DFT calculations predicted that a TP neutral tris(dithiolene)chromium complex would comprise a Cr(IV) central ion encapsulated by a $(L_3)^{4-}$ unit; that is, two oxidized holes (221). The corresponding broken symmetry calculation produced an octahedral structure defined as Cr(III) ion bound by three monoanionic dithiolene radicals; that is, $[Cr^{III}(L^{\bullet})_3]$, a solution $20 \, kcal \, mol^{-1}$ more stable than the prismatic one.

Moving down to the second-row, neutral tris(dithiolene)molybdenum complexes have until just a few years back been cast as $[Mo^{VI}(L)_3]$. A similar position has been described for tungsten, with the metal in its highest oxidation state ($+VI$) coordinated by three dithiolate ligands. A reformulation was prompted by two definitive results: the $\nu(C-S^{\bullet})$ stretch at $1106 \, cm^{-1}$ and a low-energy pre-edge peak in the S K-edge spectrum of $[Mo(tbbdt)_3]$ (227). The electronic structure description of these neutral species is the most ambiguous, and so the resonance description $\{[M^{IV}(L_3^{4-})] \leftrightarrow [M^V(L_3^{5-\bullet})]\}$, where $M = Mo$ or W, is clearly more appropriate. Neutral complexes have at least one hole in their ligand unit, almost two in some cases. The weighting of the resonance structures is guided by the magnitude of the fold angle. As previously encountered for Nb, these compounds have the correct electron configuration and suitably small HOMO–LUMO energy gap to benefit from a $D_{3h} \rightarrow C_{3h}$ distortion. Here, the HOMO is metal-based ($3a_1{}'$) and the LUMO ligand-centered ($2a_2{}'$), both free to mix in lower symmetry. Unlike vanadium, the distortion introduces ligand character into the HOMO and vice versa into the LUMO. This outcome generates a metal ion somewhere between $+IV$ and $+V$ depending on the extent of mixing, and this will vary with ligand type and metal ion, Mo or W.

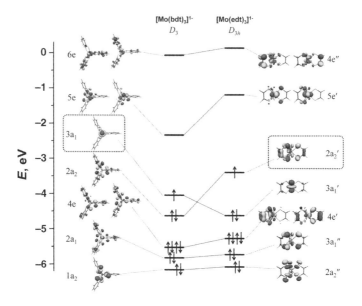

Figure 65. Molecular orbital scheme for $[Mo(bdt)_3]^{1-}$ (D_3 symmetry labels) and $[Mo(edt)_3]^{1-}$ (D_{3h} symmetry labels) with the SOMO identified in each case. Neutral complexes resemble the manifold for $[Mo(edt)_3]^{1-}$, but with C_{3h} point symmetry and one less electron in the $2a_2'$ MO. [Adapted from (232).] (See color version of this figure in the color plates section.)

As already mentioned, alkene-1,2-dithiolates have smaller fold angles than their aromatic counterparts (Table II).

The influence of dithiolene ligand substituents reaches its zenith for the monoanionic members of the series. Both S K-edge and EPR data confirm the compounds are categorized into two classes. Arene-1,2-dithiolates are more difficult to oxidize than alkene-1,2-dithiolates and their electronic structures are defined as $[Mo^V(L_3^{6-})]^{1-}$ (Fig. 65). These species have larger [95,97]Mo hyperfine coupling constants and absorption bands of weak intensity in the low-energy region of the visible spectrum. In contrast, the alkene-1,2-dithiolates are preferentially oxidized over Mo to give a $[Mo^{IV}(L_3^{5-\bullet})]^{1-}$ electronic structure (Fig. 65). The latter are distorted TP ($\Theta < 20°$), whereas the aromatic ligands prefer distorted octahedral geometries ($\Theta > 30°$). The W analogues are exceedingly more complicated. A straightforward $[W^V(L_3^{6-})]^{1-}$ can be made from those compounds with aryl-1,2-dithiolates, but the EPR data suggest $\{[W^{IV}(L_3^{5-\bullet})]^{1-} \leftrightarrow [W^V(L_3^{6-})]^{1-}\}$ resonance structures more appropriate for the olefinic dithiolenes (232). This finding is also in keeping with the intraligand bond distances. Although the electron configuration is not the most favorable, DFT calculations infer that a second-order Jahn–Teller distortion is operative, mixing the HOMO-1 and SOMO, which precludes the assignment of an integer oxidation state.

The di-, tri- and, in the case of Mo, tetra-anionic members of the series possess three closed-shell ligands and metal ions in the +IV, +III, and +II oxidation states respectively. The monocation, only seen irreversibly by electrochemistry, is highly unstable because the electron-deficient metals are no longer able to support oxidized ligands.

D.   Group 7 (VII B)

Manganese compounds all possess a +IV central ion and three closed-shell dithiolates. The heavier members of the group operate differently. From the modest spectroscopic and structural data collected for Tc compounds, it is evident these are identical to their Re homologues. The electronic structures of the Re series are visualized in the qualitative MO scheme provided in Fig. 66. Like tris(dithiolene) vanadium compounds, the first three members differ in the occupation of the $2a_2'$ MO—ligand-centered redox events, after which the $5e'$ level begins to be filled. The occupancy of these orbitals leads to a twisting of the geometry from TP for the monocation and neutral species, to distorted trigonal for the monoanion, and finally to a distorted octahedral polyhedron for the dianionic species; $\Theta = 38°$ in $[Re^{IV}(mnt)_3]^{2-}$ (248). The monocation is isoelectronic with neutral Mo and W species, and monoanions of V and Nb, but does not optimize with a C_{3h} structure (248). This result relates to the substantial energy gap between the beneficiaries of the geometric distortion, the $3a_1'$ and $2a_2'$ MOs, with the d_{z^2} orbital of Re the most stabilized across the considered metals (V, Nb, Mo, W). The more reduced members, dianion through tetra-anion, are octahedral and are interrelated by metal-based redox processes.

E.   Group 8 (VIII B)

Octahedral $[Fe(mnt)_3]^{3-}$ is a simple Werner-type coordination complex with a low-spin ferric center and three closed-shell ligands (249). The orbital manifold shows a $(t_{2g})^5$ electron configuration (Fig. 67); one unpaired electron is mainly located on the metal with some distribution onto the $(mnt)^{2-}$ ligands by virtue of covalency (Fig. 67). The facile oxidation of this compound generates what can be regarded as an Fe^{IV} species as demonstrated by Fe K-edge and Mössbauer spectroscopy.

It is with DFT calculations that the true nature of the Fe(IV) ion is expressed. *In silico* one-electron oxidation to the dianion afforded two possibilities, either a simple $S = 1$ species, or alternatively a ligand radical ($S_L = 1/2$) bound to an intermediate-spin ferric ion ($S_{Fe} = 3/2$) to give the net spin-triplet ground state. The calculations confirm the former is the more plausible description, with the electron removed from a d orbital. The consequence is a slight twist away from octahedral ($\Theta = 46.6°$) in order for the dithiolene π orbitals to overlap

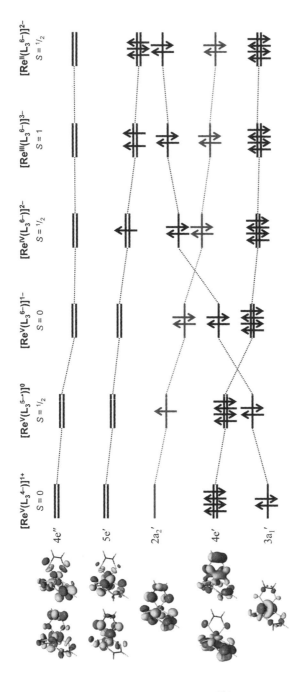

Figure 66. Qualitative MO scheme depicting the ordering of the frontier orbitals for the $[Re(L_3)]^z$ ($z = 1+, 0, 1-, 2-, 3-, 4-$) electron-transfer series. The MOs are annotated with D_{3h} symmetry labels and the redox-active $2a_2'$ is highlighted in gray. (See color version of this figure in the color plates section.)

$[Fe^{III}(mnt)_3]^{3-}$ $[Fe^{IV}(mnt)_3]^{2-}$
$S = ^1/_2$ $S = 1$

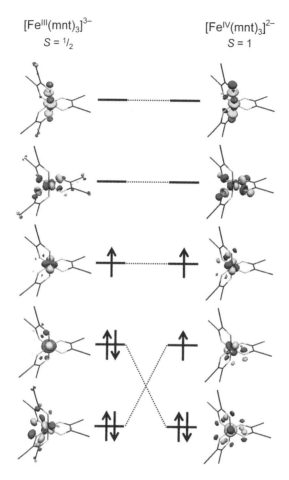

Figure 67.   Qualitative MO scheme for $[Fe(mnt)_3]^{3-}$ and $[Fe(mnt)_3]^{2-}$. (See color version of this figure in the color plates section.)

more effectively with the Fe d orbitals. As a result, the d_{z^2} (a_1) orbital is stabilized relative to the $d_{xz,yz}$ (e) orbitals as the latter attain more π^* character (Fig. 67). Additionally, the increased covalency in this $\{Fe^{IV}S_6\}$ polyhedron is evident in the spin population map, where $+1.77$ spins are found on Fe and $+0.27$ on the sulfur atoms (Fig. 68b). This process stabilizes a high oxidation state for Fe and one very different from a ferryl center, $[Fe^{IV}=O]^{2+}$. Ruthenium and osmium complexes are regarded as having closed-shell ligands in all cases; the intraligand bond distances in $[Ru(bdt)_3]^{2-}$ and $[Os(bdt)_3]^{2-}$ are consistent with this view (Table II). However, the bonding in these dianions, and the monoanionic and neutral members of the

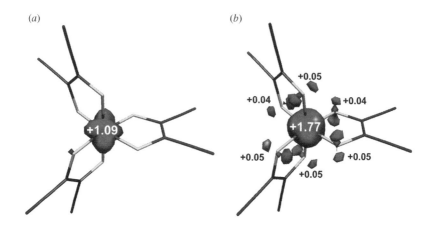

Figure 68.   Mulliken spin density plots and populations for (*a*) $[Fe^{III}(mnt)_3]^{3-}$, and (*b*) $[Fe^{IV}(mnt)_3]^{2-}$. (See color version of this figure in the color plates section.)

Os series, will be highly covalent. Their rich electronic spectra are testament to this description.

F.   Group 9 (VIII B) and Beyond

The number of tris(dithiolene) complexes diminishes as we approach the late transition metals. Several Co complexes are known, but only with highly electron-withdrawing substituents on the dithiolene ligands, such as $(mnt)^{2-}$, $(qdt)^{2-}$, and $(dtcr)^{2-}$. The central ion in each case is low-spin Co(III) d^6 and the ligands are closed-shell. Their geometry is clearly octahedral as seen in the crystal structure of $[Co(dtcr)_3]^{3-}$ (103). Rhodium is similarly described as $[Rh^{III}(L)_3]^{3-}$. Interestingly, the Co trianions are reversibly oxidized to their dianionic form. Based on the Fe chemistry, it can be assumed the event is metal centered, which infers these strong π-donating ligands are capable of stabilizing an electron-deficient Co(IV) ion. It has been shown that the oxidation of $[Co^{III}(dtc)_3]$ is metal centered giving $[Co^{IV}(dtc)_3]^{1+}$, confirmed by EPR (357). The Co(IV) entity has a finite lifetime of <2 h after which it will cannibalize itself by stripping an electron from the supposedly redox-inert $(dtc)^{1-}$ ligand, leading to the formation of $[Co_2(dtc)_5]^{1+}$ and tetrathiuramdisulfide—the sulfur-bridged dimer of $(dtc)^{1-}$. Whether a dithiolate ligand can exist attached to the highly oxidizing Co(IV) center is not known. Higher oxidation states for Rh are more accessible and the electronic structures of the di- and monoanion are anticipated to follow the group 8 (VIII B) metals as $[Rh^{IV}(L)_3]^{2-}$ and $[Rh^{V}(L)_3]^{1-}$, respectively.

Main group tris(dithiolenes) have unremarkable electronic structures as there are no d orbitals to stabilize dithiolene radicals. The single example of an oxidized ligand attached to a main group metal is $(eTTFdt)^{2-}$, which has redox chemistry independent of the metal (257). Successive one-electron oxidation of $[Sn^{IV}(eTTFdt)_3]^{2-}$ generates $[Sn^{IV}(eTTFdt)_2(eTTFdt^{\bullet})]^{1-}$, and $[Sn^{IV}(eTTFdt)(eTTFdt^{\bullet})_2]$. The monoanion has an $S = 1/2$ ground state and an EPR signal consistent with a spin localized on the TTF unit, though it is highly probable that the neutral compound is also EPR active with a low-lying triplet ($S = 1$) excited state generated by weak interligand spin coupling.

X. CONCLUSIONS

Tremendous scientific strides have been made in the last 5 years that have resolved many of the outstanding questions surrounding the molecular and electronic structures of tris(dithiolene) complexes. The most salient explanation centers on the unique geometries inherent to certain members of this class of compound. We can now predict whether a specific compound is TP, paddle wheel, or distorted octahedral, and the reasons why this particular topology is adopted. Not surprisingly, this relates entirely to the bonding between the metal ion and its dithiolene ligands. The TP structure is held together by a strong π interaction that is lost upon twisting to octahedral. The trigonal prism is favored in systems with oxidized ligands. Interligand repulsion is maximized with a fully reduced tris(dithiolene) unit, though this does not necessarily shift the geometry away from TP, as seen in many structures with heavy transition metals. The propensity of dithiolene ligands to be oxidized or reduced stems from the inherent nature of the substituents, with electron-withdrawing or electron-donating groups appended to either an aromatic or aliphatic carbon backbone. Ward and McCleverty (358) refined Jørgensen's original concept of an innocent ligand—one that allowed the metal oxidation state of a complex to be defined (359, 360)—by adding when both metal and ligand frontier orbitals are potentially redox-active and close to each other in energy, then noninnocence can occur. So it can be taken as fact that all ligands are noninnocent, but whether they will act the part depends entirely on the circumstance in which they are found. Metallodithiolene chemistry is an exploration of the redox interplay that exists between metals and their organic counterparts. Specifically, it has inspired a re-evaluation of the fundamental concepts of structure and bonding in coordination compounds, spawned the notion of delocalized orbitals and metal–ligand covalency, and challenged the applicability of the oxidation state formalism. What started 50 years ago as an investigation into the intricacies of dithiolenes has since escalated into the larger field of redox-active ligands, and demonstrates how application-driven research evolves from the humble origins of pure scientific inquiry.

ACKNOWLEDGMENTS

I am indebted to my colleagues Dr. James Walsh and Dr. George Whitehead for generous assistance in producing many of the figures in this chapter, and Dr. Joseph Sharples for thoroughly proofreading the manuscript. Support from the EPSRC National UK EPR Facility and Service at The University of Manchester is gratefully acknowledged.

ABBREVIATIONS

α	Dithiolene fold angle
abt	Aminobenzenethiolate
acac	Acetylacetonate
adt	Anisylethene-1,2-dithiolate
andt	1,2-Dianilylethene-1,2-dithiolate
anpdt	1-Anilyl-2-phenylethene-1,2-dithiolate
apdt	1-Anisyl-2-phenylethene-1,2-dithiolate
bddt	1,4-Butanediyldithioethene-1,2-dithiolate
bdt	Benzene-1,2-dithiolate
bdtCl$_2$	3,6-Dichlorobenzene-1,2-dithiolate
bdtCl$_4$	3,4,5,6-Tetrachlorobenzene-1,2-dithiolate
bdtCl$_4$H$_2$	3,4,5,6-Tetrachlorobenzene-1,2-dithiol
bdtF$_4$	3,4,5,6-Tetrafluorobenzene-1,2-dithiolate
bdtF$_4$H$_2$	3,4,5,6-Tetrafluorobenzene-1,2-dithiol
bdtH$_2$	Benzene-1,2-dithiol
bn	Benzene
bn-bdt	1,3,5-Tris(amidomethylbenzenedithiolate)benzene
bsdt	*p*-Bromophenylethene-1,2-dithiolate
btdt	1,2-Bis(butylthio)ethene-1,2-dithiolate
15C5-bdt	2,3,5,6,8,9,11,12-Octahydro-1,4,7,10,13-benzopentaoxacyclopenta-decine-15,16-dithiolate
18C6	18-Crown-6
18C6-bdt	2,3,5,6,8,9,11,12,14,15-Decahydro-1,4,7,10,13,16-benzohexaoxa-cyclooctadecine-18,19-dithiolate
Bz	Benzyl
Calcd	Calculated
cat	Catecholate
Cp	Cyclopentadienyl
csdt	*p*-Chlorophenylethene-1,2-dithiolate
cydt	Cyclohexene-1,2-dithiolate
dadt	1,2-Dianisylethene-1,2-dithiolate

dbddto	2,5-Dithioxobenzo[1,2-*d*:3,4-*d'*]bis[1,3]dithiolene-7,8-dithiolate
dbTTFdt	Di(*n*-butylthio)tetrathiafulvalenedithiolate
dbzdt	Dibenzoylethene-1,2-dithiolate
dcmdtcr	4-Dicyanomethyl-1,2-dithiocroconate
dddt	5,6-Dihydro-1,4-dithiin-2,3-dithiolate
dem	Diethylmaleate-2,3-dithiolate
deTTFdt	Dimethylthiotetrathiafulvalenedithiolate
DFT	Density functional theory
dhdndt	10,10-Dimethoxy-1,4,4aα,5,8,8aα-hexahydro-1α,4α:5β,8β-dimethanonaphthalene-2,3-dithiolate
dmdto	Dimethyldithiooxamide
DMF	*N,N'*-Dimethylformamide
dmid	2-Oxo-1,3-dithiole-4,5-dithiolate
dmit	1,3-Dithiole-2-thione-4,5-dithiolate
dmm	Dimethylmaleate-2,3-dithiolate
DMSO	Dimethyl sulfoxide
dmt	1,2-Dithiole-3-thione-4,5-dithiolate
dppa	Bis(diphenylphosphino)amine
dppm	Bis(diphenylphosphino)methane
dtc	Dithiocarbamate
dtcr	1,2-Dithiocroconate
dtdt	1,2-Ditolylethene-1,2-dithiolate
dto	Dithiooxamide
dtsq	1,2-Dithiosquarate
edt	Ethene-1,2-dithiolate
EPR	Electron paramagnetic resonance
etdt	1,2-Diethylethene-1,2-dithiolate
eTTFdt	Dimethylenetetrathiafulvalenedithiolate
EXAFS	Extended X-ray absorption fine structure
exptl	Experimental
Fc	Ferrocene
HALS	Highy anisotropic low-spin
HOMO	Highest occupied molecular orbital
IR	Infrared
ir	Irreversible
I_s	Radial transition dipole integral
IVCT	Intervalence charge transfer
L	Dithiolene
LF	Ligand field
LFT	Ligand field theory
LLCT	Ligand-to-ligand charge transfer
LMCT	Ligand-to-metal charge transfer

LUMO	Lowest unoccupied molecular orbital
mdt	1,2-Dimethylethene-1,2-dithiolate
mdto	Methyldithiooxamide
Me$_2$pipdt	1,4-Dimethylpiperazine-2,3-dithione
MeCN	Acetonitrile
MLCT	Metal-to-ligand charge transfer
mnt	1,2-Dicyanoethene-1,2-dithiolate
MO	Molecular orbital
ms	Mesitylene
ms-bdt$_3$	2,4,6-Tris(amidomethylbenzenedithiolate)mesitylene
mtdt	1,2-Bis(methylthio)ethene-1,2-dithiolate
μ_B	Bohr magneton
μ_{eff}	Effective magnetic moment
NIR	Near-infrared
NLO	Nonlinear optical
NMe$_3$Bz$^+$	Trimethylbenzylammonium
NMP	N-Methylphenazinium
NMR	Nuclear magnetic resonance
pddt	1,3-Propanediyldithioethene-1,2-dithiolate
pdt	1,2-Diphenylethene-1,2-dithiolate
PNP	Bis(triphenylphosphoranylidene)ammonium
pTTFdt	Trimethylenetetrathiafulvalenedithiolate
py	Pyridine
qdt	Quinoxaline-2,3-dithiolate
qdtH$_2$	Quinoxaline-2,3-dithiol
qpdt	1-Quinoxalyl-2-phenylethene-1,2-dithiolate
qr	Quasireversible
r	Reversible
sdt	Phenylethene-1,2-dithiolate
SCE	Saturated calomel electrode
SOMO	Singly occupied molecular orbital
tbbdt	3,5-Di-$tert$-butylbenzene-1,2-dithiolate
tbbdtH$_2$	3,5-Di-$tert$-butylbenzene-1,2-dithiol
TD–DFT	Time-dependent density functional theory
tdt	Toluene-3,4-dithiolate
tedto	Tetraethyldithiooxamide
tfd	Bis(trifluoromethyl)ethene-1,2-dithiolate
thdt	Bis(3-thienyl)ethene-1,2-dithiolate
Θ	Trigonal twist angle
THF	Tetrahydrofuran
tmdto	Tetramethyldithiooxamide
tms	3,6-Bis(trimethylsilyl)ethene-1,2-dithiolate

toldt	Tolylethene-1,2-dithiolate
TP	Trigonal prismatic
tpdt	1-Tolyl-2-phenylethene-1,2-dithiolate
tr	Triazole
tr-bdt$_3$	2,4,6-Tris(amidomethylbenzenedithiolate)-1,3,5-triazole
TTF	Tetrathiafulvalene ($C_6H_4S_4$)
UV	Ultraviolet
vdt	Veratrole-4,5-dithiolate
vdtH$_2$	Veratrole-4,5-dithiol
XANES	X-ray absorption near-edge spectroscopy
XAS	X-ray absorption spectroscopy
xdt	Xylene-4,5-dithiolate
Z_{eff}	Effective nuclear charge
ZFS	Zero-field splitting

REFERENCES

1. R. B. King, *Inorg. Chem.*, *2*, 641 (1963).

2. H. B. Gray, R. Williams, I. Bernal, and E. Billig, *J. Am. Chem. Soc.*, *84*, 3596 (1962).

3. G. N. Schrauzer and V. Mayweg, *J. Am. Chem. Soc.*, *84*, 3221 (1962).

4. A. Davison, N. Edelstein, R. H. Holm, and A. H. Maki, *Inorg. Chem.*, *2*, 1227 (1963).

5. R. Eisenberg, J. A. Ibers, R. J. H. Clark, and H. B. Gray, *J. Am. Chem. Soc.*, *86*, 113 (1964).

6. H. B. Gray and E. Billig, *J. Am. Chem. Soc.*, *85*, 2019 (1963).

7. E. I. Stiefel, J. H. Waters, E. Billig, and H. B. Gray, *J. Am. Chem. Soc.*, *87*, 3016 (1965).

8. G. N. Schrauzer, H. W. Finck, and V. P. Mayweg, *Angew. Chem. Int. Ed. Engl.*, *3*, 639 (1964).

9. G. N. Schrauzer, V. P. Mayweg, H. W. Finck, U. Müller-Westerhoff, and W. Heinrich, *Angew. Chem. Int. Ed. Engl.*, *3*, 381 (1964).

10. A. Davison, N. Edelstein, R. H. Holm, and A. H. Maki, *J. Am. Chem. Soc.*, *86*, 2799 (1964).

11. J. Locke, J. A. McCleverty, E. J. Wharton, and C. J. Winscom, *Chem. Commun.*, 677 (1966).

12. J. A. McCleverty, N. M. Atherton, J. Locke, E. J. Wharton, and C. J. Winscom, *J. Am. Chem. Soc.*, *89*, 6082 (1967).

13. J. A. McCleverty, *Prog. Inorg. Chem.*, *10*, 49 (1968).

14. A. L. Balch, I. G. Dance, and R. H. Holm, *J. Am. Chem. Soc.*, *90*, 1139 (1968).

15. C. Faulmann and P. Cassoux, *Prog. Inorg. Chem.*, *52*, 399 (2004).

16. R. Hille, *Chem. Rev.*, *96*, 2757 (1996).

17. M. K. Johnson, D. C. Rees, and M. W. W. Adams, *Chem. Rev.*, *96*, 2817 (1996).

18. M. K. Chan, S. Mukund, A. Kletzin, M. W. W. Adams, and D. C. Rees, *Science*, *267*, 1463 (1995).

19. J. McMaster, J. M. Tunney, and C. D. Garner, *Prog. Inorg. Chem.*, *52*, 539 (2004).

20. G. C. Tucci, J. P. Donahue, and R. H. Holm, *Inorg. Chem.*, *37*, 1602 (1998).

21. C. Lorber, J. P. Donahue, C. A. Goddard, E. Nordlander, and R. H. Holm, *J. Am. Chem. Soc.*, *120*, 8102 (1998).

22. C. A. Goddard and R. H. Holm, *Inorg. Chem.*, *38*, 5389 (1999).

23. B. S. Lim, J. Donahue, and R. H. Holm, *Inorg. Chem.*, *39*, 263 (2000).

24. R. Eisenberg, *Coord. Chem. Rev.*, *255*, 825 (2011).

25. R. Eisenberg and H. B. Gray, *Inorg. Chem.*, *50*, 9741 (2011).

26. S. Sproules and K. Wieghardt, *Coord. Chem. Rev.*, *255*, 837 (2011).

27. E. Hoyer, W. Dietzsch, H. Hennig, and W. Schroth, *Chem. Ber.*, *102*, 603 (1969).

28. E. Hoyer, W. Dietzsch, and H. Müller, *Z. Chem.*, *7*, 354 (1967).

29. W. Schroth and J. Peschel, *Chimia*, *18*, 171 (1964).

30. M. J. Baker-Hawkes, E. Billig, and H. B. Gray, *J. Am. Chem. Soc.*, *88*, 4870 (1966).

31. R. Adams and A. Ferretti, *J. Am. Chem. Soc.*, *81*, 4927 (1959).

32. R. Adams, W. Reifschneider, and A. Ferretti, *Org. Synth.*, *42*, 22 (1965).

33. A. Ferretti, *Org. Synth.*, *42*, 54 (1965).

34. L. Testaferri, M. Tiecco, M. Tingoli, D. Chianelli, and M. Montanucci, *Synthesis*, 751 (1983).

35. R. Sato, T. Ohyama, T. Kawagoe, M. Baba, S. Nakajo, T. Kimura, and S. Ogawa, *Heterocycles*, *55*, 145 (2001).

36. G. D. Figuly, C. K. Loop, and J. C. Martin, *J. Am. Chem. Soc.*, *111*, 654 (1989).

37. D. M. Giolando and K. Kirschbaum, *Synthesis*, 451 (1992).

38. N. D. Lowe and C. D. Garner, *J. Chem. Soc., Dalton Trans.*, 2197 (1993).

39. A. Callaghan, A. J. Layton, and R. S. Nyholm, *Chem. Commun.*, 399 (1969).

40. W. Schroth and U. Schmidt, *Z. Chem.*, *4*, 270 (1964).

41. J. S. Pap, F. L. Benedito, E. Bothe, E. Bill, S. DeBeer George, T. Weyhermüller, and K. Wieghardt, *Inorg. Chem.*, *46*, 4187 (2007).

42. T. Kreickmann and F. E. Hahn, *Chem. Commun.*, 1111 (2007).

43. J. A. McCleverty and E. J. Wharton, *J. Chem. Soc. A*, 2258 (1969).

44. D. Sellmann, G. Freyberger, R. Eberlein, E. Böhlen, G. Huttner, and L. Zsolnai, *J. Organomet. Chem.*, *323*, 21 (1987).

45. A. M. Richter, V. Engels, N. Beye, and E. Fanghänel, *Z. Chem.*, *29*, 444 (1989).

46. A. M. Richter, N. Beye, and E. Fanghänel, *Synthesis*, 1149 (1990).

47. D. C. Morrison and A. Furst, *J. Org. Chem.*, *21*, 470 (1956).

48. C. G. Krespan, *J. Am. Chem. Soc.*, *83*, 3434 (1961).

49. J. L. Hencher, Q. Shen, and D. G. Tuck, *J. Am. Chem. Soc.*, *98*, 899 (1976).

50. K. Wang, J. M. McConnachie, and E. I. Stiefel, *Inorg. Chem.*, *38*, 4334 (1999).

51. N. Jacobsen, P.de Mayo, and A. C. Weedon, *Nouv. J. Chim.*, *2*, 331 (1978).

52. P. C. Guha and M. N. Chakladar, *Quart. J. Indian Chem. Soc.*, *2*, 318 (1925).

53. L. Field, W. D. Stephens, and E. L. Lippert, *J. Org. Chem.*, *26*, 4782 (1961).

54. C. M. Bolinger and T. B. Rauchfuss, *Inorg. Chem.*, *21*, 3947 (1982).

55. T. Shimizu, H. Murakami, Y. Kobayashi, K. Iwata, and N. Kamigata, *J. Org. Chem.*, *63*, 8192 (1998).

56. W. Kusters and P.de Mayo, *J. Am. Chem. Soc.*, *96*, 3502 (1974).

57. E. Fanghänel, R. Ebisch, and B. Adler, *Z. Chem.*, *13*, 431 (1973).

58. G. N. Schrauzer and V. Mayweg, *Z. Naturforsch.*, *19b*, 192 (1964).

59. D. Coucouvanis, A. Hadjikyriacou, A. Toupadakis, S. M. Koo, O. Ileperuma, M. Draganjac, and A. Salifoglou, *Inorg. Chem.*, *30*, 754 (1991).

60. M. Draganjac and D. Coucouvanis, *J. Am. Chem. Soc.*, *105*, 139 (1983).

61. A. Mallard, C. Simonnet-Jégat, H. Lavanant, J. Marrot, and F. Sécheresse, *Trans. Met. Chem.*, *33*, 143 (2008).

62. K. Umakoshi, E. Nishimoto, M. Sokolov, H. Kawano, Y. Sasaki, and M. Onishi, *J. Organomet. Chem.*, *611*, 370 (2000).

63. M. A. Ansari, J. Chandrasekaran, and S. Sarkar, *Inorg. Chim. Acta*, *130*, 155 (1987).

64. C. L. Soricelli, V. A. Szalai, and S. J. N. Burgmayer, *J. Am. Chem. Soc.*, *113*, 9877 (1991).

65. A. Davison and R. H. Holm, *Inorg. Synth.*, *10*, 8 (1967).

66. J. Berger and I. Uldall, *Acta Chem. Scand.*, *18*, 1353 (1964).

67. A. K. Bhattacharya and A. G. Hortmann, *J. Org. Chem.*, *39*, 95 (1974).

68. P. Falaras, C.-A. Mitsopoulou, D. Argyropoulos, E. Lyris, N. Psaroudakis, E. Vrachnou, and D. Katakis, *Inorg. Chem.*, *34*, 4536 (1995).

69. A. M. Celli, D. Donati, F. Ponticelli, S. J. Roberts-Fleming, M. Kalaji, and P. J. Murphy, *Org. Lett.*, *3*, 3572 (2001).

70. C. Mitsopoulou, J. Konstantatos, D. Katakis, and E. Vrachnou, *J. Mol. Cat.*, *67*, 137 (1991).

71. P. Chandrasekaran and J. P. Donahue, *Org. Synth.*, *86*, 333 (2009).

72. P. Chandrasekaran, K. Arumugam, U. Jayarathne, L. M. Pérez, J. T. Mague, and J. P. Donahue, *Inorg. Chem.*, *48*, 2103 (2009).

73. N. Robertson and L. Cronin, *Coord. Chem. Rev.*, *227*, 93 (2002).

74. M. L. Mercuri, P. Deplano, L. Pilia, and F. Artizzu, *Coord. Chem. Rev.*, *254*, 1419 (2010).

75. J. M. Williams, J. R. Ferraro, R. J. Thorn, K. D. Carlson, U. Geiser, H. H. Wang, A. M. Kini, and M. H. Whangbo, *Organic Superconductors (Including Fullerene)*, Prentice Hall, Englewood, NJ, 1992.

76. M. Bendikov, F. Wudl, and D. F. Perepichka, *Chem. Rev.*, *104*, 4891 (2004).

77. P. Cassoux, L. Valada, H. Kobayashi, A. Kobayashi, R. A. Clark, and A. E. Underhill, *Coord. Chem. Rev.*, *110* (1991).

78. A. E. Pullen and R.-M. Olk, *Coord. Chem. Rev.*, *188*, 211 (1999).

79. G. Steimecke, H.-J. Sieler, R. Kirmse, and E. Hoyer, *Phosphorus Sulfur*, *7*, 49 (1979).

80. T. K. Hansen, J. Becher, T. Jørgensen, K. S. Varma, R. Khedekar, and M. P. Cava, *Org. Synth.*, *73*, 270 (1996).

81. L. Valada, J.-P. Legros, M. Bousseau, P. Cassoux, M. Garbauskas, and L. V. Interrante, *J. Chem. Soc., Dalton Trans.*, 783 (1985).

82. J. G. Breitzer and T. B. Rauchfuss, *Polyhedron*, *19*, 1283 (2000).

83. J. G. Breitzer, A. I. Smirnov, L. F. Szczepura, S. R. Wilson, and T. B. Rauchfuss, *Inorg. Chem.*, *40*, 1421 (2001).

84. K. Hartke, T. Kissel, J. Quante, and R. Matusch, *Chem. Ber.*, *113*, 1898 (1980).

85. R.-M. Olk, W. Dietzsch, R. Kirmse, J. Stach, E. Hoyer, and L. Golič, *Inorg. Chim. Acta*, *128*, 251 (1987).

86. G. Steimecke, J. Sieler, R. Kirmse, W. Dietzsch, and E. Hoyer, *Phosphorus Sulfur*, *12*, 237 (1982).

87. R.-M. Olk, B. Olk, W. Dietzsch, R. Kirmse, and E. Hoyer, *Coord. Chem. Rev.*, *117*, 99 (1992).

88. C. T. Vance, R. D. Bereman, J. Bordner, W. E. Hatfield, and J. H. Helms, *Inorg. Chem.*, *24*, 2905 (1985).

89. G. Chung, R. Bereman, and P. Singh, *J. Coord. Chem.*, *33*, 331 (1994).

90. J. Becher, J. Lau, P. Leriche, P. Mørk, and N. Svenstrup, *J. Chem. Soc., Chem. Commun.*, 2715 (1994).

91. T. Yoneda, Y. Kamata, K. Ueda, T. Sugimoto, T. Tada, M. Shiro, H. Yoshino, and K. Murata, *Synth. Met.*, *135–136*, 573 (2003).

92. G. N. Schrauzer and H. W. Finck, *Angew. Chem. Int. Ed. Engl.*, *3*, 133 (1964).

93. G. N. Schrauzer, H. W. Finck, and V. Mayweg, *Z. Naturforsch.*, *19b*, 1080 (1964).

94. G. N. Schrauzer, V. P. Mayweg, and W. Heinrich, *Inorg. Chem.*, *4*, 1615 (1965).

95. K. Arumugam, J. E. Bollinger, M. Fink, and J. P. Donahue, *Inorg. Chem.*, *46*, 3283 (2007).

96. G. N. Schrauzer and V. P. Mayweg, *J. Am. Chem. Soc.*, *88*, 3235 (1966).

97. K. Tatsumi, I. Matsubara, Y. Sekiguchi, A. Nakamura, and C. Mealli, *Inorg. Chem.*, *28*, 773 (1989).

98. S. Friedle, D. V. Partyka, M. V. Bennett, and R. H. Holm, *Inorg. Chim. Acta*, *359*, 1427 (2006).

99. D. V. Partyka and R. H. Holm, *Inorg. Chem.*, *43*, 8609 (2004).

100. S. Grosyman, J.-J. Wang, R. Tagore, S. C. Lee, and R. H. Holm, *J. Am. Chem. Soc.*, *130*, 12794 (2008).

101. Y. Gareau and A. Orellana, *Syn. Lett.*, 803 (1997).

102. R. F. X. Williams, *Phosphorus Sulfur*, *2*, 141 (1976).

103. W. B. Heuer and W. H. Pearson, *Polyhedron*, *15*, 2199 (1996).

104. D. Coucouvanis, F. J. Hollander, R. West, and D. Eggerding, *J. Am. Chem. Soc.*, *96*, 3006 (1974).

105. M. Gay-Lussac, *Ann. Chim. (Paris)*, *95*, 136 (1815).

106. J. Formanek, *Berchte.*, *22*, 2655 (1889).

107. J. R. Barceló, *Spec. Acta*, *10*, 245 (1958).

108. R. N. Hurd and G. DeLaMater, *Chem. Rev.*, *61*, 45 (1961).

109. M. R. Green, N. Jubran, B. E. Bursten, and D. H. Busch, *Inorg. Chem.*, *26*, 2326 (1987).

110. R. N. Hurd, G. DeLaMater, G. C. McElheny, and L. V. Peiffer, *J. Am. Chem. Soc.*, *82*, 4454 (1960).

111. S. P. Perlepes, M. Bellaihou, and H. O. Desseyn, *Spec. Lett.*, *26*, 751 (1993).

112. M. Abboudi, A. Mosset, and J. Galy, *Inorg. Chem.*, *24*, 2091 (1984).

113. R. Veit, J. J. Girerd, O. Kahn, F. Robert, and Y. Jeannin, *Inorg. Chem.*, *25*, 4175 (1986).

114. R. N. Hurd, G. DeLaMater, G. C. McElheny, R. J. Turner, and V. H. Wallingford, *J. Org. Chem.*, *26*, 3980 (1961).

115. H. O. Desseyn, A. J. Aarts, E. Esmans, and M. A. Herman, *Spec. Acta*, *35A*, 1203 (1979).

116. D. M. Hart, P. S. Rolfs, and J. M. Kessinger, *J. Inorg. Nucl. Chem.*, *32*, 469 (1970).

117. A. C. Fabretti, G. C. Pellacani, and G. Peyronel, *J. Inorg. Nucl. Chem.*, *36*, 1751 (1974).

118. G. Peyronel, G. C. Pellacani, A. Pignedoli, and G. Benetti, *Inorg. Chim. Acta*, *5*, 263 (1971).

119. R. Isaksson, T. Liljefors, and J. Sandström, *J. Chem. Res. (S)*, 43 (1981).

120. D. J. A. De Ridder, *Acta Crystallogr. Sect.*, *C49*, 1975 (1993).

121. P. C. Servaas, D. J. Stufkens, A. Oskam, P. Vernooijs, E. J. Baerends, D. J. A. De Ridder, and C. H. Stam, *Inorg. Chem.*, *28*, 4104 (1989).

122. U. T. Mueller-Westerhoff, B. Vance, and D. I. Yoon, *Tetrahedron*, *47*, 909 (1991).

123. P. Deplano, L. Pilia, D. Espa, M. L. Mercuri, and A. Serpe, *Coord. Chem. Rev.*, *254*, 1434 (2010).

124. L. Pilia, F. Artizzu, D. Espa, L. Marchió, M. L. Mercuri, A. Serpe, and P. Deplano, *Dalton Trans.*, *39*, 8139 (2010).

125. J. L. Martin and J. Takats, *Inorg. Chem.*, *14*, 73 (1975).

126. M. J. Bennett, M. Cowie, J. L. Martin, and J. Takats, *J. Am. Chem. Soc.*, *95*, 7504 (1973).

127. R. L. Melen, M. McPartlin, and D. S. Wright, *Dalton Trans.*, *40*, 1649 (2011).

128. J. Jones and J. Douek, *J. Inorg. Nucl. Chem.*, *43*, 406 (1981).

129. B. Birkmann, A. W. Ehlers, R. Fröhlich, K. Lammertsma, and F. E. Hahn, *Chem. Eur. J.*, *15*, 4301 (2009).

130. B. Birkmann, W. W. Seidel, T. Pape, A. W. Ehlers, K. Lammertsma, and F. E. Hahn, *Dalton Trans.* 7350 (2009).

131. F. E. Hahn, B. Birkmann, and T. Pape, *Dalton Trans.* 2100 (2008).

132. F. Hupka and F. E. Hahn, *Chem. Commun.*, *46*, 3744 (2010).

133. T. Kreickmann, C. Diedrich, T. Pape, H. V. Huynh, S. Grimme, and F. E. Hahn, *J. Am. Chem. Soc.*, *128*, 11808 (2006).

134. J. K. Yandell and N. Sutin, *Inorg. Chem.*, *11*, 448 (1972).

135. J. A. McCleverty, J. Locke, E. J. Wharton, and M. Gerloch, *J. Chem. Soc. A*, 816 (1968).

136. E. Billig, H. B. Gray, S. I. Shupack, J. H. Waters, and R. Williams, *Proc. Chem. Soc.*, 110 (1964).

137. C. H. Langford, E. Billig, S. I. Shupack, and H. B. Gray, *J. Am. Chem. Soc.*, *86*, 2958 (1964).

138. S. Sproules and K. Wieghardt, *Coord. Chem. Rev.*, *254*, 1358 (2010).

139. D. C. Olson, V. P. Mayweg, and G. N. Schrauzer, *J. Am. Chem. Soc.*, *88*, 4876 (1966).

140. E. J. Rosa and G. N. Schrauzer, *J. Phys. Chem.*, *73*, 3132 (1969).

141. D. Katakis, C. Mitsopoulou, and E. Vrachnou, *J. Photochem. Photobiol. A: Chem.*, *81*, 103 (1994).

142. A. Davison, N. Edelstein, R. H. Holm, and A. H. Maki, *Inorg. Chem.*, *4*, 55 (1965).

143. P. J. Baricelli and P. C. H. Mitchell, *Inorg. Chim. Acta*, *115*, 163 (1986).

144. M. Kawashima, M. Koyama, and T. Fujinaga, *J. Inorg. Nucl. Chem.*, *38*, 801 (1976).

145. F. E. Hahn and W. W. Seidel, *Angew. Chem. Int. Ed.*, *34*, 2700 (1995).

146. S. Boyde, C. D. Garner, J. A. Joule, and D. J. Rowe, *Chem. Commun.*, 800 (1987).

147. G. N. Schrauzer, V. P. Mayweg, and W. Heinrich, *J. Am. Chem. Soc.*, *88*, 5174 (1966).

148. D. Argyopoulos, E. Lyris, C. A. Mitsopoulou, and D. Katakis, *J. Chem. Soc., Dalton Trans.*, 615 (1997).

149. D. J. Harrison, A. J. Lough, N. Nguyen, and U. Fekl, *Angew. Chem. Int. Ed.*, *46*, 7644 (2007).

150. G. N. Schrauzer and H. N. Rabinowitz, *J. Am. Chem. Soc.*, *91*, 6522 (1969).

151. A. Davison, N. Edelstein, R. H. Holm, and A. H. Maki, *Inorg. Chem.*, *3*, 814 (1964).

152. A. L. Balch and R. H. Holm, *Chem. Commun.*, 552 (1966).

153. G. N. Schrauzer, V. P. Mayweg, H. W. Finck, and W. Heinrich, *J. Am. Chem. Soc.*, *88*, 4604 (1966).

154. J. H. Enemark and W. N. Lipscomb, *Inorg. Chem.*, *4*, 1729 (1965).

155. X. Yang, G. K. W. Freeman, T. B. Rauchfuss, and S. R. Wilson, *Inorg. Chem.*, *30*, 3034 (1991).

156. C. G. Pierpont and C. W. Lange, *Prog. Inorg. Chem.*, *41*, 331 (1994).

157. C. G. Pierpont, *Coord. Chem. Rev.*, *219–221*, 415 (2001).

158. C. Keller, D. Walther, J. Reinhold, and E. Hoyer, *Z. Chem.*, *28*, 410 (1988).

159. C. Schulze Isfort, T. Pape, and F. E. Hahn, *Eur. J. Inorg. Chem.*, 2607 (2005).

160. N. G. Connelly and W. E. Geiger, *Chem. Rev.*, *96*, 877 (1996).

161. G.-E. Matsubayashi, M. Nakano, H. Tamura, and K. Natsuaki, *Mol. Cryst. Liq. Sect. Cryst.*, *296*, 245 (1997).

162. T. B. Rauchfuss, *Prog. Inorg. Chem.*, *52*, 1 (2004).

163. G. Matsubayashi, T. Maikawa, and M. Nakano, *J. Chem. Soc., Dalton Trans.*, 2995 (1993).

164. C. Ma, Y. Han, and L. Dacheng, *Polyhedron*, *23*, 1207 (2004).

165. G. M. Allan, R. A. Howie, J. M. S. Skakle, J. L. Wardell, and S. M. S. V. Wardell, *J. Organomet. Chem.*, *627*, 189 (2001).

166. G. N. Schrauzer and V. P. Mayweg, *J. Am. Chem. Soc.*, *87*, 1483 (1965).

167. M. Nomura and M. Fourmigué, *Inorg. Chem.*, *47*, 1301 (2008).

168. E. Cerrada, E. J. Fernandez, P. G. Jones, A. Laguna, M. Laguna, and R. Terroba, *Organometallics*, *14*, 5537 (1995).

169. C. S. Velazquez, T. F. Baumann, M. M. Olmstead, H. Hope, A. G. M. Barrett, and B. M. Hoffman, *J. Am. Chem. Soc.*, *115*, 9997 (1993).

170. C. L. Beswick, J. M. Schulman, and E. I. Stiefel, *Prog. Inorg. Chem.*, *52*, 55 (2004).

171. R. Eisenberg and J. A. Ibers, *Inorg. Chem.*, *4*, 605 (1965).

172. R. Eisenberg and J. A. Ibers, *J. Am. Chem. Soc.*, *87*, 3776 (1965).

173. R. Eisenberg and J. A. Ibers, *Inorg. Chem.*, *5*, 411 (1966).

174. R. G. Dickinson and L. Pauling, *J. Am. Chem. Soc.*, *45*, 1466 (1923).

175. A. E. Smith, G. N. Schrauzer, V. P. Mayweg, and W. Heinrich, *J. Am. Chem. Soc.*, *87*, 5798 (1965).

176. R. Eisenberg and W. W. Brennessel, *Acta Crystallogr. Sect.*, *C62*, m464 (2006).

177. R. Eisenberg, E. I. Stiefel, R. C. Rosenberg, and H. B. Gray, *J. Am. Chem. Soc.*, *88*, 2874 (1966).

178. R. Eisenberg and H. B. Gray, *Inorg. Chem.*, *6*, 1844 (1967).

179. S. Sproules, T. Weyhermüller, S. DeBeer, and K. Wieghardt, *Inorg. Chem.*, *49*, 5241 (2010).

180. E. I. Stiefel and H. B. Gray, *J. Am. Chem. Soc.*, *87*, 4012 (1965).

181. E. I. Stiefel, R. Eisenberg, R. C. Rosenberg, and H. B. Gray, *J. Am. Chem. Soc.*, *88*, 2956 (1966).

182. R. D. Shannon, *Acta Crystallogr. Sect.*, *A32*, 751 (1976).

183. B. Cordero, V. Gómez, A. E. Platero-Prats, M. Revés, J. Echeverría, E. Cremades, F. Barragán, and S. Alvarez, *Dalton Trans.* 2832 (2008).

184. R. Eisenberg, *Prog. Inorg. Chem.*, *12*, 295 (1970).

185. R. Williams, E. Billig, J. H. Waters, and H. B. Gray, *J. Am. Chem. Soc.*, *88*, 43 (1966).

186. J. D. Forrester, A. Zalkin, and D. H. Templeton, *Inorg. Chem.*, *3*, 1500 (1964).

187. J. D. Forrester, A. Zalkin, and D. H. Templeton, *Inorg. Chem.*, *3*, 1507 (1964).

188. R. Franzi, M. Geoffroy, M. V. V. S. Reddy, and J. Weber, *J. Phys. Chem.*, *91*, 3187 (1987).

189. R. Benassi and F. Taddei, *J. Phys. Chem. A*, *102*, 6173 (1998).

190. C. G. Pierpont and R. Eisenberg, *J. Chem. Soc. A*, 2285 (1971).

191. E. I. Stiefel and G. F. Brown, *Inorg. Chem.*, *11*, 434 (1972).

192. E. I. Muetterties and L. J. Geuggenberger, *J. Am. Chem. Soc.*, *96*, 1748 (1974).

193. K. R. Dymock and G. J. Palenik, *Inorg. Chem.*, *14*, 1220 (1975).

194. D. L. Kepert, *Inorg. Chem.*, *11*, 1561 (1972).

195. D. L. Kepert, *Prog. Inorg. Chem.*, *23*, 1 (1977).

196. H. Abrahamson, J. R. Heiman, and L. H. Pignolet, *Inorg. Chem.*, *14*, 2070 (1975).

197. A. Avdeef and J. P. Fackler, Jr., *Inorg. Chem.*, *14*, 2002 (1975).

198. M. Cowie and M. J. Bennett, *Inorg. Chem.*, *15*, 1595 (1976).

199. E. I. Stiefel, Z. Dori, and H. B. Gray, *J. Am. Chem. Soc.*, *89*, 3353 (1967).

200. M. Gerloch, S. F. A. Kettle, J. Locke, and J. A. McCleverty, *Chem. Commun.*, 29 (1966).

201. A. Sequeira and I. Bernal, *J. Cryst. Mol. Struct.*, *3*, 157 (1973).

202. G. F. Brown and E. I. Stiefel, *Inorg. Chem.*, *12*, 2140 (1973).

203. N. M. Atherton and C. J. Winscom, *Inorg. Chem.*, *12*, 383 (1973).

204. W.-L. Kwik and E. I. Stiefel, *Inorg. Chem.*, *12*, 2337 (1973).

205. F. W. B. Einstein and R. D. G. Jones, *J. Chem. Soc. A*, 2762 (1971).

206. M. Könemann, W. Stüer, K. Kirschbaum, and D. M. Giolando, *Polyhedron*, *13*, 1415 (1994).

207. B. Birkmann, R. Fröhlich, and F. E. Hahn, *Chem. Eur. J.*, *15*, 9325 (2009).

208. C. Schulze Isfort, T. Kreickmann, T. Pape, R. Fröhlich, and F. E. Hahn, *Chem. Eur. J.*, *13*, 2344 (2007).

209. F. E. Hahn, C. Schulze Isfort, and T. Pape, *Angew. Chem. Int. Ed.*, *43*, 4807 (2004).

210. F. E. Hahn, T. Kreickmann, and T. Pape, *Dalton Trans.* 769 (2006).

211. J. H. Welch, R. Bereman, and P. Singh, *Inorg. Chem.*, *29*, 68 (1990).

212. M. Kondo, S. Minakoshi, K. Iwata, T. Shimizu, H. Matsuzaka, N. Kamigata, and S. Kitagawa, *Chem. Lett.*, 489 (1996).

213. T. Okubo, R. Maeda, M. Kondo, and S. Kitagawa, *Chem. Lett.*, *35*, 34 (2006).

214. G.-E. Matsubayashi, K. Akiba, and T. Tanaka, *Inorg. Chim. Acta*, *157*, 195 (1989).

215. W. E. Broderick, E. M. McGhee, M. R. Godfrey, B. M. Hoffman, and J. A. Ibers, *Inorg. Chem.*, *28*, 2902 (1989).

216. G.-E. Matsubayashi, K. Akiba, and T. Tanaka, *Inorg. Chem.*, *27*, 4744 (1988).

217. C. Livage, M. Fourmigué, P. Batail, E. Canadell, and C. Coulon, *Bull. Soc. Chim. Fr.*, *130*, 761 (1993).

218. J. H. Welch, R. D. Bereman, and P. Singh, *Inorg. Chem.*, *27*, 2862 (1988).

219. M. Cowie and M. J. Bennett, *Inorg. Chem.*, *15*, 1589 (1976).

220. J. L. Martin and J. Takats, *Inorg. Chem.*, *14*, 1358 (1975).

221. P. Banerjee, S. Sproules, T. Weyhermüller, S. DeBeer George, and K. Wieghardt, *Inorg. Chem.*, *48*, 5829 (2009).

222. R. R. Kapre, E. Bothe, T. Weyhermüller, S. DeBeer George, N. Muresan, and K. Wieghardt, *Inorg. Chem.*, *46*, 7827 (2007).

223. G. R. Lewis and I. Dance, *J. Chem. Soc., Dalton Trans.*, 3176 (2000).

224. M. Cowie and M. J. Bennett, *Inorg. Chem.*, *15*, 1584 (1976).

225. A. Cervilla, E. Llopis, D. Marco, and F. Pérez, *Inorg. Chem.*, *40*, 6525 (2001).

226. H. Sugimoto, Y. Furukawa, M. Tarumizu, H. Miyake, K. Tanaka, and H. Tsukube, *Eur. J. Inorg. Chem.*, 3088 (2005).

227. R. R. Kapre, E. Bothe, T. Weyhermüller, S. DeBeer George, and K. Wieghardt, *Inorg. Chem.*, *46*, 5642 (2007).

228. F. E. Hahn, M. Offermann, C. Schulze Isfort, T. Pape, and R. Fröhlich, *Angew. Chem. Int. Ed.*, *47*, 6794 (2008).

229. S. Boyde, C. D. Garner, J. H. Enemark, M. A. Bruck, and J. G. Kristofzski, *J. Chem. Soc., Dalton Trans.*, 2267 (1987).

230. S. Boyde, C. D. Garner, J. H. Enemark, and R. B. Ortega, *J. Chem. Soc., Dalton Trans.*, 297 (1987).

231. D. Fomitchev, B. S. Lim, and R. H. Holm, *Inorg. Chem.*, *40*, 645 (2001).

232. S. Sproules, P. Banerjee, T. Weyhermüller, Y. Yan, J. P. Donahue, and K. Wieghardt, *Inorg. Chem.*, *50*, 7106 (2011).

233. H.-W. Xu, Z.-N. Chen, and J. G. Wu, *Acta Crystallogr. Sect.*, *E58*, m631 (2002).

234. W.-W. Fu, Z.-L. Xie, and Z.-N. Chen, *Acta Crystallogr. Sect.*, *E63*, m509 (2007).

235. W.-W. Fu and Z.-N. Chen, *Acta Crystallogr. Sect.*, *E63*, m842 (2007).

236. H.-Y. Xu, J.-G. Wu, and Z.-N. Chen, *Trans. Met. Chem.*, *33*, 17 (2008).

237. G. Matsubayashi, K. Douki, and H. Tamura, *Chem. Lett.*, *21*, 1251 (1992).

238. G. Matsubayashi, K. Douki, H. Tamura, M. Nakano, and W. Mori, *Inorg. Chem.*, *32*, 5990 (1993).

239. H. Huynh, T. Lugger, and F. E. Hahn, *Eur. J. Inorg. Chem.*, 3007 (2002).

240. F. Knoch, D. Sellmann, and W. Kern, *Z. Kristallogr.*, *202*, 326 (1992).

241. T. E. Burrow, R. H. Morris, A. Hills, D. L. Hughes, and R. L. Richards, Acta Cryst., *C49*, 1591 (1993).

242. F. Knoch, D. Sellmann, and W. Kern, *Z. Kristallogr.*, *205*, 300 (1992).

243. W.-W. Fu and Z.-N. Chen, *Acta Crystallogr. Sect.*, *E63*, m482 (2007).

244. C.-H. Lin, C.-G. Chen, M.-L. Tsai, G.-H. Lee, and W.-F. Liaw, *Inorg. Chem.*, *47*, 11435 (2008).

245. S. F. Colmanet, G. A. Williams, and M. F. MacKay, *J. Chem. Soc., Dalton Trans.*, 2305 (1987).

246. S. F. Colmanet and M. F. MacKay, *Aust. J. Chem.*, *41*, 1127 (1988).

247. S. Sproules, F. L. Benedito, E. Bill, T. Weyhermüller, S. DeBeer George, and K. Wieghardt, *Inorg. Chem.*, *48*, 10926 (2009).

248. S. Sproules, T. Weyhermüller, R. Goddard, and K. Wieghardt, *Inorg. Chem.*, *50*, 12623 (2011).

249. C. Milsmann, S. Sproules, E. Bill, T. Weyhermüller, S. DeBeer George, and K. Wieghardt, *Chem. Eur. J.*, *16*, 3628 (2010).

250. D. Sellmann, K. Hein, and F. W. Heinemann, *Inorg. Chim. Acta*, *357*, 3739 (2004).

251. R. Maiti, M. Shang, and A. G. Lappin, *Chem. Commun.*, 2349 (1999).

252. D. Sellmann, A. C. Hennige, and F. W. Heinemann, *Eur. J. Inorg. Chem.*, 819 (1998).

253. B. Wenzel, B. Wehse, U. Schilde, and P. Strauch, *Z. Anorg. Allg. Chem.*, *630*, 1469 (2004).

254. R. O. Day, J. M. Holmes, S. Shafieezad, V. Chandrasekhar, and R. R. Holmes, *J. Am. Chem. Soc.*, *110*, 5377 (1988).

255. F.de Assis, Z. H. Chohan, R. A. Howie, A. Khan, J. N. Low, G. M. Spencer, J. L. Wardell, and S. M. S. V. Wardell, *Polyhedron*, *18*, 3533 (1999).

256. T. Sheng, X. Wu, P. Lin, Q. Wang, W. Zhang, and L. Chen, *J. Coord. Chem.*, *48*, 113 (1999).

257. T. Akasaka, M. Nakano, H. Tamura, and G. Matsubayashi, *Bull. Chem. Soc. Jpn.*, *75*, 2621 (2002).

258. N. M. Comerlato, G. B. Ferreira, R. A. Howie, and J. L. Wardell, *Acta Crystallogr. Sect.*, *E60*, m1781 (2004).

259. Y.-L. Wang, Y.-J. Wang, J. Yao, X.-T. Zhao, and R.-H. Wang, *Acta Crystallogr. Sect.*, *E63*, m1147 (2007).

260. N. M. Comerlato, G. B. Ferreira, W. T. A. Harrison, R. A. Howie, and J. L. Wardell, *Acta Crystallogr. Sect.*, *C61*, m139 (2005).

261. J. Wegener, K. Kirschbaum, and D. M. Giolando, *J. Chem. Soc., Dalton Trans.*, 1213 (1994).

262. G. M. Spencer, J. L. Wardell, and J. H. Aupers, *Polyhedron*, *15*, 2701 (1996).

263. M. Roger, T. Arliguie, P. Thuéry, M. Fourmigué, and M. Ephritikhine, *Inorg. Chem.*, *44*, 584 (2005).

264. M. Roger, T. Arliguie, P. Thuéry, M. Fourmigué, and M. Ephritikhine, *Inorg. Chem.*, *44*, 594 (2005).

265. J. S. Gancheff and F. E. Hahn, *Spec. Acta*, *98A*, 62 (2012).

266. S. Alvarez, R. Vicente, and R. Hoffmann, *J. Am. Chem. Soc.*, *107*, 6253 (1985).

267. H. K. Joshi, J. J. A. Cooney, F. E. Inscore, N. E. Gruhn, D. L. Lichtenberger, and J. H. Enemark, *Proc. Natl. Acad. Sci. USA*, *100*, 3719 (2003).

268. B. Domercq, C. Coulon, and M. Fourmigué, *Inorg. Chem.*, *40*, 371 (2001).

269. S. Campbell and S. Harris, *Inorg. Chem.*, *35*, 3285 (1996).

270. S. Meskaldji, L. Belkhiri, T. Arliguie, M. Fourmigué, M. Ephritikhine, and A. Boucekkine, *Inorg. Chem.*, *49*, 3192 (2010).

271. C. G. Pierpont and R. M. Buchanan, *Coord. Chem. Rev.*, *38*, 45 (1981).

272. C. G. Pierpont, *Coord. Chem. Rev.*, *216–217*, 99 (2001).

273. H. Wunderlich and D. Mootz, *Acta Crystallogr. Sect.*, *B27*, 1684 (1971).

274. A. L. Macdonald and J. Trotter, *J. Chem. Soc., Perkin Trans. 2*, 476 (1973).

275. S. Kokatam, K. Ray, J. Pap, E. Bill, W. E. Geiger, R. J. LeSuer, P. H. Rieger, T. Weyhermüller, F. Neese, and K. Wieghardt, *Inorg. Chem.*, *46*, 1100 (2007).

276. K. Ray, A. Begum, T. Weyhermüller, S. Piligkos, J.van Slageren, F. Neese, and K. Wieghardt, *J. Am. Chem. Soc.*, *127*, 4403 (2005).

277. K. Ray, T. Petrenko, K. Wieghardt, and F. Neese, *Dalton Trans.* 1552 (2007).

278. K. Ray, T. Weyhermüller, A. Goossens, M. W. Crajé, and K. Wieghardt, *Inorg. Chem.*, *42*, 4082 (2003).

279. K. Ray, T. Weyhermüller, F. Neese, and K. Wieghardt, *Inorg. Chem.*, *44*, 5345 (2005).

280. C. G. Pierpont, *Inorg. Chem.*, *50*, 9766 (2011).

281. A. L. Tenderholt, R. K. Szilagyi, R. H. Holm, K. O. Hodgson, B. Hedman, and E. I. Solomon, *Inorg. Chem.*, *47*, 6382 (2008).

282. G. N. Schrauzer and V. P. Mayweg, *J. Am. Chem. Soc.*, *87*, 3585 (1965).

283. A. Davison, N. Edelstein, R. H. Holm, and A. H. Maki, *J. Am. Chem. Soc.*, *85*, 2029 (1963).

284. A. H. Maki, N. Edelstein, A. Davison, and R. H. Holm, *J. Am. Chem. Soc.*, *86*, 4580 (1964).

285. A. H. Al-Mowali and A. L. Porte, *J. Chem. Soc., Dalton Trans.*, 250 (1975).

286. G. Gritzner and J. Kuta, *Pure Appl. Chem.*, *56*, 461 (1984).

287. K. Wang, *Prog. Inorg. Chem.*, *52*, 267 (2004).

288. E. J. Wharton and J. A. McCleverty, *J. Chem. Soc. A*, 2258 (1969).

289. P. Vella and J. Zubieta, *J. Inorg. Nucl. Chem.*, *40*, 613 (1978).

290. Z.-Q. Tian, J. P. Donahue, and R. H. Holm, *Inorg. Chem.*, *34*, 5567 (1995).

291. N. G. Connelly, C. J. Jones, and J. A. McCleverty, *J. Chem. Soc. A*, 712 (1971).

292. K. K. Ganguli, G. O. Carlisle, H. J. Hu, L. J. Theriot, and I. Bernal, *J. Inorg. Nucl. Chem.*, *33*, 3579 (1971).

293. B. J. McCormick and D. S. Rinehart, *J. Inorg. Nucl. Chem.*, *42*, 928 (1980).

294. N. G. Connelly and J. A. McCleverty, *J. Chem. Soc. A*, 1621 (1970).

295. N. D. Lowe and C. D. Garner, *J. Chem. Soc., Dalton Trans.*, 3333 (1993).

296. S. Boyde and C. D. Garner, *J. Chem. Soc., Dalton Trans.*, 713 (1991).

297. S. S. Isied, G. Kuo, and K. N. Raymond, *J. Am. Chem. Soc.*, *98*, 1763 (1976).

298. K. N. Raymond, S. S. Isied, L. D. Brown, F. R. Fronczek, and J. H. Nibert, *J. Am. Chem. Soc.*, *98*, 1767 (1976).

299. S. R. Sofen, D. C. Ware, S. R. Cooper, and K. N. Raymond, *Inorg. Chem.*, *18*, 234 (1979).

300. R. M. Buchanan, J. Clafin, and C. G. Pierpont, *Inorg. Chem.*, *22*, 2552 (1983).

301. C. G. Pierpont and H. H. Downs, *J. Am. Chem. Soc.*, *98*, 4834 (1976).

302. H.-C. Chang and S. Kitagawa, *Angew. Chem. Int. Ed.*, *41*, 130 (2002).

303. H.-C. Chang, H. Myasaka, and S. Kitagawa, *Inorg. Chem.*, *40*, 146 (2001).

304. C. C. Scarborough, S. Sproules, C. J. Doonan, K. S. Hagen, T. Weyhermüller, and K. Wieghardt, *Inorg. Chem.*, *51*, 6969 (2012).

305. C. C. Scarborough, K. M. Lancaster, S. DeBeer, T. Weyhermüller, S. Sproules, and K. Wieghardt, *Inorg. Chem.*, *51*, 3718 (2012).

306. C. C. Scarborough, S. Sproules, T. Weyhermüller, S. DeBeer, and K. Wieghardt, *Inorg. Chem.*, *50*, 12446 (2011).

307. D. Petridis, D. Niarchos, and B. Kanellakopulos, *Inorg. Chem.*, *18*, 505 (1979).

308. D. Niarchos and D. Petridis, *Chem. Phys.*, *41*, 97 (1979).

309. C. W. Allen, R. O. Fields, and E. S. Bretschneider, *J. Inorg. Nucl. Chem.*, *35*, 1951 (1973).

310. G. B. Ferreira, N. M. Comerlato, J. L. Wardell, and E. Hollauer, *Spec. Acta*, *61A*, 2663 (2005).

311. G. B. Ferreira, E. Hollauer, N. M. Comerlato, and J. L. Wardell, *Spec. Acta*, *62A*, 681 (2005).

312. M. K. Johnson, *Prog. Inorg. Chem.*, *52*, 213 (2004).

313. C. W. Schläpfer and K. Nakamoto, *Inorg. Chem.*, *14*, 1338 (1975).

314. T. Petrenko, K. Ray, K. Wieghardt, and F. Neese, *J. Am. Chem. Soc.*, *128*, 4422 (2006).

315. G. H. Spikes, S. Sproules, E. Bill, T. Weyhermüller, and K. Wieghardt, *Inorg. Chem.*, *48*, 10935 (2008).

316. C. G. Pierpont, *Inorg. Chem.*, *40*, 5727 (2001).

317. S. P. Best, S. A. Ciniawsky, and D. G. Humphrey, *J. Chem. Soc., Dalton Trans.*, 2945 (1996).

318. S. P. Best, R. J. H. Clark, R. C. S. McQueen, and J. R. Walton, *Inorg. Chem.*, *27*, 884 (1988).

319. P. S. Santos, *J. Mol. Struct.*, *220*, 137 (1990).

320. R. Sarangi, S. DeBeer George, D. Jackson Rudd, R. K. Szilagyi, X. Ribas, C. Rovira, M. Almeida, K. O. Hodgson, B. Hedman, and E. I. Solomon, *J. Am. Chem. Soc.*, *129*, 2316 (2007).

321. W. H. Mills and R. E. D. Clark, *J. Chem. Soc.*, 175 (1936).

322. S. Sproules, unpublished result.

323. E. I. Stiefel, L. E. Bennett, Z. Dori, T. H. Crawford, C. Simo, and H. B. Gray, *Inorg. Chem.*, *9*, 281 (1970).

324. E. Lyris, D. Argyopoulos, C.-A. Mitsopoulou, D. Katakis, and E. Vrachnou, *J. Photochem. Photobiol. A: Chem.*, *108*, 51 (1997).

325. E. S. Bretschneider, C. W. Allen, and J. H. Waters, *J. Chem. Soc. A*, 500 (1971).

326. R. J. H. Clark and P. C. Turtle, *J. Chem. Soc., Dalton Trans.*, 1714 (1978).

327. H. B. Gray, *Transition Metal Chem.*, *1*, 239 (1965).

328. G. Matsubayashi, K. Takahashi, and K. Tanaka, *J. Chem. Soc., Dalton Trans.*, 967 (1988).

329. J. M. Tunney, A. J. Blake, E. S. Davies, J. McMaster, C. Wilson, and C. D. Garner, *Polyhedron*, *25*, 591 (2006).

330. J. Stach, R. Kirmse, W. Dietzsch, I. N. Marov, and V. K. Belyaeva, *Z. Anorg. Allg. Chem.*, *466*, 36 (1980).

331. U. Fekl, B. Sarkar, W. Kaim, M. Zimmer-Du Iuliis, and N. Nguyen, *Inorg. Chem.*, *50*, 8685 (2011).

332. S. C. Drew, J. Baldas, and J. F. Boas, *Appl. Magn. Reson.*, *40*, 427 (2011).

333. R. E. DeSimone, *J. Am. Chem. Soc.*, *95*, 6238 (1973).

334. J. K. Gardner, N. Pariyadath, J. L. Corbin, and E. I. Stiefel, *Inorg. Chem.*, *17*, 897 (1978).

335. B. A. Goodman and J. B. Raynor, *Adv. Inorg. Chem. Radiochem.*, *13*, 135 (1970).

336. F. E. Mabbs and D. Collison, *Electron Paramagnetic Resonance of d Transition Metal Compounds*, Elsevier, Amsterdam, The Netherlands, 1992.

337. J. L. Shaw, J. Wolowska, D. Collison, J. A. K. Howard, E. J. L. McInnes, J. McMaster, A. J. Blake, C. Wilson, and M. Schröder, *J. Am. Chem. Soc.*, *128*, 13827 (2006).

338. N. G. Connelly, D. J. H. Emslie, P. Klangsinsirikul, and P. H. Rieger, *J. Phys. Chem. A*, *106*, 12214 (2002).

339. D. L. Liczwek, R. L. Belford, J. R. Pilbrow, and J. S. Hyde, *J. Phys. Chem.*, *87*, 2509 (1983).

340. U. Abram, M. Braun, S. Abram, R. Kirmse, and A. Voigt, *J. Chem. Soc., Dalton Trans.*, 231 (1998).

341. U. Abram, A. Hagenbach, A. Voigt, and R. Kirmse, *Z. Anorg. Allg. Chem.*, *627*, 955 (2001).

342. U. Abram, B. Schmidt-Brücken, A. Hagenbach, M. Hecht, R. Kirmse, and A. Voigt, *Z. Anorg. Allg. Chem.*, *629*, 838 (2003).

343. E. Stiefel, Ed., *Prog. Inorg. Chem.*, *52*, 1 (2004).

344. R. K. Szilagyi, B. S. Lim, T. Glaser, R. H. Holm, B. Hedman, K. O. Hodgson, and E. I. Solomon, *J. Am. Chem. Soc.*, *125*, 9158 (2003).

345. K. Ray, S. DeBeer George, E. I. Solomon, K. Wieghardt, and F. Neese, *Chem. Eur. J.*, *13*, 2783 (2007).

346. H. H. Downs, R. M. Buchanan, and C. G. Pierpont, *Inorg. Chem.*, *18*, 1736 (1979).

347. T. Glaser, B. Hedman, K. O. Hodgson, and E. I. Solomon, *Acc. Chem. Res.*, *33*, 859 (2000).

348. B. Hedman, K. O. Hodgson, and E. I. Solomon, *J. Am. Chem. Soc.*, *112*, 1643 (1990).

349. E. I. Solomon, B. Hedman, K. O. Hodgson, A. Dey, and R. K. Szilagyi, *Coord. Chem. Rev.*, *249*, 97 (2005).

350. T. Glaser, K. Rose, S. E. Shadle, B. Hedman, K. O. Hodgson, and E. I. Solomon, *J. Am. Chem. Soc.*, *123*, 442 (2001).

351. J. E. Huyett, S. B. Choudhury, D. M. Eichhorn, P. A. Bryngelson, M. J. Maroney, and B. M. Hoffman, *Inorg. Chem.*, *37*, 1361 (1998).

352. P. Gütlich, E. Bill, and A. X. Trautwein, *Mössbauer Spectroscopy and Transition Metal Chemistry*, Springer, Berlin/Heidelberg, 2011.

353. T. Birchall, N. N. Greenwood, and J. A. McCleverty, *Nature (London)*, *215*, 625 (1967).

354. J. Blomquist, U. Helegeson, B. Folkesson, R. Larsson, and T. Weyhermüller, *Chem. Phys.*, *76*, 71 (1983).

355. D. Petridis and B. W. Fitzsimmons, *Inorg. Chim. Acta*, *11*, 105 (1974).

356. C. W. Allen and D. B. Brown, *Inorg. Chem.*, *13*, 2020 (1974).

357. R. D. Webster, G. A. Heath, and A. M. Bond, *J. Chem. Soc., Dalton Trans.*, 3189 (2001).

358. M. D. Ward and J. A. McCleverty, *J. Chem. Soc., Dalton Trans.*, 275 (2002).

359. C. K. Jørgensen, *Coord. Chem. Rev.*, *1*, 164 (1966).

360. C. K. Jørgensen, *Oxidation Numbers and Oxidation States*, Springer, Heidelberg, Germany, 1969.

How to Find an HNO Needle in a (Bio)-Chemical Haystack

FABIO DOCTOROVICH, DAMIAN E. BIKIEL, JUAN
PELLEGRINO, SEBASTIÁN A. SUÁREZ, AND MARCELO A.
MARTÍ

INQUIMAE-CONICET, DQIAQF, Universidad de Buenos Aires, Argentina

CONTENTS

Progress in Inorganic Chemistry, Volume 58, First Edition. Edited by Kenneth D. Karlin.
© 2014 John Wiley & Sons, Inc. Published 2014 by John Wiley & Sons, Inc.

I. INTRODUCTION

A. Azanone and Its Elusive Nature

Azanone (also called nitroxyl or simply HNO) is a highly reactive compound with very interesting chemical properties, and whose role in several chemical and biochemical reactions has not been completely elucidated to date. Its high reactivity, mainly toward itself, originates its intrinsic elusive nature. Nitroxyl dimerizes at a very fast rate ($k = 8 \times 10^6 \, M^{-1} \, s^{-1}$), which is second order in its concentration. Thus its maximum concentration and lifetime are severely limited. Azanone also reacts fast with its sibling NO ($k = 5.6 \times 10^6 \, M^{-1} \, s^{-1}$), and at a moderate rate ($k = 3 \times 10^3 \, M^{-1} \, s^{-1}$) with oxygen (1). Thus, its reactivity is mostly studied in an anaerobic environment. As in the case of NO, the biological targets, of HNO have been shown to be mainly thiols (1, 2), and metalloproteins, especially heme proteins (3–7). As shown later, these reactions are fast and thus can compete with dimerization, allowing their use as trapping agents.

II. CHEMICAL AND BIOLOGICAL RELEVANCE OF HNO

A. Chemical Relevance of HNO as a Reaction Intermediate

Both ^{1}HNO and ^{3}NO$^-$ are very reactive molecules. As already mentioned, HNO reacts at a fast rate with itself and at a slower one with O_2, while the reaction of ^{3}NO$^-$ with dioxygen is nearly diffusion controlled. Nevertheless, HNO is a thermodynamically stable molecule, so much as that it has been detected in outer space, in interstellar clouds (8). However, it was only recently isolated. The salt LiNO was obtained by deprotonation of N-hydroxybenzenesulfonamide with phenyllithium (9). This lithium salt of azanone exhibits antiferromagnetic paramagnetism, consistent with the triplet ground state of NO$^-$. As expected, it produces N_2O by protonation, due to HNO dimerization to produce $H_2N_2O_2$, which loses a water molecule. It reacts with O_2 to produce peroxynitrite and eventually nitrite. It interacts with HNO/NO$^-$ traps, such as iron tetraphenylporphyrin chloride, Fe(TPP)Cl and potassium tetracyanonickelate, $K_2Ni(CN)_4$, in the same way as other HNO/NO$^-$ donors. It was determined that it is oxidized at about +0.80 V, a value consistent with the reduction potential of NO (< -0.8 V vs NHE) (10).

In previous years, a number of molecules that can release ^{1}HNO or ^{3}NO$^-$ under certain conditions (azanone donors) have been developed. Recently, a review describing azanone donors among other topics related to HNO has been published (1, 11). In order to show a complete panorama of the reactions in which azanone is involved, we will briefly describe most of the known donors with updated references. After that, the focus will be on other reactions in which HNO has been postulated to be an intermediate.

1. HNO Donors

Compounds **1–11** have been described as azanone donors and they are able to release HNO under a variety of conditions. As shown in Scheme 1, they can be classified into the three main groups: hydroxylamine and its derivatives (i.e., R−NHOH compounds), NONOates [i.e., compounds with an N(O)=N(O) moiety], and specifically designed nitroso compounds. The azanone production mechanisms, as briefly described below, are also diverse. Cyanamide (R_2NCN), for example, is considered in the hydroxylamine group since it liberates

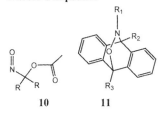

Scheme 1.   Azanone donors, grouped in the three proposed categories. (*a*) Hydroxylamine and its derivates, (*b*) NONOates, and (*c*) nitroso compounds and derivatives.

HNO in the presence of catalases, after reaction with hydrogen peroxide, producing the hydroxylamine derivative HN(OH)CN, which is the actual spontaneous azanone donor. Compounds **2** and **3** are activated by deprotonation, and like hydroxylamine itself disproportionate to produce N_2, N_2O, and NH_3. In these reactions, HNO is postulated to be an intermediate species, which could be associated with enzymes of the nitrogen cycle. Azanone also can be obtained from other hydroxylamine derivatives (e.g., hydroxyurea) through oxidation with periodic acid (12). Compounds holding the general structure **2** are derived from N-hydroxybenzenesulfonamide (with $R_1 = R_2 = R_3 = H$), commonly known as Piloty's acid, which decomposes producing HNO through deprotonation in alkaline environment (pH > 9.3). Recently, a series of derivatives holding the general structure **2** that can release HNO in aqueous media at lower pH values (4.0 and above) were described (13). These derivatives include either electron-withdrawing and/or -donating substituents like methyl, nitro, fluoro, triisopropyl, trifluoromethyl, and methoxy groups. Also recently, Toscano and co-workers (14) developed the barbiturate **5** and the pirazolone derivative **6**. Both of them release HNO at physiological pH with half-lives in the order of ~1 and 10 min, respectively, while at pH 4 they are stable for hours.

Regarding NONOates, HNO is generally released thermally or by adjusting the pH usually in the acidic region (15). For example, the NONOate sodium 1-(isopropylamino)diazene-1-ium-1,2-diolate (IPA/NO) (**8**, with R = H and i-propyl) has been shown to release both HNO and NO in a pH dependent manner, similarly as the commonly azanone donor known as Angeli's salt (AS), or trioxodinitrate (**7**) (16). More recently, a novel NONOate, compound **9**, was obtained through the reaction of IPA/NO with $BrCH_2OAc$ (17). This compound is hydrolyzed to generate HNO by two different mechanisms. In the presence of esterase, dissociation to acetate, formaldehyde, and IPA/NO is proposed to be the dominant path, while in the absence of enzyme, isopropylamine (IPA) is not produced. Nitric oxide is not observed as a product. A mechanism was proposed [Fig. 1(*a*)] in which base-induced removal of the N$-$H proton of **9** leads to acetyl group migration from the oxygen to the neighboring nitrogen atom, followed by cleavage to isopropanediazoate ion and the known acyl nitroso HNO precursor **7**.

Acyloxy nitroso compounds, like compound **10**, liberate HNO by hydrolysis in a pH dependent manner [Fig. 1(*b*)] (18). More recently, water-soluble acyloxy nitroso compounds were synthesized, and it was observed that pig liver esterase catalyzes their decomposition and favors the HNO release (19).

Finally, compounds of the general formula **11** act as photocontrollable HNO donors via retro-Diels–Alder reactions, although they can also release NO depending on the solvent [Fig. 1(*c*)] (20, 21). A review focused on these compounds has been published recently (22).

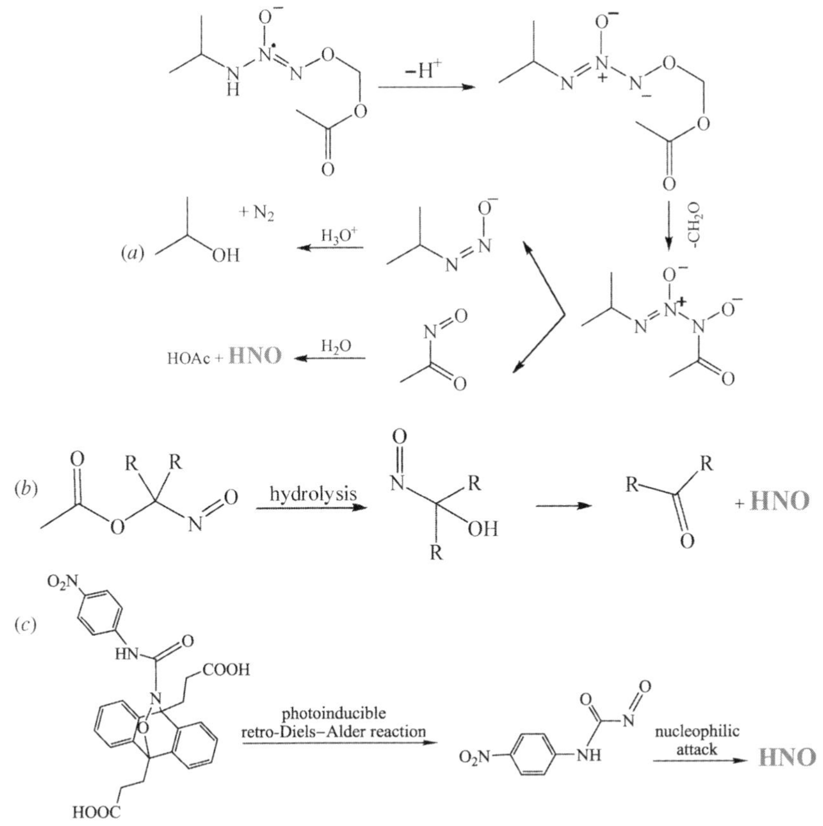

Figure 1. (*a*) Proposed mechanism of hydrolytic HNO generation from **7** (17). (*b*) Hydrolysis of acyloxy nitroso compounds (*c*) HNO release from a photocontrollable donor via a retro-Diels–Alder reaction.

2. Reactions in Which Azanone Has Been Proposed As an Intermediate

Due to its high reactivity, azanone has a very short life and its nitrogen and/or oxygen atoms end up as gaseous products (i.e., N_2O, N_2), a nitrosocompound, an inorganic salt (e.g., nitrite), or an NO/HNO coordination compound. We will try to summarize the reactions in which azanone or its anion has been postulated as an intermediate in chemical reactions, mainly in free form.

a. Hydroxylamine-Related Reactions. Hydroxylamine and its derivatives are undoubtedly the most extended source of HNO. Apart from the hydroxylamine-derived HNO donors described before, hydroxylamine itself disproportionates in alkaline media to produce variable amounts of N_2, N_2O (via the formation of hyponitrite), and NH_3 as final products (23). Surprisingly enough, the decomposition of hydroxylamine has been a matter of controversy that has not been completely solved to date. The decomposition products of hydroxylamine in basic medium are ammonia, nitrogen, nitrous oxide, and hyponitrite. Nast and Foppl (24) suggested that the reaction proceeded via HNO, since evidence for the presence of nitroxyl in the reaction mixture was found when tricyanonitrosonickelate(II) (a bent NO complex) appeared by addition of tetracyanonickelate. They proposed that the primary process is Eq. 1 (shown below), and that the distribution of gaseous products is dependent on pH: Reduction of HNO to N_2 predominates in alkaline solution, but dimerization to N_2O predominates at acidic pH.

$$2NH_2OH \rightleftharpoons NH_3 + HNO + H_2O \qquad (1)$$

However, years later Luňák and Vepřek-Šiška (23) pointed out that a putative intermediate in the formation of tricyanonitrosonickelate is not nitroxyl, but a complex $[Ni(CN)_3NH_2OH]^-$, by replacement of a cyanide ligand in tetracyanonickelate by a molecule of hydroxylamine. The mechanism shown in Eqs. 2–4 was proposed for the formation of tricyanonitrosonickelate(II):

$$[Ni(CN)_4]^{2-} + NH_2OH \rightleftharpoons [Ni(CN)_3NH_2OH]^- + CN^- \qquad (2)$$

$$[Ni(CN)_3NH_2OH]^- + NH_2OH + OH^- \rightarrow [Ni(CN)_3NO]^{2-} + NH_3 + H_2O \quad (3)$$

$$2[Ni(CN)_3NH_2OH]^- + OH^- \rightarrow [Ni(CN)_3NO]^{2-} + [Ni(CN)_3NH_3]^- + H_2O \quad (4)$$

Moreover, in another somewhat neglected work of 1974, Luňák and Vepřek-Šiška (23) showed that the decomposition is catalyzed by metal ions, such as Fe(II), Ni(II), and Cu(II). Anderson (25, 26) studied the Cu(II) catalyzed oxidation of hydroxylamine in the presence of air in the mid-1960s, finding N_2O as a product, and proposing the intermediacy of nitroxyl but not hyponitrite. This idea is contradicting, since HNO dimerizes rapidly to produce hyponitrite, which decomposes to N_2O. As established by Luňák and Vepřek-Šiška (23), the proportion of the products in these metal-catalyzed reactions depends not only on pH and concentration of hydroxylamine, but also on the concentration of the catalyzing cations, with $N_2 : N_2O$ having a value of 0.6 at pH 7. While addition of Fe(II), Co(II), and Cu(II) invariably reduced the ratio, and the use of ultrapure KOH as a base raised it up to 7. The rate equation was found to be third order in hydroxide and first order in hydroxylamine, with $k(20\,^\circ C) = 7 \times 10^{-6}\,M^{-3}\,s^{-1}$. The intermediacy of nitroxyl was objected based on the above results: (1) finding of the complex $[Ni(CN)_3NH_2OH]^-$, (2) metal ion catalysis, and (3) increasing the $N_2 : N_2O$ ratio at increasing pH. The authors proposed that the decomposition intermediates are hydroxylamine complexes with the metal ions. Which decompose to a nitrosyl complex. This complex could react by two different pathways: reaction with hydroxylamine would produce dinitrogen, while dimerization of two nitrosyl ligands would produce hyponitrite and N_2O. In any case, the complexes proposed by Luňák and Vepřek-Šiška (23) are bent NO complexes, or in other words, nitroxyl complexes. Many years later the reaction of a hydroxylamine complex coordinated to a relatively robust $[Fe^{II}(CN)_5]^{3-}$ fragment was studied (27). Spectroscopic and kinetic evidence for the coordination of NH_2OH was found, and N_2, N_2O, NH_3, and $[Fe^{II}(CN)_5(NO)]^{2-}$ (nitroprusside) were found as reaction products. Azanone was proposed as an intermediate, produced by Eqs. 5–7, in a radical-chain process:

$$[Fe^{II}(CN)_5O(H)NH_2]^{2-} + H^+ \rightarrow [Fe^{III}(CN)_5H_2O]^{2-} + NH_2 \qquad (5)$$

$$NH_2 + NH_2OH \rightarrow NH_3 + NHOH \qquad (6)$$

$$NHOH + NH_2OH \rightarrow HNO + H_2O + NH_2 \qquad (7)$$

More recently, it was found that carbon in different allotropic forms (Mogul L, graphite, Black Pearls L) also catalyzes the decomposition of hydroxylamine (28). Dinitrogen, NO, N_2O, and NH_3 were identified as products. Hydrogen atom transfer from one molecule of hydroxylamine to another to produce HNO, NH_3, and H_2O was proposed as a plausible reaction mechanism.

Another reaction involving hydroxylamine, for which nitroxyl has been proposed as an intermediate, is the reduction of NO. Cooper et al. (29) reported equimolar quantities of N_2 and N_2O as products in basic solution. A tracer experiment indicated the appearance of one N atom of NO origin in each of the two product molecules. The authors proposed initial H atom abstraction from hydroxylamine by NO, followed by formation of N-nitrosohydroxylamine, which would decompose to N_2O (Eqs. 8, 9). The formation of N_2 was rationalized as an attack of nitroxyl anion to NH_2OH (Eq. 10). Nitrous oxide (N_2O) is also formed by HNO dimerization (Eq. 11). However, given the relative $N-H$ bond strengths in HNO vs hydroxylamine, this may seem unlikely.

$$NO + NH_2OH \rightarrow HNO \text{ (or NOH)} + NHOH \text{ (or } H_2NO) \qquad (8)$$

$$NHOH + NO \rightarrow ON-NHOH \rightarrow N_2O + H_2O \qquad (9)$$

$$NO^- + NH_2OH \rightarrow N_2 + H_2O + OH^- \qquad (10)$$

$$2\,HNO \rightarrow N_2O + H_2O \qquad (11)$$

A few years later, this mechanism was reanalyzed and confirmed by Bonner et al. (30), by using a combination of tracer and kinetic experiments, and a detailed study of the stoichiometry of the final products. In the same work, Bonner found that the nitroxyl intermediate encountered in trioxodinitrate (AS) decomposition had properties significantly different from those of the intermediate of the NO–NH_2OH reaction. Whereas the first nitroxyl was found to be much more reactive toward NH_2OH than toward itself at pH 8, the second one is consumed almost exclusively by self-reaction at the same pH. This result indicated that the first form is fully deprotonated, but the second one is not. Bonner rationalized this behavior as the presence of a different tautomer of azanone in each case: HNO for Angeli's Salt decomposition and NOH in the NO–NH_2OH reaction (31).

Bonner also studied the reactions of N- and O-substituted hydroxylamines with NO. He proposed that in the case of MeNHOH, NH_2OMe, and MeNHOMe the reaction proceeds by N-bound H atom abstraction to form HNO. In the case of the dialkyl compounds $(Me)_2NOH$ and $(Et)_2NOH$, there is no N bound H atom, and abstraction is postulated to occur at the α-carbon (32).

However, these works involving NO and azanone should be reinvestigated in view of the fact that both HNO and NO^- react with NO at fast rates (33), producing N_2O and nitrite as final products.

In biological systems, hydroxylamine was found to exert a mutagenic effect on deoxyribonucleic acid (DNA) at high concentrations, and an inactivating effect at low concentrations (34). The inactivating effect was ascribed to nitroxyl and hydrogen peroxide (H_2O_2) produced by reaction between NH_2OH and O_2. The combination of hydroxylamine and H_2O_2 produced a much more rapid inactivation than either of the two compounds separately or even the additive effect that would result if the two agents acted separately. Formation of HNO and H_2O_2 was suggested to occur by reaction of hydroxylamine with O_2 (Eq. 12).

$$O_2 + NH_2OH \rightarrow H_2O_2 + HNO \qquad (12)$$

Erlenmeyer et al. (35) found that the peroxidase-like reaction between H_2O_2 and NH_2OH is catalyzed both by Cu^{2+} and by the $Cu^{2+}(2,2'$-bpy) complex where bpy = $2,2'$-bpyridyl. The authors proposed coordination of both H_2O_2 and NH_2OH to Cu^{2+} (Eq. 13), and hydrogen-atom transfer from hydroxylamine to produce HNO.

$$\qquad (13)$$

The catalytic reaction between H_2O_2 and NH_2OH (or N-hydroxy-L-arginine) was more recently studied by Donzelli et al. (36) in the presence of heme proteins, in which the iron metal center acts as the catalytic site. Formation of HNO was evaluated with a selective assay in which the release of HNO was indicated by formation of glutathione sulfinamide, $GS(O)NH_2$, when glutathione, GSH, was added (Eq. 14).

$$HNO + GSH \rightarrow GSNHOH \rightarrow GS(O)NH_2 (detected\ by\ mass\ spectrometry, MS)$$
(14)

Sulfinamide was observed upon oxidation of NH_2OH, whereas N-hydroxy-L-arginine (NOHA), the primary intermediate in the oxidation of L-arginine by NO synthase, was resistant to oxidation by the heme proteins utilized. The highest yields of $GS(O)NH_2$ (and therefore HNO) were observed with proteins in which the heme was coordinated to a histidine (horseradish peroxidase, myoglobin, and other proteins), in contrast to a tyrosine (catalase) or cysteine (cytochrome P450). It was proposed that oxidation of the metal center by H_2O_2 to produce compound I, which in turn would oxidize hydroxylamine to HNO (Scheme 2).

In another work, it was shown that nitroxyl could be produced by nitric oxide synthase (NOS) under certain conditions (37). Nitric oxide synthase is a rather complex enzyme that utilizes multiple redox-active cofactors and substrates to catalyze the five-electron oxidation of its substrate L-arginine to citrulline and NO. Two flavins, a cysteine-coordinated heme cofactor and a rare tetrahydrobiopterin cofactor are used to deliver electrons from reduced nicotinamide adenine dinucleotide phosphate (NADPH) to molecular oxygen. Neuronal NOS free of the cofactor tetrahydrobiopterin (H4B) catalyzes arginine oxidation to NOHA and citrulline in both NADPH and H_2O_2 driven reactions. It was suggested that a ferrous heme–NO complex (formally $Fe^{II}NO$ or $Fe^{III}NO^-$) was built up after initiating catalysis in both NADPH and H_2O_2 reactions, consistent with formation of nitroxyl anion as a product. For the H4B-replete enzyme, an NO releasing $Fe^{III}NO$ heme complex was produced. A model for heme-dependent oxygen activation and NO/NO^- synthesis was proposed (Scheme 3). A heme–peroxo intermediate forms and can react with NOHA in the presence or absence of H4B. If the reductase domain provides the electron to the ferrous–dioxy species (path a),

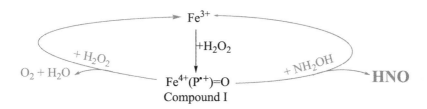

Scheme 2.   Azanone production by catalytic reaction of heme proteins with hydroxylamine and H_2O_2.

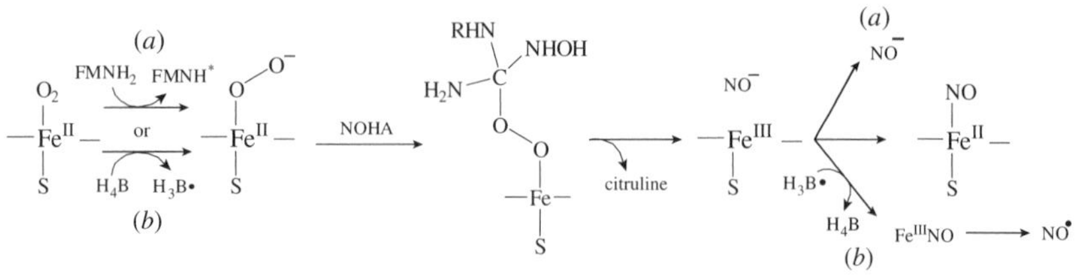

Scheme 3.　Heme-dependent oxygen activation and NO/NO⁻ synthesis from NOS and N-hydroxy-L-arginine (NOHA).

then the reaction generates NO⁻. However, if H4B provides this electron (path b), then the H4B radical can accept an electron and NO is produced.

The previous mechanism was revised 10 years later (38). The peroxide-driven reaction with both Mn and Fe containing heme domain constructs of NOS was used to characterize the formation of HNO as the initial inorganic product formed when O_2 activation occurs without H4B radical formation. Since H_2O_2 can directly form the ferric–peroxo intermediate, generation of a pterin-centered radical does not occur under these conditions, and path (b) shown in Scheme 3 does not occur, leading to the formation of the Fe^{II}–NO complex that either releases HNO or is oxidized by peroxide to generate nitrite and nitrate. In the presence of the pterin-centered radical generated by preturnover of the iron-containing enzyme with L-arginine, NO was the observed product. A modified mechanism was proposed for the NOS oxidation of NOHA in which an Fe^{III}–peroxo complex is formed initially (Scheme 4), instead of the Fe^{II}–peroxo as proposed in Scheme 3. Activation

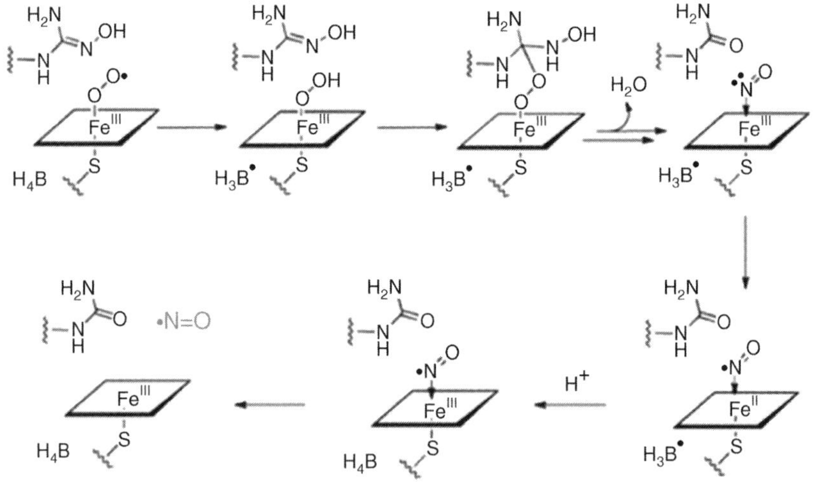

Scheme 4.　Updated mechanism for the NOS mediated oxidation of NOHA.

of O_2 occurs with one electron derived from NADPH and one from H4B, generating a pterin-centered radical $H_3B^•$. Nucleophilic attack of NOHA by this peroxo-intermediate leads to the formation of an addition complex, which decomposes to generate an $Fe^{III}-NO^-$ and citrulline. At this point, note that $Fe^{III}-NO^-$ and $Fe^{II}-NO^•$ are in fact the same species (or resonance Lewis structures). Therefore the release of nitroxyl is probably aided by protonation, producing azanone, which in the free form (noncoordinated) is expected to be protonated at physiological pH, (i.e., as HNO). Since the nitric oxide ligand in $Fe^{II}-NO^•$ is not labile, one could expect that this species could be alive long enough to be protonated. Once Fe(II) is oxidized by $H_3B^•$ to $Fe^{III}-NO^•$ (Scheme 4), nitric oxide becomes labile and is rapidly released as the final product.

In another biologically relevant reaction, Hollocher (39) suggested, based on isotope tracer experiments, that at least in the case of the denitrifying bacteria *Paracoccus denitrificans* and *Pseudomonas stutzeri*, nitroxyl could be an intermediate in the denitrification pathway (39).

In regard to the hydroxylamine derivatives, apart from the HNO donors already mentioned, hydroxylamine-*N*-sulfonate was postulated to decompose in alkaline media (40), producing HNO or NOH, depending on the involved tautomers of the deprotonated reagent, Eq. 15:

$$HONH-SO_3^- + OH^- \rightarrow HNO(or\ HON) + SO_3^{2-} + H^+ \qquad (15)$$

Related to the abovementioned reaction, the reaction of sulfamic acid with nitric acid, which produces N_2O (Eq. 16), was postulated to proceed via an intermediate nitramide (H_2NNO_2) that could decompose to HNO by cleavage of the N–N bond (30).

$$HNO_3 + NH_2SO_3H \rightarrow N_2O + H_2O + H_2SO_4 \qquad (16)$$

b. Nitric Oxide Reduction. In principle, NO reduction by an electron donor or by reaction of NO with H atoms (free or abstracted from a bond), should be the simplest way to get $^3NO^-$ or HNO, chemically or physiologically. As mentioned above, HNO has been detected in interstellar clouds (8). It could be thought that it was produced by reaction of NO with H radicals, and that under high-vacuum conditions prevailing in outer space, the fast reactions of HNO with itself or with NO do not occur. The reduction potential of NO is negative (Eqs. 17 and 18, standard reduction potentials vs NHE) (10, 41), and the −0.14-V value suggested in Eq. 17 drops to even more negative values at physiological pH due to the proton dependency on the left side of the equation.

$$NO + H^+ + e^- \rightarrow {}^1HNO \qquad E^0 \approx -0.14\ V \qquad (17)$$

$$NO + e^- \rightarrow {}^3N\,O^- \qquad E^0 < -0.8\ V \qquad (18)$$

Bonner and Pearsall (42, 43) showed that NO can be reduced to nitroxyl by Fe(II) ($E^{\circ}_{Fe(III)/Fe(II)}$ = 0.771 V), by performing detailed kinetic and isotope tracer experiments. Nitrite is reduced by Fe(II) sequentially to NO and N_2O (42). HNO was demonstrated to be a primary product of the Fe(II) reduction of NO, when $^{15}N^{18}O$ was reduced by Fe(II) in the presence of $HN_2O_3^-$ (natural abundance) producing N_2O whose isotopic composition showed both self-dimerization and dimerization of HNO from two sources. The reactions proposed, which could involve free or bound HNO and NO_2^- ligands, are the initial reduction of nitrite to produce NO (Eq. 19), and subsequent NO reduction, which renders HNO (Eq. 20), followed by HNO dimerization to form the end-product N_2O.

$$NO_2^- + Fe^{2+} + H^+ \rightarrow Fe^{3+} + NO + \frac{1}{2}H_2O \qquad (19)$$

$$NO + Fe^{2+} + H^+ \rightarrow Fe^{3+} + HNO \qquad (20)$$

Kinetic evidence suggested that an iron dinitrosyl complex should be an intermediate in the reduction pathway (43). An octahedral Fe(II) complex was proposed, bearing cis nitrosyls, one linear and the other one bent (formally NO^+ and NO^-, respectively). The NO^+ of one complex could be subject to nucleophilic attack by the oxygen atom of a nitrosyl ligand of another Fe(II) complex in the following process:

$$(21)$$

Nitrite is also reduced to HNO by trimethylamine-borane, as suggested by kinetic evidence and by the formation of N_2O as a final product (44). The stoichiometric reaction is shown in Eq. 22.

$$(Me)_3N \cdot BH_3 + 2HNO_2 + H_3O^+ \rightarrow (Me)_3NH^+ + B(OH)_3 + N_2O + H_2 + H_2O \qquad (22)$$

The proposed mechanism involves protonation of nitrous acid as a first step, followed by rate-limiting hydride attack on H_2ONO^+ or free NO^+ to produce azanone as a reactive intermediate (Eqs. 23, 24). A substrate isotope effect was found, consistent with an activated complex involving B$-$H bond scission through attack of amine-borane on H_2ONO^+ or NO^+

$$HNO_2 + H^+ \leftrightarrow H_2NO_2^+ (fast) \qquad (23)$$

$$H^- [from(Me)_3N^\bullet BH_3] + H_2NO_2^+ \rightarrow HNO + H_2O \qquad (24)$$

In regard to the reaction of NO with H atoms, it was studied by the *in situ* mercury-sensitization of H_2 in the presence of NO, and by photolysis of CH_2O, admixed with NO (45). Dinitrogen (N_2) and N_2O were observed as products of the reaction, and the ratio $[N_2]/[N_2O]$ increased with increasing P_{NO} and decreasing intensity level of radiation. Based on the observed product distribution, quantum yield data, and kinetic behavior of the system, a mechanism based on the initial formation of nitroxyl followed by dimerization or reaction with NO was proposed (Eqs. 25–27).

$$H + NO + M \rightarrow HNO + M \qquad (25)$$

$$2\,HNO + M \rightarrow N_2O + H_2O + M \qquad (26)$$

$$HNO + 2NO \rightarrow N_2 + HNO_3 \qquad (27)$$

Nowadays, it is accepted that the reaction of HNO with NO (Eq. 27) produces N_2O and nitrite as end-products (33).

In another work, the reactions D + NO and H + NO were studied by mercury photosensitization of D_2–NO and H_2–NO mixtures, Nitroxyl (HNO and DNO at $m/z = 31$ and 32, respectively) was observed as a transient intermediate by mass spectrometry (46, 47).

With regard to HNO formation by H-atom abstraction from nitric oxide, we have already described the reaction of hydroxylamine with NO (29, 30).

Bonner and co-workers (48) also suggested H-atom abstraction in the reaction of NO with hiponitrous acid (HON=NOH), based on isotope tracer experiments. The resulting $^\bullet$ONNOH radical would decompose to the observed product N_2O and OH$^\bullet$. The latter species is trapped by ethanol added as a radical-chain inhibitor. Under conditions where the radical chain is operative, a complex mechanism involving reactions of $^\bullet$ONNOH with NO and OH$^\bullet$ were proposed, among others.

An H atom from thiols could in principle be another likely pathway for HNO formation. Anaerobic aqueous solutions of thiophenol (PhSH) exposed to NO results in quantitative formation of diphenylsulfide (PhSSPh) (49). However, hydrogen-atom abstraction was ruled out by the observation of base catalysis. Nucleophilic addition of the thiol anion to NO, followed by protonation and radical coupling was proposed instead (Eqs. 28, 29).

$$RS- + {}^\bullet NO \rightarrow [RS-N-O]^{\bullet-} \rightarrow RS-N-OH^\bullet (-H^+) \qquad (28)$$

$$2RS-N-OH^\bullet \rightarrow RSN(OH)-N(OH)SR \rightarrow RSSR + HON=NOH \qquad (29)$$

c. Miscellaneous Reactions.   Azanone has been obtained by photolysis of nitromethane or methyl nitrite in an argon matrix. I was identified by infrared (IR) spectroscopy (50). The suggested mechanism of formation is shown in Eqs. 30 and 31, nitromethane isomerizes by irradiation to methyl nitrite, which decomposes photochemically to produce HNO and formaldehyde. In regards to the IR spectrum, the observed vibrations were $N-H$ weak stretching, $N=O$ stretching, and HNO bending at 3300, 1570, and $1110\,cm^{-1}$, respectively.

$$MeNO_2 + h\nu \rightarrow MeONO \tag{30}$$

$$cis\text{-}MeONO + h\nu \rightarrow H_2CO + HNO \tag{31}$$

Nitroxyl has been postulated as an intermediate in NH_3 doped low-pressure $H_2/N_2O/Ar$ flames by molecular beam mass spectroscopy (51). Its $\Delta H^{\circ}{}_f$ (298 K) was estimated to be $25.4\,kcal\,mol^{-1}$ by calculations. The reactions proposed for its formation involve N-centered free radicals, as shown in Eqs. 32 and 33.

$$NH_2{}^{\bullet} + O^{\bullet} \rightarrow HNO + H^{\bullet} \tag{32}$$

$$NH + OH^{\bullet} \rightarrow HNO + H^{\bullet} \tag{33}$$

iso-Propoxyl radicals produced by thermal decomposition of diisopropyl peroxide react readily with nitric oxide by cross-disproportionation to form acetone and nitroxyl (Eq. 34) (52).

$$(Me)_2CHO + NO \rightarrow (Me)_2CO + HNO \tag{34}$$

Condensed-phase thermal decomposition of ammonium perchlorate at temperatures of 95 °C and above produces HNO, as observed by high-resolution time-of-flight mass spectrometry (TOFMS) (53). As in the case of the low-pressure flames, HNO is postulated as a product of the reaction of NH_2 free radicals with oxygen, this time in its molecular form (Eq. 35):

$$NH_2{}^{\bullet} + O_2 \rightarrow HNO + OH^{\bullet} \tag{35}$$

The NH_2 radicals, also detected by mass spectrometry, were suggested to be produced from ammonia (formed by decomposition of NH_4ClO_4), by reactions with $Cl^{\bullet}$ and $ClO_3{}^{\bullet}$ radicals formed *in situ*.

The reaction of ascorbate with *S*-nitrosoglutathione produces NO, N_2O, and glutathione as end-products, through the intermediacy of nitroxyl (54). The yield of HNO from the reaction at physiological pH was determined to be 60%, as shown by trapping with the aid of a Mn(III)–porphyrin (6), while NO was produced in 48% yield. The same HNO yield was observed with S-nitrosothiol bound to a

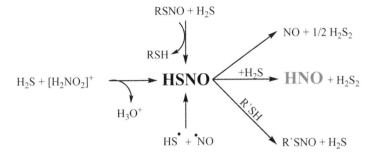

Scheme 5.   Synthetic pathways for HSNO obtention.

protein (i.e., *S*-nitrosopapain). The mechanism proposed for HNO formation involves a transnitrosation reaction between the nitrosothiol and ascorbate (ASC) (Eq. 36), followed by attack of a proton to *O*-nitrosoascorbate to yield dehydroascorbic acid (DHA) and nitroxyl (Eq. 37).

$$RSNO + ASC^- \rightarrow RSH + NOASC^- \qquad (36)$$

$$NOASC^- + H^+ \rightarrow DHA + HNO \qquad (37)$$

The smallest $S-NO$ compound, HSNO, has been obtained and characterized very recently by mass spectrometry, Fourier transform infrared (FTIR), and nitrogen nuclear magnetic resonance (^{15}N NMR) (55). Although it has been described as the "smallest nitrosothiol", formally it is not a thiol derivative since there are no carbon atoms present in the molecule. There are three routes to obtain HSNO, as shown in Scheme 5. (1) Generation of $HS^\bullet$ radicals by flash photolysis in the presence of NO. (2) Thermal reaction of H_2S with nitrous acid, or a transnitrosation reaction between H_2S and nitrosothiols. (3) Formation of HNO from HSNO was shown by reductive nitrosylation of methemoglobin, which produced nitrosylhemoglobin in the presence of both GSNO and H_2S. When human umbilical vein endothelial cells loaded with the fluorescent HNO probe CuBOT1 (56) were incubated with both GSNO and H_2S, increased fluorescence, which indicates the presence of HNO, was observed. These results demonstrate that, at the cellular level, HSNO can be metabolized to afford azanone. Given the abundance of H_2S and RSNOs in biological systems (57), transnitrosation could be in principle a route for HSNO, and therefore HNO endogenous formation.

B.   Azanone Biological Relevance: Friend or Foe?

Azanone biological relevance has two important aspects. The first one concerns the possibility of its endogenous production, either as a reaction intermediate (as proposed for nitrite and nitric oxide reducing enzymes) (58) or as a specific signaling molecule, like its sibling NO; or as an unwanted product of an enzyme (1). For example, one of the most studied *in vivo* relevant azanone sources has been proposed to result from the activity of the usually NO producing enzyme, Nitric oxide synthase (NOS), in the absence of the redox cofactor tetrahydro-biopterin, as discussed above. However, the evidence for this HNO producing activity is indirect and remains open to other interpretations. This mechanism drives the enzyme to HNO instead of NO production and is not completely understood (37, 38, 59, 60). Another well-established enzymatic azanone source relies on the oxidation of hydroxylamine, or other aminoalcohols (e.g., hydroxy-urea) by heme proteins. These proteins are capable of stabilizing oxo ferryl species (Compound **I** and Compound **II**), like peroxidases, catalases, or even myoglobin, as recently shown by several groups (36, 61). The physiological relevance and detailed mechanism of these reactions is, however, completely unknown. Finally, concerning the presence of HNO as a reaction intermediate, it has been proposed that in those heme nitrite reductases that catalyzes the six-electron reduction of nitrite to ammonia, a key step involves the proton coupled electron transfer to a ferrous nitrosyl compound, that thus yields an iron bound HNO (58). This $Fe^{II}-HNO$ adduct, which has also been found to be stable in myoglobin (62) and isolated iron porphyrins (63), must be further reduced to ammonia, and whether there is a possible azanone leak (i.e., release) or not, is still a matter of debate. Thus, although the above mentioned proposals make a strong point for enzymatic, *in-vivo* relevant, HNO production, it should be stressed that none of the proposed mechanisms have been confirmed beyond a reasonable doubt, and that to our knowledge, there has been no report of any *in vivo*, either tissue or cell based, detected azanone production. This lack of certainty is primarily related to the difficulties with the unequivocal and quantitative detection of HNO (1).

The second biological relevance of azanone concerns the study of HNO pharmacological effects and elucidation of the coincidences and differences with the observed NO pharmacological effects. Given the mentioned instability of HNO itself, usually azanone donors are used in physiological studies (1, 64). Reported pharmacological effects by HNO donors are mostly related to the cardiovascular system. Azanone, for example, has been shown to cause vaso-relaxation, and it is readily accepted that the mechanism is through the activation of soluble guanylate cyclase (65). Other studies pointed to the possible positive effects of HNO in preventing heart failure by acting on various potential targets via diverse mechanisms. Meanwhile others also show that it may be a powerful preconditioning agent that helps to alleviate the negative consequences of an ischemic event (66). Finally, HNO donors have been proposed for the treatment of cancer.

To summarize, HNO releasing agents have interesting perspectives as potential therapeutic compounds. However, more work, where HNO detections methods are critical, is needed to understand the chemical mechanisms underlying the observed physiological effects.

III. AZANONE DETECTION METHODS

A. Trapping vs Real-Time Detection

Quantification of HNO is challenging because it is a very reactive species with a high rate for dimerization. Two different methodologies have been explored in order to quantify HNO. The first approach (Scheme 6) is to use a second molecule (R) that reacts faster with HNO than HNO itself, thus actually "trapping" it and leading to a new stable product (P), which is subsequently identified and, if possible, quantified.

The advantage of this kind of approach is that there are several different reagents that can be used to trap HNO specifically, and several techniques can be used to measure [P]. On the other hand, one of the disadvantages of this approach is that most of the times only the total (or a fraction of the total) amount of HNO produced in a given time window, during the trapping reaction course, can be obtained. Moreover, since the agent R is usually in excess in relation to the azanone concentration, it prevents the quantification of HNO during its reactions, since the trapping agent inevitably modifies or significantly interferes with the studied reaction. Most of the ultraviolet–visible (UV–vis) and mass spectrometric techniques rely at some point on this approach.

While a real-time detection is far more complicated, significant progress has been achieved in the past few years through the use of modified electrodes. Recently with the use of a porphyrin-based sensor, it was shown that it is possible to "detect" a very small amount of HNO, yielding an electrochemical signal that is proportional to the azanone concentration in solution (68). Since the amount of HNO "consumed" by the sensor is negligible, by following the signal over time it is possible to obtain the azanone concentration instantaneously, thus providing a

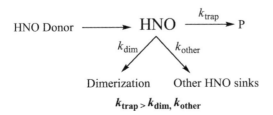

Scheme 6.   Azanone trapping and competition with other reactions.

means to study in real-time reactions, where HNO is involved. In the following sections, we review the recent advances in both types of azanone detection methods.

B. Colorimetric Methods

1. Manganese Porphyrins as Trapping Agents

Manganese porphyrins have good perspectives for the development of an azanone sensor. They are efficient HNO trapping agents displaying the following characteristics that make them a good choice for this difficult task. First, Mn(III) porphyrins are selective for azanone over nitric oxide. Second, a large ($\sim$30 nm) shift of the very intense ($\varepsilon > 1 \times 10^5 M^{-1}$ cm^{-1}) UV–vis Soret band is observed when a Mn(III) porphyrin is converted to the Mn^{II}–NO product (i.e., the nitrosyl complex). Third, the reaction rate for HNO trapping is fast ($\sim$1 $\times 10^5 M^{-1}$ s^{-1}) and thus able to compete with dimerization. However, there are also some disadvantages or inconveniences that must be taken into account when these porphyrins are used to detect HNO.

The first concerns the stability of the resulting nitrosyl complex, since it reacts with O$_2$ to reproduce the starting Mn(III) back, thus preventing accumulation of the product that must be detected. The first way to overcome this inconvenience was presented by Dobmeier et al. (69), who designed a method for quantitative detection of nitroxyl with an estimated dynamic range of 24–290 nM. This method is based on an optical sensor film obtained through encapsulation of Mn^{III}TPPS [TPPS = $meso$-tetrakis(4-sulfonato-phenyl)porphyrinate] within the anaerobic local environment of an aminoalkoxysilane xerogel membrane decorated with trimethoxysilyl-terminated poly(amidoamine-organosilicon) dendrimers, which are poorly O$_2$ permeable. This HNO sensing films were tested with the HNO donors AS and IPA/NO, and were found to provide a rapid means for determining HNO concentrations in aerobic solution (69). However, the rapid dimerization of HNO and relatively slow rate of HNO complexation in the xerogel film limit the sensor performance to environments with restricted HNO scavenging conditions. Another strategy used to avoid oxidation of the nitrosyl complex back to Mn(III) was to protect the metalloporphyrin with a protein matrix that slows down the oxidation rate (7). To accomplish this Bari and co-workers (7) recombined Mn^{III}protoporphyrinate IX into apomyoglobin and showed that it retains the desired selectivity, spectroscopic, and kinetic properties, but oxidation of the nitrosyl is significantly slower.

2. Miscellaneous Colorimetric Methods

To our knowledge, there is only one other colorimetric method developed for azanone trapping and detection. It is based on the reaction of HNO with nitrosobenzene (NB) to yield the cupferron (N-nitrosophenylhydroxylamine) (Eq. 35), which can be readily precipitated by a number of metal cations (70). The precipitates were characterized by comparison of their UV–vis spectra with those of authentic cupferron–metal complexes, and identification of their retention times by high-performance liquid chromatography (HPLC). However, this method is limited to organic solvents.

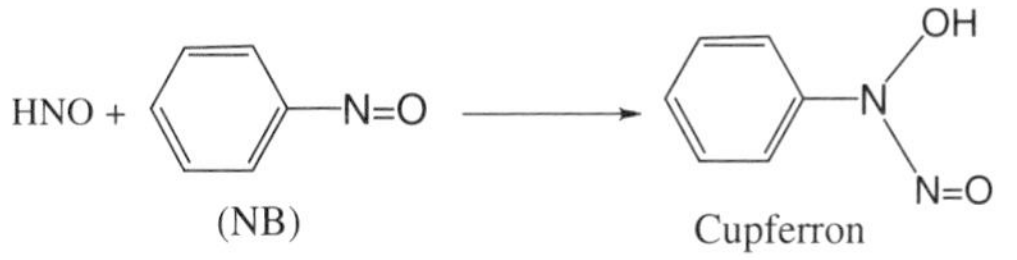

$$\text{HNO} + \underset{\text{(NB)}}{\text{C}_6\text{H}_5\text{—N=O}} \longrightarrow \underset{\text{Cupferron}}{\text{C}_6\text{H}_5\text{—N}\begin{smallmatrix}\text{OH}\\\text{N=O}\end{smallmatrix}} \tag{38}$$

As a final consideration for colorimetric methods, note that their advantages can be summarized as follows: they are fast, have fairly low detection limits, and are quantitative, provided that trapping by the reactant is more efficient than other HNO sinks in the reaction media. They have the main disadvantage that the UV–vis measurements are done in a wavelength range where biological materials (and many interesting chemical reactions) strongly absorb, thus preventing their use in most biological studies.

C. Thiol Blocking

Azanone's high reactivity toward thiols has been known for more than two decades (71, 72). Pioneering studies by Doyle et al. (71) showed that HNO could convert thiophenol to phenyl disulfide, most likely through an N-hydroxysulfena-mide intermediate. In addition Wong et al. (73) suggested that at low-thiol levels the N-hydroxysulfenamide could isomerize to sulfinamide, (see Scheme 7). Similar reactivity was also proposed for other thiols, such as glutathione (GSH) (72), N-acetyl-L-cysteine, and dithiothreitol (74). The reports of a stable sulfinamide product for the thiol reaction with azanone, which is different from the usual nitrosothiol species that results from its aerobic reaction with nitric oxide, prompted

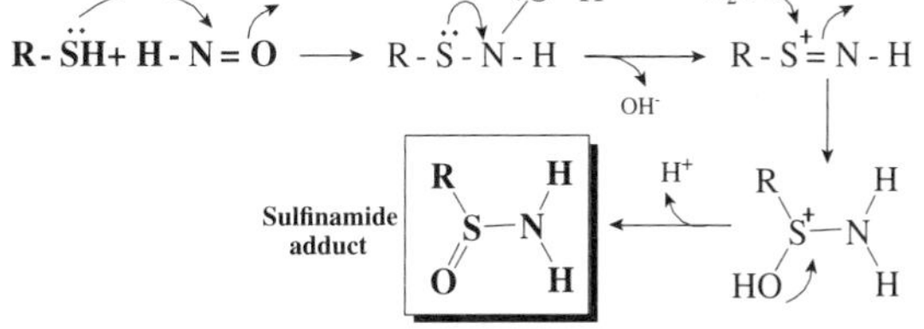

Scheme 7.   Azanone trapping and detection using thiophenol.

Wink and co-workers (36, 74) to use this reaction as a selective method for HNO detection.

Using GSH as the reactive thiol, they showed that an equimolar reaction of GSH with the common azanone donor AS, resulted in the presence of two major products, corresponding to nitrite from AS decomposition, and the glutathione sulfinamide, $GS(O)NH_2$, as identified by HPLC. The corresponding peak areas, which are proportional to the concentration of $GS(O)NH_2$, also were shown to depend significantly on the GSH/AS ratio due to competitive consumption of the hydroxylamine intermediate by a second molecule of thiol. The authors also analyzed the product distribution of the GSH reaction with NO, and undoubtedly showed that $GS(O)NH_2$ is not produced in this case. Thus, identification of the sulfinamide would provide a selective indication of the presence of HNO in a particular environment (74).

In a subsequent work (36) the above described technique was used to determine the azanone producing capacity of several heme proteins, including myoglobin (Mb) and horseradish peroxidase (HRP), due to the catalytic oxidation of hydroxylamine with hydrogen peroxide (Scheme 2).

In summary, the thiol reaction provides a reliable tool to identify, and possibly quantify, the presence of HNO in a given reaction media. However, more work needs to be done to address the method detection limit and possible interfering reactions with the RNOS besides NO. Finally, note that the present method does not detect HNO *in situ* since its presence is only revealed after purification and HPLC analysis.

D. Phosphines

Another recently developed trapping agent is based on the selectivity of phosphines for HNO over other physiologically relevant nitrogen oxides. The reaction of azanone with triarylphosphine nucleophiles is suggested first to provide equimolar amounts of phosphine oxide and the corresponding aza-ylide (Scheme 8). Second, in the presence of a properly situated electrophilic ester, the aza-ylide should undergo a Staudinger ligation to yield an amide where the nitrogen atom is derived from the HNO. These products differ from the known reaction products of triphenylphosphine with nitric oxide, which yields triphenylphosphine oxide and nitrous oxide (75, 76).

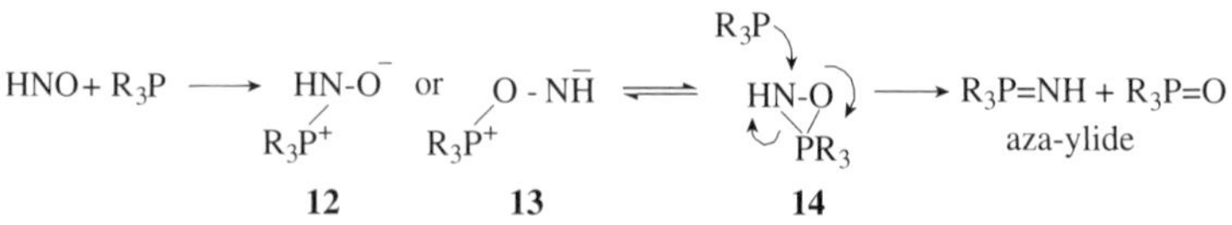

$$HNO + R_3P \longrightarrow HN\text{-}O^- \text{ or } O\text{-}N\overline{H} \rightleftharpoons HN\text{-}O \longrightarrow R_3P=NH + R_3P=O$$

12   13   14   aza-ylide

Scheme 8. Proposed mechanism for the reaction of azanone with triarylphosphines.

Reaction of the phosphine with HNO could yield a product either through P addition at the N (12) or the O atom (13), in Scheme 8. Each of these initial addition products could exist as a three-membered ring species (14) in Scheme 8. Addition of a second phosphine to 14 (or to 12 or 13) would give the corresponding aza-ylide and phosphine oxide in equal proportions.

Evidence for the proposed mechanism comes first from ^{31}P NMR experiments that reveal the reaction of Angeli's salt with the water-soluble phosphine tris(4,6-dimethyl-3-sulfonatophenyl)phosphine trisodium salt hydrate (TXPTS) at pH 7.6 produces two new phosphorous-containing products in a 1:1 ratio. The same experiment also indicates the presence of aza-ylide along with increasing amounts of TXPTS phosphine oxide in this reaction mixture, even after 6 days. Most importantly, TXPTS quenches nitrous oxide formation (>90%) during the aqueous decomposition of Angeli's salt indicating reactivity with HNO. Kinetic studies using TXPTS show that its reaction rate with HNO is comparable to that of glutathione, suggesting the utility of TXPTS as a trapping agent.

In another situation, treatment of the methyl ester derivates (15, Scheme 9) with AS in a mixture of acetonitrile/Tris buffer at pH 7.0 gives a 1:1 mixture of the benzamide phosphine oxide (16) and methyl ester phosphine oxide (17), Reaction of 15 with ^{15}N-12 produces ^{15}N-15 clearly demonstrating that the amide nitrogen atom derives from 12 and presumably from HNO. These results suggest the initial formation of aza-ylide (18, Scheme 9) by the reaction of HNO with 15 through a similar mechanism as the one shown in Scheme 8. Intramolecular reaction of the aza-ylide group of 18 with the adjacent ester group generates the amide (16, Scheme 9).

The utility of phosphines for HNO trapping was also demonstrated by the ability of TXPTS to trap the enzymatically generated HNO (using the heme-mediated hydroxylamine oxidation reaction already described), illustrating both, the ability of phosphines to trap azanone under various aqueous oxidative

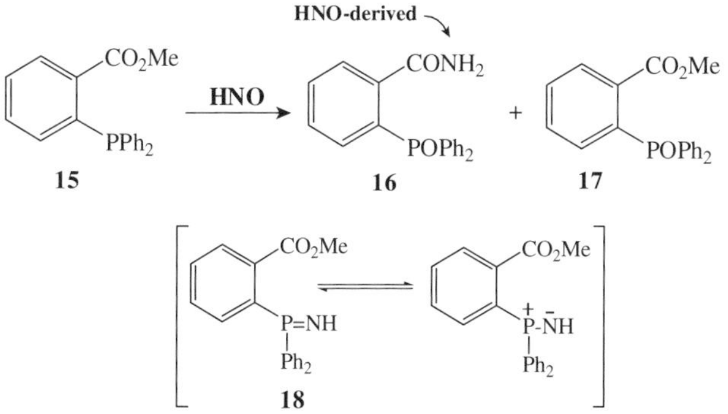

Scheme 9.   Formation of benzamide phosphine oxide (16) and methyl ester phosphine oxide (17) by reaction of the methyl ester (15) with HNO.

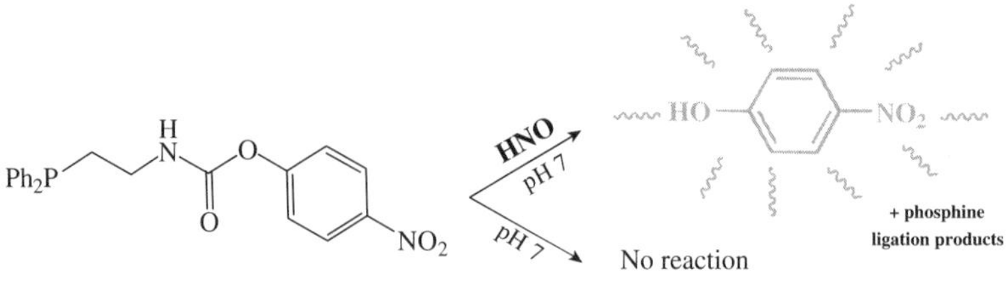

Scheme 10.   Colorimetric essay for HNO detection with derivatized phosphines.

conditions, and the stability of the resulting aza-ylide product. Moreover, with appropriately positioned intramolecular electrophiles, HNO was readily ligated to form thermodynamically stable amides and ureas, which were proposed for their use *in vivo* azanone detection. In this line of research, a biologically compatible colorimetric-based azanone detection system was developed, using a phosphine biotin derivative (Scheme 10). The trapping–signaling system is based on the design of two phosphines, equipped with carbamates as the electrophiles for the ligation reaction, and the incorporation of phenolate and *p*-nitrophenolate leaving groups, which enhance the electrophilicity of the carbonyl group, and yield a bright yellow color when released providing the basis for the colorimetric detection method. The method was tested by monitoring the increase in absorption at 400 nm in MeCN and acetate buffer (pH 7). The results show that in the absence of azanone donors the trapping agent slowly hydrolyzes over several minutes yielding a small amount of the colored species. However, addition of AS immediately yielded a bright yellow solution with a sharp increase in absorption at 400 nm indicating the complete release of *p*-nitrophenolate, through the proposed reactions. Although the described method provides a rapid colorimetric method for qualitatively indicating the presence of HNO (at the micromolar level), the complications, which result from the reactivity of the *p*-nitrophenol carbamate (i.e., its spontaneous hydrolysis), may limit its applications (61).

E.   Electron Paramagnetic Resonance

Xia et al. (77) examined the trapping specificity of different redox forms of Fe$-$MGD (*N*-methyl-D-glucamine dithiocarbamate iron). According to these authors, with Fe$^{II}-$MGD, NO generates characteristic triplet NO$^{\bullet}-$Fe$^{II}-$MGD signals, whereas HNO from AS is EPR silent. Both NO and NO^{-} give rise to NO$^{\bullet}-$Fe$^{II}-$MGD signals when Fe$^{III}-$MGD is used. The authors asseverate that "spin trapping with Fe$-$MGD can distinguish NO and NO^{-} and this depends on the redox status of the iron center". However, an electron paramagnetic resonance (EPR) signal would be observed if only NO is present, as well as if both NO and HNO are coexisting in the solution. Therefore, this method cannot discriminate

HNO from NO. In another publication, Komarov et al. (78) showed that in the presence of dioxygen Fe^{II}–MGD reacts with AS or AS derived NO^- to yield $NO^{\bullet}$–Fe^{II}–MGD, possibly by reaction of NO^- with Fe^{III}–MGD formed by aerobic oxidation of the iron center. Since the paramagnetic $NO^{\bullet}$–Fe^{II}–MGD complex is produced by AS (and/or NO^-), and also by NO, dithiocarbamate iron traps do not distinguish between NO and NO^-.

More recently, the nitronyl nitroxides 2-phenyl-4,4,5,5-tetramethylimidazoline-1-oxyl-3-oxide (PTIO) and its water soluble analogue 2-(4-carboxyphenyl)-4,4,5,5-tetramethylimidazoline-1-oxyl-3-oxide (C-PTIO) were investigated as NO/HNO discriminating agents (79, 80). Nitric oxide reacts with nitronyl nitroxides giving the corresponding iminonitroxidos plus NO_2. The reaction with HNO was studied in aerated aqueous solutions at pH 7 using AS as the HNO donor. From these experiments it was concluded that HNO also reacts with nitronyl nitroxides yielding the corresponding iminonitroxides (PTI or C–PTI) and iminohidroxylamines (PTI-H or C-PTI-H) as the ultimate products, in a relative proportion that depends on the ratio $[AS]_0/[\text{Nitronyl nitroxide}]_0$. The mechanism of the reaction of HNO with PTIO and C–PTIO involves the participation of several reactive species, including NO, NO_2, and oxoammonium cations ($PTIO^+$ and PTI^+ or $C–PTIO^+$ and $C-PTI^+$). Through a complete kinetic analysis from competitive experiments with $MbFe^{III}$ and stopped-flow techniques, the following reactions were proposed as the most important to interpret the results:

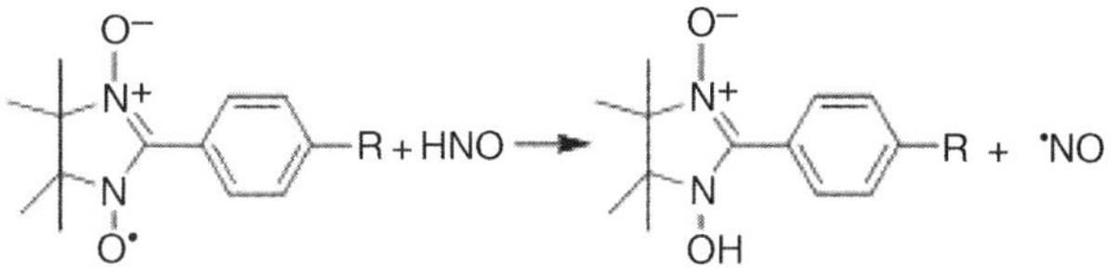

(39)

PTIO or C–PTIO   PTIO–H or C–PTIO–H

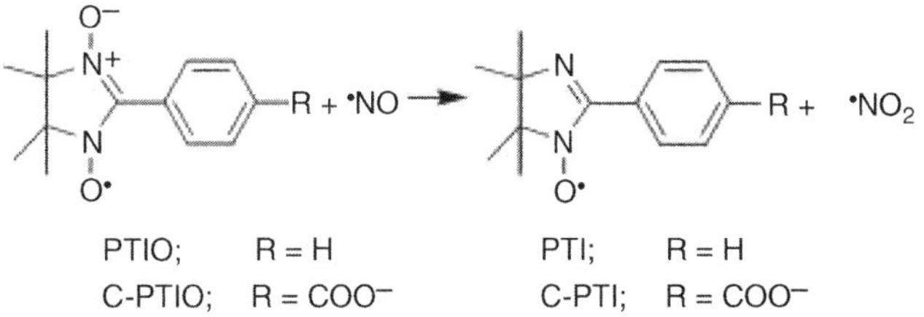

(40)

PTIO;   R = H   PTI;   R = H
C-PTIO; R = COO$^-$ C-PTI; R = COO$^-$

(41)

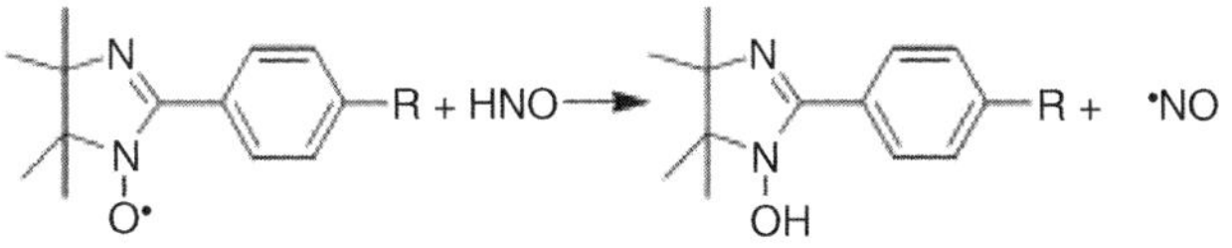

PTI or C–PTI   PTI–H or C–PTI–H

$$PTIO \text{ or } C-PTIO \qquad\qquad PTIO^+ \text{ or } C-PTIO^+ \tag{42}$$

$$PTI \text{ or } C-PTI \qquad\qquad PTI^+ \text{ or } C-PTI^+ \tag{43}$$

$$(C-)PTIO-H + (C-)PTIO^+ \rightarrow 2(C-)PTIO + H^+ \tag{44}$$

$$(C-)PTIO-H + (C-)PTI^+ \rightarrow (C-)PTIO + (C-)PTI + H^+ \tag{45}$$

The combination of reactions 39, 40, 42, 43 and 45 accounts for the observation of the iminonitroxides as the final products, but if there is excess of HNO, it will react further with the iminonitroxides (PTI or C−PTI) (Eq. 41) yielding the corresponding iminohidroxilamines (PTI−H or C−PTI−H). Since the reaction of NO with nitronyl nitroxide yields only the corresponding imino nitroxide, nitronyl nitroxide can discriminate NO from HNO only when present at a concentration much lower than the total production of HNO. Apart from this, the nitronyl nitroxides have not been used in *in vivo* experiments, where more complications can arise, as, for example, the competitive reduction by thiols, which taking into account the reduction potential of the nitronyl nitroxides, would be thermodynamically favorable.

F.   Mass Spectrometry

Using membrane-inlet mass spectrometry (MIMS), Toscano and co-workers (81) were able to detect HNO in solution. By means of different NO and HNO donors, the authors followed the intensity of the m/z 30 and 44 peaks over time (corresponding to NO^+ and N_2O^+) in donor's solutions. Due to the fact that the EI mass spectrum of HNO presents peaks at m/z 30 and 31 with relative intensities of 35:1 (47) and that the spectrum due to N_2O also presents significant amounts of the m/z 30 signal, the assignment of the origin of m/z 30 had to be analyzed using different trapping agents. Exploiting the differential reactivity of HNO and NO toward thiols and phosphines in aerobic and anaerobic conditions, the authors showed that AS and PA are not HNO pure donors at physiological conditions while the 2-bromo derivative of Piloty's acid (PA) is fundamentally a pure HNO donor.

Following this approach, this technique was used to study the formation of HNO by the reaction of *N*-hydroxyl-l-arginine with HOCl (82). The authors also found that significant amounts of HNO are produced in a similar fashion by using hydroxylamine and hydroxyurea.

While this methodology has the potential ability to detect HNO directly from solution, at this moment it allows the indirect detection of HNO by quantification of the relative products of HNO (N_2O^+ due to dimerization and NO^+ due to instrumental ionization). Further developments should be encouraged.

G. Fluorescence-Based Methods

When one wants to construct a fluorescent probe for selective nitroxyl detection under physiologically relevant conditions, it is necessary to take into account that the system should show selectivity over other reactive nitrogen species (RNS) and downstream NO oxidation products, compatibility with living biological samples, water solubility, and membrane permeability. Additionally, a signaling moiety with relatively long-wavelength absorption and emission properties is desirable to avoid unintended cellular damage by high-energy radiation and to minimize innate biological autofluorescence. On the other hand, although the fluorescent probe works well *in vitro*, the ability to image the analyte in biological systems will depend on the emission turn-on ratio and the detection limit.

Most fluorescent probes for HNO reported so far are based on the same principle: Cu(II) coordination quenches the fluorescence of the probe (off-state), but in the presence of HNO the copper is reduced to copper(I), which results in the fluorescence restoration (on-state), being the fluorescence intensity proportional to the HNO concentration. The reduction of Cu(II) by HNO has a precedent in its reaction with $SODCu^{II}$ (SOD = superoxide dismutase) to generate NO and reduced $SODCu^{I}$.

Lippard and co-workers (83) developed two compounds for fluorescent detection of nitroxyl, based on the mechanism mentioned above. The first one is a bithiophene-substituted poly(*p*-phenylene ethynylene) derivative (CP1), which forms a Cu(II) complex that becomes fluorescent upon exposure to excess NO in unbuffered solutions, but under simulated biological conditions (pH 7.4 buffer) it shows a small decrease in emission upon treatment with NO, probably as a consequence of precipitation of the probe (Scheme 11). A twofold increase in $CP1-Cu^{II}$ integrated emission occurs upon exposure to AS, indicating that $CP1-Cu^{II}$ can sense HNO selectively over NO in buffered aqueous solutions, taking into account its apparent insensitivity to NO.

The second probe, boron–dipyrromethene–triazole (BOT1), comprises a BODIPY (boron dipyrromethene) reporter site, which has optical properties that are well suited for cellular imaging experiments and a tripodal coordination environment provided by a *N*-(triazolylmethyl)-*N*,*N*-dipicolyl framework, both separated by a triazole bridge as a rigid spacer (56). This design minimizes the

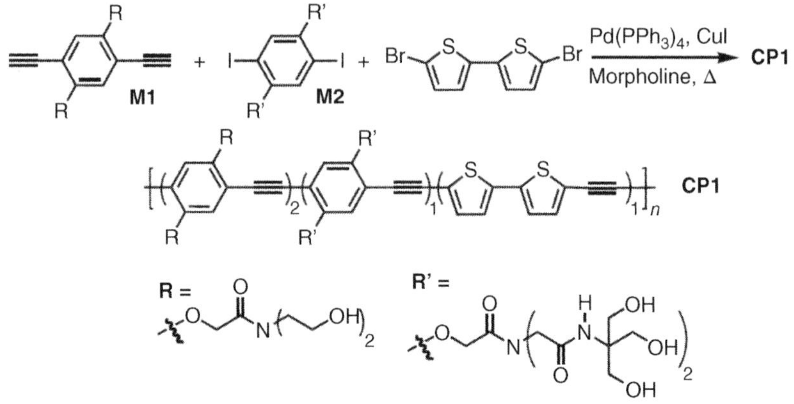

Scheme 11.   Synthesis of CP1.

distance between fluorophore and a metal-binding site, thereby assuring strong fluorescence quenching in the probe off-state.

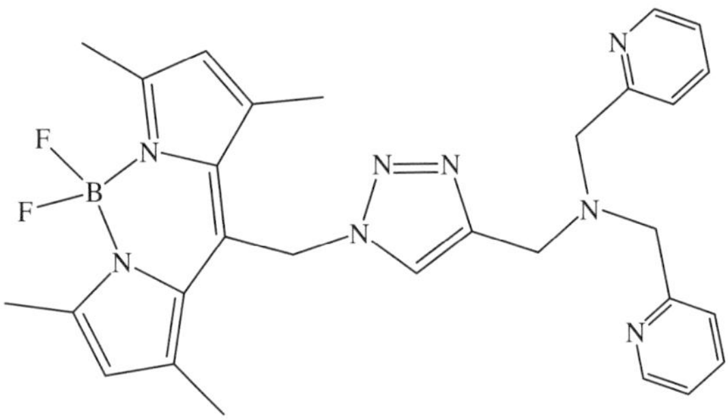

Structure of BOT1.

The photophysical properties of BOT1 were assessed under simulated physiological conditions. This probe exhibits fluorescence properties typical of a BODIPY chromophore; excitation in the maximum absorption region (518 nm) produces an emission profile with a maximum at 526 nm and $\Phi_{fl} = 0.12$. The addition of 1 equiv of CuCl$_2$ to a solution of BOT1 produces a 12-fold decrease in the fluorescence intensity and a 30-fold reduced lifetime, as a consequence of strong quenching by the paramagnetic Cu(II) ion that strongly coordinates to BOT1 (with an apparent dissociation constant $K_d = 3.0 \pm 0.1\ \mu M$). Treatment of [Cu^{II}(BOT1)Cl]Cl with 1000 equiv of cysteine that has a reduction potential lower than that of [Cu^{II}(BOT1)Cl]Cl produces the complete restoration of the fluorescence response of the free ligand BOT1.

The addition of a 1000-fold excess of AS to a solution of $[Cu^{II}(BOT1)Cl]Cl$ (pH 7.0) produced an instantaneous 10-fold fluorescent enhancement; however, progressive quenching is observed with time. Strong fluorescent enhancement also occurs when only 100 equiv of AS were added. However, addition of lower amounts results in only a weak fluorescence response, suggesting that the probe is best suited for sensing HNO in the 0.5–5-mM range, when the sensor is present at micromolar concentration levels. As mentioned aboved, the fluorescence turn-on is due to the formation of diamagnetic $[Cu^{I}(BOT1)Cl]$ by reduction. Additional evidence for this hypothesis is given by EPR and electrospray ionization–mass spectrometry (ESI–MS) and by detection of NO gas, that would be formed upon one-electron HNO oxidation.

Importantly, NO_2^- and other RNS and ROS (e.g., NO, NO_3^-, ONOO$^-$, H_2O_2, and OCl$^-$) failed to induce significant emission enhancement of the $[Cu^{II}(BOT1)Cl]Cl$ complex. The complex was also tested in HeLa cells. Incubation of HeLa cells with 3-μM $[Cu^{II}(BOT1)Cl]Cl$ for 1 h resulted in weak fluorescence that could be observed in both the green and red channels. Global citoplasmatic fluorescence was an indication that the probe is readily taken up by the cells. Then, addition of AS to these cells increased the observed intracellular fluorescence in the green (1.9-fold) and red channels (1.5-fold), consistent with an HNO induced emission response. The intensity increased by another 10% upon an incubation period of 5 min. Further incubation resulted in quenching of the fluorescence intensity. While the results *in vivo* are qualitatively consistent with the fluorescence behavior *in vitro*, the fluorescence enhancement observed in the cells was significantly lower. This lost of sensitivity can be attributed to sequestration of HNO by reactive biomolecules, (e.g., thiols) and nonspecific reduction of $[Cu^{II}(BOT1)Cl]Cl$ by intracellular reduction agents, which would produce a strong background fluorescence in the off-state. Cellular distribution of the probe was also investigated: $[Cu^{II}(BOT1)Cl]Cl$ did not penetrate the nuclear membrane and did not accumulate in mitochondria, but was found to localize in both Golgi and endoplasmatic reticulum.

To finish the description of the $[Cu^{II}(BOT1)Cl]Cl$ probe is necessary to mention that although it has been demonstrated that this compound is suitable to devise a practical protocol for fluorescence imaging of HNO in live cells, the authors recognized some weak points, emphasizing the need for improvements in the future. First, as already mentioned, thiols also reduce $[Cu^{II}(BOT1)Cl]Cl$, interfering with the selectivity of the probe for HNO. The measurement of the reduction potential of $[Cu^{II}(BOT1)Cl]Cl$ confirmed its thermodynamically favorable reaction with both azanone and thiols. So, the authors suggest that one of the interesting modifications in future probes would be to alter the Cu^{II}/Cu^{I} reduction potential so that it lies between those of azanone and thiols, making the reaction with thiols unfavorable. Another point to improve, also recognized by the authors, is the need for large excesses of HNO to induce turn-on and the gradual decrease in emission over time. This result is attributed to the bimolecular decomposition of HNO to N_2O and H_2O, with a rate constant of $\sim 10^6 \, M^{-1} \, s^{-1}$, reaction kinetically competitive with the reduction of $[Cu^{II}(BOT1)Cl]Cl$. Therefore it seems desirable to investigate how to improve the kinetics of Cu(II) reduction by HNO in future generations of this sensor family to enable more rapid detection of HNO. Finally, the authors also gave an explanation for the decrease in emission intensity over time after the initial turn-on. They attribute this to back-oxidation of Cu(I) to Cu(II) by the reactive hyponitrite radical that is known to be formed by reaction of HNO with NO [the last one is formed by Cu(II) oxidation of HNO]. The back-oxidation of the probe to the Cu(II) oxidation state was confirmed by EPR measurements of a solution of $[Cu^{II}(BOT1)Cl]Cl$ treated with 1000 equiv of AS, after overnight storage under anaerobic conditions, obtaining an identical spectrum to that of $[Cu^{II}(BOT1)Cl]Cl$.

Another fluorescent probe for HNO has been constructed by Yao and co-workers (84), following the design strategy of Lippard's $[Cu^{II}(BOT1)Cl]Cl$ sensor. Coumarin and its derivatives are well known as fluorescence-labeling reagents for their excellent photophysical properties of high-fluorescente quantum yield and efficient membrane permeability. Taking that into account, they decided to replace the BODIPY moiety in the Lippard sensor by such a derivative. Therefore, they designed a probe for nitroxyl including a coumarin chromophore and a tripodal dipicolylamine receptor, which was attached via a triazole bridge. The receptor provides a rigid Cu(II) binding-site spacer between the coumarin fluorophore and the chelating ligand (84).

Structure of COT1. (See color version of this figure in the color plates section.)

The emission profile of COT1 showed typical coumarin green fluorescence at 499 nm, with a high quantum yield ($\Phi f = 0.63$) compared to that of the free coumarin moeity ($\Phi f = 0.03$), suggesting that cycloaddition leads to an increase in the electron-donating ability in the emission of coumarin. When 1 equiv of Cu^{2+} was added to COT1 in aqueous solution, a dramatic fluorescence quenching was observed (23.6-fold). The association constant of COT1 with Cu^{2+} was determined to be $7.9 \times 10^5 \, M^{-1}$ on the basis of the fluorescence titration experiments. The fluorescence intensity increased steadily with increasing $Na_2N_2O_3$ concentration until it reached a plateau at 20-μM $Na_2N_2O_3$, which corresponded to a 17.2-fold increase in fluoresence instensity compared to the blank HNO concentration. This finding indicates that complete reduction of Cu^{II}–COT1 occurred with 20-μM $Na_2N_2O_3$. The dramatic change at 499 nm in the fluorescence intensity was much higher than the one obtained with the early reported BODIPY based sensor. The reduction of Cu^{II}–COT1 to Cu(I) was supported by EPR measurements, which showed the loss of the typical signal due to the paramagnetic Cu(II) center, and by ESI–MS, obtaining a peak due to $[COT1 + Cu^{I}]^+$ after the HNO addition, and a peak corresponding to $[COT1 + Cu^{II} + Cl]^+$ in the absence of HNO. In addition, submillimolar cysteine and sodium ascorbate could also be used to restore the typical coumarin green fluorescence because of the reduction of chelated Cu^{II}–coumarin.

On the other side, as with the previously reported $[Cu^{II}(BOT1)Cl]Cl$ probe, a much weaker response was observed with other biologically relevant ROS and RNS species, including NO_3^-, ClO^-, H_2O_2, $ONOO^-$, $ROO^{\bullet}$, and NO_2^-. With NO, a 3.2-fold increase in fluorescence intensity was observed, and this relative lack of induced fluorescence response could be used to discriminate between NO and HNO. By performing competitive experiments with $Na_2N_2O_3$ and various ROS and RNS it could also be demonstrated that the probe displays high selectivity toward HNO, since the fluorescence intensity obtained in these experiments was almost the same as the one measured in the absence of the ROS or RNS.

The probe was then applied to fluorescence imaging of HNO in living cells, using human malignant melanoma A375 cells. The incubation with the probe alone showed very faint intracellular fluorescence, while a cytotoxicity test showed 82.1% cell viability. Then, the treated cells were incubated with $Na_2N_2O_3$, observing that the fluorescence signal produced by these cells increased over time; the major fluorescence intensity in the cells was localized in the perinuclear space, which suggests that the nitrogen species did not reach the nucleolus.

In summary, a coumarin conjugate probe was developed that acted as a dual-response probe to HNO for both fluorescence and EPR detection. Compared to the Lippard BODIPY probe, it still shows the same weak points mentioned before, though it shows a higher sensitivity.

More recently, Cline and Toscano (85) reported another probe based on the restoration of fluorescence upon reaction with HNO because of one-electron reduction of the probe, giving a diamagnetic species plus NO. Although it is based on the same principle as the previous examples, instead of having a CuII complex as the redox active part of the probe, it contains a nitroxide moiety. Based on the known reactivity of HNO with nitroxide radicals (e.g., TEMPOL), see Scheme 12 (86), the authors tested the ability of a probe formed combining a TEMPO derivative with a fluorescent moiety (TEMPO-9-AC), as shown in Scheme 12 (85).

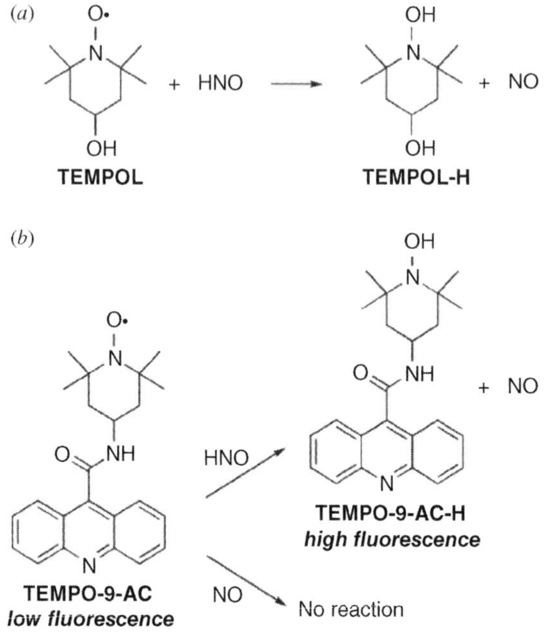

Scheme 12.   Reactions of (*a*) TEMPOL and (*b*) the fluorescent compound TEMPO-9-AC with HNO.

Though the probe was able to detect HNO with good sensitivity and selectivity, the main dawnback is the competitive reaction of HNO with the NO generated upon reduction of the probe, a reaction that occurs at a rate constant of almost two orders of magnitude higher than that estimated rate for the reaction of HNO with TEMPOL. On the other hand, the use of higher concentrations of TEMPO-9-AC to increment the rate of its reaction with HNO is problematic because of inter-molecular fluorescence quenching at concentrations $>60\,\mu M$.

Despite the above comments, TEMPO-9-AC can be used to study HNO chemistry in aqueous solution. This finding was demonstrated by the ability of the probe to work at different pH values (using different HNO donors) and a by a good estimation of the rate constant for the reaction of HNO with thiols, from the fluorescent intensity decrease in a competitive experiment with TEMPO-9-AC.

H.   Electrochemical Real-Time Detection

In situ electrochemical detection in real time is a commonly used analytical technique that allows detection and quantification of many elusive chemical species, including nitric oxide (87–91). As for the other methods mentioned, key aspects for azanone detection are selectivity (mainly against other nitrogen

TABLE I
Reactions, Rate Laws, and Rate Constants That Were Used in the Simulations for HNO Production[a]

Reaction Number	Reaction	Constant Value	
R1	Angeli's salt $\rightarrow$ HNO	$0.00089\,s^{-1}$	(k_{dec})
R2	$Co^{III}P + HNO \rightarrow Co^{III}PNO^- + H^+$	$3.1 \times 10^4\,M^{-1}\,s^{-1}$	(k_{on})
R3	$Co^{III}PNO^- \rightarrow Co^{III}PNO + e^-$	$0.089\,s^{-1}$	(k_{ox})
R4	$Co^{III}PNO \rightarrow Co^{III}P + NO$	$0.11\,s^{-1}$	(k_{off})
R5	$HNO + NO \rightarrow N_2O_2^- + H^+$	$5.8 \times 10^6\,M^{-1}\,s^{-1}$	(k_{NO-HNO})
R6	$2HNO \rightarrow N_2O + H_2O$	$8 \times 10^6\,M^{-1}\,s^{-1}$	(k_{dim})

[a] Numbered according to **Scheme 14** (93).

oxides), as well as high efficiency, since HNO concentrations are usually quite low. Metalloporphyrins, as previously described, are good candidates for achieving this task, since they are highly reactive toward azanone, and are also widely used in electrochemical devices when coupled to a surface, including gas sensors (87, 92). Ordered monolayers of metalloporphyrins can be easily built, based on the establishment of Au−S bonds that directly wire the porphyrin to the electrode. Therefore, based on the known redox state-dependent differential reactivity of CoPor (Por = porphyrin) toward azanone, we have developed an azanone amperometric sensor that works according to Scheme 14 and Table I (67, 68, 93).

The working electrode consists of a monolayer of Co^{II}-5,10,15,20-tetrakis[3-(p-acetylthiopropoxy)phenyl]porphyrin [Co(Por) (Scheme 13)], immobilized on a gold electrode through the spontaneous formation of Au−S bonds (94). Electrochemical measurements of the Co(P) electrode (with and without bound nitrosyl) confirmed the presence of the key $Co^{II}(NO)/Co^{III}(NO)$ redox couple, and showed interestingly, that in the unbound state the obtained $E_{1/2}$ value is shifted ~400 mV to lower potentials (compared to the $E_{1/2}$ in solution), strongly suggesting that Co(P) adsorption on the gold electrode facilitates Co(II) oxidation. The shift is

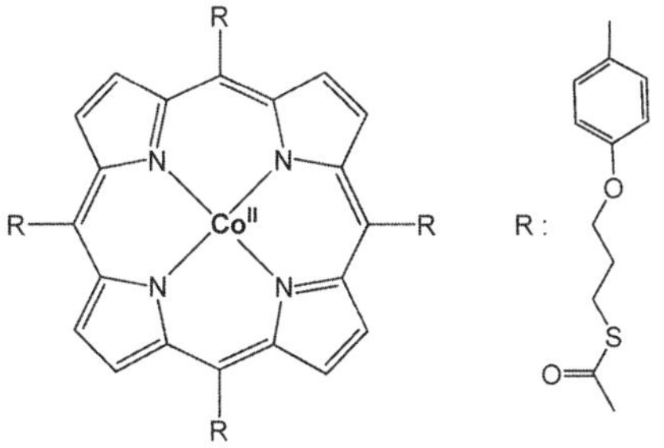

Scheme 13.   Structure of Co^{II}-5,10,15,20-tetrakis[3-(p-acetylthiopropoxy)phenyl]porphyrin.

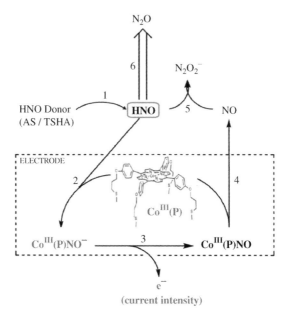

Scheme 14.   Mechanistic scheme considered for HNO production from the decomposition of AS, plus the reactions occurring at the electrode surface (TSHA = p-methyl-sulfohydroxamic acid).

significantly reduced in the nitrosyl complex. The electrode attached Co(P) also retains its selective reactivity behavior toward HNO and NO, as observed in solution, since $Co^{III}(Por)$ reacts efficiently with HNO, but does not with NO. On the contrary, NO reacts rapidly with $Co^{II}(Por)$ modified electrodes while HNO does not. The azanone selectivity and different values for the redox couple (~400 mV vs SCE for Co^{II}/Co^{III} and 800 mV for $Co^{III}-NO/Co^{II}-NO$) thus allowed us to design the sensing electrode that works according Scheme 14. Briefly, the resting state electrode potential is set to 0.8 V, a value where the porphyrin is stable in the $Co^{III}(Por)$ state and no current flow is observed. Reaction with HNO yields, according to the above mentioned observations, the $Co^{III}(Por)$ NO^- complex, which under the described conditions is oxidized to $Co^{III}(Por)NO$. The resulting $Co^{III}(Por)NO$ complex releases the NO ligand in a fast manner and yields $Co^{III}(Por)$, which allows the catalytic cycle to start again. In this scheme, the current intensity is proportional to the amount of azanone that binds the $Co^{III}(Por)$ (67, 68).

The azanone sensing electrode was tested first using AS as the donor (67, 68, 93). Figure 2 shows the corresponding current vs time plot. As expected, setting the electrode potential at 0.8 V, where $Co^{III}(Por)$ is stable, does not show any measurable current. However, a few seconds after the addition of AS the current

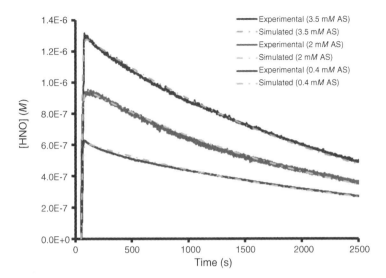

Figure 2. Current intensity vs time measured for three different concentrations of AS. Dotted lines show the corresponding simulations for the current intensity obtained from $Co^{III}PNO^-$ oxidation, by using the model shown in Scheme 14. (See color version of this figure in the color plates section.)

intensity increases, and is maintained for several minutes, due to the catalytic cycle, which is sustained by continuous HNO production from the donor (see Scheme 13). The figure also shows that the signal decreases as the presence of azanone decreases due to donor consumption, which is consistent with a real-time detection. The peak intensity is proportional to the concentration of the initial donor (i.e., HNO) (67, 68).

The electrode can be calibrated by comparing the peak current intensity with the peak azanone concentration produced by a given amount of donor, as determined by the method utilized by Schoenfisch and co-workers (69), rendering a practically linear function for the peak current intensity vs [HNO] (93). By using this method, HNO can be quantified in a concentration range of ~ 1–$200\,nM$.

The presented electrochemical method has several advantages compared to most of the previous described methods. First, the signal is detected in real time without the need of product separation and characterization by analytical techniques (e.g., HPLC). Second, it has one of the lowest detection limits (low-nanomolar range), which considering the elusive nature of azanone is of outmost importance. Compared to optical UV–vis methods, it avoids the interference that would be presented by many systems of interest that have intense absorption (e.g., heme proteins) and can be used in combination with them to follow the different participants in interesting chemical reactions. Moreover, the electrode signal reflects the azanone concentration in real-time, thus allowing us to measure HNO production–consumption on time, extracting relevant kinetic information from these data. Last, but not least, since the HNO reaction with the modified electrode is very efficient, and a small amount of current can be detected, the "real" amount of azanone that reacts with the electrode is usually negligible compared to the total amount of HNO in the system ($<1\%$). It also does not significantly perturb the HNO producing and consuming reactions, thus becoming a powerful tool in mechanistic and kinetic studies where azanone plays a key role.

IV. CONCLUSIONS AND FUTURE PERSPECTIVES

Azanone is an intrinsic elusive molecule, with a proven history of chemically and biologically relevant reactivity. After a bit more of a decade of intense research, many HNO donors, potential HNO production reactions, and azanone potential biological effects have been described. Azanone donors are based mainly on hydroxylamine and its derivatives, NONOates, and C-nitroso compounds. Nitroxyl has been suggested to be a reactive intermediate in many reactions, most of them involving hydroxylamine oxidation (from the -1 to $+1$ oxidation state) or nitric oxide reduction (from the $+2$ to $+1$ oxidation state).

However, to fully understand the role played by azanone as a reaction intermediate during chemical reactions or as a potential in vivo (or endogenously) produced metabolite, unequivocal, efficient, and easy HNO detection at the low-nanomolar level is required. As described in this chapter, research toward this goal has steadily moved forward, exploring a diverse set of trapping and detection strategies including UV–vis spectrophotometry, fluorescence, MS, EPR and electrochemical methods, among others. Although, many of these have promising perspectives to become the method of choice or benchmark for azanone detection and quantification, their response in a wider variety of conditions including lower and higher putative azanone concentrations, presence of potential interfering agents, and competition with HNO sink reactions is needed. However, we are confident that a reliable, sensitive, selective, and physiologically compatible HNO sensing device should be available in the near future. A definitive answer regarding HNO endogenous production in biological systems will be obtained.

ACKNOWLEDGMENTS

This work was financially supported by UBA through UBACYT 20020100100583 and UBACyT 2010-12, MinCyT (PICT-07-1650) and CONICET PIP1207 and 112 201001 00125. Thanks to the John Simon Guggenheim Memorial Foundation (FD).

ABBREVIATIONS

AS	Angeli's salt
ASC$^-$	Ascorbate
BODIPY	Boron dipyrromethene
BOT1	Boron–dipyrromethene–triazole
bpy	2,2-Bipyridyl
CP1	Phenylene ethylene
Co(Por)	Co^{II}-5,10,15,20-tetrakis[3-(p-acetylthiopropoxy)phenyl] porphyrin
C-PTIO	2-(4-Carboxyphenyl)-4,4,5,5-tetramethylimidazoline-1-oxyl-3-oxide
DHA	Dehydroascorbic acid
DNA	Dioxyribonucleic acid
EPR	Electron paramagnetic resonance
ESI–MS	Electrospray ionization–mass spectrometry
Fe–MGD	N-Methyl-D-glucamine ditliiocarbaniate iron
FTIR	Fourier transform infrared
GS(O)NH$_2$	Glutathione sulfinamide
GSH	Glutathione
^{1}H NMR	Proton nuclear magnetic resonance
H4B	Tetrahydrobiopterin
HNO	Nitrosyl azanone
HPLC	High-performance liquid chromatography
HRP	Horseradish peroxidase
IPA	Isopropylamine
IPA/NO	Sodium 1-(isopropylamino)diazene-1-ium-1,2-diolate
IR	Infrared
MIMS	Membrane-inlet mass spectrometry
Mb	Mioglobin
MS	Mass spectrometry
NADPH	Nicotinamide adenine dinucleotide phosphate
NB	Nitrosobenzene
NHE	Normal hydrogen electrode
NMR	Nuclear magnetic resonance
NOHA	N-Hydroxy-L-arginine
NOS	Nitric oxide synthase
PA	Piloty's acid
Por	Porphyrin
PTIO	2-phenyl-4,4,5,5-tetramethylimidazoline-1-oxyl-3-oxide
RNS	Reactive nitrogen species
SOD	Superoxide dismutase

TEMPO	2,2,6,6-(Tetramethylpiperidin-1-yl)oxyl
TOF MS	Time-of-flight mass spectrometry
TPP	Tetraphenylporphyrin
TPPS	*meso*-Tetrakis(4-sulfonato-phenyl)porphyrinate
TSHA	*p*-methyl-sulfohydroxamic acid (*p*-methyl derivative of Piloty's acid)
TXPTS	Tris(4-6-dimethyl-3-sulfonatophenyl)phosphine trisodium hydrate
UV–vis	Ultraviolet–visible

REFERENCES

1. F. Doctorovich, D. Bikiel, J. Pellegrino, S. A. Suárez, A. Larsen, and M. A. Martí, *Coord. Chem. Rev.* 255, 2764 (2011).

2. M. D. Bartberger, J. M. Fukuto, and K. N. Houk, *Proc. Natl. Acad. Sci. USA* 98, 2194 (2001).

3. P. C. Ford, *Inorg. Chem.* 49, 6226 (2010).

4. M. Hoshino, L. Laverman, and P. C. Ford, *Coord. Chem. Rev.* 187, 75 (1999).

5. S. A. Suárez, M. A. Martí, P. M. De Biase, D. A. Estrin, S. E. Bari, and F. Doctorovich, *Polyhedron* 26, 4673 (2007).

6. M. A. Martí, S. E. Bari, D. A. Estrin, and F. Doctorovich, *J. Am. Chem. Soc.* 127, 4680 (2005).

7. I. Boron, S. A. Suárez, F. Doctorovich, M. A. Martí, and S. E. Bari, *J. Inorg. Biochem.* 105, 1044 (2011).

8. J. B. Pickles and D. A. Williams, *Nature (London)* 271, 335 (1978).

9. C. H. Switzer, T. W. Miller, P. J. Farmer, and J. M. Fukuto, *J. Inorg. Biochem.* 118, 128 (2013).

10. V. A. Benderskii, A. G. Krivenko, and E. A. Ponomarev, *Sov. Electrochem. Engl. Tr.* 25, 154 (1989).

11. J. DuMond and S. King, *Antioxid. Redox Signal.* 14, 1637 (2011).

12. B.-B. Zeng, J. Huang, M. W. Wright, and S. B. King, *Bioorg. Med. Chem. Lett.* 14, 5565 (2004).

13. K. Sirsalmath, S. a Suárez, D. E. Bikiel, and F. Doctorovich, *J. Inorg. Biochem.* 118, 134 (2013).

14. D. A. Guthrie, N. Y. Kim, M. A. Siegler, C. D. Moore, and J. P. Toscano, *J. Am. Chem. Soc.* 134, 1962 (2012).

15. D. Thomas, K. Miranda, M. Graham, D. C. Espey, D. Jourd'heuil, N. Paolocci, S. J. Hewett, C. A. Colton, M. B. Grisham, M. Feelisch, and D. A. Wink, *Meth. Enzymol.* 359, 84 (2002).

16. D. J. Salmon, C. L. Torres de Holding, L. Thomas, K. V. Peterson, G. P. Goodman, J. E. Saavedra, A. Srinivasan, K. M. Davies, L. K. Keefer, and K. M. Miranda, *Inorg Chem* 50, 3262 (2011).

17. D. Andrei, D. J. Salmon, S. Donzelli, A. Wahab, J. R. Klose, M. L. Citro, J. E. Saavedra, D. A. Wink, K. M. Miranda, and L. K. Keefer, *J. Am. Chem. Soc. 132*, 16526 (2010).

18. X. Sha, T. S. Isbell, R. P. Patel, C. S. Day, and S. B. King, *J. Am. Chem. Soc. 128*, 9687 (2006).

19. J. F. Dumond, M. W. Wright, and S. B. King, *J. Inorg. Biochem. 118*, 140 (2013).

20. Y. Adachi, H. Nakagawa, K. Matsuo, T. Suzuki, and N. Miyata, *Chem. Commun.* 5149 (2008).

21. K. Matsuo, H. Nakagawa, Y. Adachi, E. Kameda, H. Tsumoto, T. Suzuki, and N. Miyata, *Chem. Commun. 46*, 3788 (2010).

22. H. Nakagawa, *J. Inorg. Biochem. 118*, 187 (2013).

23. S. Luňák and J. Vepřek-Šiška, *Collect. Czech. Chem. Commun. 39*, 391 (1974).

24. R. Nast and I. Foppl, *Z. Anorg. Allg. Chem. 263*, 310 (1950).

25. J. H. Anderson, *Analyst (London) 89*, 357 (1964).

26. J. H. Anderson, *Analyst (London) 91*, 532 (1966).

27. G. E. Alluisetti, A. E. Almaraz, V. T. Amorebieta, F. Doctorovich, and J. A. Olabe, *J. Am. Chem. Soc. 126*, 13432 (2004).

28. J. W. Larsen, J. Jandzinski, M. Sidovar, and J. L. Stuart, *Carbon NY 39*, 473 (2001).

29. J. N. Cooper, J. E. Chilton, and R. E. Powell, *Inorg. Chem. 9*, 2303 (1970).

30. F. T. Bonner, L. S. Dzelzkalns, and J. A. Bonucci, *Inorg. Chem. 17*, 2487 (1978).

31. N. Y. Wang and F. T. Bonner, *Inorg. Chem. 25*, 1858 (1986).

32. N. Y. Wang and F. T. Bonner, *Inorg. Chem. 25*, 1863 (1986).

33. S. V. Lymar, V. Shafirovich, and G. A. Poskrebyshev, *Inorg Chem 44*, 5212 (2005).

34. E. Freese, E. B. Freese, and E. Rutherford, Biochemistry, *4* (1965).

35. H. Erlenmeyer, C. Flierl, and H. Sigela, *J. Am. Chem. Soc. 91*, 1065 (1969).

36. S. Donzelli, M. G. Espey, W. Flores-Santana, C. H. Switzer, G. C. Yeh, J. Huang, D. J. Stuehr, S. B. King, K. M. Miranda, D. A. Wink, and M. Graham, *Free Radic. Bio. Med. 45*, 578 (2008).

37. S. Adak, Q. Wang, and D. J. Stuehr, *J. Biol. Chem. 275*, 33554 (2000).

38. J. J. Woodward, Y. Nejatyjahromy, R. D. Britt, and M. A. Marletta, *J. Am. Chem. Soc. 132*, 5105 (2010).

39. T. C. Hollocher, *Antonie Van Leeuwenhoek 48*, 531 (1983).

40. M. N. Ackermann and R. E. Powell, *Inorg. Chem. 5*, 1334 (1966).

41. V. Shafirovich and S. V. Lymar, *Proc. Natl. Acad. Sci. USA 99*, 7340 (2002).

42. F. T. Bonner and K. A. Pearsall, *Inorg. Chem. 21*, 1973 (1982).

43. F. T. Bonner and K. A. Pearsall, *Inorg. Chem. 21*, 1978 (1982).

44. K. E. Bell and H. C. Kelly, *Inorg Chem 35*, 7225 (1996).

45. O. P. Strausz and H. E. Gunning, *Trans. Faraday Soc. 60*, 347 (1964).

46. F. C. Kohout and F. W. Lampe, *J. Am. Chem. Soc. 87*, 5795 (1965).

47. R. M. Lambert, *Chem. Commun.(London)* 850 (1966).

48. M. J. Akhtar, F. T. Bonner, and M. N. Hughes, *Inorg. Chem. 24*, 1934 (1985).

49. W. A. Pryor, D. F. Church, C. K. Govindan, and G. Crank, *J. Org. Chem. 47*, 156 (1982).

50. H. W. Brown and G. C. Pimentel, *J. Chem. Phys. 29*, 883 (1958).

51. R. C. Sausa, G. Singh, G. W. Lemire, and W. R. Anderson, *Flame Structure Studies of Neat and NH_3-Doped Low-Pressure $H_2/N_2O/Ar$ Flames by Molecular Beam Mass Spectroscopy*, Weapons and Materials Research Directorate, ARL, July 1997.

52. M. J. Y. Quee and J. C. J. Thynne, *Trans. Faraday Soc. 64*, 1296 (1968).

53. E. E. Hackman, H. H. Hesser, and H. C. Beachell, *J. Phys. Chem. 76*, 3545 (1972).

54. M. Kirsch, A.-M. Büscher, S. Aker, R. Schulz, and H. de Groot, *Org. Biomol. Chem. 7*, 1954 (2009).

55. M. R. Filipovic, J. L. Miljkovic, T. Nauser, M. Royzen, K. Klos, T. Shubina, W. H. Koppenol, S. J. Lippard, and I. Ivanović-Burmazović, *J. Am. Chem. Soc. 134*, 12016 (2012).

56. J. Rosenthal and S. J. Lippard, *J. Am. Chem. Soc. 132*, 5536 (2010).

57. X. Shen, C. B. Pattillo, S. Pardue, S. C. Bir, R. Wang, and C. G. Kevil, *Free Radic. Biol. Med. 50*, 1021 (2011).

58. O. Einsle, A. Messerschmidt, R. Huber, P. M. H. Kroneck, and F. Neese, *J. Am. Chem. Soc. 124*, 11737 (2002).

59. H. H. H. W. Schmidt, H. Hofmann, U. Schindler, Z. S. Shutenko, D. D. Cunningham, and M. Feelisch, *Proc. Natl. Acad. Sci. USA 93*, 14492 (1996).

60. S. Stoll, Y. NejatyJahromy, J. J. Woodward, A. Ozarowski, M. A. Marletta, and R. D. Britt, *J. Am. Chem. Soc. 132*, 11812 (2010).

61. J. A. Reisz, C. N. Zink, and S. B. King, *J. Am. Chem. Soc. 133*, 11675 (2011).

62. R. Lin and P. J. Farmer, *J. Am. Chem. Soc. 122*, 2393 (2000).

63. J. Pellegrino, S. E. Bari, D. E. Bikiel, and F. Doctorovich, *J. Am. Chem. Soc. 132*, 989 (2010).

64. J. M. Fukuto, M. D. Bartberger, A. S. Dutton, N. Paolocci, D. A. Wink, and K. N. Houk, *Chem. Res. Toxicol. 18*, 790 (2005).

65. X. L. Ma, F. Gao, G. L. Liu, B. L. Lopez, T. A. Christopher, J. M. Fukuto, D. A. Wink, and M. Feelisch, *Proc. Natl. Acad. Sci. USA 96*, 14617 (1999).

66. X. L. Ma, A. S. Weyrich, D. J. Lefer, and A. M. Lefer, *Circ. Res. 72*, 403 (1993).

67. S. A. Suárez, M. H. Fonticelli, A. A. Rubert, E.de La Llave, D. Scherlis, R. C. Salvarezza, M. A. Martí, and F. Doctorovich, *Inorg Chem 49*, 6955 (2010).

68. S.A. Suárez, D.E. Bikiel, D.E. Wetzler, M.A. Martí, and F. Doctorovich, *Anal. Chem., 85*, 10262 (2013).

69. K. P. Dobmeier, D. A. Riccio, and M. H. Schoenfisch, *Anal. Chem. 80*, 1247 (2008).

70. D. W. Shoeman and H. T. Nagasawa, *Nitric Oxide 2*, 66 (1998).

71. M. P. Doyle, S. N. Mahapatro, R. D. Broene, and J. K. Guy, *J. Am. Chem. Soc. 110*, 593 (1988).

72. S. I. Liochev and I. Fridovich, *Free Radic. Bio. Med. 34*, 1399 (2003).

73. P. Wong, J. Hyun, and J. Fukuto, *Biochemistry 37*, 5362 (1998).

74. S. Donzelli, M. G. Espey, D. D. Thomas, D. Mancardi, C. G. Tocchetti, L. A. Ridnour, N. Paolocci, S. B. King, K. M. Miranda, G. Lazzarino, J. M. Fukuto, and D. A. Wink, *Free Radic. Bio. Med. 40*, 1056 (2006).

75. M. B. Soellner, B. L. Nilsson, and R. T. Raines, *J. Am. Chem. Soc. 128*, 8820 (2006).

76. M. D. Lim, I. M. Lorkovic, and P. C. Ford, *Inorg. Chem. 41*, 1026 (2002).

77. Y. Xia, A. Cardounel, A. F. Vanin, and J. L. Zweier, *Free Radic. Bio. Med. 29*, 793 (2000).

78. A. M. Komarov, D. A. Wink, M. Feelisch, and H. H. H. Schmidt, *Free Radic. Bio. Med. 28*, 739 (2000).

79. U. Samuni, Y. Samuni, and S. Goldstein, *J. Am. Chem. Soc. 132*, 8428 (2010).

80. A. Bobko, A. Ivanov, and V. Khramtsov, *Free Radic. Res. 47*, 74 (2013).

81. M. R. Cline, C. Tu, D. N. Silverman, and J. P. Toscano, *Free Radic. Bio. Med. 50*, 1274 (2011).

82. M. R. Cline, T. A. Chavez, and J. P. Toscano, *J. Inorg. Biochem. 118*, 148 (2013).

83. A. G. Tennyson, L. Do, R. C. Smith, and S. J. Lippard, *Nitric Oxide 26*, 4625 (2007).

84. Y. Zhou, K. Liu, J.-Y. Li, Y. Fang, T.-C. Zhao, and C. Yao, *Org. Lett. 13*, 2357 (2011).

85. M. R. Cline and J. P. Toscano, *J. Phys. Org. Chem. 24*, 993 (2011).

86. K. M. Miranda, N. Paolocci, T. Katori, D. D. Thomas, E. Ford, M. D. Bartberger, M. G. Espey, D. A. Kass, M. Feelisch, J. M. Fukuto, and D. A. Wink, *Proc. Natl. Acad. Sci. USA 100*, 9196 (2003).

87. I. R. Davies and X. Zhang, *Meth. Enzymol. 436*, 63 (2008).

88. J. H. Shin and M. H. Schoenfisch, *Analyst (London) 131*, 609 (2006).

89. J. Vitecek, J. Petrlova, J. Petrek, V. Adam, D. Potesil, L. Havel, R. Mikelova, L. Trnkova, and R. Kizek, *Electrochim. Acta 51*, 5087 (2006).

90. S. Kudo, J. L. Bourassa, S. E. Boggs, Y. Sato, and P. C. Ford, *Anal. Biochem. 247*, 193 (1997).

91. Y. C. Boo, S. L. Tressel, and H. Jo, *Nitric Oxide 16*, 306 (2007).

92. G. B. Richter-Addo, S. J. Hodge, G. B. Yi, M. A. Khan, T. Ma, E. Van Caemelbecke, N. Guo, and K. M. Kadish, *Inorg. Chem 35*, 6530 (1996).

93. J. L. Heinecke, C. Khin, J. C. M. Pereira, S. A. Suárez, A. V. Iretski, F. Doctorovich, and P. C. Ford, *J. Am. Chem. Soc. 135* (10), 4007 (2013).

94. D. G. Whitten, E. W. Baker, and A. H. Corwin, *J. Org. Chem. 28*, 2363 (1963).

Photoactive Metal Nitrosyl and Carbonyl Complexes Derived from Designed Auxiliary Ligands: An Emerging Class of Photochemotherapeutics

BRANDON J. HEILMAN, MARGARITA A. GONZALEZ, AND
PRADIP K. MASCHARAK

Department of Chemistry and Biochemistry, University of California, Santa Cruz, CA 95060

CONTENTS

Progress in Inorganic Chemistry, Volume 58, First Edition. Edited by Kenneth D. Karlin.
© 2014 John Wiley & Sons, Inc. Published 2014 by John Wiley & Sons, Inc.

I. INTRODUCTION

Among the simple diatomic gases, nitric oxide (NO) and carbon monoxide (CO) are of interest mostly because of their harmful effects. Both gases are toxic enough so that CO earned the moniker the "silent killer" and NO is identified as the major source of harmful "smog" in urban areas. The noxious property of these "toxic twins" arises primarily from their high affinity toward heme proteins. Research during the past few decades has, however, demonstrated that NO and CO are formed endogenously in mammals at very low concentrations and both molecules play critical roles in various biological signaling pathways (1). For example, a nanomolar concentration of NO, generated from arginine by the enzyme nitric oxide synthase (NOS) (2), mediates key processes such as blood pressure regulation and neurotransmission (3–5). During chronic infection, macrophages produce higher (micromolar and above) concentrations of NO through up-regulation of the inducible NOS and provide defense against the invading pathogen as part of our innate immune response (6, 7). Such levels of NO also induce programmed cell death (apoptosis) through inhibition of cellular respiration, activation of caspases, and deoxyribonucleic acid (DNA) damage (8–10). In mammals, CO is generated through catabolism of heme by the enzyme heme oxygenase (HO) (11). Present in almost all tissues, this enzyme causes ring opening of the porphyrin macrocycle with the elimination of CO (and biliverdin). The constitutive isoform HO-2 takes care of the heme processing in the liver and spleen, while the inducible form of HO, namely, HO-1, catalyzes elevated levels of localized heme degradation in case of tissue injury. Although a major portion of CO produced endogenously bind hemoglobin in the blood and is carried away to the lungs where it is exhaled as CO gas, small amounts of this toxic gas is utilized by the body as a cytoprotective agent to thwart oxidative damage (12). The salutary effects of CO at low concentrations (15–250 ppm) have recently been demonstrated in preclinical animal models of disease including shock, organ transplantation, enterocolitis, and ischemia–reperfusion injury (IRI) (12, 13). Carbon monoxide inhibits various proinflammatory mediators through the mitogen-activated protein kinase (MAPK) pathways in these pathological conditions (14). Taken together, it is now clear that (a) NO and CO are produced in controlled amounts on demand and utilized effectively without exerting any apparent toxicity in mammalian tissues and (b) pathological outcomes are observed with both molecules only at elevated (and uncontrolled) levels.

The discovery of the unusual beneficial effects of NO and CO has prompted research toward development of NO and CO releasing molecules (NORMs and CORMs, respectively) that could be employed to modulate the levels of these two signaling molecules within the body. Although drugs like glyceryl trinitrate (GTN, Fig. 1) for angina pectoris and sodium nitroprusside (SNP, Fig. 1) to tackle hypertensive episodes have been in use for quite sometime, their identity as NO

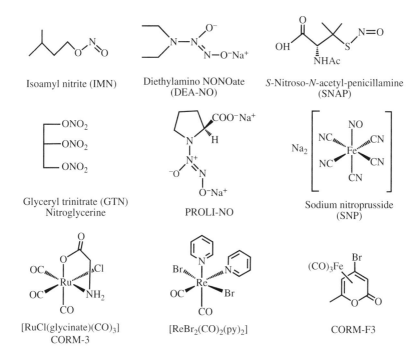

Figure 1. Examples of NORMs and CORMs (py = pyridine).

donors has been realized only recently (15, 16). To date, a number of NO donating compounds have been synthesized and several of them have found wide use as pharmaceuticals (17–19). In recent years, a few CORMs also have been synthesized and attempts have been made to establish their utility as exogenous CO donors (20–22). Most of the NORMs in Fig. 1 afford NO upon exposure to heat, oxidants, or thiols, and in some cases through enzymatic reactions. The diazeniumdiolates (e.g., DEA-NO and PROLI-NO) (Fig. 1) generate NO with changes in pH and the rate of NO release depends on the substituents. The CORMs like [RuCl(glycinate)(CO)$_3$] (CORM-3, Fig. 1) and CORM-F3 release CO upon solvation in biological media.

Despite excellent NO and CO releasing properties, the NORMs and CORMs of Fig. 1 all suffer from one common problem, namely, the lack of control on the NO or CO delivery; the drug goes everywhere in the body and NO (or CO) is released through enzymatic and non-enzymatic (e.g., heat, pH) pathways. This inexorability of NO (or CO) release and the unpredictable rates of NO (or CO) loss in different parts of the body become major issues in many biomedical applications. For example, although the ability of higher doses of NO to eradicate pathogens or tumor cells of different grade and origin has been known for quite sometime, use of a systemic NO donor (e.g., GTN) is strictly prohibited for such a purpose due to the severe hypotensive effect of NO. Delivery of NO or CO to a specific biological target requires both site-specific localization of the drug and strict control on the release of the gaseous molecule. Since control on the enzymatic pathways or pH within a selected part of the body is extremely difficult, researchers in recent years have relied on *light as the trigger* for the release of NO and CO from designed NORMs and CORMs. Over the past few years, a variety of photoactive metal NO complexes and carbonyls (CO complexes) have been synthesized and their capacities of NO/CO delivery have been determined (23–29). In addition to one or more NO or CO as ligands, these metal complexes contain designed organic frames as auxiliary ligand(s). These auxiliary ligands have been carefully tailored so as to promote release of NO or CO from the resulting nitrosyls and carbonyls upon exposure to light of selected wavelengths and with desired effeciencies (30–32). In many cases, established chemical principles and theoretical calculations have aided the design of such ligands (33, 34). Once isolated, the NORMs and CORMs have been subjected to a wide range of *in vitro* and *in vivo* biological experiments to evaluate the utility of the photoactive NORMS and CORMs as photochemotherapeutics. This chapter focuses on the results from these pursuits by various groups during the past few years.

II. METAL NITROSYL AND CARBONYL COMPLEXES AS NITRIC OXIDE AND CARBON MONOXIDE DONORS

The occurrence of NO and CO in human physiology and the potential of metal nitrosyl and carbonyl complexes as pharmaceuticals have added new paradigms in modern biology and medicine. Despite the enormous success of cis-platin (*cis*-$[PtCl_2(NH_3)_2]$) as an anticancer drug (35), metal complexes are often shun by the pharmaceutical research community. This approach is somewhat surprising since metal complexes can carry drug molecules as ligands and release them under controlled conditions. Both NO and CO serve as ligands to many metals and collectively constitute a major part of organometallic chemistry (36–39). Metal complexes of NO (metal nitrosyls) and CO (metal carbonyls), as well as mixed-metal carbonyl–nitrosyl complexes have been studied over several decades and their electronic structures have provided important insight into the so-called "18-electron rule" (40). Although the Lewis basicity of these two molecules hardly qualifies them as good ligands, CO binds to metals in low oxidation states quite readily through the carbon atom and the bond between the metal center and CO is strengthened through back-donation of electron density from the out-of-plane d orbitals to the π^* MO. In the case of NO, there is one more electron in the π^* orbital (compared to CO) that alters its behavior as a ligand. In the case of metals in a higher oxidation state, NO transfers this extra electron to the partially empty out-of-plane d orbital on the metal in addition to the lone pair that constitutes the coordinate bond between the metal and the nitrogen atom (*linear* M−NO bond). Formally, NO is considered as a three-electron donor in such cases and acts more as an NO^+ unit (isoelectronic with CO). With metals in a low oxidation state, NO forms nitrosyls with a *bent* M−NO bond. Formally, NO is considered as a one-electron donor (the other coming from the metal to form the M−NO bond) in such cases. The bent NO ligand is an analogue of an organic nitroso group or the NO group in Cl−N=O, where a lone pair stays on the N atom in sp^2 hybrid orbital. The different extent of transfer of electron density from NO to the metal center often raises the difficulty of assigning oxidation states to both NO and the metal center in metal nitrosyls. A special notation of $\{M-NO\}^n$ was therefore devised in 1974 by Enemark and Feltham (41) to denote a metal–NO bond where $n =$ the total number of electrons in the metal d plus the NO π^* orbital. For example, $\{Ru-NO\}^6$ could represent one of the two possible combinations of formal oxidation states of the metal center and NO, namely, $Ru(III)-NO^{\bullet}$ or $Ru(II)-NO^+$.

The nature of M−NO and M−CO bonds as described above provides hints toward photolabilization of NO and CO lignds from metal nitrosyls and carbonyls. For example, transfer of electron density from the low-valent metal center of a designed carbonyl to the π^* MO of the auxiliary ligand through metal-to-ligand charge-transfer (MLCT) transition could diminish the extent of metal-to-CO back-bonding and labilization of the M−CO bond (42, 43). Similarly, transfer of electron density from the metal center of a designed nitrosyl to low-lying MOs with strong metal−NO antibonding character through MLCT could weaken the M−NO bond and promote NO photorelease (33). As described in the following sections, both these expectations have been realized in metal nitrosyl and carbonyl complexes derived from designed auxiliary ligands.

III. PHOTOACTIVE METAL NITROSYL COMPLEXES

A. Metal Nitrosyl Complexes With Monodentate Ligands

Sodium nitroprusside ($Na_2[Fe(CN)_5(NO)]$, SNP, Fig. 1) is one of the first documented photoactive metal nitrosyl complexes. This $\{Fe-NO\}^6$ nitrosyl has been in clinical use as a vasodilatory drug (Nitropress^{TM}) for quite sometime (FDA approval in 1974) (44). The mechanism of *in vivo* NO release is a two-step process beginning with a one-electron reduction of the Fe(III) center by endogenous thiols like glutathione to afford a semistable Fe(II) species that readily loses a CN^- ion to relieve excess negative charge at the metal center (45). The pentacoordinated intermediate $[Fe(CN)_4(NO)]^{2-}$ eventually decomposes to afford NO in addition to

CN⁻. Cogeneration of CN^- often limits the amount of SNP employed in intravenous administration (as a dilute solution in sterile 5% dextrose solution) during a hypertensive episode (46). Severe toxicity is observed in the case of patients with compromised liver and renal function.

The photodissociation of NO from SNP upon absorption of 400-nm light proceeds through a more direct mechanism (47, 48). Absorption of 400-nm photon excites SNP to a metastable state MS1, a linkage isomer with O-bonded NO. Upon absorption of a second photon, MS1 gives rise to another metastable state MS2 in which the NO ligand is bonded sideways to the Fe center (48). This latter transient species MS2 rapidly dissociates into $[Fe(CN)_5]^{2-}$ and NO in solution. Solvent-assisted decomposition of $[Fe(CN)_5]^{2-}$ also generates toxic CN^- (and Prussian blue) during photodissociation from SNP in solution. The loss of ancillary ligands thus remains a challenge with the use of SNP in both the "dark" and the "light-assisted" mode of application.

Synthetic iron–sulfur cluster nitrosyl complexes like Roussin's red salt (RRS, $[Fe_2S_2(NO)_4]^{2-}$), Roussin's black salt (RBS, $[Fe_4S_3(NO)_7]^-$), and tetranitrosyl-tetra-μ_3-sulfido-tetrairon ($[(FeNOS)_4]^-$) (Fig. 2) resemble iron–sulfur clusters found in biological systems. The bright color of these complexes in solutions results from strong absorption bands in the visible region. Photorelease of NO from these cluster nitrosyl complexes has been studied extensively by Ford et al. (27). Photolysis (350–450 nm) of RRS (quantum yield values $\phi = 0.14$–0.40) in aerobic aqueous solution leads to quantitative production of the RBS with release of NO. The less-photoactive RBS ($\phi = 0.001$–0.004) undergoes photodecomposition to give Fe(III) precipitates and NO. Both RBS and $[(FeNOS)_4]^-$ release NO upon illumination at 515 (and 458) nm and induce vasorelaxation in rat aorta muscle rings in tissue bath experiments (49). In addition, photodelivery of NO from RRS greatly sensitized γ-radiation-induced damage to hypoxic cultures of Chinese hamster V79 cells in an early experiment (50). Attempts to use these cluster nitrosyl complexess as biologically acceptable NO donors in radiation oncology have, however, met with limited success due to their severe toxicity.

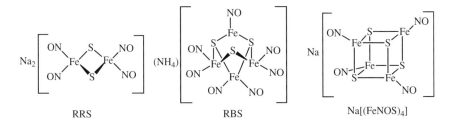

Figure 2.   Structures of iron–sulfur nitrosyl complexes, RRS, RBS, and Na[(FeNOS)₄].

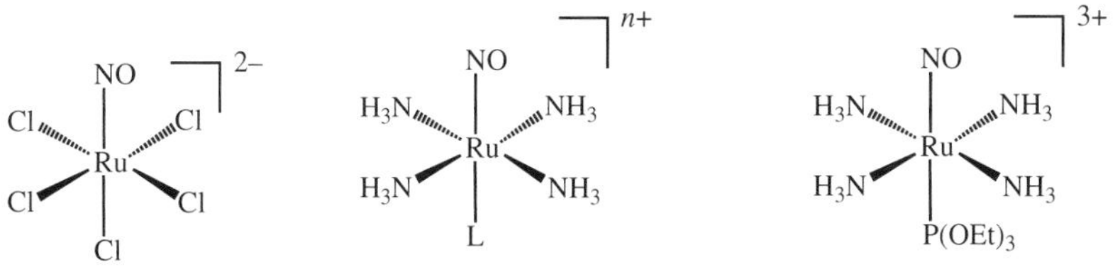

L = Cl, H_2O, im, py, 4-pic, NO_2^-

(1)
$[Ru(NH_3)_4(NO)(P(OEt)_3)](PF_6)_3$

Figure 3. Photoactive $\{Ru-NO\}^6$ nitrosyl complexes with monodentate ligands (im = imidazole).

The instability of iron nitrosyl complexes in biological media (loss of ancillary ligands, formation of insoluble precipitates) has prompted researchers to focus more on photoactive ruthenium nitrosyl complexes that in most cases exhibit superior stability. Photosensitivity of $\{Ru\text{-}NO\}^6$ nitrosyl complexes was noted as early as 1971, when Cox and Wallace (51) observed that an acidic aqueous solution of $K_2[RuCl_5(NO)]$ (Fig. 3) turned brown over a period of hours when exposed to light, yet it was thermally stable under dark conditions. Careful experiments showed that such activity is limited to ultraviolet (UV) light and results from the loss of NO. Ruthenium nitrosyl ammine complexes of the type $[Ru(NH_3)_5(NO)]X_3$ and $[Ru(NH_3)_4(NO)L]X_3$ $[X = BF_4^-, ClO_4^-; L = ligands$ e.g., py, 4-pic, $P(OEt)_3$; Fig. 3] also exhibit similar photorelease of NO upon exposure of their aqueous solution to UV light (300–370 nm), and afford $[Ru(NH_3)_5(H_2O)]^{3+}$ and $[Ru(NH_3)_4(H_2O)L]^{3+}$, respectively, as the photoproducts (25, 26). A $d_\pi(Ru)$ to $\pi^*(NO)$ MLCT transition promotes NO release in these $[Ru(NH_3)_4(NO)L]X_3$ complexes and the rate of NO photorelease depends on the nature of L (the ligand trans to NO). The coordination frame of $[Ru(NH_3)_4(NO)L]^{n+}$ thus allows some freedom in the design of photoactive NORMs as the stability, solubility, and quantum yield values for NO release can all be tuned with change in L. For example, the quantum yield values (ϕ) for NO release in aqueous solution (pH 4, $\lambda_{irr} = 330$ nm) of $[Ru(NH_3)_4(NO)L]^{n+}$ changes from 0.12×10^{-3} to 130×10^{-3} with change of L from 4-picoline (4-pic) to py (26). These Ru nitrosyl complexes are soluble in water and exhibit low toxicity, both desirable properties for pharmaceuticals. Consequently, they have been subjected to a wide variety of biological studies to evaluate their chemotherapeutic potential (52, 53). Franco and co-workers (54) showed that $[Ru(NH_3)_4(NO)(P(OEt)_3)](PF_6)_3$ (**1**) rapidly releases NO upon activation with biological reductants (e.g., glutathione) *in vivo* and induces strong vasodilatory effects in hypertensive rats. Despite such activity, the lack of strong absorption bands in the visible region restricts use of these ammine–nitrosyl complexes in phototherapeutic applications because of the detrimental effects of UV light in general.

B. Metal Nitrosyl Complexes Derived from Polydentate Ligands
With Extended Structure

Alteration of the ancillary ligands of $\{Ru-NO\}^6$ nitrosyl complexes to impart further stability and sensitivity toward vis light (i.e., lights of longer wavelength) has been attempted by several groups. In such pursuits, macrocyclic ligands have been employed to alleviate the problems associated with loss of ancillary ligands. The biological utility of the photoactive nitrosyl *trans*-[RuCl(cyclam)(NO)]X$_2$ (**2**, cyclam = 1,4,8,11-tetraazacyclotetradecane; X = Cl$^-$, PF$_6^-$, ClO$_4^-$; Fig. 4), *trans*-[RuCl([15]ane)(NO)]X$_2$ (**3**, [15]ane = 1,4,8,12-tetraazacyclopentadecane, Fig. 4), and related species have been explored by Franco and co-workers (53). Both **2** and **3** are photoactive and induce vasorelaxation in hypertensive rats through activation by biological reductants. In addition, **3** can be activated by UV light (355 nm, $\phi = 0.60$) and such activation causes relaxation effects faster than the thiol-induced pathway (55). The lower quantum yield value of **2** ($\phi = 0.16$) fails to produce NO fast enough to cause vasorelaxation in rat aorta. Both Ru and Fe nitrosyls derived from more conjugated macrocycles, such as porphyrins, display strong absorption bands in the visible region and also exhibit NO

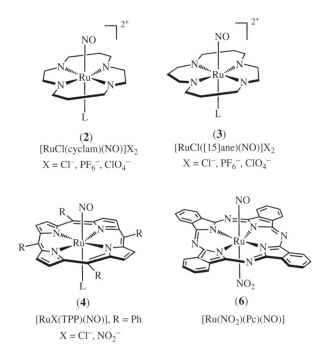

(**2**)
[RuCl(cyclam)(NO)]X$_2$
X = Cl$^-$, PF$_6^-$, ClO$_4^-$

(**3**)
[RuCl([15]ane)(NO)]X$_2$
X = Cl$^-$, PF$_6^-$, ClO$_4^-$

(**4**)
[RuX(TPP)(NO)], R = Ph
X = Cl$^-$, NO$_2^-$

(**6**)
[Ru(NO$_2$)(Pc)(NO)]

Figure 4.   Photoactive Ru nitrosyls with macrocyclic ligands.

photolability. Ford and co-workers (56, 57) studied the photochemistry of [MX(TPP)(NO)] (**4**, Fig. 4) and [MX(OEP)(NO)] (**5**) (M = Ru, Fe; TPP = *meso*-tetraphenylporphyrin, OEP = octaethylporphyrin; X = Cl$^-$, NO$_2^-$) in various solvents under different conditions to evaluate their potential as NO donors upon illumination with UV and visible light. Exposure of both [RuCl(TPP)(NO)] and [RuCl(OEP)(NO)] to light leads to photolysis of the Ru–NO bond. The quantum efficiency depends on the nature of the solvent, temperature, and the wavelength of irradiation. Because of the high affinity of NO toward metal–porphyrin centers, capture of the photoreleased NO within the solvation sphere poses a problem with these porphyrin–nitrosyls. Because of competing rates of NO release (k_{off}) and capture (k_{on}), efficient NO delivery by these porphyrin–nitrosyls is observed only under high-power illumination. As the macrocyclic ligand frames of **2–6** allow multiple sites for substitutions, addition of groups (e.g., –SO$_3$H and –COOH) to the periphery of the ligands can improve their solubility in aqueous solutions. Substitutions on the ligand frame, however, do not alter the overall photosensitivity of these porphyrin–nitrosyls because such alterations bring minor shifts in the major absorption bands of these complexes. Sensitivity to oxygen is another disadvantage in the case of Fe porphyrin–nitrosyls. Despite these drawbacks, preferential accumulation of porphyrin–containing species in dermal tissues (58) provides impetus to develop metal nitrosyl complexes derived from porphyrin-based ligands for the treatment of topical malignancies through site-selective delivery of NO under illumination.

Phthalocyanines have also been tried as ligands to afford metal nitrosyl complexes with strong absorption in the visible range. The phthalocyanine complex [Ru(NO$_2$)(Pc)(NO)] (**6**, Pc = phthalocyanine, Fig. 4) displays an intense band ($\varepsilon = 2.75 \times 10^4 \, M^{-1} \, cm^{-1}$) at 690 nm in its absorption spectrum. Because of poor solubility, this nitrosyl complex has been encapsulated in poly(lactic-*co*-glycolic acid) nanoparticles and the polymer–nitrosyl composite material rapidly releases NO upon irradiation with 660-nm light (59). In general, the poor solubility and biocompatibility of Ru (and Fe) nitrosyl complexes derived from macrocyclic ligands with strong absorptions require additional strategies to convert them into potential photochemotherapeutics. The recent report on a topical formulation of Ru nitrosyl-loaded lipid microemulsion for the treatment of skin malignancies by Lopez and co-workers (60) is another excellent example along this direction. The stearic acid–sodium taurodeoxycholate microemulsion with entrapped [Ru(bdqi) (terpy)](PF$_6$)$_3$ (**7**, bdqi = 1,2-benzoquinonediimine, terpy = terpyridine) exhibits an excellent tolerance to skin and efficient NO releasing kinetics upon illumination with visible light.

In more recent studies, light-harvesting antennas have been used as an artifice to promote photoinduced release of NO, especially by irradiation with visible and near-IR (NIR) light (600–1100 nm), which corresponds to the desired window (61) for greater penetration in the skin and safe noninvasive phototherapy. Ford and

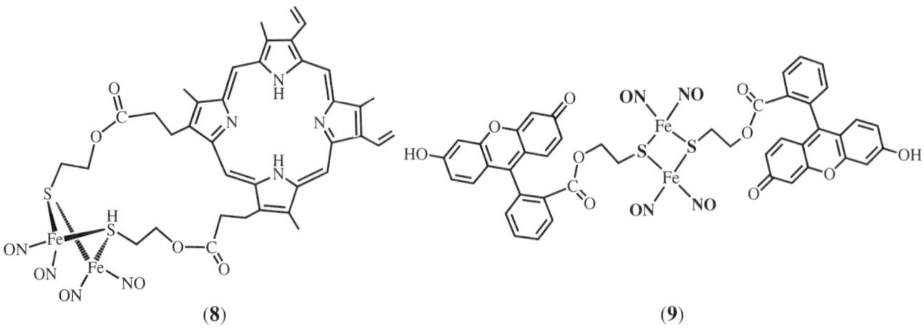

(8)   (9)
PPIX-RSE   Fluo-RSE

Figure 5.   Structure of PPIX-RSE and Fluo-RSE.

co-workers (62) adopted peripheral attachment of light-harvesting chromophores like protophorphyrin-IX (PPIX) and Fluorescein (63) to sensitize Roussin's red salt ester $[Fe_2(\mu\text{-}SR)_2(NO)_4]$ (RSE), which releases NO only upon irradiation at 365 nm. The strong absorption in the red by PPIX dominates the absorption spectrum of the PPIX–RSE (**8**, Fig. 5) and upon exposure to 546-nm light (in CHCl$_3$), **8** releases NO with a quantum yield value ~50% higher than that of RSE alone. In the case of the water-soluble Fluo-RSE conjugate (**9**, Fig. 5), the quantum yield values for NO release upon illumination at 436 nm showed modest improvement ($\phi = 0.0036$). Two-photon excitation with a 800-nm laser pulse (450 mW) also promotes moderate extent of NO release from Fluo-RSE (64). As discussed in Section III.C, such peripheral attachment of light-harvesting antennas does not improve the NO releasing capacities since most of the light energy absorbed by the nitrosyl-dye conjugates is lost through the fluorescence pathway. Also, NO release invariably leads to decomposition of these conjugates in solution.

The problem associated with fast recombination of NO to its precursor in the case of Ru nitrosyl complexes derived from porphyrins can be partially avoided with complexes derived from the tetradentate Schiff base ligand frame with a N_2O_2 donor set, namely, salen N,N'-ethylenebis(salicylideneiminato) dianion, and the related derivatives (65, 66). In aqueous solution, the quantum yield values for NO photo-release (365-nm illumination) from [Ru(R−salen)(NO)(X)] (**10**, Fig. 6) follow the order X = Cl$^-$ > ONO$^-$ > H$_2$O ($\phi = 0.13, 0.067$, and 0.005, respectively) showing the effect of the σ-donor strength of X (trans to NO). At longer wavelengths of light (500–600 nm), the quantum yield values drop significantly. A similar trend is observed with [Ru(R−salophen)(NO)(X)] (**11**, Fig. 6 salophen $= N,N'$-1,2-phenylenebia(salicylideneiminato) dianion. The quantum yield values also change significantly upon different substitution on the ligand frame of these {Ru−NO}^{6} nitrosyl complexes. Clean isosbestic points in the changing absorption spectra under illumination clearly suggest the formation of the Ru(III) aqua species

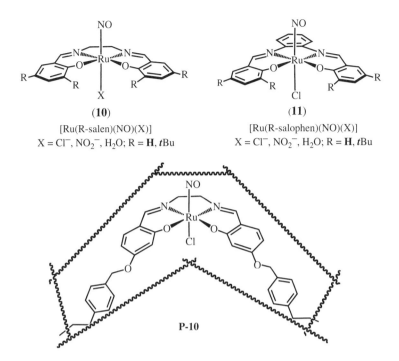

Figure 6.   Structures of Ru nitrosyls with Schiff base ligands and **P-10**.

$[Ru(ligand)(H_2O)(X)]^{n+}$ upon photorelease of NO. Despite slower NO recombination kinetics of these salen-based Ru nitrosyl complexes in comparison to the porphyrin-derived analogues, NO recombination does impede the efficiency of NO photorelease from **10** and **11**. Borovik and co-workers (67) utilized styryloxy unit as appendages to the phenolato ring of salen to entrap the resulting Ru complex [RuCl(P−salen)(NO)] in a methacyralate matrix upon copolymerization of the vinyl group and methacrylate. Despite slower NO release kinetics, the composite material (**P-10**, Fig. 6) exhibits no recombination of NO even in coordinating solvents (e.g., MeCN or H_2O). In phosphate buffer, **P-10** readily delivers NO to reduced myoglobin upon exposure to visible light. The pharmacological potential of these NO donors remains unexplored at this time.

C.   Metal Nitrosyl Complexes Derived from Polydentate Ligands With Carboxamide Groups

While attempts to design photoactivate metal nitrosyl complexes based on the Fe−porphyrin prosthetic group of heme proteins have been unsuccessful due to rapid NO recombination, the remarkable photorelease of NO from the Fe(III) center of the microbial enzyme *nitrile hydratase* (Fe—NHase) (68, 69) has inspired Mascharak and his research group in their design of a library of metal-chelating ligands that coordinate to Fe, Ru, and Mn centers to afford UV, vis, and NIR light activated NORMs. The first series of photoactive nitrosyl complexes reported by this group was derived from the pentadentate ligand, $PaPy_3H$ (*N,N*-bis(2-pyridylmethyl)amine-*N*-ethyl-2-pyridine-2-carboxamide), which incorporates several important features of the metal-binding locus of the Fe—NHase active site including (1) a unique deprotonated carboxamide N-donor group that serves to stabilize the Fe(III) oxidation state observed in the enzyme and (2) a negatively charged donor (thiolate in the case of Fe—NHase) trans to NO that could promote photorelease through the trans-labilization effect. Reaction of $PaPy_3^-$ with Fe(III) starting salts afforded a purple Fe(III) complex that rapidly bound $NO_{(g)}$ to afford the low-spin, diamagnetic $\{Fe-NO\}^6$ complex, $[Fe(PaPy_3)(NO)](ClO_4)_2$ (**12**, Fig. 7). Illumination of **12** in MeCN with low-power, visible light resulted in rapid release of NO ($\phi_{500} = 0.185$) while under dark conditions loss of NO was not observed even after 48 h (70, 71). However, as revealed by the high energy of $\nu_{NO} \sim 1919\,cm^{-1}$ of **12**, considerable transfer of electron density from the $\pi^*(NO)$ antibonding orbitals to Fe renders the NO unit susceptible to nucleophilic attack by OH^- during storage in aqueous solutions limiting the therapeutic potential of **12**.

In order to isolate a stable photoactive metal nitrosyl for the delivery of NO to biological targets, Patra and Mascharak (72) synthesized the $\{Ru-NO\}^6$ nitrosyl complex $[Ru(PaPy_3)(NO)](BF_4)_2$ (**13**, Fig. 7). Although **13** demonstrates indefinite stability in aqueous solutions at a physiologically relevant pH range of 5–9, it requires low-power UV light (5–10 mW, 300–450 nm) for release of NO in solution as detected by an NO sensitive electrode ($\phi_{410} = 0.05$). The photoproduct, the corresponding Ru(III) solvato (solv) species $[Ru(PaPy_3)(solv)](BF_4)_2$, has been characterized by spectroscopic techniques. When one of the py donors in

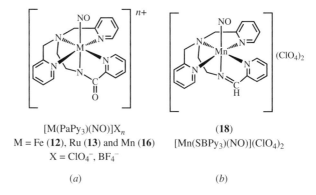

[M(PaPy_3)(NO)]X_n
M = Fe (**12**), Ru (**13**) and Mn (**16**)
X = ClO_4^-, BF_4^-

(*a*)

(**18**)
$[Mn(SBPy_3)(NO)](ClO_4)_2$

(*b*)

Figure 7.   Structures of $\{M-NO\}^6$ complexes derived from $PaPy_3H$ ligand (*a*) and the $\{Mn-NO\}^6$ complex (*b*) of the Schiff base ligand $SBPy_3$.

the ligand frame was substituted with a quinoline (Q) ring, the ligand $PaPy_2QH$ (*N,N*-bis(2-pyridylmethyl)amine-*N*-ethyl-2-quinaldine-2-carboxamide), afforded the isostructural nitrosyl complex $[Ru(PaPy_2Q)(NO)](BF_4)_2$ (**14**) which revealed a shortening of the $Ru-N_{amido}$ bond (1.988 Å for **14** vs 1.999 Å for **13**) as a result of increased σ-bonding interaction between N_{amide} and the Ru center in the case of **14** (73). The increase in σ-bonding of the trans ligand to NO shifted the λ_{max} of **14** to 420 nm from 410 nm for **13** and dramatically improved the quantum yield of NO photorelease ($\phi_{410} = 0.20$ for **14** vs 0.05 for **13**) through enhancement of the $d_{\pi}(Ru) \rightarrow \pi^*(NO)$ transition that increases NO photodissociation. To date, **13** has been employed to deliver NO to various biological targets (e.g., myoglobin and reduced cytochrome *c* oxidase) (74), papain (a cysteine protease) (75), and soluble guanylate cyclase (sGC) (76) under the control of light. The activation of sGC allowed vasorelaxation of rat aorta muscle rings in tissue bath experiments and indicated the potential of its utility as a light-activated vasodilator (76).

The low-spin configuration of the Fe and Ru centers of **12–14** suggests that stabilization of the "hole" in the $(t_{2g})^5$ subshell of the metal center by the auxiliary ligand plays a key role during the reversible binding of NO. The strong trans effect of the carboxamide group of $PaPy_3^-$ dramatically decreases the affinity of metals with a $(t_{2g})^6$ electron configuration toward NO as revealed by the strictly anaerobic conditions required for binding of NO by the $[Fe^{II}(PaPy_3)(solv)]^+$ species to afford $[Fe(PaPy_3)(NO)](ClO_4)$ (**15**) and the rapid conversion of **15** to the nitrito complex in the presence of dioxygen (71). In addition, **15** exhibits no NO photolability. Comparison of the structural parameters of **15** (an $\{Fe-NO\}^7$ complex) with the corresponding diamagnetic $\{Fe-NO\}^6$ complex (**12**) reveals considerable alteration of the $N_{amido}-Fe-NO$ axial vector including lengthening of the Fe−NO bond [1.751(1) Å] and dramatic bending of the Fe−N−O bond angle (142.2°) indicating a Fe(II)−NO• formulation of **15** (confirmed by an $S = \frac{1}{2}$ spin state as measured by superconducting quantum interference device, SQUID).

The excellent NO photolability of the $\{Fe-NO\}^6$ nitrosyl complex (**12**) prompted Mascharak and co-workers (77) to isolate the $\{Mn-NO\}^6$ complex $[Mn(PaPy_3)(NO)](ClO_4)$ (**16**). The soft nature of the Mn(II) center increases the covalency of the Mn−NO bond leading to excellent stability of **16** in biological media while exposure of **16** in aqueous solution ($\lambda_{max} = 635$ nm) to low-power visible light results in rapid labialization of the Mn−NO bond ($\phi_{550} = 0.38$). The photoproduct of **16** undergoes rapid oxidation under aerobic conditions to generate the Mn(III) solvato species $[Mn(PaPy_3)(solv)]^{2+}$, which exhibits no affinity toward NO. Isolation of the $\{Mn-NO\}^5$ species was only possible through electro-chemical oxidation ($E_{1/2} = 0.9$ V vs SCE in MeCN) or chemical oxidation of **16**. While **16** demonstrates excellent photoactivity in the far red region of the visible spectrum, the effectiveness of the designed metal nitrosyl complexes as photo-chemotherapeutics could be greatly enhanced by sensitization of the NO releasing photoband to 700–900-nm NIR light (the diagnostic window). Minimal absorption of light in this region by tissue and body fluids results in increased photoactivity of the nitrosyl during treatment of deep tissue neoplasms (78). In order to achieve the desired NIR light sensitivity, the $PaPy_2QH$ ligand was employed in the synthesis of the $\{Mn-NO\}^6$ nitorysl complex $[Mn(PaPy_2Q)(NO)]^+$ (17), which was the first metal nitrosyl complex to demonstrate biological stability and photoactivation by low-power light (2 mW) in the NIR region up to 900 nm (79). The 100% conversion of 17 into the colorless solvato Mn(II), species $[Mn(PaPy_2Q)(solv)]^+$ upon exposure to 810 nm (4 mW) was indicated by a clean isosbestic point (at 850 nm) in H_2O. The efficient photorelease of NO from the Mn nitrosyl complexes of the $PaPy_3^-/PaPy_2Q^-$ ligands upon exposure to visible light ($\phi_{550} = 0.38$ and 0.70 for 16 and 17, respectively) and NIR light activity of 17 are quite in contrast to the modest photoactivity of the corresponding Ru nitrosyl complexes (13 and 14) that is limited to the UV region.

Density functional theory (DFT) and time-dependent DFT (TD–DFT) studies on the $[M(PaPy_3)(NO)]^{n+}$ (M = Fe, Ru, Mn) nitrosyl complexes indicate that photorelease of NO upon 350–500-nm light irradiation of these $\{M-NO\}^6$ species results from a *direct mechanism* involving excitation of symmetry-allowed electronic transition(s) originating from predominantly $d_{xz}(M)$-$\pi_x^*(NO)$ and $d_{yz}(M)$-$\pi_y^*(NO)$ bonding orbitals [denoted as $(d_\pi(M)$-$\pi^*(NO)]$ with a partial admixture of $\pi_y(N_{amide})$ character (more significant for the Fe and Mn complexes) to orbitals with $d_\pi(M)$-$\pi^*(NO)$ antibonding character (33, 34, 80, 81). In the case of the Mn nitrosyl complexes 16 and 17, an additional *indirect method* for NO photorelease occurs through an initial low-energy (visible and NIR light activated) symmetry-allowed MLCT transition from the nonbonding $d_{xy}(Mn)$ orbital to orbitals with predominantly equatorial ligand π-(Py/Q) character. These orbitals are energetically matched with the T_2 state allowing intersystem crossing and excitation of $d_{xy}(M) \rightarrow d_\pi(M)$-$\pi^*(NO)$ transition(s) that are inaccessible through photoexcitation due to their symmetry-forbidden nature (81). The nature of the indirect mechanism of NO photorelease from these Mn nitrosyl complexes therefore allows tuning of the photoactivity at the desired wavelength through judicious modification of the ligand frame. This binding has been confirmed further through isolation of the $\{Mn-NO\}^6$ complexess of the analogous Schiff base ligands with pyridine $([Mn(SBPy_3)(NO)]^{2+}$, 18, Fig. 7) and quinoline donors $([Mn(SBPy_2Q)(NO)]^{2+}$, 19) (82). Spcectroscopic analysis of the Schiff base derived nitrosyl complexes demonstrates further red shifting of the NO releasing photoband to 720 nm (for 18) and 785 nm (for 19) and an increase in molar absorptivity $\varepsilon_{720} = 750\,M^{-1}\,cm^{-1}$ (18) and $\varepsilon_{785} = 1200\,M^{-1}\,cm^{-1}$ (19). The TD–DFT analysis of the photoactivity of 18 and 19 has revealed a decrease in the energy gap between the first singlet excited state S_1 and the dissociative triplet excited state T_2 $[E(S_1 \rightarrow T_2)]$ in comparison to the carboxamide analogues (16 and 17) due to the decreased ligand field strength of the N_{imine} donor of the Schiff base ligands (83). This reduction facilitates intersystem crossing to the dissociative T_2 state, especially in the case of **19** with $E(S_1 \rightarrow T_2) = 0.04\,\text{eV}$, which displays sensitivity to low-power light up to 1000 nm (82). As described in Section III.D, the robust $\{Mn-NO\}^6$ nitrosyl complex **16** has been extensively employed in delivering NO to biological targets. In such applications, the NORM has been incorporated in a variety of polymeric matrices for site-specific NO delivery.

Mascharak and his research group also employed a designed tetradentate ligand with two carboxamide donors to develop a second set of robust and photoactive $\{Ru-NO\}^6$ nitrosyl complexes that could be used to deliver NO to biological targets. This design opens up a coordination site at the Ru center (with one bound NO) available for attachment of different groups for specific purposes. The ligand is based on the 1,2-bis(pyridine-2-carboxamido)benzene (abbreviated as H$_2$bpb where Hs are dissociable carboxamide protons, Fig. 8). In their first attempt, nitrosyl complexes of the type [Ru(R$_2$bpb)(NO)X]$^{n+}$ (R = H, Me; X = Cl$^-$, py, im, OH$^-$, Fig. 8) were isolated and structurally characterized (84). The photolability of NO from this designed set established a clear trend in the influence of the trans coligand on the photorelease of NO from these $\{Ru-NO\}^6$ nitrosyl complexes of planar dianionic tetradentate ligands. For example, [Ru(Me$_2$bpb)(NO)(py)]$^+$ (**20**) releases NO at an apparent rate of $0.004\,\text{s}^{-1}$ under 302-nm illumination, while [RuCl(Me$_2$bpb)(NO)] (**21**) with a negatively charged ligand Cl$^-$ exhibits enhanced NO photorelease of $0.012\,\text{s}^{-1}$ (84). This result indicated that the ligation of a strong

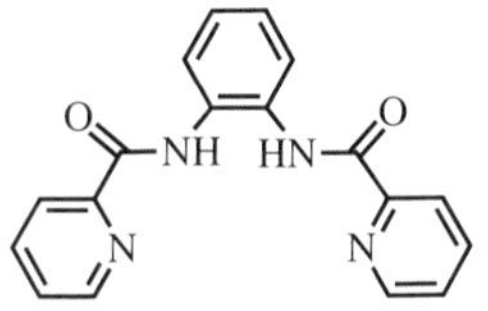

H$_2$bpb

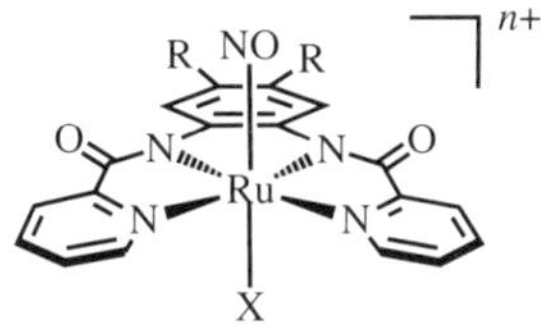

[Ru(R$_2$bpb)(NO)X]$^{n+}$
R = H, Me; X = Cl$^-$, im, py, OH$^-$

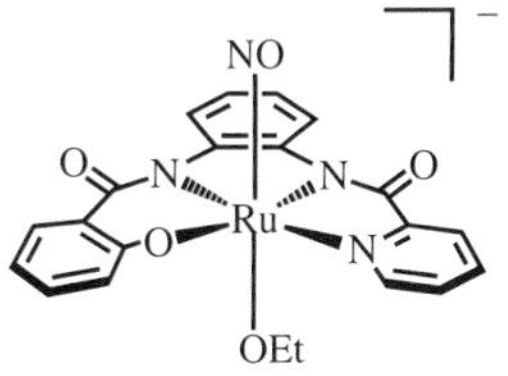

(23)
[Ru(hypyb)(OEt)(NO)]$^-$

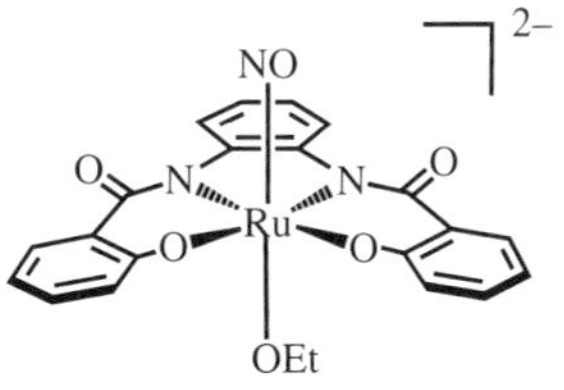

(24)
[Ru(hybeb)(OEt)(NO)]$^{2-}$

Figure 8. Structures of H$_2$bpb and Ru nitrosyl complexes derived from various anionic tetradentate ligands.

trans-labilizing ligand at the sixth site would enhance NO photorelease from this type of complex.

In later attempts, the tetradentate ligand frame was modified to achieve enhancement in photoactivity through an increase in the number of negatively charged donor atoms in the equatorial ligand field. A series of ligands containing a combination of neutral pyridine-N donors and charged phenolato-O donors was synthesized and the photoactivity of the corresponding Ru nitrosyl complexes, namely, [Ru(bpb)(OEt)(NO)] (22), (PPh$_4$)[Ru(hypyb)(OEt)(NO)] (23, H$_3$hypyb = 1-(2-hydroxybenzamido)-2-(2-pyridinecarboxamido)benzene), and (NEt$_4$)$_2$[Ru(hybeb) (OEt)(NO)] (24, H$_4$hybeb = 1,2-bis(2-hydroxybenzamido)benzene) was investigated (Fig. 8) (85). Interestingly, the quantum yield of NO photorelease from this series of nitrosyl complexes follows the order 24 (ϕ_{300} = 0.025) < 22 (ϕ_{300} = 0.051) < 23 (ϕ_{300} = 0.067). In another approach, appendage of peripheral groups with increasing electron-donor strength (in the order H < Me < OMe) to the phenylenedicarboxamide (PDA) portion of the symmetric H$_2$bpb ligand was utilized to improve the NO releasing capacity of the {Ru−NO}^{6} nitrosyl complexes of such tetradentate dicarboxamide ligands (84). Spectroscopic and photochemical analysis of the nitrosyl complexes [Ru(bpb)(Cl)(NO)] (25), [Ru(Me$_2$bpb)(Cl)(NO)] (21), and [Ru((OMe)$_2$bpb)(Cl)Ru(NO)] (26) (86) reveals that the photoactivity is enhanced by increasing electron donation from the PDA portion of the ligand frame to the [M−NO] moiety. In addition to a red shift in λ_{max}, the extinction coefficient and quantum yield values at 500 nm (ϕ_{500}) both increase in the order H < Me < OMe.

Since adequate NO donating capacity under visible and NIR radiation is the primary requirement of a metal nitrosyl complex as an NO donor in photo-therapy (76), Mascharak and co-workers investigated synthetic strategies to move the photoband of {Ru−NO}^{6} nitrosyl complexes into the vis–NIR region. The first approach involves introduction of additional conjugation in the equatorial ligand frame. In such attempts, this group synthesized a set of Ru nitrosyl complexes, namely, [Ru((OMe)$_2$bpb)(Cl)(NO)] (26), [Ru((OMe)$_2$bQb)(Cl)(NO)] (27, H$_2$(OMe)$_2$bQb = 1,2-bis(quinoline-2-carboxamido)-4,5-dimethoxybenzene, Fig. 9) and [Ru((OMe)$_2$IQ1)(Cl)(NO)] (28, H$_2$(OMe)$_2$IQ1 = 1,2-bis(isoquino-line-1-carboxamido)-4,5-dimethoxybenzene, Fig. 9) in which the extent of conju-gation in the equatorial ligand frame was increased in a systematic manner (86). Scrutiny of the absorption characteristics of 26–28 revealed a red shift in the λ_{max} of the photoband of 70 nm upon incorporation of quinoline rings (in 27, λ_{max} = 490 nm) in place of the pyridine (in 26, λ_{max} = 420 nm), while the corresponding isoquinoline-substituted nitrosyl complex (28) displayed an intermediate shift of the photoband with λ_{max} at 475 nm. The twist in the equatorial plane of 27 (dihedral angle of 30°) due to steric interactions between two cis-coordinated quinoline groups was relieved in 28 (effectively planar) upon exchange of the quinolone units with isoquinoline. Interestingly, the planarity in the equatorial

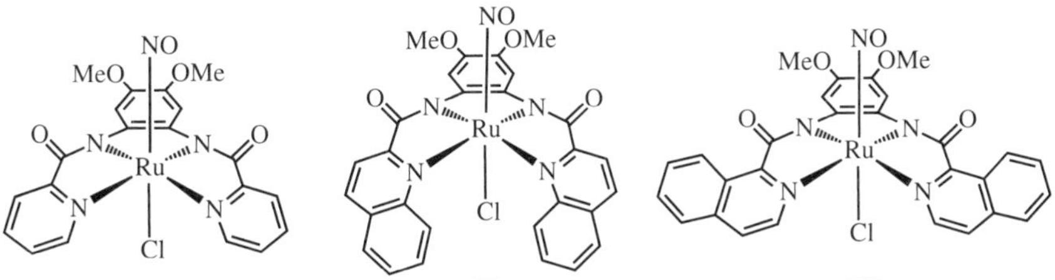

(26)   (27)   (28)

[Ru((OMe)₂bpb)(Cl)(NO)]   [Ru((OMe)₂bQb)(Cl)(NO)]   [Ru((OMe)₂IQ1)(Cl)(NO)]

Figure 9.   Structures of $\{Ru-NO\}^6$ nitrosyl complexes with increasing conjugation in the equatorial ligand frame.

plane enhanced the ϕ_{500} value of **28** to 0.035 compared to the ϕ_{500} value of **27** (0.025). In addition, the molar absorptivity of **28** was increased significantly.

The second approach by Mascharak and his research group, to move the photoband of $\{Ru-NO\}^6$ nitrosyl complexess to the vis/NIR region, involves *direct attachment* of a light-harvesting dye molecule as the sixth ligand to the Ru center. This approach is distinct from that employed by Ford and his research group (*peripheral attachment* of light antennas) and affords dye–nitrosyl conjugates that exhibit superior NO photolability upon illumination with low-power visible light. The first set of dye–nitrosyl conjugates utilized the fluorescent dye resorufin (Resf) that absorbs strongly in the visible region ($\varepsilon_{600} = 105,000 \, M^{-1}$ cm^{-1}). Direct ligation of the phenolato end of Resf to the Ru center (trans to NO) in [Ru(Me₂bpb)(Resf)(NO)] (**29**), [Ru(Me₂bQb)(Resf)(NO)] (**30**), [Ru((OMe)₂bQb) (Resf)(NO)] (**31**), and [((OMe)₂IQ1)Ru(Resf)(NO)] (**32**) (Fig. 10) was achieved by replacing the chloride ligand of the corresponding $\{Ru-NO\}^6$ complexes with Ag$^+$ in MeCN (87, 88). These dye–nitrosyl conjugates exhibit intense absorption bands in the ~500-nm region and enhanced NO photorelease with quantum yield values at 500 nm (ϕ_{500}) in the range of 0.20–0.30. Results of DFT and TD–DFT studies indicate that the dye-based molecular orbitals (MOs) are intercalated within the MOs of the Ru−NO moiety and a favorable transition from a π(dye) orbital to a mixed $\pi^*(dye) \| d_\pi(Ru)-\pi^*(NO)$ MO enhances the photolability of NO in these nitrosyl–dye conjugates (88). The mixing of the dye orbital with orbitals of Ru−NO antibonding character allows efficient energy transfer through the Dexter pathway (89). The extent of photosensitization by direct binding of the light-harvesting dye depends on the extent of spectral overlap of the dye $\pi \rightarrow \pi^*$ transition(s) with the photobands of the parent $\{Ru-NO\}^6$ nitrosyls. This mechanism has been verified through heavy atom substitution for the O atom of Resf with S and Se and determining the degree of enhancement of NO photosensitivity of the corresponding dye–nitrosyl conjugates. Change of the Resf dye of

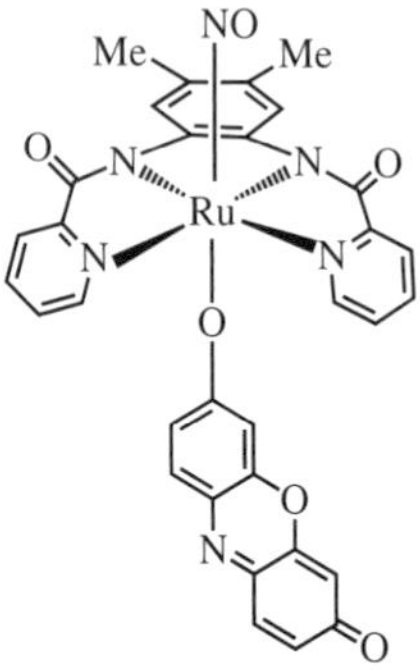

(29)
[Ru(Me₂bpb)(Resf)(NO)]

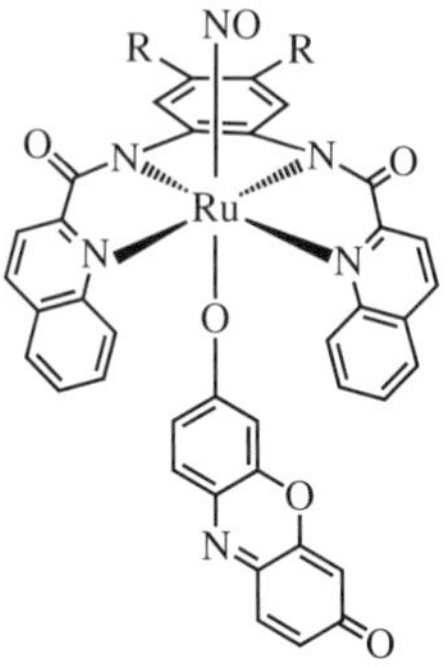

[Ru(R₂bQb)(Resf)(NO)]
R = Me (30), OMe(31)

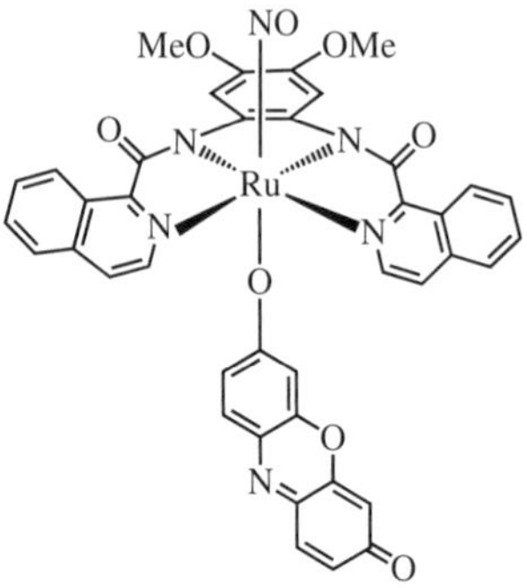

(32)
[Ru((OMe)₂IQ1)(Resf)(NO)]

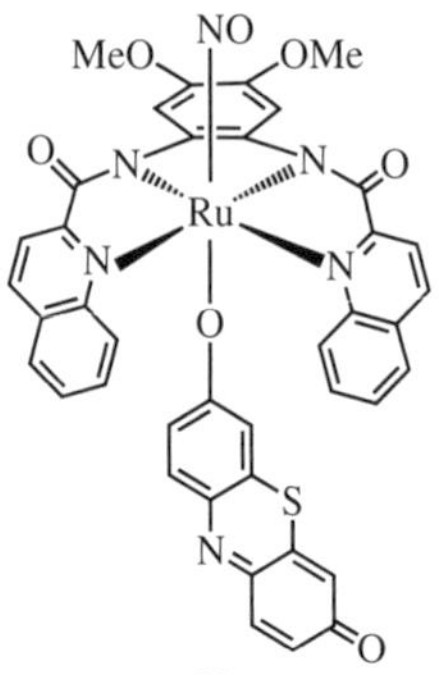

(33)
[Ru((OMe)₂bQb)(Thnl)(NO)]

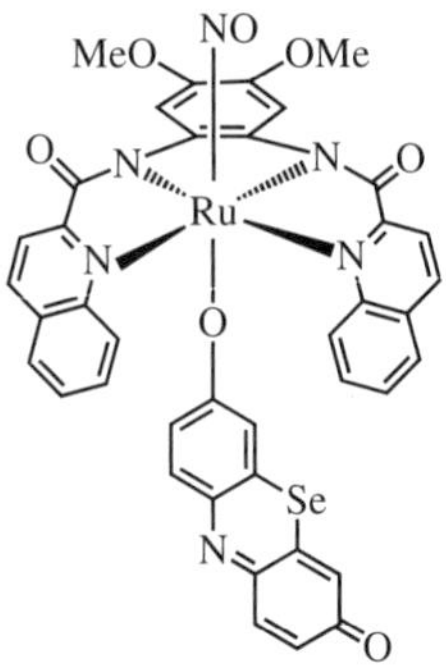

(34)
[Ru((OMe)₂bQb)(Seln)(NO)]

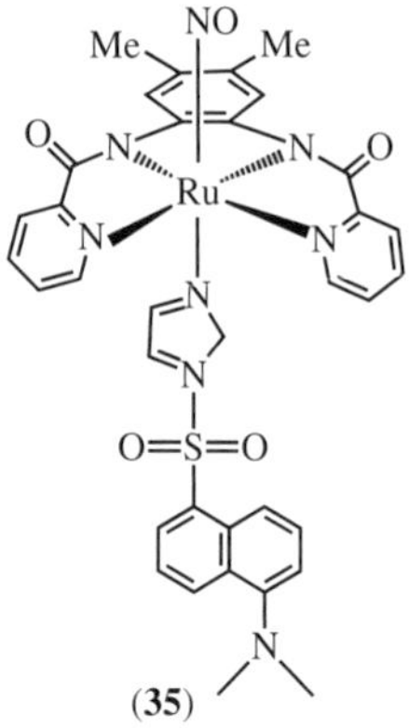

(35)
[Ru(Me₂bpb)(Ds-im)(NO)]BF₄

Figure 10.   Dye-tethered {Ru$-$NO}^{6} nitrosyl complexes that exhibits NO photolability upon exposure to visible light.

[Ru((OMe)₂bQb)(Resf)(NO)] (**31**) with Thnl (Thnl = thionol; Resf with O replaced by S) and Seln (Seln = selenophore; Resf with O replaced by Se) in [Ru((OMe)₂bQb)(Thnl)(NO)] (**33**) and [Ru((OMe)₂bQb)(Seln)(NO)] (**34**) caused red shift of the photoband from 500 to 525 to 535 nm (90). This result demonstrates that direct binding of dye molecules can sensitize {Ru$-$NO}^{6} nitrosyls toward visible light quite readily. Indeed, **33** exhibits modest NO photolability at 600 nm.

The dye–nitrosyl conjugates designed by Mascharak and co-workers (88) are excellent candidates as phototherapeutic agents for promoting NO induced apoptosis in cellular targets. For example, **29** has been employed to deliver NO to malignant cells under the control of low-power visible light (450–600-nm) light. In one experiment, light-triggered NO delivery from **29** caused apoptosis (as indicated by DNA fragmentation and cellular morphology) in human breast cancer cells (MDA-MB-231) within 6–8 h following illumination for 1 min. More experiments are, however, required before these nitrosyl–dye conjugates could find use as photopharmaceuticals in clinical settings. Another advantage of the dye–tethered nitrosyls lies in the fact that the NO delivery can be easily *tracked* in cellular matrices due to the strong fluorescence of these diamagnetic $\{Ru-NO\}^6$ nitrosyl complexes. Since low-spin Ru(III) paramagnetic photoproducts are generated in all cases, the fluorescence of the dye–nitrosyl conjugates is quenched upon NO photorelease. The kinetics of NO photorelease correlates with that of fluorescence quenching quite closely. Consequently, these dye–tethered Ru nitrosyls can be used as "NO donors with a fluorometric on–off light switch" (91). An example of trackable NO donor is exemplified by [Ru(Me₂bpb)(Ds-im)(NO)]BF₄ (**35**, Ds-im = Dansyl-imidazole Fig. 10). When MDA-MB-231 human breast cancer cells were incubated with **35**, entry of the dye–nitrosyl conjugate inside the cells was indicated by a strong green fluorescence. Rapid loss of this fluorescence upon illumination of the loaded cells with low-power visible light (for 1 min) clearly indicated photorelease of NO within the cells prior to NO induced apoptosis noted within the next 4–6 h.

The dye–nitrosyl conjugates **29–35** are all "turn-off" NO donors because of the loss of their fluorescence upon NO delivery. Recently, Mascharak and co-workers (92) also reported a set of dye-tethered $\{Ru-NO\}^6$ nitrosyl complexes that act as "turn on" NO donors. In this set, Fluorescein ethyl ester (FlEt) has been ligated to the Ru center through the phenolato-O end of the dye. The two dye–nitrosyl conjugates, namely, [Ru(Me₂bpb)(FlEt)(NO)] [**36**, Fig. 11)] and [Ru((OMe)₂IQ1)(FlEt)(NO)] (**37**) absorbs strongly in the 450–600-nm region. Exposure to visible light promotes rapid loss of NO from these nitrosyls to

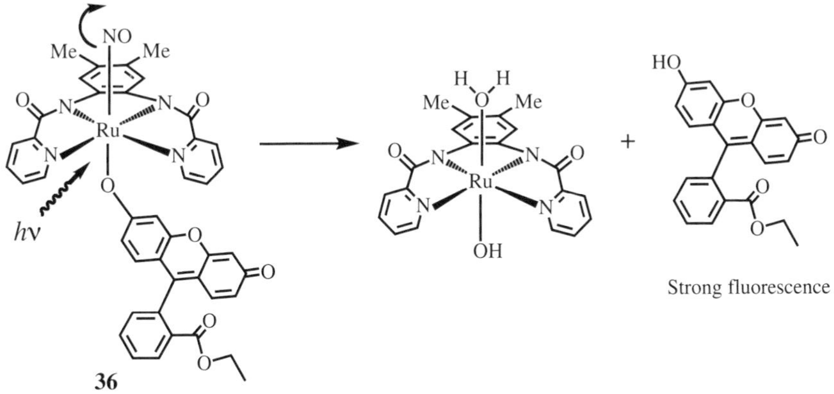

Figure 11. "Turn on" fluorescence upon NO photorelase from [Ru(Me₂bpb)(FlEt)(NO)] (**36**) in aqueous media.

generate the usual solvent bound Ru(III) photoproducts in dry aprotic solvents. The FlEt⁻ ligand remains bound to the paramagnetic Ru(III) center in these solutions and very weak fluorescence from the dye unit is noted. However, in the presence of water the Ru(III) photoproducts undergo further aquation and loss of FlEt⁻ ligand through protonation (Fig. 11). Consequently, a dramatic rise in fluorescence is observed (due to free FlEt) upon NO release from **36** and **37** in aqueous solution. This increase in fluorescence could provide a visual signal of the NO photorelease from **36** and **37** in cellular matrices.

D.   Polymer Matrices With Incorporated Metal Nitrosyl Complexes

The toxicity of the transition metal nitrosyl complexes and their photoproducts often raises concern related to their utility as pharmaceuticals. In addition, site-specific NO delivery requires confinement of these NO donors within the biological targets. In order to address both these issues, metal nitrosyl complexes have been encapsulated or immobilized within polymer matrices (31, 93). An example of such confinement is **P-10** (Fig. 6), reported by Borovik and co-workers (67). Recently, Mascharak and his research group have synthesized a variety of nitrosyl–polymer conjugates that have been employed to combat chronic infection of bacterial and fungal origin in site-specific fashion. To begin with, this group utilized sol–gel (SG) chemistry (employing silicon alkoxides) for the development of a diverse set of composite materials. The mild reaction conditions of SG processing and the ability to control important properties (e.g., the optical transparency, biocompatibility, chemical and thermal stability, and morphology) afforded several composites with the desired properties (94). The highly efficient NO donors [Mn(PaPy₃)(NO)](ClO₄) (**16**) and [Mn(PaPy₂Q)(NO)](ClO₄) (**17**) have been encapsulated within a silica based SG to afford transparent and flexible green **16·SG** and maroon **17·SG** patches that rapidly released NO ($\phi_{532} = 0.25$ and 0.55, respectively) upon exposure to low-power (5–10 mW) visible (in the case of **16·SG**) or NIR light (in the case of **17·SG**) (79, 95). A thin coat of polyurethane, a biocompatible and NO permeable polymer, was applied to the exterior of these patches to inhibit leaching of the incorporated nitrosyl and its photoproduct(s) while permitting $NO_{(g)}$ diffusion. Storage at low temperatures hinders solvent evaporation from these hydrogels and extends the shelf life of **16·SG** and **17·SG** to weeks. Both **16·SG** and **17·SG** deliver NO to biological targets (e.g., heme-proteins) under the complete control of light. The NO releasing properties of **16·SG** and **17·SG** demonstrate that the extent of NO release from such materials is proportional to the duration of light exposure and that the rate of NO release can be controlled by (1) varying the intensity of light and (2) raising the concentration of the Mn nitrosyls. These features confirm that antibiotic quantities of NO could be delivered to microbial pathogens in a site-specific fashion with such nitrosyl-impregnated SG composites.

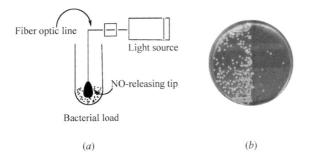

Fiber optic line

Light source

NO-releasing tip

Bacterial load

(a)   (b)

Figure 12. (a) Schematic representation of remote NO delivery to bacterial culture with the aid of **16-FO**. (b) Growth of *S. aureus* in the absence of NO treatment (left side) and lack of growth upon treatment with photoreleased NO from the catheter tip (right side).

The antibiotic capabilities of photoreleased NO from the nitrosyl–SG composites has been tested with a remote delivery platform constructed by casting a **16·SG** tip at the end of a fiber optic (FO) line to afford the endoscopic devise **16·FO** (Fig. 12) (96). The FO line guided light to the tip of **16·FO** upon coupling to a light source and initiated NO photorelease from **16**. When this endoscope was immersed into narrow cavities containing clinically significant (10^5 CFU mL^{-1}, CFU = colony-forming unit) bacterial loads of *Staphylococcus aureus, Pseudomonas aeruginosa*, and *Escherichia coli* and the light was turned on, rapid activation of **16** released μM levels of NO for a period of 5–10 min. Such NO flux eradicated the bacterial loads quite efficiently (Fig. 12). These results suggest that remote activation of these nitrosyl–polymer composites by sending light through fiber optic cable could greatly expand the applicability of photoactive metal nitrosyls to combat chronic bacterial infections in body cavities (e.g., deep inside urethra and in ventilator-related infections).

Evaporation of solvent from the pores of the SG materials (e.g., **16·SG** or **17·SG**) affords a brittle, glass-like xerogel that is highly susceptible to fragmentation. Mascharak and co-workers (97) employed such xerogels to develop a new NO donating platform. This group has dispersed nitrosyl-containing xerogel particles (up to 3 mol% of **16**) within a silicate matrix in liquid polyurethane swelled with excess water and then spin-casted the thick mixture onto a glass slide and allowed it to cure for 16 h. The procedure affords a flexible green film (**PUX-NO**) that rapidly releases NO under visible light. When the PUX–NO films containing various concentrations of **16** were placed over nutrient-rich agar plates inoculated with 10^5 CFU mL^{-1} of bacteria common to skin and soft-tissue infections (including a strain of methicillin-resistant *S. aureus*) and then illuminated for 1–2 h, a dose-dependent decrease in bacterial colonies was observed. It is evident that antibacterial films of this type could find use as linings of hospital stretchers and beds for patients with severely infected wounds. Another hydrogel formulation, Pluoronic® F127 gel, with thermosetting properties, has been employed to dissolve **16** in its liquid state at $0\,°C$, and then transferred to a polydimethylsiloxane (PDMS) mold at room temperature to solidify (98). The high moisture content of the Pluoronic gel reservoir ensures adequate solvation of the metal nitrosyl while the PDMS shell provides structural integrity to this hybrid material. Upon illumination, this material releases NO that has been shown to reduce bacterial loads quite efficiently.

Microbial infection in burn tissue is often difficult to treat due to severe tissue damage and loss of vasculature. Such infections also hinder wound repair to a significant extent. In many cases, the wound healing process can be greatly facilitated by treatment with NO, which has been shown to improve wound healing by increasing angiogenesis and the production of extracellular matrix in addition to its antimicrobial properties (99). With this in mind, Mascharak and co-workers (100) incorporated two $\{Mn-NO\}^6$ nitrosyl complexes, namely, **16** and [Ru(Me$_2$bpb)(NO)(4-vpy)]BF$_4$ (**38**, 4-vpy = 4-vinyl pyridine) into a poly(2-hydroxyethyl methacrylate) (pHEMA) hydrogel, which is widely used as a wound dressings due to its excellent swelling capacity, oxygen permeability, and bio-compatibility. When patches of pHEMA with incorporated **16** and methylene blue (a singlet oxygen generator) were illuminated in cultures of *P. aeruginosa* and *E. coli*, the combination of NO and singlet oxygen produced a potent antibiotic effect resulting in dramatic clearing of the bacterial loads (101). The permeability of **16·pHEMA** requires coating of the material with polyurethane in order to retain the metal nitrosyl and its photoproducts inside the dressing material. The need for such coatings has been circumvented in a formulation employing **38**, which contains a polymerizable vinyl group as the axial ligand. This axial ligand of the nitrosyl complex copolymerizes with HEMA and the cross-linker, ethylene glycol dimethacrylate, during radical induced polymerization and firmly attaches it with the polymer backbone through a covalent bond. The composit **38·pHEMA** demonstrates no leaching of the covalently attached nitrosyl complex and rapidly releases NO upon exposure to low-power UV light (5 mW) while the photoproduct remains bonded to the polymer frame (102). Among other biocompatible dressing materials, polylactide-based hosts also have been employed to immobilize metal nitrosyls. For example, Schiller and co-workers (103) reported incorporation of a $\{Ru-NO\}^6$ nitrosyl complex derived from the *N*-methyl imidazole analogue of H$_2$bpb, namely, [RuCl(impb)(NO)] (**39**) into poly(L-lactide-*co*-D,L-lactide) nano-fibers with the aid of electrospinning. Non woven scaffolds derived from such fibers release NO upon illumination with UV light (366 nm) and exhibit low toxicity against 3T3 mouse fibroblasts.

Silica-based nanostructured materials have drawn attention as hosts for metal nitrosyls in the current design of NO donating materials. Mascharak and co-workers (104) incorporated the Mn nitrosyl complex (**16**) into a mesoporous silicate material Al-MCM-41 with hexagonally packed unilateral pore structure.

The Al substituted sites give rise to negative point charges throughout the porous framework and trap the cationic nitrosyl (**16**) through strong electrostatic interaction to afford a loading of 25 wt% **16** within the Al-MCM-41 host. Elemental dispersive X-ray spectroscopy and inductively coupled plasma mass spectrometry (MS) data demonstrate the homogeneity of nitrosyl loading throughout the porous structure. Exposure of **16**@Al-MCM-41 to visible light (sunlight on a bright day) releases high fluxes of NO that effectively eradicates the extensively drug resistant Gram-negative bacterium *Acenatobacter baumannii* (XDR-AB), commonly observed in wounds of soldiers involved in recent conflicts in the Persian Gulf region. Clearly, this powdery material shows potential as a first-line antibiotic treatment for battlefield wounds infected with endemic pathogens.

The quest for NIR promoted NO delivery in recent years has recruited innovative modes of photoexcitation in addition to new materials. Ford and co-workers (105) employed $NaYF_4$ nanocrystals doped with lanthanide cations to upconvert 980-nm laser light (from a NIR diode laser) into the UV and visible range. These upconverting nanocrystals (UPNCs) were first coated with silica, then functionalized with terminal amines, and finally conjugated to **RBS** (Fig. 2) to afford UPNCs@SiO_2-**RBS**, a composite material that releases NO upon irradiation with 980-nm NIR light in aqueous solution (pH 5). Taken together, these results indicate that novel composite materials with incorporated metal nitrosyl complexes and new techniques of phototriggering could eventually merge into a new treatment modality in NO based phototherapy.

IV. PHOTOACTIVE METAL CARBONYL COMPLEXES

The photosensitivity of metal carbonyls is known for quite sometime and the mechanisms of CO release from them upon illumination have been extensively studied (106). The principles previously established are, however, being revisited due to renewed interest in the design of metal carbonyls as CORMs. Metal carbonyls are known to undergo CO loss via several pathways, with excitation of ligand field (LF) and MLCT states being the most prominent factors contributing to CO photodissociation. Using group 6 (VI B) hexacarbonyls as archetypical compounds, the classical interpretation of photoinduced CO ligand dissociation has been attributed to ligand field excitations. Single electron excitation into the e_g^* orbital of $Cr(CO)_6$ reduces the number of bonding electrons in the t_{2g} orbitals, while populating a strongly antibonding $d_z^2(Cr)$ -$e_g^*(5\sigma$-CO) orbital (107). The net effect of LF excitation is loss of an antibonding CO, with concomitant lowering of the d_z^2 orbital energy and loss of symmetry in the formation of $Cr(CO)_5$. Theoretical results are consistent with the original assignment of ligand field transitions in $Cr(CO)_6$ corresponding to dissociative CO loss (108). This standard model of LF excitation leading to photoinduced CO dissociation was employed thereafter to account for CO loss in various metal carbonyl complexes.

A. Homoleptic Metal Carbonyls

The first compounds to be employed as CORMs were commercially available homoleptic metal carbonyls because of the well-established photoactivity of this class of compounds. Motterlini et al. (109) selected the mono and dinuclear metal carbonyls $Fe(CO)_5$ and $Mn_2(CO)_{10}$ (Fig. 13) and demonstrated that irradiation with a cold light source leads to CO release. The ability of such compounds to mimic endogenously produced CO was also confirmed by their incorporation in vasorelaxation studies involving precontracted rat aortic rings. Mechanisms concerning CO loss can vary depending on the nature of the starting carbonyl and the incoming ligand and may even be further complicated by recombination of the photoproduct(s). For $Fe(CO)_5$, quantum yield measurements by Burkey and co-workers (110) previously accounted for the formation of both the mono and disubstituted carbonyl species $Fe(CO)_4L$ and $Fe(CO)_3L_2$ upon UV irradiation. The compound $Fe(CO)_4$ has been the common intermediate leading to each distinct photoproduct. Indeed, $Fe(CO)_5$ is a unique example wherein substitution of two CO groups occurs per photon absorbed. In the case of $Mn_2(CO)_{10}$, CO loss has been recognized to proceed via a dissociative process wherein the transient species $Mn_2(CO)_9$ is generated, as opposed to the typical metal–metal bond cleavage observed during continuous photolysis of dinuclear metal carbonyls (111). While effective in fostering prolonged vasodilation, these homoleptic metal carbonyls are mostly insoluble in aqueous solutions and have limited applicability. An impetus in the development of metal carbonyls for controlled CO release has therefore centered on the smart design of ligand frames to not only impart the desired solubility in biologically relevant media, but to also sensitize the resulting carbonyls to low-energy light.

Recent theoretical studies have assigned a low-energy transition in $Cr(CO)_6$ as a symmetry forbidden charge transfer (CT) state (as opposed to ligand field) with ^1MLCT character (T_{2u}), the potential energy curve of which crosses over to LF at increasing metal−CO bond lengths (112). Upon irradiation, the single electron excitation initially populates a low energy ^1MLCT ($d_\pi \rightarrow \pi^*$) state that, upon

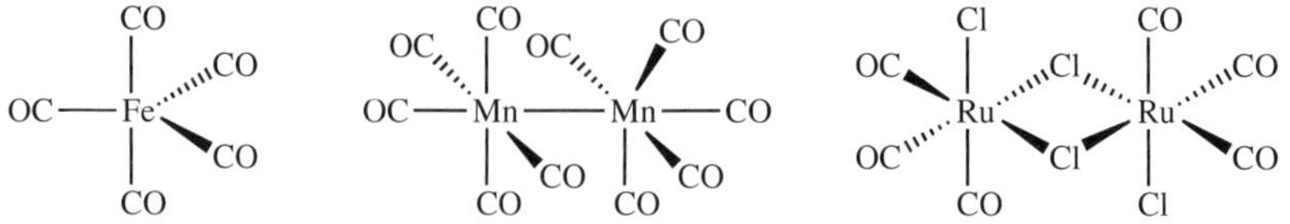

Figure 13.   Homoleptic CORMs employed in early biological studies.

further mixing with an excited ^{1}LF ($d_\pi \rightarrow d_{\sigma*}$) state, contributes to the observed metal–CO bond labilization. This updated model of a mixed ^{1}(MLCT/LF) state explaining metal–CO bond lability has now allowed for the tuning of CO release via incorporation of *designed ligands* with low-lying π^* orbitals. Lowering of CT states has been noted with the incorporation of ligands containing low-lying orbitals, such as the α-diimines, bpy (2,2$^\prime$-bipyridine) and R-DAB (substituted 1,4-diaza-1,3-butadiene) in a number of metal centers (43). Initial excitation onto a low-lying MLCT state has been known to cause thermal population of LF excited states resulting in photosubstitution, as has been observed for [Ru(bpy)$_3$]$^{2+}$, [W(CO)$_5$(py-x)], and [Ru(NH$_3$)$_5$(py-x)] (113). Ford and co-workers (114) reported the photochemical reaction of a water soluble salt Na$_3$[W(CO)$_5$(TPPTS)] (TPPTS^{3-} = tris(sulfonatophenyl)phosphine trianion) under UV irradiation and coined the term "photoCORM" to describe this type of light-triggered CO releasing compound. Aerobic oxidation following the initial photolysis leads to a total yield of 1.2–1.6 equiv CO per molecule of Na$_3$[W(CO)$_5$(TPPTS)], suggesting the formation of various photoproduct(s) in the photolyzed solution.

B. Metal Carbonyl Complexes With Amino Acid Ligands

Darensbourg et al. (115) investigated the use of amino acids and their derivatives as ligands in promoting CO dissociation from Cr and W amino acid complexes. Rates of CO loss from complexes of the type [W(O$_2$CCH(R)NHR$^\prime$)(CO)$_4$]$^+$ (where R = H, *t*-Bu; R$^\prime$ = H, Me$_3$) were found to be unaffected by substituents on the methylene carbon, but were significantly enhanced when R$^\prime$ = H. This mechanism points to NH hydrogen abstraction in solution to afford a CO labilizing amido functionality (as in **40**, Fig. 14), which then enhances CO dissociation rates. Metal complexes bearing amino acid, amino ester, or non-natural amino acid moieties have found renewed utility in the area of targeted CO delivery via solvent-assisted CO release mechanisms. The complex [RuCl(glycinate)(CO)$_3$], developed by Motterlini and co-workers (116) (CORM-3, Fig. 1), is the prototypical water-soluble CORM in

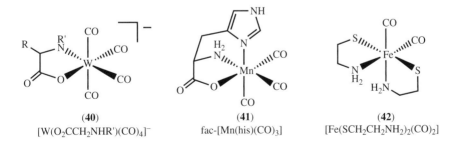

(40)
[W(O$_2$CCH$_2$NHR')(CO)$_4$]$^-$

(41)
fac-[Mn(his)(CO)$_3$]

(42)
[Fe(SCH$_2$CH$_2$NH$_2$)$_2$(CO)$_2$]

Figure 14.   The CORMs with amino acid as ligand.

this area. Facile CO loss from CORM-3 is pH dependent and arises from base (hydroxide) attack on the carbonyl leading to the formation of either $[Ru(CO_2)Cl$ (glycinate)$(CO)_2]^{2-}$ or $[Ru(CO_2H)(glycinate)(OH)(CO)_2]^-$. Solvent-assisted CO release from CORM-3 has been shown to (1) induce vasodilation in precontracted rat aorta muscle rings (117), (2) confer protection against cardiac allograft rejection in mice (118), and (3) provide beneficial effects during postischemic myocardial recovery (119). No light-induced CO release has, however, been reported for CORM-3. Interestingly, the histidinato (his) complex fac-[Mn(his)$(CO)_3$] (41, Fig. 14 where his = histidinato) is both water-soluble and undergoes light-induced CO release as evidenced by a myoglobin assay (120). Its sensitivity to UV radiation (365 nm) renders it a photoCORM best suited for slow CO release applications due to its half-life of ~93 min. While typical amino acid metal complexes have either mono or bidentate amino acid donors, the deprotonated histidine in fac-[Mn(his)$(CO)_3$] acts as a tridentate ligand through the carboxylate oxygen and nitrogen atoms of the amine and imidazole groups (Fig. 14).

Although not strictly incorporating an amino acid framework, the compound dicarbonyl-bis(cysteamine)iron(II) (42, Fig. 14, referred to as CORM-S1) features an N/S ligand donor set (121). The combination of low-valent iron centers and coordinated thiolates has been known to result in metal carbonyls exhibiting reversible CO desorption properties (122), making this particular motif suitable in the design of CO carriers. The CORM-S1 compound exhibits CO photorelease upon exposure to visible light ($\lambda = 400$ nm) yielding two CO molecules per CORM-S1. This compound stimulates voltage and Ca^{2+} activated potassium channels in a light-dependent manner making it a viable photoCORM for physiological applications. Interestingly, no CO photorelease has been noted with the Ru analogue of CORM-S1 (123).

C. Manganese(I) Tricarbonyl Complexes

The facile coordination of a variety of Lewis bases including arenes, phosphines, thioethers, and other heterocyclic moieties onto the $Mn(CO)_3^+$ fragment has led to the characterization of a series of fac-Mn(I) tricarbonyl complexes, which exhibit CO photodissociation when irradiated with UV and/or visible light. Revisiting the work of Trofimenko, Schatzschneider and co-workers (124) reported the first in a series of fac-Mn(I) tricarbonyl complexes bearing tripodal ligand frames in the preparation of fac-[Mn(CO)$_3$(tpm)]PF$_6$ [43, Fig. 15, tpm = tris (pyrazolyl)methane]. This carbonyl was shown to release CO upon exposure to 365-nm light and it demonstrated a potential to be a phototherapy agent. When HT29 colon cancer cells were incubated with 43, a light-dependent reduction in cell biomass was observed and the result was comparable to the effect of the known cytotoxic agent 5-fluorouracil. Unlike previous applications of CORMs in cyto-protective studies, 43 has so far been utilized mostly as a cytotoxic agent that

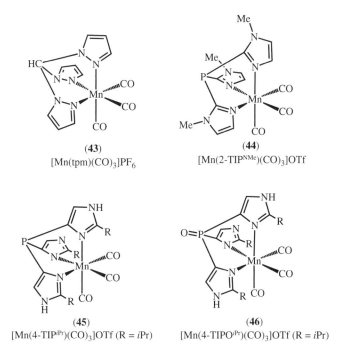

(43)
[Mn(tpm)(CO)₃]PF₆

(44)
[Mn(2-TIP^NMe)(CO)₃]OTf

(45)
[Mn(4-TIP^iPr)(CO)₃]OTf (R = iPr)

(46)
[Mn(4-TIPO^iPr)(CO)₃]OTf (R = iPr)

Figure 15.   Structures of the Mn(I) photoCORMs with *fac*-Mn(CO)₃ unit.

delivers CO to malignant sites. In such attempts, **43** has been employed in a variety of CO delivery systems including conjugation with model peptides (125), nano-particles (126), and nanodiamonds (127).

To elucidate the factors that dictate CO photorelease from the *fac*-Mn(I) tricarbonyl complexes, Schatzschneider and co-workers (128, 129) expanded their series of tripodal ligands to include imidazole moieties [Fig. 16(a)] . Within the subset of the tricarbonyl[tris(imidazolyl)phosphine]manganese(I) complexes (**44–46**, Fig. 15), the substitution pattern on the imidazolyl moiety was found to influence the CO releasing efficiency and stoichiometry of the resulting photo-active *fac*-Mn(I) tricarbonyls. In compounds bearing tris(imidazol-2yl)phosphine ligand (2-TIP^H) frames 2 equiv of CO per mole of complex were released, while only 1 equiv of CO was released per molecule when tris(imidazol-4yl)phosphane (4-TIPO^iPr) was employed as the ligand (128). In addition to the substitution pattern on the imidazolyl group, steric bulk of the substituents was also found to significantly influence CO release kinetics for this series of photoCORMs (129).

Mascharak and co-workers (130) reported a series of photoactive manganese tricarbonyl complexes derived from tripodal amine ligands, namely, *fac*-

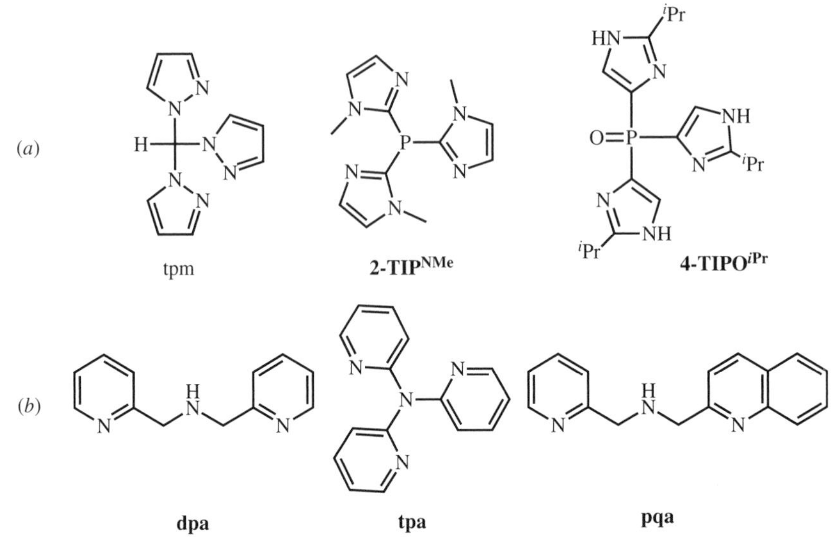

Figure 16.   Examples of ligands employed to isolate *fac*-Mn(I) tricarbonyl complexes.

[Mn(tpa)(CO)₃]ClO₄ [**47**, tpa = tris(2-pyridyl)amine], *fac*-[Mn(dpa)(CO₃)]ClO₄ [**48**, dpa = *N,N*-bis(2-pyridylmethyl)amine], and *fac*-[Mn(pqa)(CO)₃]ClO₄, [**49**, pqa = (2-pyridylmethyl)(2-quinolylmethyl)amine] with varying degrees of conjugation in the ligand framework (Fig. 17). An increase in the number of pyridine N donors and/or extent of conjugation in the ligand frame was found to enhance the extent of CO photolabilization under low-energy light. The designed carbonyl (**49**) with both pyridine and quinolone moieties in the ligand frame exhibits the most red-shifted absorption maximum (360 nm) in this series. This carbonyl is stable in aqueous buffers and promotes rapid CO induced vasorelaxation of mouse aortic muscle rings in tissue bath experiments under the control of visible light (130). Kunz and co-workers (131) incorporated **47** (and one of its analogues) in biocompatible 2-hydroxypropyl methacrylamide (HPMA) based polymer matrix and studied the utility of the carbonyl–polymer conjugate in CO delivery to malignant targets. Although the carbonyl–polymer conjugate releases CO upon illumination at 365-nm light (as evident by myoglobin assay), it exhibits marginal toxicity to human colon carcinoma and HepG2 human hepatoma cells.

As discussed in Section IV.A, the affinity of CO toward low-valent metal centers arises from back-donation of electrons from the metal center to the π^* orbital of CO. It is therefore expected that transfer of electron density from the metal center into orbitals of predominantly ligand character through MLCT transition will disrupt such back-bonding leading to CO photorelease. This

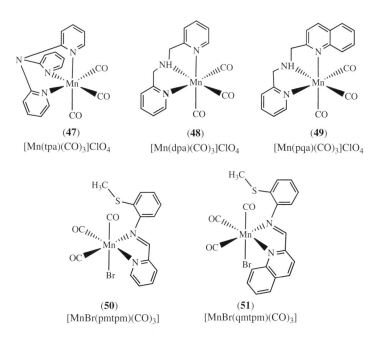

(47)
[Mn(tpa)(CO)₃]ClO₄

(48)
[Mn(dpa)(CO)₃]ClO₄

(49)
[Mn(pqa)(CO)₃]ClO₄

(50)
[MnBr(pmtpm)(CO)₃]

(51)
[MnBr(qmtpm)(CO)₃]

Figure 17.   The Mn(I) photoCORMS with N-donor ligands.

hypothesis has been tested by Mascharak and co-workers (132) who employed Schiff base ligands with extended conjugation to isolate a set of Mn(I) carbonyls that strongly absorb in the 400–550-nm region. In this set, two carbonyls, namely, *fac*-[Mn(pmtpm)(CO)₃Br] [**50**, pmtpm = 2-pyridyl-*N*-(2′-methylthiophenyl)methyleneimine), Fig. 17] and *fac*-[Mn(qmtpm)(CO)₃Br] [**51**, qmtpm = 2-quinoline-*N*-(2′-methylthiophenyl)methyleneimine, Fig. 17] exhibit MLCT absorption bands at 500 and 535 nm, respectively. Both release CO (ϕ_{509} = 0.34 and 0.37, respectively) upon exposure to low-power (5-mW) visible light. Results of DFT and TD–DFT calculations indicate that transition from a predominantly Mn−CO bonding orbital to ligand orbitals (of mostly imine and pyridine–quinoline character) contributes heavily to the photobands of these photoCORMs. Transfer of electron density from the Mn−CO bonding orbital(s) to orbitals associated with the ligand frame could therefore reduce the affinity of the Mn(I) center toward CO in photoCORMs and promotes photorelease of CO. In a recent review, Ford and co-workers (133) discussed this issue in detail in relation to development of photoCORMs that are sensitive to visible light.

Carbon monoxide loss from photoCORMs containing the *fac*-Mn(I) tricarbonyl unit has been investigated by Berends and Kurz (134) using [Mn(CO)₃(tpm)](PF₆) (**43**) and [Mn(CO)₃(bpzaa)] [**52**, bpzaa = bis(pyrazolyl) acetic acid]. A generalized scheme detailing CO loss starts with two CO releasing steps followed by oxidation of the manganese center to Mn(II) and finally to Mn(III), with μ-O-di-Mn(III) complexes purported to be the final products. These findings contrast mechanisms outlined for those of tungsten carbonyls, wherein exhaustive photolysis yields a mixture of photoproducts containing bound CO moieties due to recombination reactions (114). Formation of bridged μ-oxo-di-Mn(III) species is consistent with observations of complete loss of ν_{CO} bands and an EPR active photoproduct when photoCORMs, such as [Mn(dpa)(CO)$_3$]ClO$_4$ (**48**) and [Mn(pqa)(CO)$_3$]ClO$_4$ (**49**), are completely photolyzed (131).

D. Metal Carbonyl Complexes Derived from Polydentate Ligands

Kodanko and co-workers (135) reported a stable iron carbonyl complex, featuring a pentadentate polypyridine ligand, *N,N*-bis(2-pyridylmethyl)-*N*-bis(2-pyridyl) methylamine (N4Py). The resulting carbonyl complex, [Fe(N4Py)(CO)](ClO$_4$)$_2$ (**52**, Fig. 18) is stable in aerobic aqueous solutions, exhibits fast CO release upon irradiation with UV light (365 nm), and shows potent photoinduced cytotoxicity against PC-3 prostate cancer cells. This photoCORM is amenable to being attached to small peptides and is expected to be applicable for tissue- or cell-specific CO delivery. Interestingly, the Fe(II) carbonyl derived from the structurally similar PaPy$_3$H ligand, namely, [Fe(PaPy$_3$)(CO)]ClO$_4$ (**53**, Fig. 18) does not exhibit CO

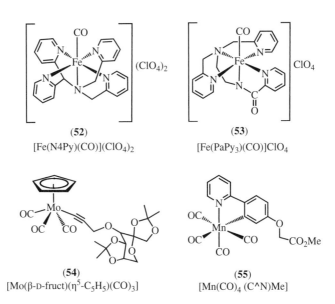

(**52**)
[Fe(N4Py)(CO)](ClO$_4$)$_2$

(**53**)
[Fe(PaPy$_3$)(CO)]ClO$_4$

(**54**)
[Mo(β-D-fruct)(η^5-C$_5$H$_5$)(CO)$_3$]

(**55**)
[Mn(CO)$_4$ (C^N)Me]

Figure 18.   Miscellaneous photoCORMs derived from polydentate ligands.

photolability (136). Instead, this water-soluble designed carbonyl complex releases CO upon dissolution in aqueous solutions and promotes concentration-dependent vasorelaxation in precontracted mouse aorta muscle rings.

Metal–alkynyl complexes typically feature strong metal–carbon bonds and variation of the substituents on the alkynyl group can modulate the properties of such complexes. Incorporation of CO ligands in such metal alkynyl complexes could generate photoCORMs for therapeutic applications. The sugar substituted molybdenum complex, [Mo(C≡C-β-D-fruct)(η^5-C$_5$H$_5$)(CO)$_3$] (**54**, Fig. 18) was synthesized via the copper-promoted coupling of [Mo(η^5-C$_5$H$_5$)Cl(CO)$_3$] with β-D-fructopyranose (fruct) (137). When exposed to UV light (325 nm, 6 W), this Mo carbonyl undergoes rapid CO release, yielding 2.2 equiv of CO per photo-CORM. Manganese-based tetracarbonyl complexes of the type [Mn(CO)$_4$(C^N)], where C^N = ortho-metalated 2-phenylpyridine or benzoquinoline have also been found to exhibit CO photoactivity in the presence of two-electron donor ligands, such as dimethyl sulfoxide (DMSO) or PPh$_3$ (138). In the case of C^N = methyl 4-(pyridin-2-yl)phenyl carbonate, the resulting Mn tetracarbonyl complex (**55**, Fig. 18) exhibits UV light-dependent CO release (365 nm). Irradiation of **55** yields approximately three CO ligands per molecule in a stepwise manner. This photoCORM shows excellent cell viability before and after irradiation, suggesting that its related degradation product(s) are minimally toxic.

Upon replacement of the halide ligands with phosphines, luminescent rhenium carbonyl complexes of the type *fac*-[Re(bpy)(CO)$_3$(X)] (X = Cl, Br) exhibits CO photolability (139). Ford and co-workers (140) recently examined CO release from *fac*-[Re(bpy)(CO)$_3$(thp)](CF$_3$SO$_3$) [**56**, thp = tris(hydroxymethyl)phosphine] upon exposure to 405 nm. When incorporated in human prostatic carcinoma cell line PPC-1, this water-soluble and air-stable photoCORM exhibits no apparent cytotoxicity. Since **56** and its photoproduct(s) are both luminescent, the researchers were able to monitor the light-induced loss of CO within the various compartments of the cell by changes in emission wavelengths with the aid of confocal microscopy. These results clearly indicate that inherent photoluminescence of designed metal carbonyl complexes of this kind could be exploited in tracking photoCORMs in cellular matrices.

V. CONCLUSION

The acute toxicity of CO and NO severely restricts their direct use to combat illness and disorders in human patients. Consequently, NO and CO donating compounds are necessary for the safe delivery of these signaling molecules in humans. At present, GTN, IMN, and SNP are used regularly in hospitals as NO-donors. Research over the past few years, however, indicates that innovative approaches   to   the   design   of   photoactive   metal   nitrosyl   complexes (photonorms) (32–34) and effective delivery systems (141) could expand the repertoire of NO donating pharmaceuticals to a significant extent. Such additions will enjoy specific advantages over conventional organic NO-donors as the photosensitivity of the metal nitrosyls will allow more control on the amount of NO delivery and smart design of the delivery systems will make site-specific NO delivery more of a reality.

The quest for CORMs has gained momentum only recently following their initial success in animal experiments. Selected CORMs have been incorporated into polymeric matrices for safer delivery to biological targets. Fluorescence sensors to detect CO in living cells have also been developed during the past few years (142, 143). These sensors detect CO within cells quite selectively. Despite such progress, to date no CORM has been subjected to human clinical trial. The lag in the discovery of the physiological roles of CO (cf. to NO) is the principal reason for this delay in applications of CORMs in modern medicine. Nevertheless, it is now apparent that CO donating drugs could be synthesized on the basis of smart design and photoCORMs could find use as novel photochemotherapeutics.

ACKNOWLEDGMENTS

Research efforts in the Mascharak group are supported by grants from the National Science Foundation (CHE-0957251 and DMR-1105296). M. G. acknowledges support from the IMSD Grant GM-58903 and the NHGRI Grant HG002371.

ABBREVIATIONS

Al-MCM-41	Al-Substituted MCM-41 microporous material
[15]ane	1,4,8,12-Tetraazacyclopentadecane
bdqi	1,2-Benzoquinonedimine
bpy	2,2′-Bipyridine
bpzaa	Bis(pyrazolyl) acetic acid
CFU	Colony-forming unit
C^N	Methyl 4-(pyridin-2-yl)phenyl carbonate
CORM	CO-Releasing molecule
CORM-3	$[RuCl(glycinate)(CO)_3]$
CT	Charge transfer
cyclam	1,4,8,11-Tetraazacyclotetradecane
DEA-NO	Diethylamine NONOate
DFT	Density functional theory

DMSO	Dimethyl sulfoxide
DNA	Deoxyribonucleic acid
dpa	N,N-Bis(2-pyridylmethyl)amine
Ds-im	Dansyl-imidazole
EPR	Electron paramagnetic resonance
FlEt	Fluorescein ethyl ester
FO	Fiber optic
fruct	Fructopyranose
GTN	Glyceryl trinitrate
H_2bpb	1,2-Bis(pyridine-2-carboxamido)benzene
his	Histidine
H_3hypyb	1-(2-Hydroxybenzamido)-2-(2-pyridinecarboxamido) benzene
H_4hybeb	1,2-Bis(2-hydroxybenzamido)-benzene
HO	Heme oxygenase
H_2(OMe)$_2$bQb	1,2-Bis(quinoline-2-carboxamido)-4,5-dimethoxybenzene
H_2(OMe)$_2$IQ1	1,2-Bis(isoquinoline-1-carboxamido)-4,5-dimethoxybenzene
HPMA	2-Hydroxypropyl methacrylamide
im	Imidazole
IMN	Isoamyl nitrite
IR	Infrared
IRI	Ischemia–reperfusion injury
LF	Ligand field
MAPK	Mitogen-activated protein kinase
MLCT	Metal-to-ligand charge transfer
MO	Molecular orbital
MS	Mass spectrometry
NIR	Near-infrared
NOS	Nitric oxide synthase
NORM	NO Releasing molecule
N4Py	N,N-Bis(2-pyridylmethyl)-N-bis(2-pyridyl)methylamine
OEP	Octaethylporphyrin
PaPy$_3$H	N,N-Bis(2-pyridylmethyl)amine-N-ethyl-2-pyridine-2-carboxamide
PaPy$_2$QH	N,N-Bis(2-pyridylmethyl)amine-N-ethyl-2-quinaldine-2-carboxamide
Pc	Phthalocyanine
PDA	Phenylenedicarboxamide
PDMS	Polydimethoxysilane
pHEMA	Poly(2-hydroxyethyl methacrylate)
photoCORM	Photoactive carbon monoxide releasing molecule
4-pic	4-Picoline
pmtpm	2-Pyridyl-N-(2′-methylthiophenyl)methyleneimine

PPIX	Protophorphyrin-IX
pqa	(2-Pyridylmethyl)(2-quinolylmethyl) amine
py	Pyridine
qmtpm	2-Quinoline-N-(2'-methylthiophenyl)methyleneimine
RBS	Roussin's black salt
R-DAB	Substituted 1,4-diaza-1,3-butadiene
Resf	Resorufin
RRS	Roussin's red salt
RSE	Roussin's red salt ester
salenH$_2$	N,N'-Bis(salicylidene)ethylenediamine
salophen	N,N'-Bis(salicylaldehydo)-1,2-phenylenediimine
Seln	Selenophore
SG	Sol–gel
SNAP	S-Nitroso-N-acetyl-penicillamine
SNP	Sodium nitroprusside
solv	Solvent
SQUID	Superconducting quantum interference device
TD-DFT	Time-dependent DFT
terpy	Terpyridine
Thnl	Thionol
thp	Tris(hydroxymethyl)phosphine
2-TIP^H	Tris(imidazol-2yl)phosphine
TIPO^{iPr}	Tris(imidazol-4yl)phosphane
tpa	Tris(2-pyridyl)amine
tpm	Tris(pyrazolyl)methane
TPP	$meso$-Tetraphenylporphyrin
TPPTS	Tris(sulfonatophenyl)phosphine
UPNC	Upconverting nanocrystal
UV	Ultraviolet
vis	Visible
4-vpy	4-Vinyl pyridine

REFERENCES

1. R. Wang, Ed., *Signal Transduction and the Gasotransmitters: NO, CO and H$_2$S in Biology and Medicine*, Humana Press, Totowa, NJ, 2004.
2. T. Li and T. L. Poulos, *J. Inorg. Biochem.*, **99**, 293 (2005).
3. L. J. Ignarro, Ed., *Nitric Oxide: Biology and Pathobiology*, Academic Press, San Diego, 2000.
4. S. Kalsner, Ed., *Nitric Oxide and Free radicals in Peripheral Neurotransmission*, Birkhauser, Boston, MA, 2000.

5. A. R. Butler and R. Nicholson, *Life, Death and Nitric Oxide*, The Royal Society of Chemistry, Cambridge, UK, 2003.

6. F. C. Fang, Ed., *Nitric Oxide and Infection*, Kluwer Academic/Plenum Publishers, New York, 1999.

7. S. Moncada, E. A. Higgs, and G. Bagetta, Eds., *Nitric Oxide and Cell Proliferation, Differentiation and Death*, Portland Press, London, 1998.

8. S. B. Abramson, A. R. Amin, R. M. Clancy, and M. Attur, *Best Pract. Res. Clin. Rheumatol.*, *15*, 831 (2001).

9. C. S. Boyd and E. Cadenas, *Biol. Chem.*, *383*, 411 (2002).

10. C. Q. Li and G. N. Wogan, *Cancer Lett.*, *226*, 1 (2005).

11. G. Kikuchi, T. Yoshida, and M. Noguchi, *Biochem. Biophys. Res. Commun.*, *338*, 558 (2005).

12. H. P. Kim, S. W. Ryter, and A. M. K. Choi, *Ann. Rev. Pharmacol. Toxicol.*, *46*, 411 (2006).

13. D. J. Kaczorowski and B. S. Zuckerbraun, *Curr. Med. Chem.*, *14*, 2720 (2007).

14. L. E. Otterbein, F. H. Bach, J. Alam, M. Soares, H. T. Lu, M. Wysk, R. J. Davis, R. A. Flavell, and A. M. K. Choi, *Nature Med.*, *6*, 422 (2000).

15. C. Napoli and L. J. Ignarro, *Ann. Rev. Pharmacol. Toxicol.*, *43*, 97 (2003).

16. P. G. Wang, M. Xian, X. Tang, X. Wu, Z. Wen, T. Cai, and A. J. Janczuk, *Chem. Rev.*, *102*, 1091 (2002).

17. P. G. Wang, T. B. Cai, and N. Taniguchi, Eds., *Nitric Oxide Donors for Pharmaceutical and Biological Applications*, Wiley–VCH, Weinheim, Germany, 2005.

18. C. S. Degoute, *Drugs*, *67*, 1052 (2007).

19. P. K. Mascharak, *Ind. J. Chem.*, *51(A)*, 99 (2012).

20. T. R. Johnson, B. E. Mann, J. E. Clark, R. Foresti, C. J. Green, and R. Motterlini, *Angew. Chem. Intl. Ed.*, *42*, 3722 (2003).

21. B. E. Mann and R. Motterlini, *Chem. Commun.*, 4197 (2007).

22. C. R. Romão, W. A. Blättler, J. D. Seixas, and G. J. L. Bernardes, *Chem. Soc. Rev.*, *41*, 3571 (2012).

23. A. D. Ostrowski and P. C. Ford, *Dalton Trans.* 10660 (2009).

24. M. J. Rose and P. K. Mascharak, *Coord. Chem. Rev.*, *252*, 2093 (2008).

25. J. C. Toledo, B. D. S. L. Neto, and D. W. Franco, *Coord. Chem. Rev.*, *249*, 419 (2005).

26. E. Tfouni, M. Krieger, B. R. McGarvey, and D. W. Franco, *Coord. Chem. Rev.*, *236*, 57 (2003).

27. P. C. Ford, J. Bourassa, K. Miranda, B. Lee, I. Lorkovic, S. Boggs, S. Kudo, and L. Laverman, *Coord. Chem. Rev.*, *171*, 185 (1998).

28. R. D. Rimmer, A. E. Pierri, and P. C. Ford, *Coord. Chem. Rev.*, *256*, 1509 (2012).

29. U. Schatzschneider, *Inorg. Chim. Acta*, *374*, 19 (2011).

30. U. Schatzschneider, *Eur. J. Inorg. Chem.*, 1451 (2010).

31. A. A. Eroy-Reveles and P. K. Mascharak, *Future Med. Chem.*, *1*, 1497 (2009).

32. M. J. Rose and P. K. Mascharak, *Curr. Opin. Chem. Biol.*, *12*, 238 (2008).

33. N. L. Fry and P. K. Mascharak, *Dalton Trans.*, *41*, 4726 (2012).

34. N. L. Fry and P. K. Mascharak, *Acc. Chem. Res.*, *44*, 289 (2011).

35. D. Wang and S. J. Lippard, *Nature Rev. Drug Discov.*, *4*, 307 (2005).

36. J. S. Anderson, *Qu. Rev. Chem. Soc.*, *1*, 331 (1947).

37. E. W. Abel and F. G. A. Stone, *Qu. Rev. Chem. Soc.*, *23*, 325 (1969).

38. P. J. Dyson and J. S. McIndoe, *Transition Metal Carbonyl Cluster Chemistry*, CRC Press, Boca Raton, FL, 2000.

39. G. B. Richter-Addo and P. Legzdins, *Metal Nitrosyls*, Oxford University Press, New York, 1992.

40. N. V. Sidgwick, *The Electronic Theory of Valency*, Cornell University, Ithaca, NY, 1927.

41. J. H. Enemark and R. D. Feltham, *Coord. Chem. Rev.*, *13*, 339 (1974).

42. M. A. Gonzales, S. J. Carrington, N. L. Fry, J. L. Martinez, and P. K. Mascharak, *Inorg. Chem.*, *51*, 11930 (2012).

43. E. J. Baerends and A. Rosa, *Coord. Chem. Rev.*, *177*, 97 (1998).

44. J. H. Tinker and J. D. Michenfelder, *Anesthesiology*, *15*, 340 (1976).

45. J. N. bates, M. T. Baker, R. GuerraJr., and D. G. Harrison, *Biochem. Pharmacol.*, *42*, S157 (1991).

46. A. Lockwood, J. Patka, M. Rabinovich, K. Wyatt, and P. Abraham, Open Access J. *Clin. Trials*, *2*, 133 (2012).

47. P. Coppens, I. Novozhilova, and A. Kovalevsky, *Chem. Rev.*, *102*, 861 (2002).

48. M. S. Lynch, M. Cheng, B. E.van Kuiken, and M. Khalil, *J. Am. Chem. Soc.*, *133*, 5255 (2011).

49. F. W. Flitney, I. L. Megson, J. L. M. Thomson, G. D. Kennovin, and A. R. Butler, *Br. J. Pharmacol.*, *117*, 1549 (1996).

50. J. B. Mitchell, D. A. Wink, W. DeGraff, J. Gamson, L. K. Keefer, and M. C. Krishna, *Cancer Res.*, *53*, 5845 (1993).

51. A. B. Cox and R. M. Wallace, *Inorg. Nucl. Chem. Lett.*, *7*, 1191 (1971).

52. E. Tfouni, D. R. Truzzi, A. Tavares, A. J. Gomes, L. E. Figueiredo, and D. W. Franco, *Nitric Oxide*, *26*, 38 (2012).

53. E. Tfouni, F. G. Doro, L. E. Figueiredo, J. C. M. Pereira, G. Metzker, and D. W. Franco, *Curr. Med. Chem.*, *17*, 3643 (2010).

54. P. G. Zanichelli, H. F. G. Estrela, R. C. Spadari-Bratfisch, D. M. Grassi-Kassisse, and D. W. Franco, *Nitric Oxide*, *16*, 189 (2007).

55. C. Z. Ferezin, F. S. Oliveira, R. S.da Silva, A. R. Simioni, A. C. Tedesco, and L. M. Bendhack, *Nitric Oxide*, *13*, 170 (2005).

56. M. Hoshino, L. Laverman, and P. C. Ford, *Coord. Chem. Rev.*, *187*, 75 (1999).

57. P. C. Ford and I. M. Lorkovic, *Chem. Rev.*, *102*, 993 (2002).

58. A. P. Castano, T. N. Demidova, and M. R. Hamblin, *Photodiagn. Photodyn.*, *2*, 92 (2005).

59. Z. N.da Rocha, R. G.de Lima, F. G. Doro, E. Tfouni, and R. S.da Silva, *Inorg. Chem. Commun.*, *11*, 737 (2008).

60. F. Marquele-Oliveira, D. C. D. A. Santana, S. F. Taveira, D. M. Vermeulen, A. R. M. de Oliveira, R. S.da Silva, and R. F. V. Lopez, *J. Pharmaceut. Biomed.*, *53*, 843 (2010).

61. Z. Huang, *Technol. Cancer Res. Treat.*, *4*, 283 (2005).

62. C. L. Conrado, S. Wecksler, C. Egler, D. Magde, and P. C. Ford, *Inorg. Chem.*, *43*, 5543 (2004).

63. S. R. Wecksler, J. Huchinson, and P. C. Ford, *Inorg. Chem.*, *45*, 1192 (2006).

64. S. R. Wecksler, A. Mikhailovsky, D. Korystov, and P. C. Ford, *J. Am. Chem. Soc.*, *128*, 3831 (2006).

65. C. F. Works, C. J. Jocher, G. D. Bart, X. Bu, and P. C. Ford, *Inorg. Chem.*, *41*, 3728 (2002).

66. J. Bordini, D. L. Hughes, J. D.da Motto Neto, and C. J.da Cunha, *Inorg. Chem.*, *41*, 5410 (2002).

67. J. T. Mitchell-Koch, T. M. Reed, and A. S. Borovik, *Angew Chem. Int. Ed.*, *43*, 2806 (2004).

68. S. Nagashima, M. Nakasako, N. Dohmae, M. Tsujimora, K. Takio, M. Odaka, M. Yohda, N. Kamiya, and I. Endo, *Nat. Struct. Biol.*, *5*, 347 (1998).

69. L. Song, M. Z. Wang, J. J. Shi, Z. Q. Xue, M. X. Wang, and S. J. Qian, *Biochem. Biophys. Res. Commun.*, *362*, 319 (2007).

70. A. K. Patra, R. K. Afshar, M. M. Olmstead, and P. K. Mascharak, *Angew. Chem. Intl. Ed.*, *41*, 2512 (2002).

71. A. K. Patra, J. M. Rowland, D. S. Marlin, E. Bill, M. M. Olmstead, and P. K. Mascharak, *Inorg. Chem.*, *42*, 6812 (2003).

72. A. K. Patra and P. K. Mascharak, *Inorg. Chem.*, *42*, 7363 (2003).

73. M. J. Rose, M. M. Olmstead, and P. K. Mascharak, *Polyhedron*, *26*, 4713 (2007).

74. I. Szundi, M. J. Rose, I. Sen, A. A. Eroy-Reveles, P. K. Mascharak, and Ó. Einarsdóttir, *Photochem. Photobiol.*, *82*, 1377 (2006).

75. R. K. Afshar, A. K. Patra, and P. K. Mascharak, *J. Inorg. Biochem.*, *99*, 1458 (2005).

76. M. Madhani, A. K. Patra, T. W. Miller, A. A. Eroy-Reveles, A. J. Hobbs, J. M. Fukuto, and P. K. Mascharak, *J. Med. Chem.*, *49*, 7325 (2006).

77. K. Ghosh, A. A. Eroy-Reveles, B. Avila, T. R. Holman, M. M. Olmstead, and P. K. Mascharak, *Inorg. Chem.*, *43*, 2988 (2004).

78. R. Weissleder and V. Ntziarchristos, *Nat. Med.*, *9*, 123 (2003).

79. A. A. Eroy-Reveles, Y. Leung, C. M. Beavers, M. M. Olmstead, and P. K. Mascharak, *J. Am. Chem. Soc.*, *130*, 4447 (2008).

80. N. L. Fry, X. P. Zhao, and P. K. Mascharak, *Inorg. Chim. Acta*, *367*, 194 (2011).

81. A. C. Merkle, N. L. Fry, P. K. Mascharak, and N. Lehnert, *Inorg. Chem.*, *50*, 12192 (2011).

82. C. G. Hoffman-Luca, A. A. Eroy-Reveles, J. Alvarenga, and P. K. Mascharak, *Inorg. Chem.*, *48*, 9104 (2009).

83. W. Zheng, S. Wu, S. Zhao, Y. Geng, J. Jin, Z. Su, and Q. Fu, *Inorg. Chem.*, *51*, 3972 (2012).

84. A. K. Patra, M. J. Rose, K. M. Murphy, M. M. Olmstead, and P. K. Mascharak, *Inorg. Chem.*, *43*, 4487 (2004).

85. N. L. Fry, M. J. Rose, D. L. Rogow, C. Nyitray, M. Kaur, and P. K. Mascharak, *Inorg. Chem.*, *49*, 1487 (2010).

86. N. L. Fry, B. J. Heilman, and P. K. Mascharak, *Inorg. Chem.*, *50*, 317 (2011).

87. M. J. Rose, M. M. Olmstead, and P. K. Mascharak, *J. Am. Chem. Soc.*, *129*, 5342 (2007).

88. M. J. Rose, N. L. Fry, R. Marlow, L. Hink, and P. K. Mascharak, *J. Am. Chem. Soc.*, *130*, 8834 (2008).

89. D. L. Dexter, *J. Chem. Phys.*, *21*, 836–850 (1953).

90. M. J. Rose and P. K. Mascharak, *Inorg. Chem.*, *48*, 6904 (2009).

91. M. J. Rose and P. K. Mascharak, *Chem. Commun.*, *7*, 3933 (2008).

92. N. L. Fry, J. Wei, and P. K. Mascharak, *Inorg. Chem.*, *50*, 9045 (2011).

93. G. M. Halpenny and P. K. Mascharak, *Anti-Infect. Agents Med. Chem.*, *9*, 187 (2010).

94. T. Coradin, M. Boissière, and J. Livage, *Curr. Med. Chem.*, *13*, 99 (2006).

95. A. A. Eroy-Reveles, Y. Leung, and P. K. Mascharak, *J. Am. Chem. Soc.*, *128*, 7166 (2006).

96. G. M. Halpenny, K. R. Gandhi, and P. K. Mascharak, *ACS Med. Chem. Lett.*, *1*, 180 (2010).

97. B. J. Heilman, G. M. Halpenny, and P. K. Mascharak, *J. Biomed. Mater. Res. Part B*, *99B*, 328 (2011).

98. G. M. Halpenny, B. J. Heilman, and P. K. Mascharak, *Chem. Biodivers.*, *9*, 1829 (2012).

99. A. B. Shekhter, V. A. Serezhenkov, T. G. Rudenko, A. V. Pekshev, and A. F. Vanin, *Nitric Oxide*, *12*, 210 (2005).

100. N. A. Peppas, Ed., *Hydrogels in Medicine and Pharmacy, vols. I–III.* CRC Press, Boca Raton, FL, 1987.

101. G. M. Halpenny, R. C. Steinhardt, K. A. Okialda, and P. K. Mascharak, *J. Mater. Sci: Mater. Med.*, *20*, 2353 (2009).

102. G. M. Halpenny, M. M. Olmstead, and P. K. Mascharak, *Inorg. Chem.*, *46*, 6601 (2007).

103. C. Bohlender, M. Wolfram, H. Goerls, W. Imhof, R. Menzel, A. Baumgaertel, U. S. Schubert, U. Mueller, M. Frigge, M. Schnabelrauch, R. Wyrwa, and A. Schiller, *J. Mater. Chem.*, *22*, 8785 (2012).

104. B. J. Heilman, J. St. John, S. J. Oliver, and P. K. Mascharak, *J. Am. Chem. Soc.*, *134*, 11573 (2012).

105. J. V. Garcia, J. Yang, D. Shen, C. Yao, X. Li, R. Wang, G. D. Stucky, D. Zhao, P. C. Ford, and F. Zhang, *Small*, *8*, 3800 (2012).

106. G. L. Geoffroy and M. S. Wrighton, *Organometallic Photochemistry*, Academic Press, New York, 1979.

107. C. Pollak, A. Rosa, and E. J. Baerends, *J. Am. Chem. Soc.*, *119*, 7324 (1997).

108. K. Pierloot, J. Verhulst, P. Verbeke, and L. G. Vanquickenborne, *Inorg. Chem.*, *28*, 3059 (1989).

109. R. Motterlini, J. E. Clark, R. Foresti, P. Sarathchandra, B. E. Mann, and C. J. Green, *Circ. Res.*, *90*, e17 (2002).

110. S. K. Nayak, G. J. Farrell, and T. J. Burkey, *Inorg. Chem.*, *33*, 2236 (1994).

111. R. S. Herrick and T. L. Brown, *Inorg. Chem.*, *23*, 4550 (1984).

112. N. Ben Amor, S. Villaume, D. Maynau, and C. Daniel, *Chem. Phys. Lett.*, *421*, 378 (2006).

113. A. Vlček, Jr., *Coord. Chem. Rev.*, *177*, 219 (1998).

114. R. D. Rimmer, H. Richter, and P. C. Ford, *Inorg. Chem.*, *49*, 1180 (2010).

115. D. J. Darensbourg, J. D. Draper, and J. H. Reibenspies, *Inorg. Chem.*, *36*, 3648 (1997).

116. T. R. Johnson, B. E. Mann, I. P. Teasdale, H. Adams, R. Foresti, C. J. Green, and R. Motterlini, *Dalton Trans.* 1500 (2007).

117. R. Foresti, J. Hammad, J. E. Clark, T. R. Johnson, B. E. Mann, A. Friebe, C. J. Green, and R. Motterlini, *Br. J. Pharmacol.*, *142*, 453 (2004).

118. J. E. Clark, P. Naughton, S. Shurey, C. J. Green, T. R. Johnson, B. E. Mann, R. Foresti, and R. Motterlini, *Circ. Res.*, *93*, e2 (2003).

119. J. Varadi, I. Lekli, B. Juhasz, I. Bacskay, G. Szabo, R. Gesztelyi, L. Szendrei, E. Varga, I. Bak, R. Foresti, R. Motterlini, and A. Tosaki, *Life Sci.*, *80*, 1619 (2007).

120. F. Mohr, J. Niesel, U. Schatzschneider, and C. W. Lehmann, *Z. Anorg. Allg. Chem.*, *638*, 543 (2012).

121. R. Kretschmer, G. Gessner, H. Görls, S. H. Heinemann, and M. Westerhausen, *J. Inorg. Biochem.*, *105*, 6 (2011).

122. A. Szakács-Schmidt, J. Kreisz, L. Markó, Z. Nagy-Magos, and J. Takács, *Inorg. Chim. Acta*, *200*, 401 (1992).

123. V. P. L. Velâsquez, T. M. A. Jazzai, A. Malassa, H. Görls, G. Gessner, S. H. Heinemann, and M. Westerhausen, *Eur. J. Inorg. Chem.*, 1072 (2012).

124. J. Niesel, A. Pinto, H. W. Peindy N'Dongo, K. Merz, I. Ott, R. Gust, and U. Schatzschneider, *Chem. Commun.*, 1798 (2008).

125. H. Pfeiffer, A Rojas, J. Niesel, and U. Schatzschneider, *Dalton Trans.* 4292 (2009).

126. G. Dördelmann, H. Pfeiffer, A. Birkner, and U. Schatzschneider, *Inorg. Chem.*, *50*, 4362 (2011).

127. G. Dördelmann, T. Meinhardt, T. Sowik, A. Krueger, and U. Schatzschneider, *Chem. Commun.*, *48*, 11528 (2012).

128. P. C. Kunz, W. Huber, A. Rojas, U. Schatzschneider, and B. Spingler, *Eur. J. Inorg. Chem.*, 5358 (2009).

129. W. Huber, R. Linder, J. Niesel, U. Schatzschneider, B. Spingler, and P. C. Kunz, *Eur. J. Inorg. Chem.*, 3140 (2012).

130. M. A. Gonzalez, M. A., M. A. Yim, S. Cheng, A. Moyes, A. J. Hobbs, and P. K. Mascharak, *Inorg. Chem.*, *51*, 601 (2012).

131. N. E. Brückmann, M. Wahl, G. J. Reiß, M. Kohns, W. Wätjen, and P. C. Kunz, *Eur. J. Inorg. Chem.* 4571 (2011).

132. M. A. Gonzalez, S. J. Carrington, N. L. Fry, J. L. Martinez, and P. K. Mascharak, *Inorg. Chem.*, *51*, 11930 (2012).

133. R. D. Rimmer, A. E. Pierri, and P. C. Ford, *Coord. Chem. Rev.*, *256*, 1509 (2012).

134. H. M. Berends and P. Kurz, *Inorg. Chim. Acta*, *380*, 141 (2012).

135. C. S. Jackson, S. Schmitt, Q. P. Dou, and J. J. Kodanko, *Inorg. Chem.*, *50*, 5336 (2011).

136. M. A. Gonzalez, N. L. Fry, R. Burt, R. Davda, A. J. Hobbs, and P. K. Mascharak, *Inorg. Chem.*, *50*, 3127 (2011).

137. W. Zhang, A. J. Atkin, I. J. S. Fairlamb, A. C. Whitwood, and J. M. Lynam, *Organometallics*, *30*, 4643 (2011).

138. J. S. Ward, J. M. Lynam, J. W. B. Moir, D. E. Sanin, A. P. Mountford, and I. J. S. Fairlamb, *Dalton Trans.*, *41*, 10514 (2012).

139. K. Koike, N. Okoshi, H. Hori, K. Takeuchi, O. Ishitani, H. Tsubaki, I. P. Clark, M. W. George, F. P. A. Johnson, and J. J. Turner, *J. Am. Chem. Soc.*, *124*, 11448 (2002).

140. A. E. Pierri, A. Pallaoro, G. Wu, and P. C. Ford, *J. Am. Chem. Soc.*, *134*, 18197 (2013).

141. M. C. Jen, M. C. Serrano, R.van Lith, and G. A. Ameer, *Adv. Funct. Mater.*, *22*, 239 (2012).

142. J. Wang, J. Karpus, B. S. Zhao, Z. Luo, P. R. Chen, and C. He, *Angew. Chem. Intl. Ed.*, *51*, 9652 (2012).

143. L. Yuan, W. Lin, L. Tan, K. Zheng, and W. Huang, *Angew. Chem. Intl. Ed.*, *52*, 1628 (2013).

Metal−Metal Bond-Containing Complexes as Catalysts for C−H Functionalization

KATHERINE P. KORNECKI AND JOHN F. BERRY

Department of Chemistry, University of Wisconsin – Madison, Madison, WI53706

DAVID C. POWERS AND TOBIAS RITTER

Department of Chemistry and Chemical Biology, Harvard University, Cambridge, MA02138

CONTENTS

Progress in Inorganic Chemistry, Volume 58, First Edition. Edited by Kenneth D. Karlin.
© 2014 John Wiley & Sons, Inc. Published 2014 by John Wiley & Sons, Inc.

I. INTRODUCTION

A. Overview of Metal−Metal Multiple Bonds

Metal–metal (M−M) multiple bonds in coordination compounds were first described in anionic Re halide clusters, such as $[Re_3Cl_{12}]^{3-}$ and $[Re_2Cl_8]^{2-}$, 50 years ago (1, 2). Since then, M−M bonds based on virtually every transition element have been prepared (3). By the 1980s M−M bonding in Mo_2, (4, 5) Tc_2 (6), Ru_2 (7), Rh_2 (8), Cr_2 (8), Pt_2 (9), W_2 (10), Os_2 (11), and Ir_2 compounds (12), mostly second- or third-row transition metal complexes, had been described. The M−M bond-containing coordination compounds of Nb(II) (13) and Pd(III) (14, 15) were first synthesized in the late 1990s. Aside from quadruply bonded Cr_2 compounds, which were first reported in 1844 (16, 17), M−M bond-containing complexes based on first-row transition metals historically have been less well known. Dicobalt compounds were first-synthesized in 1987 (18), followed by V_2 and Fe_2 compounds, which were first synthesized in the 1990s (19–21).

The synthesis and study of compounds containing M–M bonds has continued to provide both new structural motifs as well as provide new opportunities to study the reactivity of multiple-metal-containing complexes. Recent developments include the preparation of Cr(I) or Mo(I) dimers that contain formal quintuple bonds (22, 23), incorporation of M_2 units into larger polynuclear clusters (24–27), and extended metal-atom chain (EMAC) compounds containing delocalized M–M bonding in linear M_n chains with $n > 2$ (28). Ongoing research efforts are being directed toward both the synthesis of new M–M bond-containing complexes, as well as application of M–M bond-containing complexes and materials. For example, theoretical studies have predicted the stability of multiple bonds between actinide atoms, (e.g., uranium) (29), though stable compounds have yet to be prepared (30). In addition, M–M bond-containing units can play an important structural role in metal-organic framework (MOF) coordination polymers, which are highly porous materials of current interest in part for their potential application in catalysis (31–37).

While many M–M bond-containing coordination complexes have been prepared, the potential role of these species in catalysis has received relatively less attention by the inorganic chemistry community. Though Rh_2 complexes have long standing importance in catalysis, recent synthetic and mechanistic investigations have revealed catalytic roles for both Ru_2 and Pd_2 complexes as well. In this chapter, we will explore examples in which binuclear complexes are involved in C–H functionalization reactions and explore the effect of M–M bonds in catalysis.

B. Early Examples of M–M Bond-Containing Complexes in Catalysis

One of the first reactions reported to be catalyzed by M–M bond-containing compounds was the decomposition of organic diazo compounds. In 1973, Teyssié and co-workers (38) reported that treatment of ethyldiazoacetate with $Rh_2(OAc)_4$ (1) in the presence of olefinic or alcoholic substrates resulted in the formation of either cyclopropanes or ethers, respectively (see Scheme 1; OAc = acetate). The observed reactivity was rationalized as arising from the formation of a carbenoid

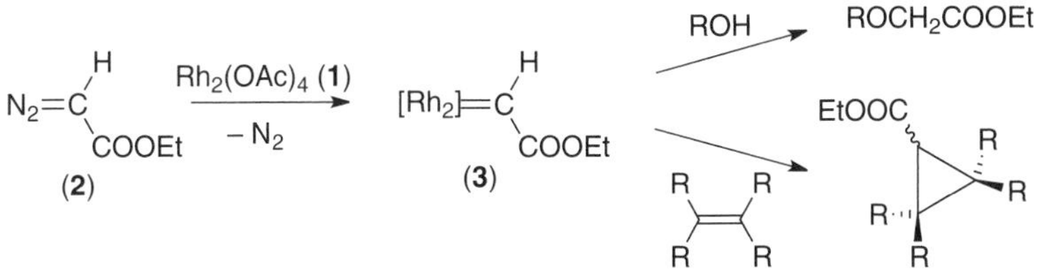

Scheme 1.   Decomposition of ethyl diazoacetate (2) with $Rh_2(OAc)_4$ (1) generates an Rh_2 carbenoid intermediate (3), which was shown in early studies to react with alcohol and alkene substrates.

intermediate (**3**), formed by expulsion of N_2 from ethyldiazoacetate (**2**) (38, 39). Since these initial reports, the use of Rh_2 catalysts in the decomposition of diazo compounds has found several applications, including $C-H$ functionalization via insertion of the carbene species into $C-H$ bonds, which will be discussed in more detail in Section II.

During the early 1980s, ditungsten complexes were investigated in alkyne metathesis reactions (40). Ditungsten complexes of the form $W_2(OR)_6$ (e.g., **4**) have a $W\equiv W$ triple bond and a trigonal, staggered D_{3d} core structure and have been extensively studied by Chisholm and co-workers (41–43). Related mononuclear tungsten carbyne complexes, such as (tBuO)$_3$W$\equiv$C–tBu (e.g., **5**), have been investigated as catalysts for alkyne metathesis reactions. In 1982, Schrock and co-workers (40) hypothesized that a potential mechanism for catalyst deactivation during alkyne metathesis could be a bimolecular reaction between two W$\equiv$C species (i.e., **5**) to form such a triply bonded W$\equiv$W complex (**4**) and the homo-coupled C$\equiv$C alkyne product. The hypothesized dimerization reaction is represented as the reverse of the top reaction in Scheme 2. Subsequent investigations indicated that this potential dimerization pathway was not operative. Instead, it was shown that W_2 complexes react cleanly with alkynes or nitriles to yield either 2 equiv of the mononuclear W alkylidyne complex (**5**) in the former case, or one equivalent each of a W–alkylidyne and a W–nitride complex (Scheme 2) (40). Thus dimerization does not constitute a deactivation pathway, but instead, dimer cleavage has been used to liberate active catalysts from binuclear precatalysts for alkyne metathesis (44–46), alkyne polymerization (47, 48), as well as nitrile metathesis (49, 50). Computational studies indicate that the metathetical reaction depicted in Scheme 2 proceeds via a ditungstacyclobutadiene-like intermediate (**6**) having a planar W_2C_2 or W_2CN core (51–53).

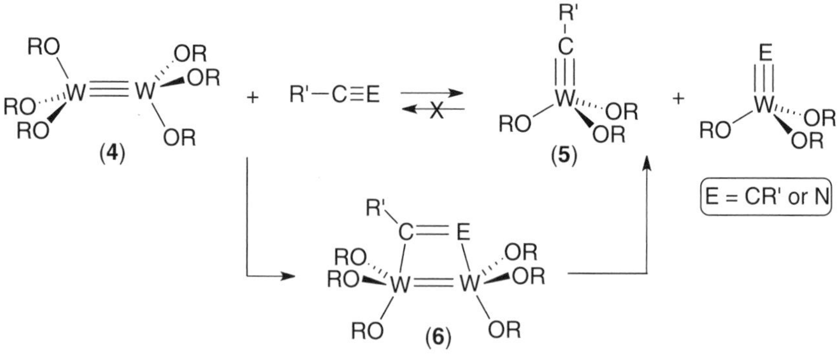

Scheme 2.   Ditungsten complexes (e.g., **4**) can react with alkynes, thus acting as precursors to alkyne metathesis catalysts (e.g., a W–alkylidyne complex, **5**).

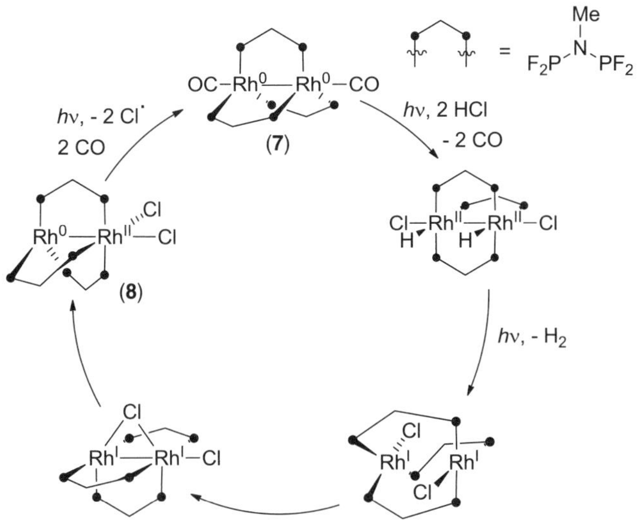

Scheme 3. Stoichiometric photocycle for hydrogen generation from dirhodium complexes.

Another early example of catalysis using M−M bond-containing compounds stems from the observation by Gray and co-workers (54) in 1977 that Rh(I) dimers spontaneously reduce HCl to molecular hydrogen upon irradiation into the lowest-energy $\sigma^*(4d) \rightarrow \sigma(5s)$ electronic transition. Further breakthroughs in this area have been made by the Nocera group using dirhodium phosphazane complexes, such as 7 (Scheme 3). The overall reaction, using light to formally split HCl into H_2 and Cl_2, features unusual two-electron mixed valency (i.e., Rh(0)/Rh(II)) in intermediate (8) (55–57). Approximately 80 turnovers have been achieved with these Rh systems for proton reduction. A similar Rh_2–phosphazane complex has been employed in the stoichiometric reduction of O_2 to water (58). Also of note are reactions promoted by "platinum blues", polynuclear mixed-valent Pt(II)/Pt(III) species having Pt–Pt bonds, which were first described in 1908 (59) and first characterized by Lippard and co-workers (60) in 1977. Matsumoto and co-workers (61–63) demonstrated the ability of Pt blue species to catalyze photo-chemical hydrogen production and water oxidation reactions (64), as well as a number of other catalytic transformations of olefins, including both addition reactions and oxidations (65–72).

This chapter will focus on the emerging role of binuclear transition metal complexes in C−H functionalization chemistry. Functionalization of C−H bonds is a field of current interest, and many methodologies for achieving selective C−H functionalization continue to appear in the literature (73–86). These can generally be classified into one of two mechanistic regimes, as shown in Scheme 4 (87).

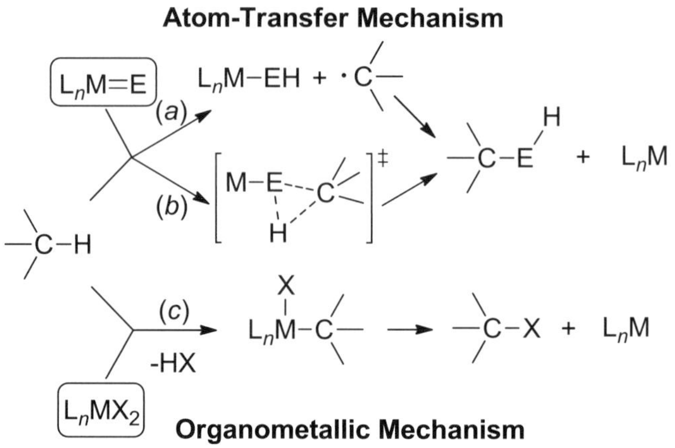

Atom-Transfer Mechanism

Scheme 4. The two general mechanistic pathways involved in C−H functionalization. (*a*) Radical action–recombination ATM. (*b*) Concerted insertion (ATM). (*c*) Inner-sphere C−H insertion (OM = organometallic mechanism). Here $L_n = n$ supporting ligands.

First, in the atom-transfer mechanism (ATM), which is reminiscent of C−H functionalization reactions that take place in cytochrome P450 s (88), a reactive metal complex having a metal-ligand multiple bond, denoted M=E, reacts to insert E directly into the substrate C−H bond to form product. In cytochrome P450, the reactive metal complex (M) is the thiolate-ligated heme iron cofactor and E is an oxygen atom. Conceptually, the ATM may be thought to occur either in a two-step sequence, beginning with hydrogen-atom abstraction to form an iron hydroxo complex and an organic radical, followed by radical recombination to form the hydroxlyated product (Scheme 4*a*), or by a single-step, concerted insertion of E into the C−H bond (Scheme 4*b*). Evidence for the stepwise mechanism has been found in cytochrome P450 s (88). Importantly, E does not have to be an oxygen atom: It can be other fragments that can form a metal–ligand multiple bond.

The second mechanistic paradigm for C−H functionalization involves the intermediacy of species with M−C bonds. Therefore we call it the organometallic mechanism (OM) (Scheme 4*c*). The C−H metalation to generate the requisite M−C bonds can proceed via various pathways: Oxidative addition of a C−H bond to a coordinatively unsaturated transition metal fragment (89), electrophilic substitution reactions, and concerted metalation–deprotonation (CMD) reactions are all pathways for C−H metallation (90). In most of the Pd catalyzed reactions presented in this chapter, C−H metalation is proposed to proceed by a CMD pathway in which electrophilic metalation and C−H deprotonation are coupled (91–98). The product-forming step in the OM is frequently reductive elimination of a C−X bond-containing product (Scheme 4*c*). In any given transformation proceeding by the OM pathway, it may be necessary for ligand metathesis reactions or oxidation steps to occur between metalation and reductive elimination in order to install the desired X ligand on the metal center and to facilitate C–X reductive elimination.

Despite the major differences between the two C−H functionalization mechanisms, M−M bond-containing complexes of exactly the same electron count [the isoelectronic Rh_2(II,II) and Pd_2(III,III) dimers, which contain a M−M σ bond], have been shown to be involved in both ATM and OM pathways. We will also discuss mixed-valent Ru_2(II,III) dimers, having an Ru−Ru bond order of 2.5. Although the topics of Rh_2 and Pd_2 C−H functionalization have been reviewed extensively (73, 86, 87, 89, 99–105), this chapter focuses on the relationship between electronic structure and reaction mechanisms. It is our goal to compare and contrast the chemistries of Rh_2 and Pd_2 complexes and discuss the ways in which the intriguing geometric and electronic structure of these species may facilitate C−H functionalization reactions.

C. Metal–Metal Bonding

The valence orbitals of the transition metals are d orbitals, and it is d-orbital overlap that must be considered when discussing M−M bonds. In a given binuclear transition metal complex in which the metal centers are within bonding distance, three types of orbital overlap can occur (3). The d_z^2 orbital of each metal atom can overlap to form σ bonding and antibonding combinations, the d_{xz} and d_{yz} orbitals can overlap forming π-bonding and π-antibonding sets, and the d_{xy} and $d_{x^2-y^2}$ orbitals can overlap to form δ bonding and antibonding orbitals. In the absence of exogenous ligands on the metal atoms, the relative energies of these orbitals would be based solely on the degree of orbital overlap; the orbitals would increase in energy as follows: $\sigma < \pi_1,\pi_2 < \delta_1,\delta_2 < \delta^*_1,\delta^*_2 < \pi^*_1,\pi^*_2 < \sigma^*$ (Fig. 1).

As ligands are added to the binuclear unit, those orbitals that engage strongly in metal–ligand (M−L) bonding will be utilized mainly for that purpose and will therefore contribute little to M−M bonding (2). For example, the compounds reviewed here generally will have four σ-donor ligands (L) attached to each metal atom in a roughly square disposition, $L_4M–ML_4$, where L represents only the atoms coordinated to the metal centers. If we allow these M−L bonds to form along the x and y vectors (Fig. 1), then the lone pairs of the L atoms will be properly oriented in such a way as to interact strongly with the $d_{x^2-y^2}$ derived orbitals. Thus, this set of δ and δ* orbitals will be used solely to make M−L bonds and will not contribute strongly to M−M bonding. In $L_4M–ML_4$ compounds that have all of the M−M bonding orbitals filled and all of the antibonding orbitals empty, we find a maximum M−M bond order of 4, corresponding to a $\sigma^2\pi^4\delta^2$ electron configuration (3). This configuration requires the presence of eight valence electrons from the metal atoms. Thus, compounds that contain quadruple bonds are formed with the following d^4 metal ions: Cr^{2+}, Mo^{2+}, W^{2+}, Tc^{3+}, and Re^{3+} (3). Both oxidation

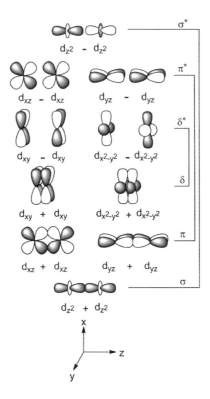

Figure 1.   Bonding and antibonding combinations of d orbitals used to construct M$-$M bonds.

or reduction of complexes with the $\sigma^2\pi^4\delta^2$ configuration decreases the bond order. For example, there are two ways to make M$-$M triple bonds in the L$_4$M$-$ML$_4$ geometry: The first is to remove the two δ electrons as in Mo$_2$(III,III) complexes yielding an "electron poor" $\sigma^2\pi^4$ triple bond; the second is to add two electrons to the δ^* antibonding orbital yielding an "electron-rich" $\sigma^2\pi^4\delta^2\delta^{*2}$ triple bond.

It is also possible to break the M$-$M bond in its entirety by filling all of the bonding and antibonding combinations of M$-$M orbitals with 16 electrons, as in d^8–d^8 dimers. However, an interesting phenomenon occurs in such systems. The nd σ^* orbital, which should generally be strongly antibonding, can mix with the (n + 1)p$_z$ orbitals in such a way as to mitigate antibonding character (106–109). This secondary orbital mixing creates a situation in which the nd σ^* orbital has less antibonding character than the σ orbital has bonding character, leading to an overall small, but nonzero, σ interaction (110). For example, the compound Pd$_2$(OAc)$_2$(phpy)$_2$ with phpy = 2-phenylpyridyl has been calculated to have a very weak Pd$\cdots$Pd σ-type interaction (calculated bond order of ~0.1, and Pd–Pd stretching frequency < 200 cm^{-1}) (110). This weak σ interaction can be increased by removal of the electrons occupying the nd σ^* orbital. Compounds in the Ni_2(II,III), Pd_2(II,III), and Pt_2(II,III) oxidation states have 15 valence electrons, and thus a $\sigma^2\pi^4\delta^2\delta^{*2}\pi^{*4}\sigma^{*1}$ configuration with an overall formal bond order of 0.5 (3). Removal of the remaining σ^* electron yields compounds with a $\sigma^2\pi^4\delta^2\delta^{*2}\pi^{*4}$ single σ bond, the most numerous of which are Rh_2(II,II) species.

When the difference in energy between the δ^* and π^* orbitals is sufficiently small, high-spin binuclear complexes can be obtained (3). For example, consider Ru_2(II,III) complexes, which have 11 valence electrons and will be discussed in Section II.B.3 in the context of C–H amination chemistry. Given the expected orbital ordering, we may anticipate that the last three electrons are added such that the δ^* orbital is filled completely before electrons are put into the π^* level to afford a low-spin electronic configuration. Because of the nature of the ligand field at each Ru center, the δ^* and π^* levels in fact are very close in energy, and the electron configuration is instead best described as $(\delta^*, \pi^*)^3$, where each orbital is occupied by one electron yielding a high-spin, $S = 3/2$ state; the Ru–Ru bond order is 2.5 in these compounds (3). The near-degeneracy of the δ^* and π^* orbitals can have implications for Rh_2 compounds as well. When Rh_2(II,II) compounds are oxidized by one electron, the electronic structure of the resulting Rh_2(II,III) species is determined by whether the electron has been removed from a δ^* or π^* orbital; both possibilities have been suggested to occur (111). In particular, Rh_2(II,III) complexes with N,N-donor ligands show an electron paramagnetic resonance (EPR) spectra consistent with a δ^{*1} ground state, whereas carboxylate analogues show a much larger g-value anisotropy, which is taken as evidence for an orbitally degenerate π^{*3} configuration (111).

Note the effects of equatorial ligands on the electronic structure of these compounds as in the example above (112). Here we will consider only the effects of changing from O,O-donors to N,O-donors to N,N-donors in anionic equatorial ligands with a three-atom bridge $[L–X–L]^-$ with L = N- or O-donors. These ligands are formally isoelectronic to carboxylate ligands. Importantly, the donor atoms in these ligands have 2p orbitals that are of the appropriate energy and symmetry to interact with the M_2 δ and δ^* orbitals. A combination of L based 2p orbitals that are perpendicular to the L–X–L plane, which are filled and serve as lone pairs on the L donor atoms, overlaps with an M_2 δ^* orbital to form a $\delta^* + L$ bonding and $\delta^* - L$ antibonding combination, and the effect on the latter is illustrated in Fig. 2 (a similar interaction between ligand-based lone pairs and the δ orbitals is neglected in Fig. 2 for simplicity). The bonding combination will be filled by the L lone-pair electrons, leaving any available M electrons to populate the $\delta^* - L$ orbital. Notably, the energy of the lone pairs of the ligand donor atom will change as we change L from O to N because of the change in effective nuclear charge. When L = O, the lone-pair orbitals are lower in energy and have weaker interactions with the metal orbitals. When L = N, the lone-pair orbitals are better energetically matched with the metal orbitals and can form strong covalent

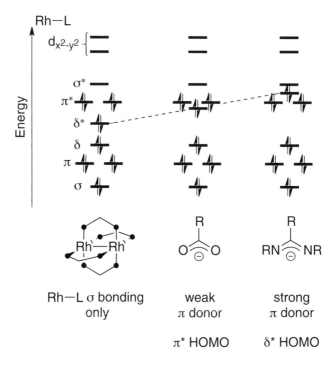

Figure 2.   Interaction of ligand-derived p orbitals with the M_2 δ^* orbital. A similar interaction with the δ orbital is neglected in the diagram.

interactions. As shown in Fig. 2, the variable electronic structure of $Rh_2(II,III)$ compounds is well explained by this bonding analysis. With carboxylate-donor ligands, the HOMO of the $Rh_2(II,II)$ dimer is the set of π^* orbitals, but when N,N-donors are used, the $\delta^* - L$ orbital is raised higher in energy and becomes the highest occupied molecular orbital (HOMO).

D.   Structural Manifestations of M–M Bonding

The theoretical framework discussed above provides a basis for examining M–M multiple bonding, but what are the physical properties of a compound that we may use to determine whether an $M-M$ bond exists or not? This question is especially important considering that the most favorable bite angle of the three-atom equatorial bridging ligands could simply hold together atoms in close proximity that do not engage in bonding. The $M-M$ bond distances, available from crystallographic data, have been an important piece of information used to invoke bonding, although they are not the only measure of M–M interactions. In general, we may expect that M–M distances will become shorter as the $M-M$ bond

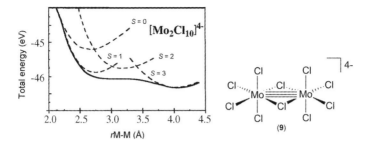

Figure 3. Potential energy surface for the $[Mo_2Cl_{10}]^{4-}$ ion in its ground state and various excited states. [Adapted from (113)].

order increases. Critical analysis, however, reveals some interesting nuances that will be examined here.

The σ, π, and δ orbitals have different shapes. Thus, σ, π, and δ overlap will be maximal at different M–M bond distances and the observed bond distance will represent a balance between the constituents of the multiple bond. For example, breaking a δ bond in a compound may be expected to have less of an effect on the observed M–M bond distance than breaking a σ bond. An illustration of the varying effects of bond length as a function of orbital overlap comes from computational work on the edge-sharing bioctahedral molybdate $[Mo_2Cl_{10}]^{4-}$ (**9**) by Stranger and co-workers (113) (Fig. 3). The geometry of this molecule precludes the formation of two Mo–Mo π bonds due to the bridging chlorides. This d^3–d^3 dimer can thus form one σ, one π, and one δ interaction between the Mo atoms, and it is possible to break these bonds one at a time *in silico*. From the triply bonded $\sigma^2\pi^2\delta^2$ $S = 0$ state, promotion of one δ electron to the δ^* orbital yields an $S = 1$ state with a $\sigma^2\pi^2\delta^1\delta^{*1}$ configuration with only a σ and π bond (i.e., a double bond). Similarly, an $S = 2$ state, generated by promotion of a π electron to a π^* orbital, with a $\sigma^2\pi^1\delta^1\delta^{*1}\pi^{*1}$ configuration has only a net σ bond, and an analogously generated $S = 3$ $\sigma^1\pi^1\delta^1\delta^{*1}\pi^{*1}\sigma^{*1}$ state has no net bond between the Mo atoms. Energies of these states as a function of the Mo–Mo distance are shown in Fig. 3. As expected, the triply bonded $S = 0$ state has the shortest equilibrium bond distance, at ~ 2.5 Å. Breaking the δ bond yields a Mo–Mo distance that is only slightly longer than in the $S = 0$ state. The π bond has a larger effect on M–M distance, as the Mo–Mo separation in the $S = 2$ state is longer by ~ 0.5 Å, and breaking the σ bond yields a minimum at > 4.0 Å.

In paddlewheel-type L_4M–ML_4 complexes in which four bridging ligands are present, the changes in M–M distance are often more subtle than in the above-described Mo_2 example. Oxidation or reduction of a complex by one electron formally changes the M–M bond order by 0.5, and these changes are usually accompanied by a change in the M–M bond distance of ~ 0.03–0.07 Å (3). For

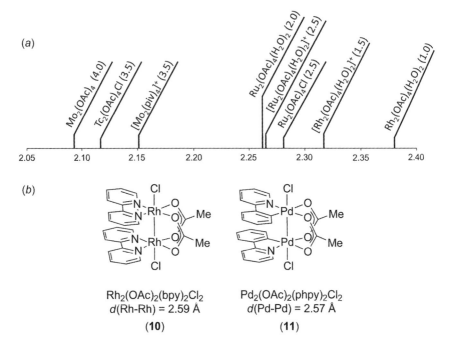

Figure 4.   (a) The M−M bond distance scale containing compounds with bond orders of 1.0 up to 4.0. (b) Comparison of the M−M bond distances in a $Rh_2(II,II)$ species and an isoelectronic and structurally analogous $Pd_2(III,III)$ species (bpy = 2,2′-bipyridine).

example, one-electron oxidation of $Rh_2(II,II)$ species to the $Rh_2(II,III)$ causes a net increase in Rh−Rh bond order of 0.5, and thus we expect a contraction of the Rh−Rh bond. Indeed, the Rh−Rh bond distance in $Rh_2(OAc)_4(H_2O)_2$, 2.39 Å (8, 114), decreases to 2.32 Å upon oxidation to the $[Rh_2(OAc)_4(H_2O)_2]^+$ monocation (115). There is generally a strong correlation between observed bond distances and the M−M bond order. For example, if we limit ourselves to compounds of second-row transition metals bridged by carboxylate donor ligands and, where possible, with either no axial ligands or pure σ-donor axial ligands (e.g., H_2O) (3), a correlation exists between bond distance and bond order, shown in Fig. 4(a).

Exceptions to the correlation between bond order and M−M distance have been observed for some binuclear complexes. For example, $Ru_2(II,II)$ complexes with a formal double bond have consistently shorter Ru–Ru distances than $Ru_2(II,III)$ compounds with a bond order of 2.5. There are two issues that contribute to this result. First, there are compounds that differ effectively by the presence or absence of a δ* electron, which is not expected to yield a major difference in the M−M bond distance. Second is the fact that the d orbitals will be more contracted in the $Ru_2(II,III)$ complex because of the added positive charge. This effect will decrease orbital overlap in general, and increase M–M electrostatic repulsion, both effects contributing to a longer Ru–Ru bond distance. Related effects lead to variability in the M–M bond length changes upon one electron oxidation of $Pd_2(II,II)$ compounds to the mixed-valent $Pd_2(II,III)$ species. There are only two compounds that have been crystallographically characterized in both the $Pd_2(II,II)$ and $Pd_2(II,III)$ states, $Pd_2(DTolF)_4$ (108), and $Pd_2(DAniF)_4$ (116). In each case, one would expect the loss of a σ^* electron via oxidation to yield a significant decrease in the Pd–Pd separation. Indeed, the Pd–Pd distance in $Pd_2(DAniF)_4$ (DAniF = N,N'-di-p-methoxyphenylformamidinate) decreases by 0.05 Å upon oxidation to the corresponding monocation (116). However, the Pd–Pd distance in the $[Pd_2(DTolF)_4]^+$ (DTolF = N, N'-di-p-tolylformamidinate) monocation, bearing less electron-donating ligands than $[Pd_2(DAniF)_4]^+$, is 0.02 Å $longer$ than in the neutral species (108). High-field EPR studies have shown that oxidation is metal-centered in both cases, and the delicate balance of M–M bonding via orbital overlap and electrostatic repulsion of the positively charged Pd(2.5+) ions appears to be affected by electronic effects of the equatorial ligands (116).

Also illustrated in Fig. 4 is the effect of axial ligation on M–M distance. Introduction of axial ligands (e.g., chloride) can cause significant deviations in M–M bond distances, as exemplified by the increase in the Ru–Ru distance by 0.03 Å on going from the $[Ru_2(OAc)_4(H_2O)_2]^+$ cation to $Ru_2(OAc)_4Cl$ [Fig. 4(a)]. This finding is rationalized as a result of the differing π-donating strengths of Cl and H_2O); the Cl^- ligand is not just a σ-donor, but is also a π-donor and can weaken M–M π interactions when it binds. Notably, in singly bonded compounds, where π-bonding interactions are not present, this effect is less significant, as exemplified by the comparison in Fig. 4(b). For compounds of higher M–M bond orders, it is not difficult to find examples of compounds having no axial ligands (3). In contrast, with the exception of d^8–d^8 dimers, compounds with M–M bond orders less than three almost exclusively are found with ligands bound to the axial sites (3). For example, it was not until 2002 that an Rh_2-tetracarboxylate compound devoid of axial ligands was characterized in the solid state and in solution (117). The low Lewis acidity of binuclear complexes of the early metals, and high Lewis acidity of binuclear complexes of the late-metals toward axial donors is contrary to our expectation in the periodic trends of mononuclear transition metal compounds. Moreover, trends in hard-soft Lewis acid-base theory would lead us to expect late-metals (e.g., Rh) to prefer soft Lewis base donor ligands (118, 119). However, $Rh_2(II,II)$ tetracarboxylates can be oxophilic and prefer hard Lewis bases, particularly when electron-withdrawing equatorial ligands are employed (120, 121).

How do we explain these observations? It is important to recognize that the binding of axial σ-donor ligands requires empty orbitals on the M_2 unit of the appropriate symmetry to accept the electron pairs of the axial donors. The M–M σ orbital is inappropriate for this interaction when it is involved in a strong M–M multiple bond. Quadruply bonded $Cr_2(II,II)$ compounds provide an interesting case study: they very strongly bind axial ligands despite a high formal $M-M$ bond order. This can be explained by looking at the electronic structure of $Cr_2(II,II)$ compounds: The $M-M$ bond in these complexes is relatively weak because the $\sigma^2\pi^4\delta^2$ electron configuration does not contribute much to the ground-state wavefunction (30, 122–124). Thus, for the second-row metals, axial σ-donor ligands must use the $M-M$ σ^* orbital to form a dative bond to the metal atoms. For early metal compounds with strong $M-M$ bonds, the σ^* orbital lies very high in energy, likely having strong mixing with the metal 5p orbitals, and this hybrid orbital can only be used to make weak $M-L$ bonds. Increasing effective nuclear charge and electron occupation of the d orbitals reduces the $M-M$ σ^* orbital energy as we go from the early metals to the later metals, thus it is not surprising that $Rh_2(II,II)$ complexes with a σ^* orbital as the lowest unoccupied molecular orbital (LUMO) have pronounced $M-L_{axial}$ bonding. The preference of $Rh_2(II,II)$ for hard Lewis bases may also be explained from an electronic structure viewpoint (121). The bonding paradigm for an octahedral mononuclear complex with soft Lewis bases is that the unoccupied e_g^* type orbitals can be used to accept σ electron density while the filled metal t_{2g} set can provide back bonding to vacant orbitals on the ligands (e.g., the empty $C-O$ π^* orbitals of CO, or $P-C$ σ^* orbitals of phosphines). In an $Rh_2(II,II)$ complex, the $Rh-Rh$ σ^* orbital can be used analogously to the e_g^* orbitals of a mononuclear complex, but the analogues of the t_{2g} orbitals, $Rh-Rh$ π and δ orbitals, are engaged in $Rh-Rh$ bonding and are therefore less available to interact with π acids.

E. Physical and Spectroscopic Properties of M–M Bond-Containing Compounds

Spectroscopic and electrochemical methods are useful for studying catalysts in solution, and the $M-M$ bonds in the compounds discussed here give rise to some unique spectral features and electrochemical properties that can be used to assess the electronic structure of the catalysts. Below, we will briefly review some of the most useful spectroscopic methods and give examples of physical features of the Rh_2, Ru_2, or Pd_2 complexes that will be discussed in the context of catalysis later.

1. Absorption Spectroscopy

One of the simplest methods for assessing the electronic structure of $M-M$ bond-containing compounds is electronic absorption spectroscopy. This method directly probes electronic transitions, which can be readily correlated to a molecular orbital picture, giving insight into bonding. Most importantly, changes in metal oxidation states are typically accompanied by major changes in the features of the electronic spectrum. Thus, absorption spectra of catalysts *in situ* allow for an assessment of metal oxidation states by comparison with the expected signature absorption features.

The $Rh_2(II,II)$ tetracarboxylates show two distinct bands in the visible region, Band I at ~600–700 nm, and Band II at ~450 nm. Although conflicting assignments of Band I appear in the literature (125–129), a preponderance of evidence favors assignment of this band as the $Rh_2 \pi^* \rightarrow Rh_2 \sigma^*$ HOMO to LUMO transition. This evidence includes the observed x,y polarization of this band (125), a 297-cm^{-1} vibronic progression observed at 15 K attributable to the Rh–Rh stretching frequency of the $\pi^{*3}\sigma^{*1}$ excited state (127), magnetic circular dichroism intensity consistent with a σ^* excited state (130), and variation of the energy of this band as a function of the axial ligands (125). An Rh_2-carboxylate compound characterized as having no axial ligands shows Band I at very low energy, 760 nm (117). Band II is assigned as an $Rh_2 \pi^* \rightarrow Rh–O \sigma^*$ transition.

Oxidation of $Rh_2(II,II)$ tetracarboxylates to the $Rh_2(II,III)$ level is accompanied by drastic changes in the electronic spectrum (e.g., oxidation of a to d in Fig. 5) (131). The most striking feature is the appearance of a new band, Band Ia, at ~800 nm. The $Rh_2(II,II)$ Bands I and II both shift to higher energy to become Band IIa and Band IIIa in the $Rh_2(II,III)$ species, and Band IIIa gains significant intensity. On the basis of early SCF–Xα–SW calculations on $[Rh_2(OAc)_4(H_2O)_2]^+$ (132), the ground state was assigned to the $\sigma^2\pi^4\delta^2\pi^{*4}\delta^{*1}$

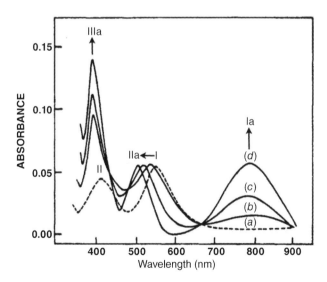

Figure 5. Changes in the absorption spectrum of $Rh_2(OAc)_4$ (a) upon electrochemical oxidation in acetonitrile. The fully oxidized species is (d), and (b) and (c) are spectra taken at intermediate points. Band assignments are as discussed in the text. [Adapted from (131)].

configuration, although subsequent EPR studies (see below) suggest a $\sigma^2\pi^4\delta^2\delta^{*2}\pi^{*3}$ configuration (111). Nevertheless, the SCF$-$X$\alpha-$SW results were used to interpret the electronic spectrum. It is suggested that Band IIIa is a Rh$_2$ $\pi^* \rightarrow$ Rh$-$O σ^* transition and that Band IIa is a Rh$_2$ $\pi^* \rightarrow$ Rh$_2$ σ^* transition, as in the Rh$_2$(II,II) species (132). Band Ia has been assigned tentatively as a $\delta \rightarrow \delta^*$ transition (132), although Band Ia is absent in Rh$_2$(II,III) mixed acetate–acetamidate compounds (131); the electronic origin of Band Ia is currently unknown. Observation of such a low-energy band has, however, been useful for assigning the catalyst resting state of Rh$_2$ mediated intermolecular C$-$H amination as an Rh$_2$(II,III) complex (see below) (133, 134).

The major ambiguity in the electronic structure of Rh$_2$(II,III) complexes is whether the compounds have a $\delta^{*2}\pi^{*3}$ or a $\pi^{*4}\delta^{*1}$ ground state. This ambiguity exists because the π^*- and δ^*-orbitals in Rh$_2$ complexes are very close in energy (see above). In ultraviolet (UV) photoelectron spectra of Rh$_2$(O$_2$CCF$_3$)$_4$, the π^* and δ^* bands appear overlapped in the spectrum, and a deconvolution of this feature yields an energy difference of only 0.2 eV (135). Thus, the nature of the ligands bound to the Rh$_2$(II,III) core can strongly influence whether a $\delta^{*2}\pi^{*3}$ or a $\pi^{*4}\delta^{*1}$ ground state is observed.

For Ru$_2$(II,III) compounds, the visible region of the absorption spectra usually contain charge-transfer bands that obscure d–d bands in this region (136). The visible absorption energies are, however, sensitive to changes in the nature of axial and equatorial ligands (137, 138), and therefore make good optical probes for chemical reactions. Some d–d bands from the $\sigma^2\pi^4\delta^2(\delta^*,\pi^*)^3$ manifold occur significantly lower in energy, in the near-IR region of the spectrum. For example, the $\delta^2\delta^{*1} \rightarrow \delta^1\delta^{*2}$ transition has been assigned at 9000 cm^{-1} ($\sim$1100 nm; ε $\sim$25–50 M^{-1} cm^{-1}) on the basis of its z polarization and the observation of vibrational progressions for the (excited state) Ru–Ru and Ru–O stretching frequencies at $\sim$300 and $\sim$400 cm^{-1}, respectively (139).

Oxidation of Pd$_2$(II,II) complexes to Pd$_2$(III,III) complexes frequently results in the appearance of a low-energy absorption band in the electronic absorption spectrum. Both tetrabridged Pd$_2$(III,III) complexes, as well as dibridged Pd$_2$(III, III) complexes display low-energy absorption bands, which have been assigned as Pd–Pd $\sigma \rightarrow \sigma^*$ excitations (15, 140–144).

2. Electron Paramagnetic Resonance Spectroscopy

Catalysts or intermediates that have unpaired electrons are amenable to study by EPR spectroscopy. For a compound with nonzero spin S, application of a magnetic field splits the ground state into $2S + 1$ non-degenerate m$_S$ states. Electron paramagnetic resonance spectroscopy uses a microwave photon to excite transitions between m$_S$ states. These transitions can be affected by coupling of the electron spin with nearby nuclear spins (hyperfine or superhyperfine coupling) to give fine structure to the observed signals. For example, the ^{103}Rh nucleus has a nuclear spin (I) of $I = 1/2$, and the larger the ^{103}Rh hyperfine interaction, the more spin density in the molecule resides near the Rh nucleus. Thus, analysis of an EPR spectrum can give detailed information about the number of unpaired electrons in a compound and their chemical environment.

Electron paramagnetic resonance spectroscopy is a useful probe of the electronic structure of Rh_2(II,III) compounds. As mentioned in Section I.E., Rh_2(II,III) complexes having mixed N,O- or N,N-donor ligands have a δ^{*1} configuration, whereas carboxylate compounds have a π^{*3} ground state. Electron paramagnetic resonance spectroscopy provides the strongest evidence for this assignment. Compounds with the δ^{*1} configuration show an axial $S = 1/2$ signal with $g_\perp > g_\parallel$, where $g_\perp$ is $\sim$2.05–2.11 and $g_\parallel$ is $\sim$1.89–1.98 (111, 131). The g-values are very close to the free electron value, indicating little contribution from spin–orbit coupling. The EPR spectra for Rh_2(II,III) carboxylate species are significantly different. They have $g_\perp < g_\parallel$, and the values of $g_\perp$ and $g_\parallel$, $\sim$1.4 and 3.7, respectively (111, 145). These parameters have been analyzed in terms of a degenerate π^{*3} ground state having a first-order spin–orbit contribution to the wavefunction (145). The unusual g-values that result from this spin–orbit splitting lead to two Kramers' doublets [$\pi^*(x)^1$ and $\pi^*(y)^1$] that are very close in energy. The observed g-values depend on the splitting between these two doublets in much the same way as in the Ru_2(II,III) case described below.

For Ru_2(II,III) complexes, the near-degeneracy of the δ^* and π^* levels leads to a high-spin $S = 3/2$ ground-state configuration, which is unusual for complexes based on second-row transition metals. The compounds have an $S = 3/2$ ground state as a consequence of their $\sigma^2\pi^4\delta^2\delta^{*1}\pi^{*2}$ electron configuration. As with all states having $S > 1/2$, the $S = 3/2$ state is subject to zero-field splitting, which splits the four fold degeneracy of the $S = 3/2$ state into $m_S = \pm 1/2$ and $m_S = \pm 3/2$ Kramers' doublets (146). The energetic splitting between these two doublets is described by the zero-field splitting tensor, which has two major components: D, the axial zero-field splitting parameter, and E, the rhombic term. It is useful to describe the symmetry of the zero-field splitting tensor in terms of its rhombicity, E/D, which may take values from $E/D = 0$ (fully axial) to $E/D = 0.33$ (fully rhombic). The molecular symmetry therefore plays an important role in the zero-field splitting tensor. The Ru_2(II,III) complexes having at least C_4 symmetry are axial, with $E/D = 0$.

The strong molecular anisotropy provided by the Ru–Ru bond leads to an unusually large zero-field splitting. From solid-state magnetic susceptibility measurements, a D value of $\sim$77 cm^{-1} is obtained for Ru_2(butyrate)$_4$Cl (147), and most other Ru_2(II,III) compounds that have been investigated by magnetic susceptibility have D values of a similar magnitude (148, 149). The D values for Ru_2(II,III) compounds are significantly larger than typical microwave quanta available for EPR spectroscopic measurements (e.g., $h\nu = 9.5$ GHz or $= 0.32$ cm^{-1},

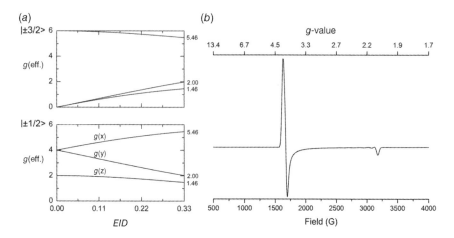

Figure 6.   (*a*) Left: Rhombogram depicting the dependence of effective *g*-values on *E/D* for the $m_S = \pm 1/2$ and $m_S = \pm 3/2$ Kramers' doublets of an $S = 3/2$ system. (*b*) Simulated X-band EPR spectrum of an $S = 3/2$ species with $D > h\nu$ and $E/D = 0$. Note the observed *g*-values of 4 and 2, which are derived theoretically from the $m_S = \pm 1/2$ rhombogram.

for a typical X-band EPR spectrum). The EPR spectra of these compounds must therefore be analyzed in the $D > h\nu$ regime using a rhombogram (150), as shown in Fig. 6(*a*), which shows how the expected EPR signals (effective *g*-values) change with rhombicity. Thus, for compounds with axial symmetry, $E/D = 0$, the lower energy $m_S = \pm 1/2$ Kramers' doublet is predicted to have effective *g*-values of $g_x = g_y = 4.00$; $g_z = 2.00$. Transitions within the $m_S = \pm 3/2$ Kramers' doublet would not be observed. A simulated $S = 3/2$ spectrum assuming X-band frequency with these parameters is given in Fig. 6(*b*), and such a spectrum is similar to those measured for many axially symmetric Ru$_2$(II,III) compounds. If the EPR line-widths are small enough, Ru hyperfine coupling can be observed (147), as the $I = 5/2$ ^{99}Ru and ^{101}Ru isotopes together are abundant in ~30 % of terrestrially occurring Ru.

Electron paramagnetic resonance spectroscopy also has played a major role in establishing the $\sigma^2\pi^4\delta^2\delta^{*2}\pi^{*4}\sigma^{*1}$ ground state of Pd$_2$(II,III) complexes. The EPR spectra of [Pd$_2$(DTolF)$_4$]$^+$ and [Pd$_2$(dpb)$_4$]$^+$ (dpb = *N,N'*-diphenylbenzamidinate), both measured in the 1980s, gave conflicting results (108, 151). The former compound showed a simple but broad isotropic signal at $g = 2$ (108), and the latter showed an axial signal with $g_\perp = 2.17$ and $g_{||} = 1.99$, with clearly visible ^{105}Pd hyperfine signals (151). Based on these results, the unpaired electron of [Pd$_2$(DTolF)$_4$]$^+$ was proposed (108) to be ligand-centered, whereas [Pd$_2$(dpb)$_4$]$^+$ was regarded to arise from M-based oxidation (151). This observation was rather unusual considering that there are only very minor changes between the two ligands employed in these studies; based on further investigations, both complexes are believed to arise from M-centered oxidation. Furthermore, an X-ray crystal structure of $[Pd_2(DTolF)_4]^+$ showed that the Pd–Pd distance was very similar to that in the neutral $Pd_2(II,II)$ compound, which seemed to support the hypothesis of ligand-centered oxidation (108). No crystal structure of the $[Pd_2(dpb)_4]^+$ cation has been reported. The $[Pd_2(DTolF)_4]^+$ species was reinvestigated recently along with the $[Pd_2(DAniF)_4]^+$ congener (116). Both species show broadened, isotropic EPR signals, though the latter shows slight g anisotropy and superhyperfine coupling to two ^{14}N nuclei when the measurement is made in butyronitrile glass. The g anisotropy of both compounds can be directly measured by the use of high-field EPR spectroscopy (W or D band). At these high fields, the EPR spectrum of $[Pd_2(DAniF)_4]^+$ resolves into a rhombic pattern with $g_1 = 2.029$, $g_2 = 2.014$, and $g_3 = 2.000$. The spectrum of $[Pd_2(DTolF)_4]^+$ is axial with $g_\perp = 2.026$ and $g_\parallel = 1.989$. In both these cases, the spectra are consistent with a metal-centered unpaired electron (116).

3. Electrochemistry

Many of the catalytic processes discussed here involve chemical steps that are accompanied by M-centered redox chemistry. Thus, it is pertinent to examine the electrochemical properties of catalysts to establish the nature of any redox activity that may influence catalytic mechanisms. Cyclic voltammetry (CV) is the most common method for determining oxidation or reduction potentials of catalysts in solution, and is useful for assessing the reversibility of redox processes.

Dirhodium complexes typically display an electrochemical wave corresponding to the $Rh_2(II,II)/Rh_2(II,III)$ redox couple (3). For carboxylate compounds, this wave typically appears at over +0.50 V vs Fc/Fc^+ (Fc = ferrocene), and the exact potential depends strongly on the nature of both the equatorial and axial ligands (3). For example, the $Rh_2(II,II)/Rh_2(II,III)$ redox couple for $Rh_2(OAc)_4(MeCN)_2$ appears at +0.77 V in acetonitrile solution (152), and substitution of the equatorial OAc^- ligands for the more electron-withdrawing trifluoroacetate (TFA^-) ligands, that is, in $Rh_2(TFA)_4(MeCN)_2$, leads to a potential of +1.40 V (153), a change of nearly 0.7 V. The effects of different axial ligands are less pronounced. The redox potential of $Rh_2(OAc)_4$ in a broad range of solvents varies from +0.77 V (in acetonitrile) to +0.58 V (in methanol) vs Fc/Fc^+ (152). Dirhodium complexes with N,O donor ligands or N,N donor ligands have been observed to display further oxidation waves to $Rh_2(III,III)$ and $Rh_2(III,IV)$ species (154, 155), as well as reduction waves to $Rh_2(I,II)$ species (3), though potential roles for this reduced state in catalysis are not yet known.

The redox chemistry of Pd_2 complexes is also of importance to the C–H functionalization catalysis discussed in this chapter. Due to the disparate geometries typical of mononuclear Pd(II) (square planar) and Pd(III) (tetragonally elongated octahedron), one-electron oxidation of mononuclear Pd(II) species is frequently irreversible, but has been achieved with mononuclear Pd complexes bearing flexible polydentate ligands that change their coordination mode upon oxidation (156). Tetrabridged $Pd_2(II,II)$ complexes with close Pd–Pd contacts can show reversible oxidation waves (108, 151, 157).

4. Metal–Metal Bond-Containing Catalyst Preparation

The complexes discussed herein are predominantly supported by carboxylate ligands or their derivatives, which frequently bridge metal centers. There are two general catalyst synthesis methods that are commonly employed in the reactions reviewed here. Binuclear catalysts can either be pre-formed or formed during the reaction. Pre-formed catalysts are synthesized on their own in separate synthetic steps and are isolated, whereas a combination of simple metal salts and ligands may be added at the beginning of a reaction to form a catalyst *in situ*. The propriety of pre-forming a binuclear catalyst versus *in situ* catalyst formation depends on the rates of ligand exchange reactions for each metal. In the cases presented in this chapter, Rh_2 and Ru_2 complexes are typically prepared *ab initio*, and Pd_2 complexes are formed *in situ*. It is important to synthesize Rh_2 catalysts ahead of time rather than *in situ* because of the large disparity of ligand exchange rates for axial versus equatorial ligands. Axial ligand exchange is very fast, but equatorial ligand exchange can be several orders of magnitude slower (158). Consequently, one cannot rely on an *in situ* equatorial ligand substitution to precede catalyst initiation. More details of catalyst preparation are given in the next sections.

There has been significant interest in immobilizing Rh_2 complexes on solid supports to aid in catalyst recovery and recycling (159). Immobilization can be done either by modifying the equatorial ligands with moieties that are readily incorporated into polymeric matrices (160–165), or by binding polymer substituted ligands via the axial position (166, 167). The latter approach is effective despite the high lability of Rh_2 axial sites.

II.   DIRHODIUM AND DIRUTHENIUM C−H FUNCTIONALIZATION CHEMISTRY

A.   Carbenoid Chemistry

Carbenes are divalent carbon-containing species that are typically highly reactive and have been studied since the early twentieth century (168). Carbenes were first proposed in 1903 in studies of the photolytic and thermolytic decomposition of ethyl diazoacetate (**2**) (169). Staudinger and Kupfer (170) demonstrated the cyclopropanation of olefins proceeding from free carbenes in 1912. Free carbenes can participate in a variety of reactions, such as cyclopropanation, C−H

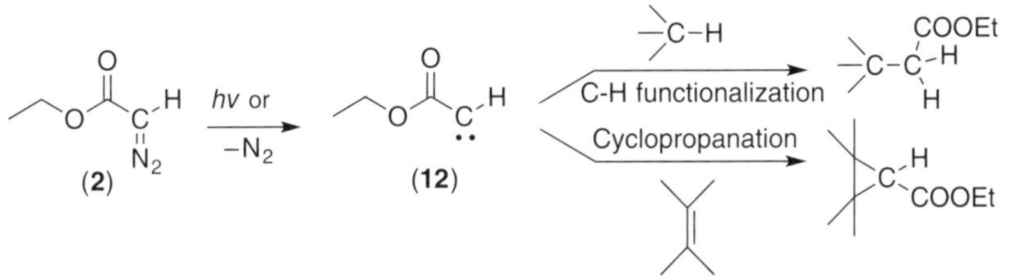

Scheme 5. Free carbenes participate in both C–H insertion and cyclopropanation reactions.

insertion, and X–H insertion (171–173), and early reports of carbene reactivity generally noted poor selectivity when more than one reaction pathway was available (Scheme 5) (173).

The first report of carbene generation from diazoalkanes by use of a transition metal was in 1952 when copper bronze and copper oxide were shown to promote insertion reactions of diazoalkanes into O–H, N–H, and C–H bonds (174). This reaction was soon followed by reports of intramolecular ring-closing C–H insertions of diazo compounds, also catalyzed by simple copper catalysts (175). The first intermolecular cyclopropanation from diazomethane was reported in 1963, also catalyzed by Cu (176). Early reports of Cu–carbenoid reativity were important in that they demonstrated that selectivity in carbene reactions could be influenced by the transient stabilization of a carbene by a metal center, referred to as a metallocarbenoid. The discovery of carbenoid-type reactivity resulting from the decomposition of diazo compounds by mononuclear Cu complexes, which is still a flourishing research field (102, 177), led to broad interest in the development of metallocarbenoid chemistry.

1. Overview of Dirhodium–Carbenoid Reactivity

Dirhodium tetracarboxylates were first used in the C–H functionalization of hydrocarbons in 1981 (178). Seminal reports by Teyssié and co-workers (178) showed intermolecular C–C bond formation from diazo compounds with $Rh_2(TFA)_2$ (13) and simple long-chain alkanes. The C–H site selectivity preferences of reactions of free photolytically generated carbenes as compared to reactions using 13 were evaluated and a greater level of chemoselectivity for 2° C–H bonds over 1° C–H bonds was observed when 13 was used (Scheme 6). As seen in this example, Rh_2–carbenoids are typically generated by Rh_2–mediated decomposition of diazo compounds and display electrophilic reactivity, participating in insertion reactions with electron-rich C–H bonds. This example further highlights the control of product selectivity that metallocarbenoids can offer as compared to free carbenes, and Rh_2 carbenoid intermediates have become topics of significant interest in the development of new synthetic methodologies (173, 179–183).

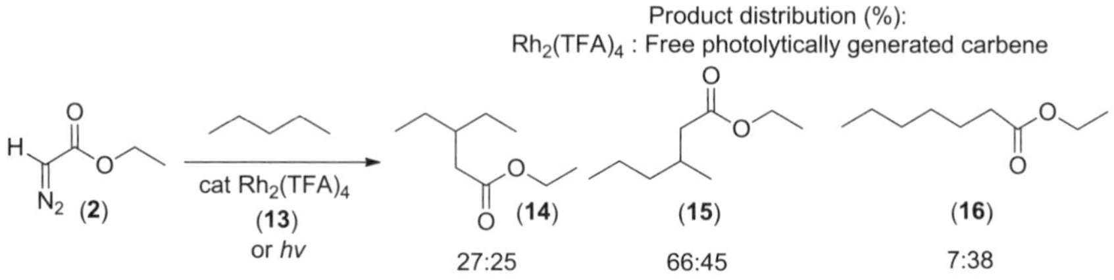

Scheme 6.   Site selectivities using **13** as a catalyst compared with photolytically generated free carbene. Insertion into primary C—H bonds is suppressed when **13** is used as a diazo decomposition catalyst (cat = catalytic).

2.   Current Mechanistic Understanding of Carbenoid Chemistry

A general mechanism for carbene generation from diazo compounds and C—H insertion of the resulting carbene into a C—H bond is shown in Scheme 7. Diazo decomposition is generally proposed to proceed by coordination of the diazo compound to the Rh₂ core followed by N₂ expulsion to generate an Rh₂

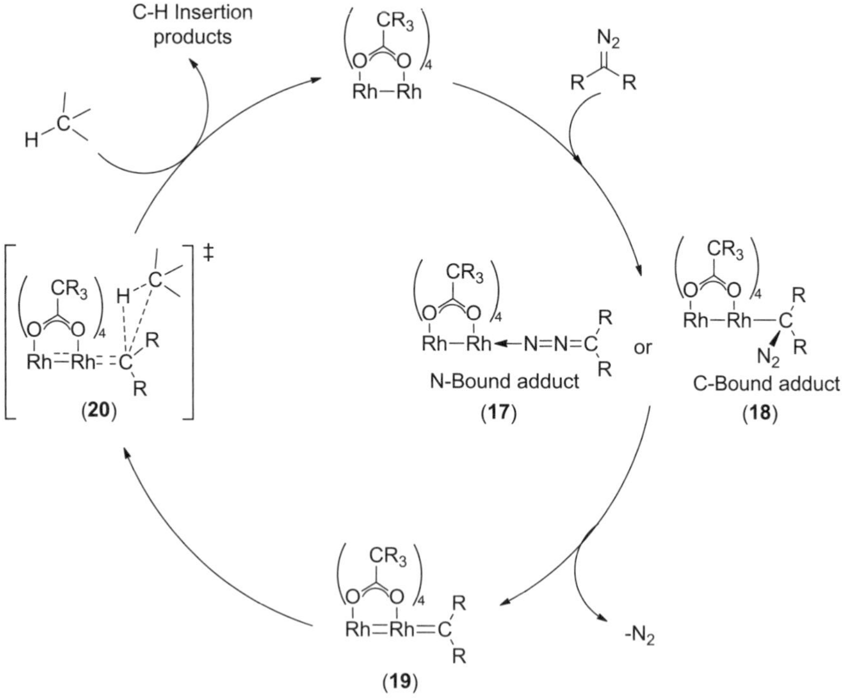

Scheme 7.   A proposed mechanism for Rh₂ catalyzed C—H functionalization with diazo compounds.

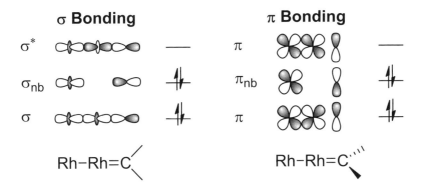

3 center-4 electron σ and π bonds

Figure 7. Depiction of σ and π bonding orbitals in a 3c–4e manifold for an Rh_2–carbenoid. The terminal Rh=C bond is lengthened by the 3c–4e interaction, thus enhancing reactivity (nb = nonbonding).

carbenoid (184). The C–H insertion proceeds from Rh_2–carbenoid **19** via a proposed concerted asynchronous transition state (**20**) (184). Dirhodium-catalyzed carbene transformations can be very efficient with turnover numbers of $>10^6$ being reported for some transformations (185). Because of the efficiency of Rh_2 catalysts in carbenoid transformations, direct spectroscopic studies of the putative Rh_2 stabilized carbene intermediate are lacking (see Note Added in Proof), although mononuclear Rh–carbene species have been characterized (186). Nevertheless, indirect mechanistic evidence from both experiment and computation is available (87).

A three-center–four-electron (3c–4e) bonding manifold involving the orbitals of σ and π symmetry on both Rh atoms and the carbene/nitrene center has been proposed as important to the reactivity of dirhodium carbenes and nitrenes (187). As shown in Fig. 7, the three σ (π) atomic orbitals combine to form three molecular orbitals: a three-center bonding orbital in which all orbitals are in phase, a nonbonding orbital having a node at the central Rh atom, and a completely out-of-phase antibonding combination. In the case of Rh_2 carbene or nitrene species, the bonding and nonbonding orbitals are filled and the antibonding orbitals are empty, leading to their description as having 3c–4e bonds of both σ and π symmetry. In comparison with conventional 2c–2e bonds, the Rh–Rh–C(N) 3c–4e bonds are weaker and the antibonding combinations lie lower in energy, leading to more reactive terminal M = E (E = CR_2, NR) groups in binuclear species (187). While we draw the Rh_2–carbenoid and nitrenoid species as having an Rh=E double bond, computed bond orders are significantly lower than 2.0 (187).

Rh_2–carbenoids have not yet been directly characterized due to the high intrinsic reactivity of these species (186). As a result, much of what we know

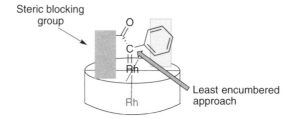

Figure 8.   An example of a steric model for predicting catalyst selectivity.

about the impact of ligand perturbation on carbenoids comes from characterization of the reactivity of these species as a function of ligand structure. An early study by Doyle and co-workers (112, 188, 189) investigated the C−H functionalization of simple hydrocarbons by various $Rh_2(OAc)_4$ derivatives and indicated that site selectivity was catalyst controlled. This study demonstrated that ligand electronic effects can effectively switch between competitive reaction pathways by influencing the relative charge on the reactive carbenoid carbon (189). Because ligand electronic effects can so readily alter reactivity patterns, investigation of the selectivity of C−H functionalization reactions has become a prominent indirect method for understanding the nature of the Rh_2–carbene intermediate.

Since carbenoid chemistry does not rely on covalently linked directing groups in either catalyst or substrate, selectivity is a function of the delicate interplay of electronics and sterics of the putative metallocarbenoid intermediates. Both electronic effects (*see above*) (112), as well as steric considerations are important factors in determining selectivity in metallocarbenoid reactions. Davies and co-workers (104, 190–192) proposed a predictive steric model for understanding regio- and enantioselectivity for intermolecular reactions using donor–acceptor carbenoids; a simplified example of such a model is depicted in Fig. 8. When a catalyst is designed to contain bulky groups that can block substrate access, Davies and co-workers (104, 190–192) argue that the least encumbered approach of the substrate can be readily predicted, thus providing a handle for predicting selectivity. A tutorial review on this subject can be found in the recent literature (75).

Early mechanistic studies examined relative rates of cyclopropanation by various Rh_2 catalysts (193). An important study by Doyle et al. (194) compared cyclopropane formation from $(CO)_5W=CHPh$ to cyclopropane formation from $PhCHN_2/Rh_2(OAc)_4$ with alkenes. Similar linear free energy relationships were observed for both the cyclopropanation of various monosubstituted alkenes catalyzed by $PhCHN_2/Rh_2(OAc)_4$ or mediated by pre-formed $(CO)_5W=CHPh$, suggesting that formation of an electrophilic Rh_2–carbenoid precedes cyclopropane production. Since tungsten–carbene complexes are well defined, this was the first evidence of an Rh_2 carbene complex as an intermediate in

Rh$_2$ catalyzed cyclopropanations (189). Subsequently, several kinetic studies have been published that outline the mechanism of N$_2$ extrusion from diazo compounds and propose mechanisms for both carbenoid formation and reactivity with C–H substrates (195). Hammett analyses, kinetic isotope effect (KIE) measurements and analysis of product distributions are generally consistent with a concerted C–H insertion mechanism, in agreement with related computational studies (195, 196).

The nature of the carbenoid or nitrenoid intermediate that performs the C–H functionalization event is controlled by a multitude of factors including, but not limited to, the electronic structures of both the catalyst and the substrate as well as the functional groups appended to the carbene or nitrene center. Small carboxylate ligands (e.g., pivalate), which functions as a small ligand in dirhodium catalysis because the sterically demanding *tert*-butyl group is directed away from the axial site of the catalyst, or strongly electron-withdrawing carboxylate ligands, (e.g., trifluoroacetate) can offer high conversions, but usually at the expense of selectivity. This occurrence is presumably because the metallocarbenoid intermediate reacts too quickly to effectively discriminate between the substrate C–H bonds (188).

In recent computational work, Davies and co-workers (190) analyzed the reaction coordinate of Rh$_2$ complexes with diazo compounds **21** and **22** to rationalize experimentally observed differences in selectivity between donor–acceptor carbenes (e.g., **21**) and acceptor carbenes. It was found that donor–acceptor carbenes substantially stabilize the Rh$_2$–carbene intermediate, causing a late transition state with substantial charge buildup, and a high potential energy barrier to C–H functionalization (17.4 kcal mol^{-1}), making C–H functionalization with donor–acceptor carbenes rate limiting (Fig. 9). The barrier to C–H functionalization is lowered to only 3.5 kcal mol^{-1} when diazo compound **22** is used. This energy barrier causes selectivity to be dependent on both the electronic structure of the substrate, as well as steric constraints. Thus, donor–acceptor carbenes discriminate well between small electronic differences in C–H bonds, resulting in highly chemoselective reactions (192). These computational results are consistent with experiment; the enhanced selectivity of donor–acceptor carbenes had previously been investigated in a Hammett study of cyclopropanation of substituted styrenes. Reactions with methyl 2-diazo-2-phenylacetate (**21**) were shown to be strongly influenced by the electronics of the styrene ($\rho = -0.9$, on a $\rho+$ scale), whereas other diazo compounds showed no selectivity (193).

3. Catalyst Design, Synthesis, and General Reactivity

Based on the perceived importance of ligand structure on selectivity of group-transfer reactions, a variety of ligands have been developed for dirhodium complexes; several common catalysts have been compiled in Fig. 10 and will

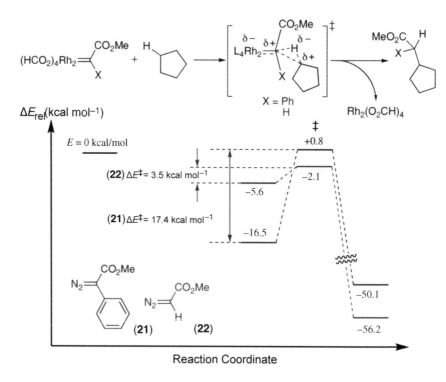

Figure 9.   Calculated reaction coordinate for C–H insertion: donor–acceptor carbenes are more chemoselective due to enhanced stability of the Rh_2–carbenoid. [Adapted from (192)].

be referred to throughout this section. Bulky ligands are frequently employed in order to enforce high levels of regioselectivity by preventing substrate access to the Rh_2–carbene intermediate in undesired conformations (188). The effect of ligands on chemo- and enantioselectivity has been reviewed elsewhere (101–103, 197), here we will focus on the impact of ligand variation on the redox behavior of dirhodium complexes and the resulting effect on reactivity. Below is a brief discussion of some of the ligand classes that have been used in dirhodium-catalyzed reactions; the ligand classes will be introduced, and subsequently, a brief discussion of reactions catalyzed by these complexes will be presented.

   a Dirhodium   Carboxylates.   The   archetypical Rh_2 tetracarboxylate, $Rh_2(OAc)_4$ (**1**), is easily prepared by treatment of $RhCl_3 \cdot 3\,H_2O$ (**37**) with sodium acetate in a mixture of ethanol and acetic acid (Scheme 8a) (198, 199). Carboxylate exchange is another useful synthetic method for accessing Rh_2-carboxylate species, whereby compounds (e.g., **1** or **13**) react with an excess of carboxylic acid to achieve ligand exchange (Scheme 8b) (200, 201). The Rh_2 tetracarboxylates can also be

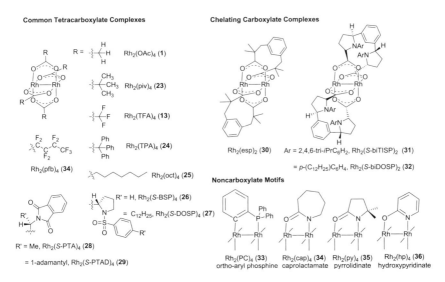

Figure 10. Common ligands and Rh$_2$ complexes used in C–H functionalization. (piv = pivalate; hp = hydroxypyridinate; pc = orthometalated phosphine; py = pyrrolidinate; pfb = perfluorobutyrate; cap = caprolactamate; oct = octanoate; TPA = triphenylacetate; S-biTISP = prolinate; S-biDOSP = (5R,5'R)-5,5'-(1,3-phenylene)bis[1-[(4-dodecylphenyl)sulfonyl]-L-prolinate; esp = l α,α,α',α'-Tetramethyl-1,3-benzenedipropanoate; S-PTA = N-napthoyl-(S)-alaninate; S-PTAD = adamantan-1-yl-(1,3-dioxo-1,3-dihydro-iso-indol-2-yl)-acetate.

prepared from K$_4$Rh$_2$(CO$_3$)$_4$ (**38**) (202), which reacts cleanly and irreversibly with carboxylic acids to yield the Rh$_2$-tetracarboxylate product and expelling CO$_2$ (Scheme 8c) (203). Chiral Rh$_2$ tetracarboxylates may also be prepared through the use of optically pure carboxylic acids, such as naturally occurring amino acids, as ligands (204–207).

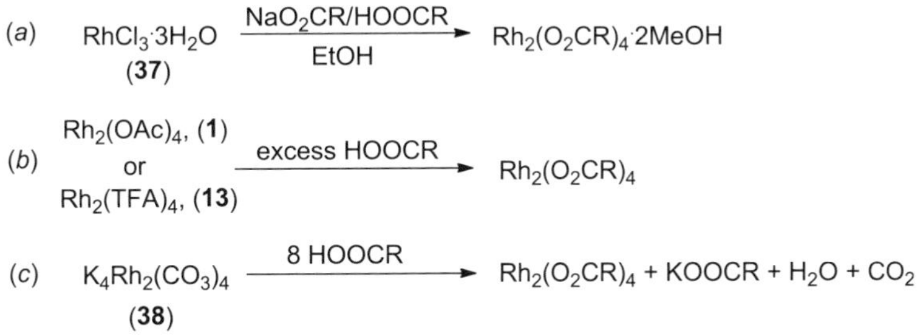

(a) RhCl$_3$·3H$_2$O $\xrightarrow[\text{EtOH}]{\text{NaO}_2\text{CR/HOOCR}}$ Rh$_2$(O$_2$CR)$_4$·2MeOH
    (**37**)

(b) Rh$_2$(OAc)$_4$, (**1**)
    or $\xrightarrow{\text{excess HOOCR}}$ Rh$_2$(O$_2$CR)$_4$
    Rh$_2$(TFA)$_4$, (**13**)

(c) K$_4$Rh$_2$(CO$_3$)$_4$ $\xrightarrow{\text{8 HOOCR}}$ Rh$_2$(O$_2$CR)$_4$ + KOOCR + H$_2$O + CO$_2$
    (**38**)

Scheme 8. General synthesis of dirhodium compounds (a) Starting from mononuclear Rh(III) (b) ligand metathesis starting from dimeric Rh$_2$(II,II), and (c) starting from Rh$_2$(II,II) carbonate.

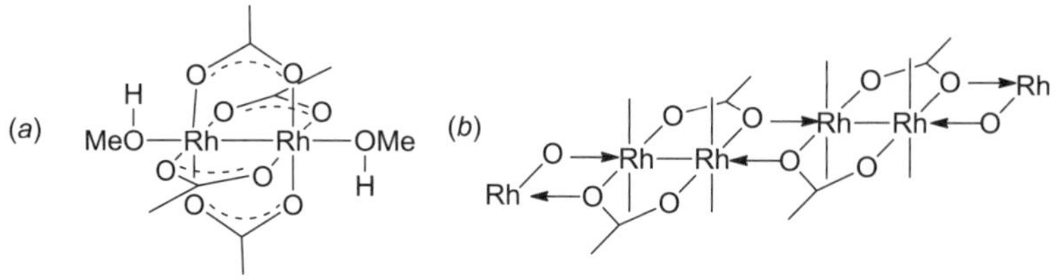

(1·2MeOH)   (39)

Scheme 9.   (a) Complex **1** with two MeOH molecules axially ligated. (b) General scheme for a coordination polymer of an Rh_2–carboxylate complex similar to **1**.

The most common preparative method for making $Rh_2(OAc)_4$ yields the deep blue–green bis-methanol adduct from a recrystallization of the crude reaction (Scheme 9a) (198). Methanol can be removed from the axial positions by heating under vacuum, and two resulting compounds have been described. Prolonged heating at 100 °C will convert **1**·2MeOH into a bright green insoluble form that is proposed to be a coordination polymer of $Rh_2(OAc)_4$ in which carboxylate oxygen atoms from one molecule can bridge to coordinate to the axial sites of the next molecule in the chain (**39**, Scheme 9b) (198). Heating **1**·2MeOH under vacuum for < 24 h at 40–50 °C reportedly leads to an olive green axial ligand free form of $Rh_2(OAc)_4$, which is more reactive than the polymeric form (208). Additionally, the axial ligands are labile and ligand exchange will typically take place when the Rh_2 complex is dissolved or recrystallized from the desired axial ligand as solvent.

A number of Rh_2 complexes with chelating dicarboxylate ligands have been prepared and used as catalysts for C−H functionalization (209–212). These compounds are also prepared by carboxylate exchange. There are two important effects of using these chelating dicarboxylate ligands in catalysis. First is that these chelating dicarboxylate ligands enhance catalyst stability via the chelate effect (211, 213). Second, chelating ligands can lower the effective symmetry of the Rh_2 complexes, breaking the fourfold axis along the Rh–Rh vector. With chiral, C_2-symmetric chelating dicarboxylate ligands, the resulting Rh_2 complexes will have D_2 symmetry (209, 210). This approach represents one strategy for influencing how chirality is conferred in asymmetric C−H functionalization.

b Dirodhium-Carboxamidates.   An important class of Rh_2 catalysts are complexes of ligands that are more basic than carboxylates. These include N,O-donors such as hydroxypyridinates or carboxamidates. Changing from O,O-donor equatorial ligands to N,O-donors has three major effects on $Rh_2(II,II)$ complexes. First, unlike an Rh_2 tetracarboxylate complex, an Rh_2 tetracarboxamidate compound can potentially have four isomers, as shown in Scheme 10. The most common isomer is the cis-(2,2) isomer, in which each Rh atom is ligated by two

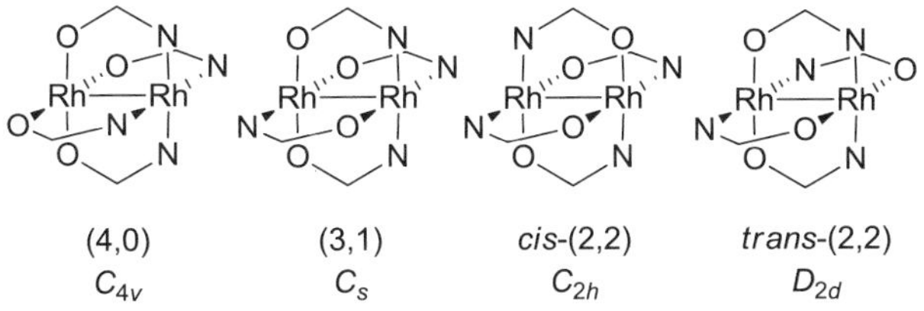

(4,0) (3,1) cis-(2,2) trans-(2,2)
C_{4v} C_s C_{2h} D_{2d}

Scheme 10. Isomers of Rh_2–carboxamidates.

N atoms and two O atoms in a cis orientation (214). Second, carboxamidate ligands are more kinetically inert than carboxylates, and, to our knowledge, there is no evidence that these isomers may interconvert in solution. Third, Rh_2-tetracarboxamidate compounds have considerably lower oxidation potentials to reach the Rh_2(II,III) state as compared with that of carboxylate compounds (214). While Doyle and co-workers (214, 215) note that this effect can be detrimental to carbene reactions, which fail if the catalyst becomes oxidized to the Rh_2(II,III) state, it has been possible to capitalize on this feature to use Rh_2–tetracarboxamides to catalyze oxidation reactions (Section III.C). Recently, the use of discrete Rh_2(II,III) dimers has been expanded to C–H amination chemistry (Section II.B.3) (216). Unlike Rh_2 carboxamidates, Rh_2 amidinate complexes, bearing highly electron-donating N,N-bridging ligands, are relatively rare and are infrequently used as group-transfer catalysts. These completes are thus not discussed in this chapter.

Although Rh_2 carboxamidates are less efficient at diazo compound decomposition due to their attenuated electrophilicity (214), they find general utility in protocols for oxidation chemistry (217, 218), as well as aziridination chemistry (219). The difference in reaction scope for Rh_2 carboxamidate complexes compared with carboxylate complexes is likely due to the fact that the former readily undergo one-electron redox processes due to lowered oxidation potential, yielding mixed-valent Rh_2(II,III) species that may be amenable to radical chemistry. Aziridination chemistry is favored by Rh_2 carboxamidate catalysts, likely due to the lower oxidation potentials observed for these species (219). Du Bois and co-worker (220) first made this observation in 2002 using Rh_2(tfacam)$_4$ (tfacam = F_3CCONH) for olefin aziridination, followed by studies by Doyle and co-workers (219) using Rh_2(cap)$_4$ (**34**) in 2005.

c Phosphine Ligands. Dirhodium complexes with ortho-metalated aryl phosphines (e.g., **33**, Fig. 10) have been developed by Lahuerta and co-workers (221–224) and tend to show a high propensity for cyclopropanation chemistry and, like tetracarboxamidates, a lesser degree of activity for diazo decomposition. Despite the decreased level of diazo activation seen in complexes with more heavily donating ligands, such as phosphines and amides, high enantiocontrol using such catalysts is possible (102).

4. *The Role of the M–M Bond in Dirhodium Carbenoid Species*

Computational results have indicated that $C-H$ functionalization occurs using only one of the two Rh atoms of the binuclear core, but the role of the second metal is indispensible to the high activity of these catalysts (184). The exact role of the second metal cannot be easily investigated due to the synthetic difficulty involved in removing this second metal atom, or in making heterobimetallic analogues for comparison. The only example of a heterobimetallic catalyst for $C-H$ functionalization is $BiRh(O_2CCF_3)_3(O_2CMe)$ (225). This complex was first reported by Dikarev and et al. (225) and later was investigated as a catalyst for diazo decomposition in collaboration with Davies and co-workers (226). The reactivity of $BiRh(O_2CCF_3)_3(O_2CMe_3)$ was compared with isostructural $Rh_2(O_2CCF_3)_3(O_2CMe)$. It was found that the bismuth-containing analogue performs favorably as a catalyst in metal–carbenoid transformations and is effective at low catalyst loadings (2 mol%). However, $BiRh(O_2CCF_3)_3(O_2CMe)$ decomposes methyl phenyldiazoacetate 1600 times slower than the analogous Rh_2 complex. Computational studies are consistent with the experimental reactivity differences, and suggest that the electronic delocalization across the binuclear core, described as a "bifunctional electron pool" (see below) (184), lowers the energy of the $C-H$ insertion transition state. Inefficient orbital overlap between Rh and Bi precludes the binuclear core of these heterometallic dimers from participating in effective delocalization, and thus the activation barriers for group transfer are higher. This observation is consistent with $M-M$ synergy during group transfer lowering the barrier to $C-H$ functionalization. The cooperativity of a heterometal with respect to $C-H$ functionalization chemistry has been systematically investigated in a computational study describing electrophilic aromatic $C-H$ amination (see Section II.D) (227).

Nakamura et al. reported (184) a computational study of the mechanism for Rh_2–carbenoid $C-H$ functionalization in 2002 in which the role of each of the Rh centers in binuclear Rh complexes during both diazo decomposition and $C-H$ insertion was investigated. This study resulted in two important conclusions: The $Rh-Rh$ interaction is capable of mediating significant electron delocalization, which results in both enhanced electrophilicity of the carbene carbon, as well as increased facility of $Rh-C$ cleavage. Nakamura et al. (184) refers to this phenomenon as a "bifunctional electron pool", alluding to an orbital manifold that is capable of extensive electron delocalization, in this case, across three centers: the two Rh atoms and the C atom of the carbene ligand. This effect bears resemblance to "bimetallic redox chemistry", which will be discussed in the Section II.A.4). The computational results substantiate the view that the mechanism of $C-H$ functionalization is fundamentally different from organometallic $C-H$ activations, as the Rh atoms do not ever come in contact with alkane $C-H$ bonds. The proposed transition state is consistent with small experimentally observed KIEs and is described as concerted but asynchronous (**20**, Scheme 7). Nakamura (184) also provided an extensive DFT comparison of monometallic Ru–carbenoid complexes (228) with Rh_2–carbenoids, focusing on the importance of the trans-effect to reactivity. It is worth noting that mononuclear Rh–porphyrin complexes have been reported to be active in cyclopropanation chemistry, presumably via the intermediacy of an Rh–carbenoid, as early as 1992 (229). It remains a possibility that certain ligand frameworks, (e.g., porphyrins) can act as a bifunctional electron pool, thus facilitating group-transfer reactivity.

B. Nitrenoid Chemistry

Functionalization of C–H bonds via nitrene intermediates, which are isoelectronic to carbenes, to generate C–N bonds is attractive for chemical synthesis due to the prevalence of biologically relevant nitrogen-containing molecules (230–232). Although the putative intermediates of Rh_2–carbene and Rh_2–nitrene chemistry have been described to be analogous in their structure and reactivity, C–H functionalization via nitenoid intermediates is far more mechanistically complex than C–H functionalization via carbenoid intermediates and currently is less well understood. Herein we will trace the historical origins of nitrene-type reactions, discuss the structure and reactivity of Rh_2 nitrenes, and draw comparisons between the mechanisms proposed for nitrenoid reactions with those of Rh_2 catalyzed carbenoid reactions (232).

1. Development of Nitrenoid-Based Reactions

Early reports of transition metal mediated nitrenoid transformations utilized Cu catalysts. Kwart and Khan (233) performed seminal work in the decomposition of organic azides and chloramine-T by mononuclear Cu complexes showing the feasibility of C–N bond-forming reactions. The use of iminoiodinanes has enabled significant progress to be made in the development of synthetic methodologies based on nitrene intermediates (234–236). The decomposition of iminoiodinanes, to afford a putative metallonitrenoid and a stoichiometric amount of aryl iodide waste, has been observed using various mononuclear complexes (Cu, Ru, Fe, Mn, Co, etc.) in addition to Rh_2 complexes (e.g., Scheme 11*b*) (232). In 1983, Breslow and Gellman (236) identified that the combination of $Rh_2(OAc)_4$ and iminoiodinanes as a nitrogen-atom source was a promising C–H amination protocol. The efficiency of this reaction prompted the investigation of $Rh_2(OAc)_4$ and its derivatives, as well as isostructural Ru_2 complexes, as catalysts for C–H amination. Method development in the field of Rh_2 catalyzed C–H amination has been largely pioneered by Du Bois and co-workers (237) in the early 2000s. Important advances included the generation of iminoiodinanes *in situ*, the identification of

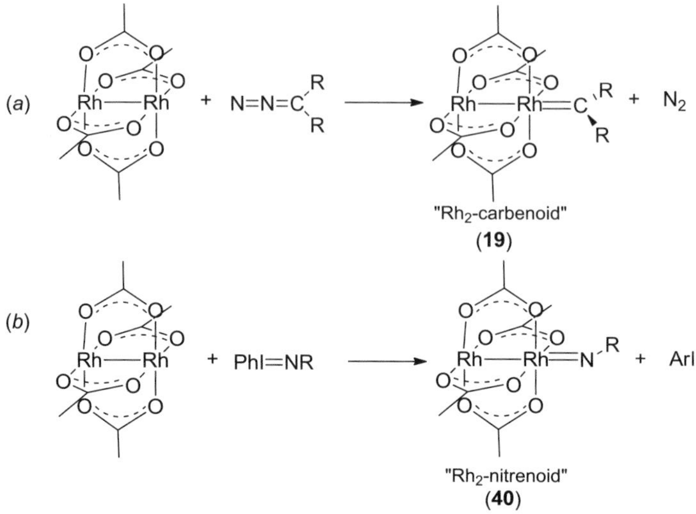

Scheme 11.   Generation of carbenoid and nitrenoid species from Rh$_2$–carboxylate complexes.

competent nitrogen-atom sources, namely, highly electron-withdrawing sulfamate esters for intermolecular aminations (e.g., **41**, Scheme 12*c*) (133, 134, 211, 212, 238–240) and the development of new catalysts, such as Rh$_2$(esp)$_2$ (**30**, Fig. 10), (211), which have allowed the development of intermolecular C−H amination reactions (Scheme 13*c*) .

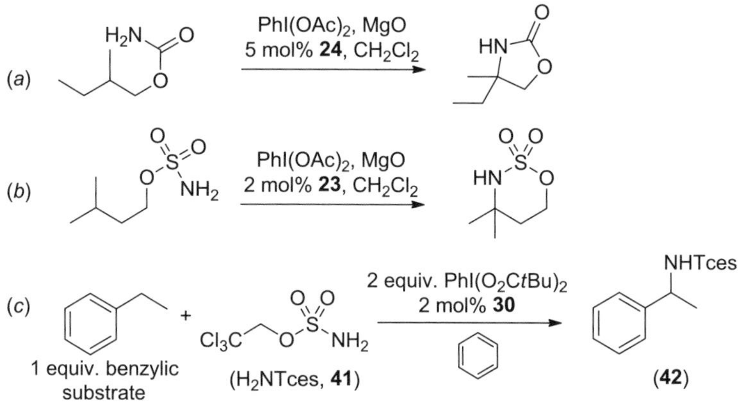

Scheme 12.   Examples of efficient Rh$_2$ catalyzed amination reactions (*a*) Intramolecular cyclization using a carbamate ester. (*b*) Intramolecular cyclization using a sulfamate ester. (*c*) Intermolecular oxidative amination of ethyl benzene using Rh$_2$(esp)$_2$ (**30**) where Tces = 2,2,2-trichloroethoxysulfonamide and esp = α,α,α',α'-tetramethyl-1,3-benzenedipropanoate.

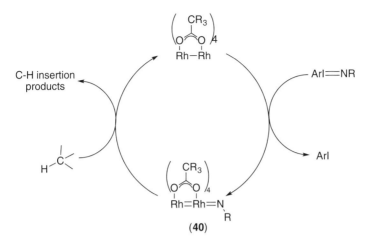

Scheme 13. Simplified mechanistic model for Rh_2 nitrene formation and concomitant C–H functionalization.

Methodology development and catalyst design for C–H amination have been heavily reliant on mechanistic observations, which is why we discuss these topics together in the following sections. Nitrene chemistry has its roots in the decomposition of preoxidized nitrogen-atom sources (e.g., iminoiodinanes or chloramine-T). Due to the similarity of these conditions with the decompostition of diazo compounds used to generate carbene intermediates, early mechanistic models of Rh_2 catalyzed nitrene reactivity were believed to proceed through two-electron steps similar to the atom-transfer mechanism discussed for carbene transfer (Scheme 13).

Both experimental and computational results show that carbene chemistry follows a mechanistic model like the one outlined for nitrene chemistry in Scheme 13 (Section II.A.2) due to the fact that Rh_2 carbenes typically form and react on a singlet energy surface (168). Nitrenes have a more complicated electronic structure than do carbenes due to the fact that the triplet state for a free nitrene is typically close to, or lower than, the corresponding singlet state (241, 242). The accessibility of a triplet potential energy surface adds a complicating factor to the simple mechanism shown in Scheme 13. Modifying the electronics of the nitrene fragment to probe the participation of the triplet surface has not been heavily investigated, however, Schomaker and co-workers (243) observed differential stereochemical results when comparing sulfamate and carbamate sources in reactions with allenes and Rh_2 catalysts, which suggests that mechanistic differences can be imparted by the nitrogen-atom substrate. Due to the nature of nitrenes, radical chemistry is often implicated in nitrene-transfer reactions. An unusual feature of Rh_2 catalyzed C–H amination is that, while intramolecular C–H amination seems to follow the mechanism set forth in Scheme 13 closely, intermolecular amination is much more mechanistically complex, presumably because the nitrene intermediate is not immediately trapped by substrate once it is formed (238). Du Bois and co-workers (211) proposed that enhanced catalyst stability should improve the yield of intermolecular amination, and developed $Rh_2(esp)_2$ (**30**), featuring chelating bis-carboxylates based on this proposal. The emerging importance of one-electron chemistry in $C-H$ amination will be further discussed in the following sections as we develop a more modern mechanistic model for $C-H$ amination.

2. Synthetic Scope

Intramolecular $C-H$ amination chemistry is becoming a powerful tool for constructing complex molecular architectures (231, 244). The scope of this chemistry is quite broad, with several examples of atypical alkene–alkane substrates as well as cascade reactions. Schomaker and co-workers reported intramolecular propargylic $C-H$ amination from carbamate-derived nitrenes (243) and aziridination at allenes (245). Blakey and co-workers (246, 247) reported the construction of polycyclic scaffolds via vinyl-cation intermediate traps. Intramolecular nitrene chemistry was recently reviewed by Dauban and co-worker (231); the reader may find several examples of cascade reactions and heterocycle synthesis therein.

Oxidative amination chemistry poses unique challenges as compared to Rh_2- carbenoid chemistry because of the need for a suitable oxidizing agent, and intermolecular $C-H$ amination is a particularly challenging transformation. Du Bois and co-workers (211, 238–240) have pioneered the development of intermolecular $C-H$ amination reactions. Dodd and co-workers (248) also developed efficient intermolecular oxyamidations of indoles. The Du Bois group has optimized intermolecular $C-H$ amination reactions catalyzed by $Rh_2(esp)_2$ (**30**) and identified that slow addition of hypervalent iodine oxidant $[PhI(O_2CtBu)_2]$ to a solution of substrate and 2,2,2-trichloroethylsulfamate ester (**41**) is key (Scheme 12c) (238). This methodology is state-of-the-art in its catalyst loading and efficient conversion of stoichiometric amounts (1 equiv) of substrate.

A strategy for building the oxidant into the nitrogen substrate was recently developed by Lebel et al. (249, 250). Unlike the conventional paradigm wherein nitrenes are formed by group transfer from iminoiodinane species, Lebel et al. (249, 250) show that intermolecular $C-H$ insertion is possible via the decomposition of preoxidized 2,2,2-trichloroethyl-*N*-tosyloxycarbamate (**43**) providing $C-H$ functionalized products in good yields (Scheme 14) (249, 250). This methodology has developed from the analogous intramolecular reactions (249). Catalyst loadings for this transformation are somewhat higher than in iminoiodinane reactions (5 mol% as compared with ~2 mol%), however, simple addition

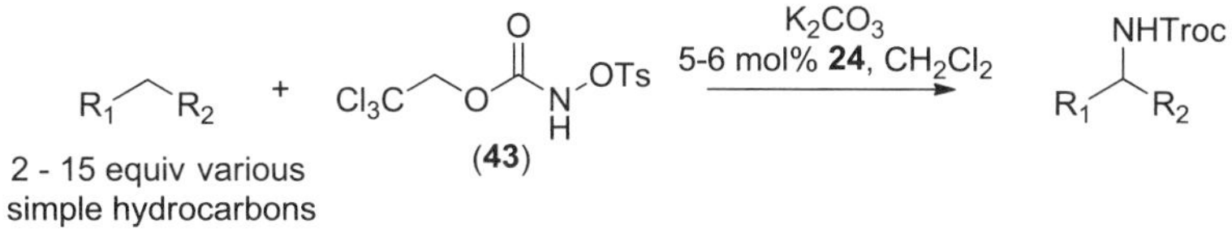

Scheme 14. Intermolecular amination from *N*-tosyloxycarbamates (Troc = 2,2,2-trichloroethoxycarbamate).

of a base is sufficient for catalytic turnover. Note that, similar to iminoiodinanes, the nitrogen-atom in *N*-tosyloxycarbamates is preoxidized, which limits the nitrogen-atom source substrate scope. Although this transformation has not been mechanistically investigated, it suggests that nitrene formation need not come from iminoiodinane sources.

Dirhodium–nitrenes can also be derived from azide precursors. Surprisingly few examples of azide activation using Rh$_2$ complexes have been reported. Recently, Driver and co-workers (251) reported intramolecular cyclization reactions from organic azides (Scheme 15). This work distinguishes itself from other examples of azide reactivity because it does not require a highly electron-withdrawing azide precursor. Isotope-labeling studies support a hydrogen-atom abstraction/radical-recombination mechanism with an intramolecular KIE of ~6. Thus, an Rh$_2$-amido radical is a potential intermediate for this transformation. Being able to choose between a concerted or stepwise (radical) nitrene insertion mechanism will have important future applications in terms of site selectivity, and careful choice of both nitrogen-atom precursor and catalyst will likely play a large role in understanding this dichotomy.

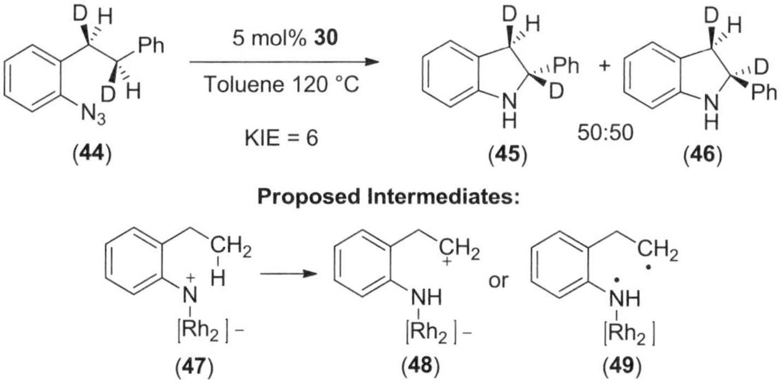

Scheme 15. Intramolecular amination using organic azides. Isolation of two diastereomers suggests the intermediacy of a radical (or cation) since scrambling of the stereocenter must occur before recombination.

3. Current Mechanistic Understanding and Its Impact on Catalyst Development

The most generally effective catalyst for intra- and intermolecular C−H amination is $Rh_2(esp)_2$ (**30**) pioneered by Du Bois and co-workers (211). This catalyst was initially developed with the idea that a chelating dicarboxylate ligand would prevent catalyst decomposition under the highly oxidizing conditions necessary for C−H amination [simple carboxylates, e.g., **1**, have low efficiency in C−H amination and typically decompose to Rh(III) species]. Du Bois and co-workers have extensively studied the mechanism of both intra- (240) and intermolecular (238) C−H amination by **30**, as well as other Rh_2 complexes. Intramolecular amination is facile (compared to intermolecular amination) and can proceed using a variety of simple catalysts (e.g., **1** and **23–25**).

The mechanism of intramolecular C−H amination was investigated using a variety of catalysts: Central reactivity patterns include the fact that the active oxidizing species is an Rh_2 bound nitrene, and that C−H insertion is likely a concerted-asynchronous event based on small KIEs (1.9 ± 0.2) and no observed ring opening in reactions using cyclopropane radical-clock substrates (240). An interesting finding in the kinetic study of the intramolecular reaction was that the initial rate of product formation was independent of catalyst concentration. Consequently, it was proposed that the condensation of sulfamate ester and oxidant to form an iminoiodinane governs the reaction rate at early reaction times (238). This finding is consistent with observations made by Kornecki and Berry (134) who proposed two separate mechanistic regimes for intermolecular C−H amination of ethylbenzene by **30** that appear to be strongly dependent on the concentration of substrate in solution (see below).

The mechanism for intermolecular C−H amination has been shown to be similar to the intramolecular mechanism based on small KIE values. In addition, Hammett analyses indicate a small but discernable cationic charge stabilization in the transition state, which supports a concerted mechanism over radical abstracation–rebound (238). Intermolecular radical clock experiments, like their intramolecular counterparts, do not show any ring opening. (Fiori and Du Bois (238) do note, however, that this does not completely discount a radical mechanism—should a radical mechanism be active, the lifetime of the radical species would have to be extremely short (on the order of 200 fs). The reason that a radical mechanism has not been discounted is that the chemoselectivity of intermolecular C−H amination is opposite to otherwise identical intramolecular conditions. Benzylic C−H bonds are strongly preferred in the intermolecular reaction, while 3° C−H centers are preferred sites for intramolecular amination (238). This discrepancy is explained by the fact that selectivity may be a function of the rate at which the Rh_2 nitrene is trapped by substrate versus the rate at which it decomposes via nonproductive pathways. A monosubstituted benzylic position has 2 equivalent C−H bonds, thus doubling the apparent rate with which it can intercept the Rh$_2$–nitrene.

An important observation made by Du Bois and co-workers (133, 238) during the study of intermolecular C−H amination was the existence of mixed-valent Rh$_2$(II,III) species under amination conditions, as exemplified by a color change from green to red. A pathway is proposed wherein carboxylic acid byproducts (from hypervalent iodine oxidant decomposition) can reduce the Rh$_2$(II,III) state back to Rh$_2$(II,II), which is the proposed active form of the catalyst (133). It was hypothesized that **30** is a better catalyst for the oxidative C−H amination reaction because the kinetic stability of the mixed-valent Rh$_2$(II,III) state is superior in **30** as compared to non-chelating Rh$_2$ tetracarboxylate complexes.

Compound **30** is indeed a more robust catalyst, and interestingly, it has been shown that the resting state of this catalyst is likely a mixed-valent Rh$_2$(II,III)– amido species, as the red color of the catalyst under the reaction conditions is not identical to electrochemically oxidized **30**$^+$ free of axial ligation (133, 134, 238). Spectroelectrochemical studies on **30** have led to the proposal that the Rh$_2$(II,III) state is accessed during catalysis via a putative proton-coupled electron transfer (PCET) reaction between a sulfamate ester and hypervalent iodine oxidant, which acts as both an oxidant and a base (outer cycle, Scheme 16) (134). A second PCET reaction from a putative Rh$_2$ amido species (**50**) results in convergence with the two-electron nitrene transfer mechanism (134). The proposed amido species (**50**) may have significant nitrogen-radical character and it is not yet clear if an Rh$_2$ amido species can itself be the active oxidant in lieu of a nitrene. This result could

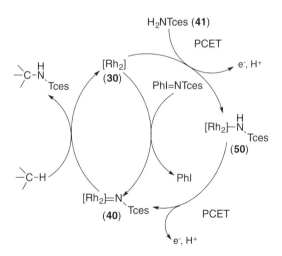

Scheme 16.   Two possible limiting mechanisms for turnover catalyzed by Rh$_2$(esp)$_2$. The outer mechanism features the spectroscopically observable catalyst resting state, Rh$_2$(II,III) amido, **50**.

explain mechanistic aspects of the latter reaction stage typically observed for $C-H$ amination vis-à-vis the fast initial reaction rates. An experiment using **49** in the presence of a pre-formed iminiodinane and ethyl benzene resulted in a lower yield than when the same sulfamate ester and appropriate oxidant were used independently (238), indicating that a purely atom-transfer mechanism (e.g., inner cycle, Scheme 16) does not sufficiently account for the total product formed under catalytic conditions. A recent study highlighting the use of desorption electrospray ionization mass spectrometry (DESI–MS) for the observation of putative reaction intermediates has shown evidence for species with masses consistent with both an Rh_2 nitrene and an Rh_2 amido in a prototypical $C-H$ amination mixture, further implying that more than one mechanistic pathway may be operative at any given time (252).

The necessity of a mixed-valent Rh_2 complex in intermolecular $C-H$ amination chemistry is further implicated in a study utilizing Rh_2 compounds supported by redox-active chelating carboxylates structurally analogous to H_2esp (Scheme 17). These complexes are shown to be less effective $C-H$ amination catalysts, presumably due to their inability to access an $Rh_2(II,III)$ state as their resorcinol-based ligands are preferentially oxidized (212). It is also worth noting that complexes such as $Rh_2(S\text{-biTISP})_2$ (**31**), are not effective at promoting intermolecular amination despite having chelating ligands (238).

A recent study has described a stable mixed-valent analogue of **30** with chelating diamidate ligands, $Rh_2(espn)_2Cl$ (**51**, Scheme 18). This $Rh_2(II,III)$ catalyst is free of axial chloride in low concentrations as measured by CV, established due to the differing oxidation potentials of the chloride-bound and unbound species. Catalyst **51** displays intramolecular selectivity preferences that are similar to **30**, providing indirect support for a mixed-valent resting state in reactions using **30**. Furthermore, cationic catalyst **51** displays enhanced longevity in simple cyclization reactions compared to **30** (Scheme 18), implying that the

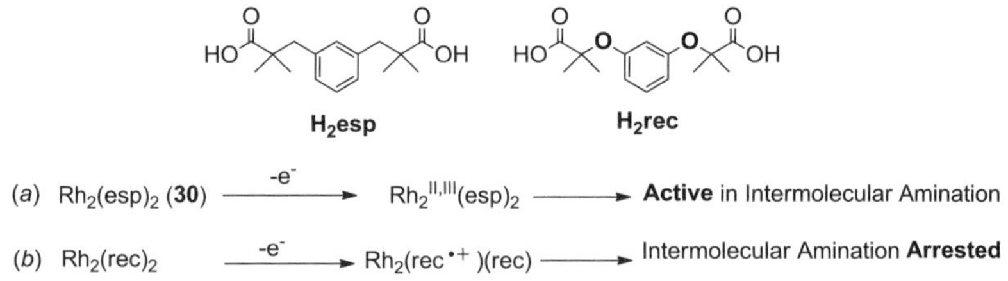

H_2esp H_2rec

(a) $Rh_2(esp)_2$ (**30**) $\xrightarrow{-e^-}$ $Rh_2^{II,III}(esp)_2$ $\longrightarrow$ **Active** in Intermolecular Amination

(b) $Rh_2(rec)_2$ $\xrightarrow{-e^-}$ $Rh_2(rec^{\bullet+})(rec)$ $\longrightarrow$ Intermolecular Amination **Arrested**

Scheme 17.   Resorcinol-based ligands result in ineffective intermolecular amination catalysts due to preferential ligand-based oxidation (e.g., *b*), where rec = 2-[3-(1-carboxy-1-methylethoxy) pheoxy]-2-methylpropanoate)

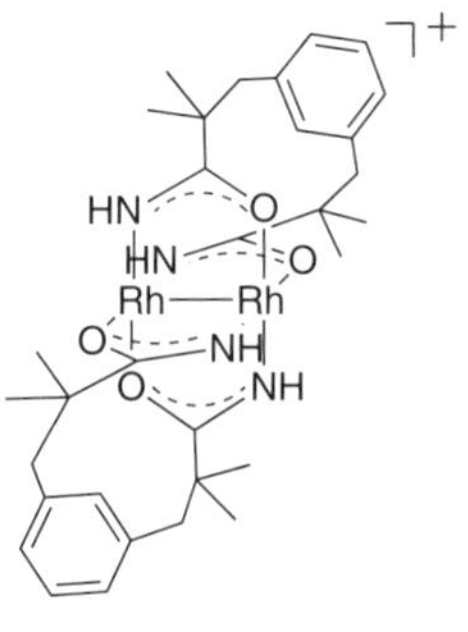

$Rh_2(espn)_2^+$ (**51**)

Substrate + $PhI(OAc)_2$ $\xrightarrow[CH_2Cl_2,\ RT]{0.05\ mol\%\ Rh_2}$ Cyclized product

Substrate	Product	TON Catalyst: (**30**)	TON (**51**)
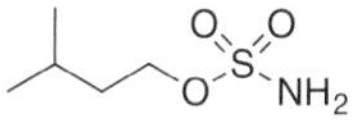	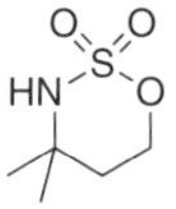	580	1400
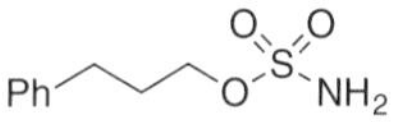	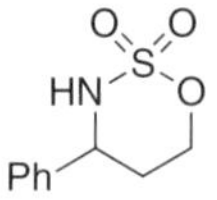	490	1450

Scheme 18. Structure of mixed-valent catalyst **51** and a comparison of maximum turnover number (TON) between catalysts **30** and **51**. Catalyst **51** is longer lived, presumably due to its superior stability in the Rh_2(II,III) state (RT = room temperature; espn = α,α,α',α'-tetramethyl-1,3-benzenedipropanamidate).

stability of the Rh_2(II,III) redox state may be key in designing robust amination catalysts (216).

More recently, diastereotopic differentiation at the benzylic methylene position using **30** as a C–H amination catalyst was studied; a density functional theory (DFT) model that reproduces experimentally observed KIEs was also reported (253). The study concluded that while *intra*molecular amination reactions proceed through a concerted mechanism (254), *inter*molecular amination likely occurs via a hydroen-atom abstraction–radical recombination on a singlet energy surface (253). While radical clock experiments on intermolecular aminations from Du Bois and coworker (238) dispute this conclusion, it is in agreement with other mechanistic discrepancies between the intra- and intermolecular reactivity of **30** described by both Du Bois and Berry.

C. Allylic Oxidations

As we have seen in a number of contexts, one-electron oxidation of $Rh_2(II,II)$ complexes can afford isolable $Rh_2(II,III)$ complexes. In the early 1980s, complex **1** was investigated in the context of *tert*-butylhydroperoxide activation and met with limited success (255). It was not until 2004 that Doyle and co-workers (217) developed practical methodology to harness *tert*-butylhydroperoxide activation by Rh_2 catalysts for oxidation chemistry (Scheme 19). Doyle notes in this first report of allylic oxidation by $Rh_2(cap)_4$ (**34**) that facile one-electron redox chemistry is generally essential to allylic oxidation protocols, thus the choice of a readily oxidized Rh_2 carboxamide catalyst. Indeed, the metal species observed upon radical generation is consistent with a mixed-valent $Rh_2(II,III)$ species, much like the data obtained for the resting state of **34** in aziridination reactions (219).

Mechanistic work shows that $Rh_2(cap)_4$ is not actually involved in the $C-O$ bond-forming step, rather, it functions only to generate *tert*-butylperoxy radicals that are themselves selective reagents for hydrogen-atom abstraction (256). More recent work has shown that the *tert*-butylperoxy radical produced from a reaction of **34** and T-HYDRO (70% *tert*-butylhydroperoxide in water) is effective at oxidizing phenols to 4-(*tert*-butyl-dioxy)cyclohexadienones and anilines to nitroarenes, representing another transformation catalyzed by Rh_2 complexes (218).

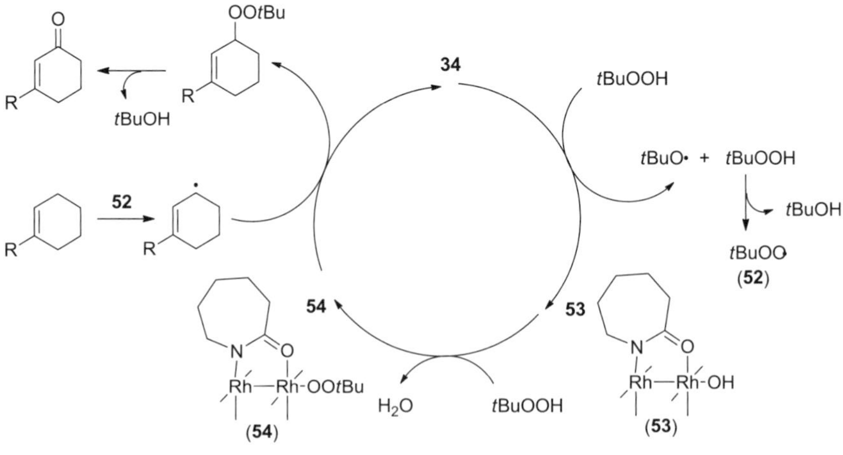

Scheme 19.   Allylic oxidation utilizing $Rh_2(cap)_4$. Mixed-valent $Rh_2(II,III)$ intermediates **53** and **54** are proposed.

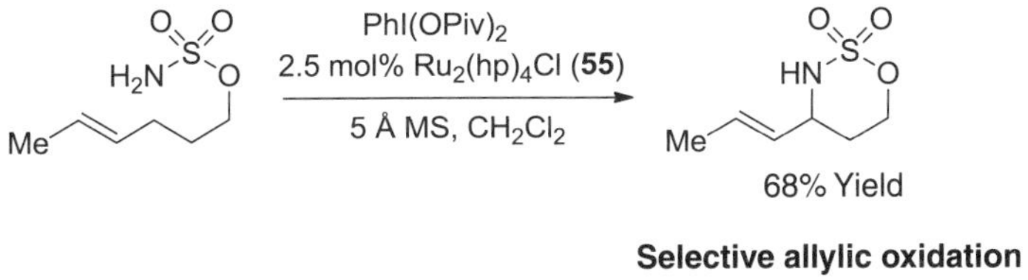

Selective allylic oxidation

Scheme 20. Diruthenium-hydroxypyridine complexes are efficient intramolecular C–H amination catalysts (MS = molecular sieves).

D. Diruthenium Chemistry

Binuclear Ru complexes also have received attention with respect to C–H amination reactions. The Ru_2 hydroxypyridine complex **55** recently has been shown to be highly effective in intramolecular allylic C–H amination (Scheme 20) (257). Du Bois and co-workers (257) suggest that hydrogen-atom abstraction/radical recombination may be the active pathway in these reactions, as a diradical intermediate is calculated to be lowest in energy on a doublet–quartet potential energy surface.

It has been proposed that as the M–M core becomes more electron deficient, radical chemistry is favored over concerted pathways, thus creating new venues for catalyst design (257). Now that there is evidence for a stepwise diradical mechanism in $Ru_2(II,III)$ catalyzed amination transformations, the idea that the Ru_2 or Rh_2 catalyst can dictate a particular mechanistic pathway (concerted vs radical) is becoming a possibility (257).

The closest experimental analogues to the proposed reactive amination intermediates that have been studied are Ru_2 nitride compounds reported by the Berry lab by cryogenic solution photolysis of an Ru_2 azide (**56**) (258). The resulting Ru_2 nitride (**57**), which is at the same oxidation state as the nitrene complexes proposed in Ru_2 catalyzed C–H amination chemistry, undergoes intramolecular aryl C–H functionalization in which the nitride attacks one of the C–H bonds on a supporting ligand (**58**, Scheme 21) (259).

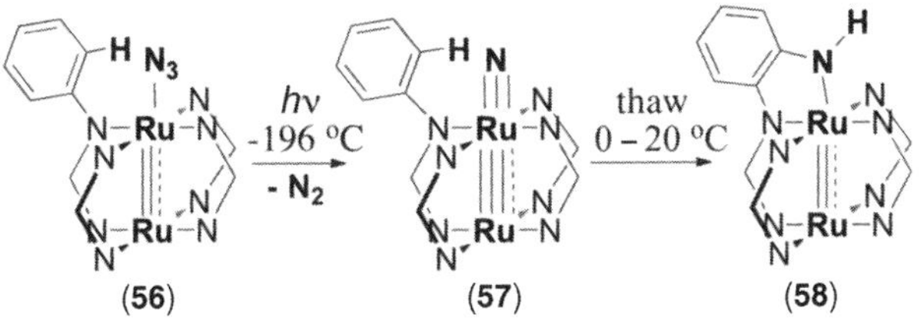

Scheme 21. Photolysis of an Ru_2 azide results in the formation of a reactive Ru_2 nitride species that rapidly undergoes an electrophilic aromatic C–H bond insertion of the supporting formamidinate ligand. [Adapted from (259)].

Kinetic analysis of the intramolecular insertion of the photolytically generated Ru_2 nitride into a ligand aryl $C-H$ bond was validated by DFT and supports an electrophilic aromatic substitution mechanism for this reaction, with the nitride N atom serving as the electrophile. Although not a functional example of catalytic reactivity, this study made it possible to benchmark experimental data with a computational model for the $C-H$ bond functionalization mechanism. This work led to an exhaustive study of $M-M-N$ electronic structure as it relates to reactivity (227). The experimentally validated Ru_2 manifold was computationally modified to an $M-Ru{\equiv}N$ ($M = Zr$, Mo, Ru, Pd) heterobimetallic structure, and reaction coordinates were compared. A relationship between ground-state electronic structure and transition state energy was described, and it is made clear that a three-center orbital manifold is best suited for $C-H$ functionalization compared with mononuclear metal nitrides. In addition, it was shown that changing the identity of heterometal M can either enhance or diminish the reactivity. Heterometallic complex $Mo-Ru{\equiv}N$ has been proposed as a synthetic target that may offer lower barriers to $C-H$ amination (227).

III. DIPALLADIUM $C-H$ FUNCTIONALIZATION CHEMISTRY

A. History of Oxidative $C-H$ Functionalization Catalyzed by Pd

Oxidative Pd catalyzed $C-H$ functionalization reactions have been developed for the preparation of both $C-C$ and $C-$heteroatom bond-containing products (86, 99, 260–264). Beginning in the 1960s, it was recognized that aromatic substrates (e.g., benzene and toluene) undergo $C-H$ palladation with Pd(II) salts to afford labile organopalladium(II) complexes (265–270). As shown in Scheme 22a, heating a

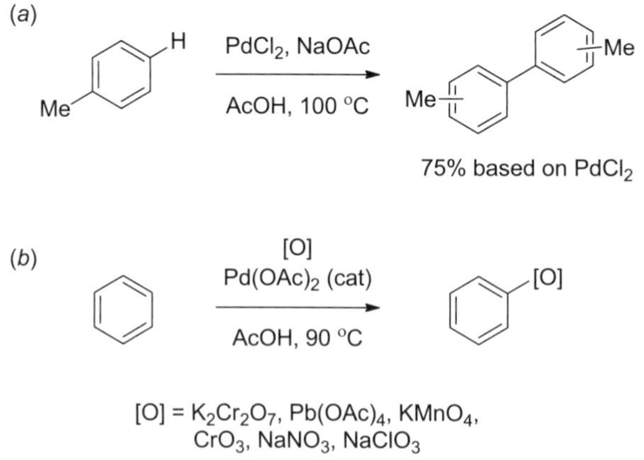

(a)

$$Me \overset{H}{\diagdown} \quad \xrightarrow[\text{AcOH, 100 °C}]{\text{PdCl}_2, \text{NaOAc}} \quad Me \diagdown \diagdown Me$$

75% based on $PdCl_2$

(b)

$$\diagdown \quad \xrightarrow[\text{AcOH, 90 °C}]{\begin{array}{c}\text{[O]}\\ \text{Pd(OAc)}_2 \text{ (cat)}\end{array}} \quad \diagdown\text{[O]}$$

[O] = $K_2Cr_2O_7$, $Pb(OAc)_4$, $KMnO_4$, CrO_3, $NaNO_3$, $NaClO_3$

Scheme 22.   (a) Thermolysis of simple arenes, (e.g., toluene), with stoichiometric amounts of Pd(II) salts affords biaryls and Pd(0) by reductive $C-H$ coupling. (b) Treatment of benzene with oxidants in the presence of Pd(OAc)$_2$ affords $C-H$ oxidation products.

mixture of PdCl$_2$ and NaOAc in toluene results in the formation of a mixture of bitolyls, as well as Pd(0) (271, 272). These reactions require stoichiometric Pd(II) salts, and likely proceed by C–C reductive elimination from diaryl palladium(II) intermediates to afford biaryl and Pd(0). Subsequently, it was found that the aryl Pd(II) intermediate in these reactions can be intercepted by external oxidants prior to C–C coupling (273–276). For example, in 1971, Henry reported that treatment of simple arenes (e.g., benzene and toluene), with oxidants in the presence of a catalytic amount of Pd(OAc)$_2$ in AcOH, resulted in the formation of aryl acetates (Scheme 22b) (273).

The generic aromatic C–H oxidation reaction depicted in Scheme 22 typically proceeds through two distinct mechanism regimes: (1) C–H palladation and (2) Pd–C bond oxidation (Scheme 23a). The mechanistic details of the C–H

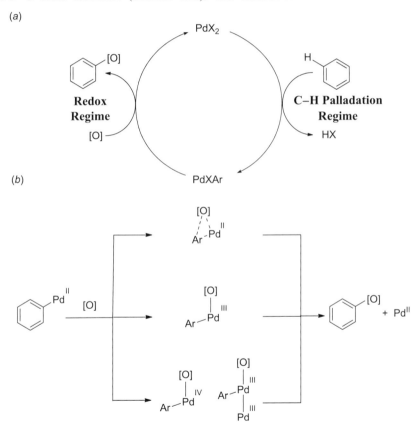

Scheme 23.   (a) Many oxidative Pd catalyzed aromatic C–H oxidation reactions proceed via C–H palladation followed by product-forming redox chemistry. Discussion of the mechanism of C–H palladation is in Section III.B. (b) The redox regime of the Pd catalyzed C–H oxidation reactions discussed herein could proceed via direct electrophilic substitution (redox neutral at Pd) or by one- or two-electron metal-based oxidation. Two-electron, metal-based oxidation of binuclear complexes has been proposed to be operative during catalysis.

palladation regime will be discussed in Section III.C. Because the C−heteroatom bond-forming reactions take place during the redox regime of catalysis, there has been significant interest in elucidation of the structures of catalytic intermediates involved in the C−X bond-forming event. In concept, a variety of pathways are available for the "redox regime" of catalysis; Pd−C bond oxidation with zero- (277, 278), one- (279–281), or two-electron (282–284) oxidation of Pd have been documented (Scheme 23b) (285).

In the oxidative Pd catalyzed C−H functionalization reactions that are the focus of this section two-electron, metal-centered oxidation is usually invoked to rationalize the observed oxidation products. Mononuclear Pd(IV) intermediates were first proposed by Henry in 1971 (273) as reaction intermediates in Pd catalyzed aromatic C−H acetoxylation reactions. Subsequent reports by Stock et al. (276) and Crabtree and co-worker (286) also entertained the possibility of Pd(IV) intermediates in aromatic acetoxylation reactions (285, 287). More recently, binuclear Pd_2(III,III) species have been proposed as intermediates in some Pd catalyzed C−H oxidation reactions (143, 285, 287–289). The potential role of Pd−Pd bond-containing structures in oxidative Pd catalyzed reactions will be discussed herein. In addition, the hypothesis that M−M redox synergy can lower activation barriers to both oxidation, as well as C−X reductive elimination, will be discussed.

B. Catalyst Structure and Metalation Reactions

The compound $Pd(OAc)_2$ is frequently employed as the catalyst in oxidative C−H functionalization reactions (285, 290). In the solid state, $Pd(OAc)_2$ is trimeric, with each of the Pd centers bridged by two acetate ligands (**59**, Scheme 24a) (291). In solution, palladium acetate can exist in a variety of aggregation states, including tri-, bi-, and mononuclear depending on the reaction medium (Scheme 24a) (292–294). Similarly, the organopalladium(II) complexes generated by C−H palladation with $Pd(OAc)_2$ can exist in various aggregation states. While acetate-bridged binuclear complexes are commonly encountered (295), higher nuclearity complexes, (e.g., trinuclear complex **62**), also have been observed and characterized (Scheme 24b) (296–298).

Several mechanisms for C−H palladation are conceivable, including (a) oxidative addition, (b) electrophilic aromatic substitution, and (c) CMD (Scheme 25) (90). Electrophilic aromatic substitution and CMD mechanisms are closely related, differing only in the relative timing of Pd−C bond formation and C−H deprotonation; in electrophilic aromatic substitution, the Pd−C bond forms prior to deprotonation, while in the CMD mechanism, Pd−C bond formation and deprotonation occur simultaneously. Experimental and theoretical investigations of the mechanisms of palladation suggest that electrophilic aromatic substitution, or CMD mechanisms, are the most common pathways (91–98).

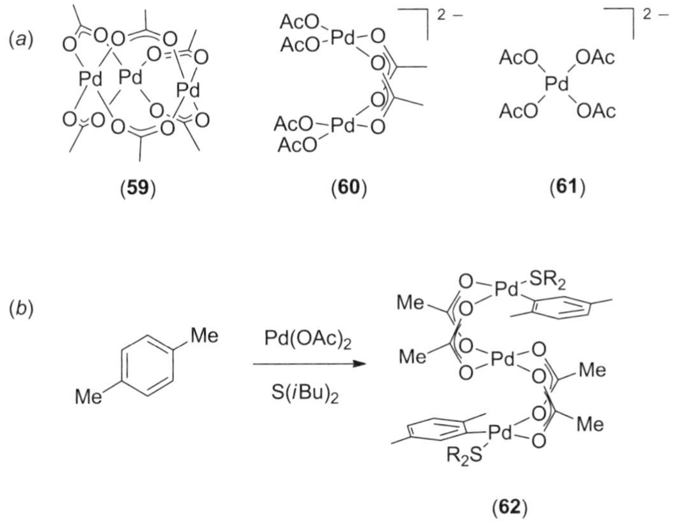

(59)   (60)   (61)

(62)

Scheme 24. (*a*) In both the solid state and in noncoordinating solvents, Pd(OAc)$_2$ exists as a trinuclear complex (**59**). Both bi- and mononuclear structures (**60** and **61**) have been observed in the presence of coordinating anions and solvents. (*b*) Similar to Pd(OAc)$_2$, acetate-ligated organopalladium complexes also exhibit a variety of nuclearities.

While both Pd(OAc)$_2$, as well as the products of C–H palladation with Pd (OAc)$_2$, are frequently polynuclear under a given set of reaction conditions, the nuclearity of the species that participate in C–H palladation need not be polynuclear. For example, in the Pd(OAc)$_2$ catalyzed acetoxylation of benzene with

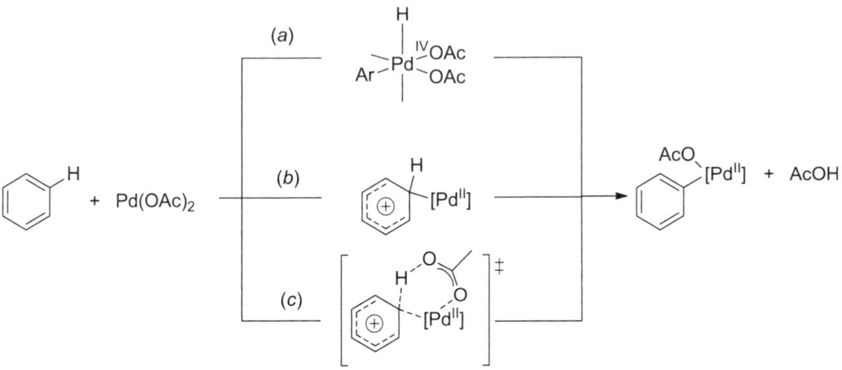

Scheme 25. Potential mechanisms for C–H palladation include (*a*) C–H oxidative addition. (*b*) Electrophilic aromatic substitution. (*c*) CMD.

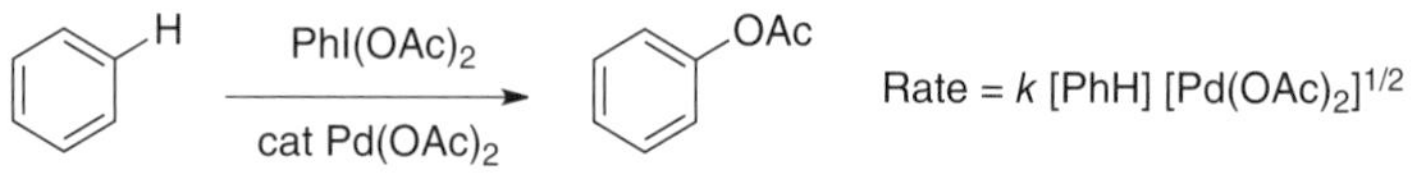

Scheme 26. The rate of $Pd(OAc)_2$ catalyzed acetoxylation of benzene with $PhI(OAc)_2$ is $\frac{1}{2}$ order dependent on $Pd(OAc)_2$ concentration, consistent with deaggregation of a binuclear catalyst resting state prior to turnover-limiting palladation.

$PhI(OAc)_2$ reported by Crabtree and co-worker (Scheme 26), a reaction order of 0.5 was observed with respect to $Pd(OAc)_2$ (286). In combination with a zero-order dependence on oxidant concentration, the observed reaction order of Pd is consistent with palladation following deaggregation of the catalyst resting state. While the catalyst resting state is this reaction is not yet known, if the catalyst resting state were binuclear (i.e., **60**), the accumulated data are consistent with $C-H$ palladation at a mononuclear Pd complex.

C. Current Mechanistic Understanding

The intermediates relevant to $C-X$ bond formation during catalysis have been difficult to elucidate because $C-H$ palladation is frequently turnover-limiting during catalysis (299). For reactions in which $C-H$ palladation is turnover limiting, measurements of reaction kinetics provide information regarding the mechanism of metalation during catalysis, but not the redox chemistry involved in catalysis. For example, efforts to elucidate structures of potential high-valent intermediates in the $Pd(OAc)_2$ catalyzed acetoxylation of 2-phenylpyridine derivatives (**63–66** in Scheme 27) have been hampered by the fact the $C-H$ metalation is turnover limiting. As such, despite the preparation of binuclear $Pd_2(III,III)$ (300) complex **65** (Scheme 27) and the observation of $C-OAc$ bond-forming reductive elimination to afford acetoxylation product **66**, information about whether binuclear $Pd_2(III,III)$ complex **65** is relevant during catalysis is not available.

Unlike $Pd(OAc)_2$ catalyzed $C-H$ acetoxylation of 2-phenylpyridine derivatives (Scheme 27), $C-H$ chlorination of 2-phenylpyridine derivatives proceeds via turnover-limiting oxidation (Scheme 28) (288, 301). Using the chlorination of benzo[h]quinoline **67**, (Scheme 28, NCS = N-chlorosuccinimide), as a tool to probe the nuclearity of the Pd complexes involved in $C-H$ chlorination, binuclear Pd intermediates were implicated in the redox chemistry (288, 301). The experimentally derived rate law, rate = k [NCS] [AcOH] [Pd], in combination with the assignment of the catalyst resting state [($Pd_2(II,II)$ complex **69**], suggest that $Pd_2(III,III)$ complex **70** may be the immediate product of oxidation during catalysis. Independent synthesis of proposed intermediate **70** confirmed the viability of chemoselective $C-Cl$ reductive elimination (Scheme 29).

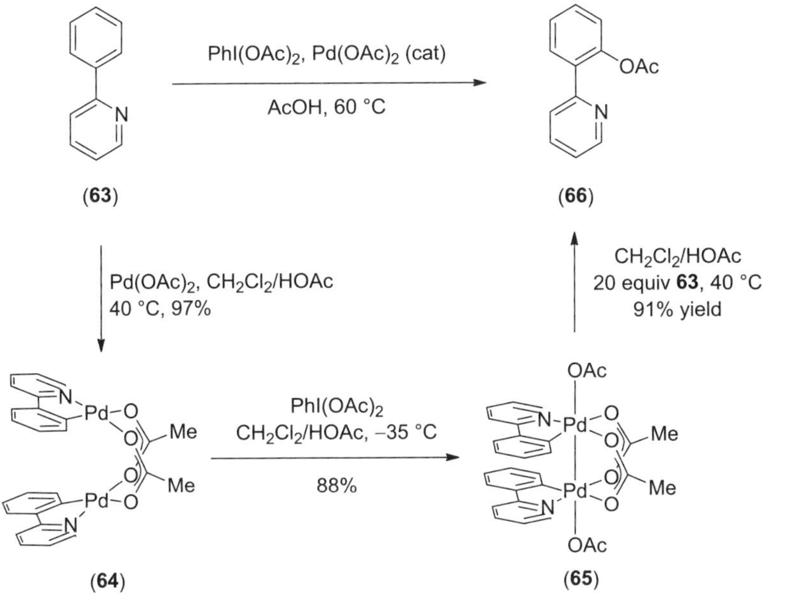

Scheme 27. A synthetic cycle for the acetoxylation of 2-phenylpyridine (**63**) has been established via binuclear Pd$_2$(III,III) complex **65**. Because C–H metalation is turnover limiting during catalysis, the nuclearity of oxidation intermediates in acetoxylation has not been established.

Similarly, a study of the Pd(OAc)$_2$-catalyzed C–H arylation of 2-phenyl-pyridine derivatives revealed that oxidation is turnover limiting during catalysis (Scheme 30) (289). Based on the rate law of catalysis (rate $= k$ [**72**] [**Pd**]^2 [**71**]$^{-3}$), as well as the spectroscopically assigned resting states of both catalyst (**73**) and oxidant (**75**), binuclear complex **76** was proposed to be the immediate product of oxidation during catalysis. Unlike binuclear Pd$_2$(III,III) complex **70**, in which the coordination sphere of each Pd is the same, one Pd center in **76** bears a strongly donating axial aryl ligand while the other Pd center has a vacant coordination site. Intermediate **76** is assigned based on kinetic data, and thus information is not available to discern whether additional ligands, (e.g., acetate), which is present in the reaction mixture, bind to the binuclear core following turnover-limiting oxidation but prior to C–C bond formation. Due to the differing ligand sets of the two Pd centers in **76**, both binuclear Pd$_2$(III,III) (**76**), as well as mixed-valent Pd$_2$(II,IV) (**77**), formulations may contribute to the description of the electronic structure. Computational investigation of Pd$_2$(III,III) complexes that bear different apical ligands, (e.g., complex **70**), has shown that increasing the sigma-donating ability of one of the apical ligands results in the population of the $4d_z{}^2$ orbital of the other Pd, which in the limit of complete electron transfer would be a mixed-valent

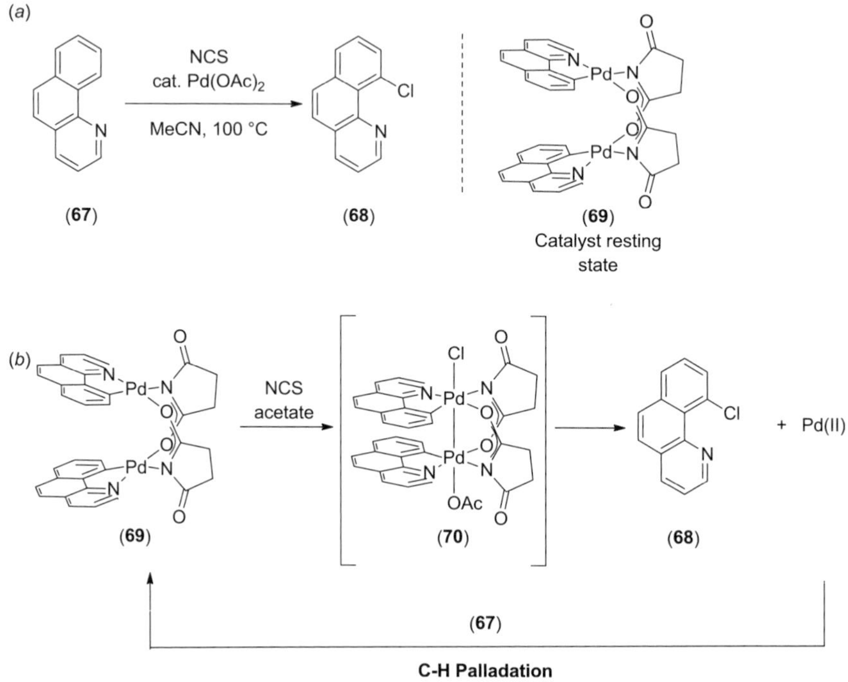

(a)

(67)   (68)   (69)
                                      Catalyst resting
                                      state

(b)

(69)   (70)   (68)

(67)

C-H Palladation

Scheme 28. (a) The resting state of Pd(OAc)$_2$ catalyzed chlorination of benzo[h]quinoline (67) is succinate-bridged binuclear Pd$_2$(II,II) complex 69. (b) Evaluation of the reaction kinetics of the reaction in (a) suggests that binuclear Pd$_2$(III,III) complex 70 is the immediate product of oxidation during catalysis.

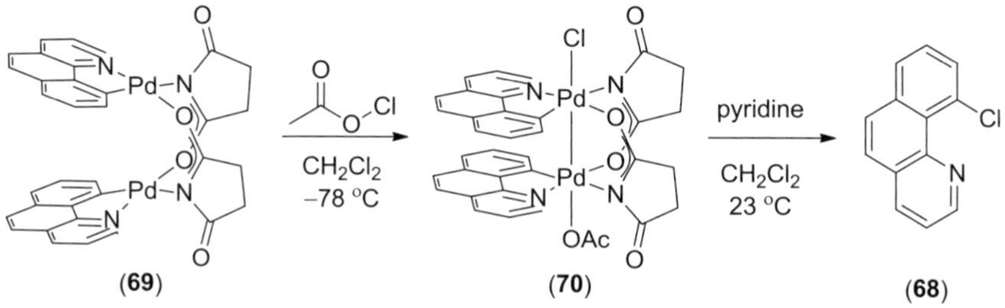

(69)   (70)   (68)

Scheme 29. Oxidation of binuclear Pd(II) complex 69 with acetyl hypochlorite affords proposed catalytic intermediate 70, which undergoes chemoselective C−Cl bond-forming reductive elimination.

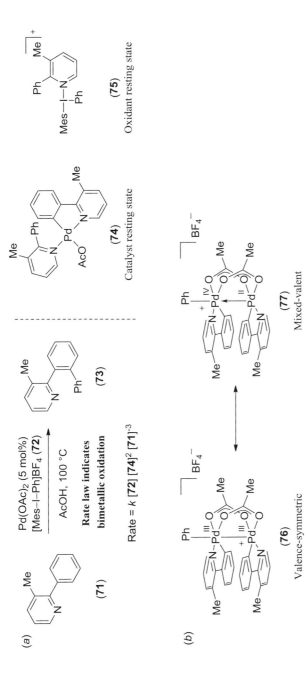

Scheme 30. (a) Oxidized binuclear Pd complexes have been implicated in Pd(OAc)$_2$ catalyzed arylation of 2-phenylpyridine. (b) Both Pd$_2$(III,III) (**76**) and mixed-valent Pd$_2$(II,IV) (**77**) formulations may contribute to the electronic structure of the oxidized intermediate in catalysis.

273

description. For further discussion of the potential impact of valence asymmetry in binuclear complexes, see Section III.G.

D. Role of Binuclear Core in C–X Bond-Forming Redox Chemistry

Binuclear oxidation intermediates have been implicated in both of the reactions for which there is information regarding the mechanism of oxidation during catalysis (Schemes 28 and 30) (289, 301). It is interesting to consider potential roles for the binuclear core during oxidation, and the potential kinetic advantage for oxidation of binuclear complexes over mononuclear complexes. Holding two square planar Pd(II) complexes in proximity enforces mixing of the $4d_z^2$ orbitals. Orbital mixing affords two filled σ-type orbitals in the M−M bonding manifold, one stabilized σ-bonding orbital, and one higher energy σ*-anti-bonding orbital, which may result in a more nucleophilic Pd complex (302). In addition, nascent Pd−Pd bonding interactions in the transition state of oxidation could lower the reorganization energy associated with electron transfer and thus facilitate oxidation. The catalyst resting state of the Pd(OAc)$_2$ catalyzed C−H arylation discussed above is a mononuclear Pd(II) complex, which in principle can undergo direct oxidation to afford mononuclear Pd(IV) intermediates (289). However, binuclear intermediates are implicated by kinetic measurements, suggesting a kinetic preference for oxidation to binuclear Pd$_2$(III,III) complexes over available mononuclear species.

Both electrochemical and structural evidence support the hypothesis that the ligand-enforced proximity of Pd centers facilitates oxidation. Isomeric complexes **78** and **79**, which differ in the connectivity of the ligands to the metal centers, display significantly different oxidation behaviors (Scheme 31a) (151). In **78**, all four ligands are bridging the two Pd centers and the Pd–Pd distance is 2.58 Å, while in **79**, two ligands are bridging and two ligands are chelating the two Pd centers and the Pd–Pd distance is 2.90 Å. The one-electron oxidation potential for the oxidation of **78** to **78**$^+$ [the Pd$_2$(II,II) to Pd$_2$(II,III) redox couple] is 370 mV lower than the potential necessary for one-electron oxidation of **79**, which is consistent with M−M bond formation stabilizing the products of oxidation. Structural support for the simultaneity of Pd−Pd bond formation and oxidation is the observation of Pd−Pd contraction upon coordination of a soft Lewis acid to binuclear Pd$_2$(II,II) complexes (302). Donor–acceptor complexes (e.g., **80**) can be viewed as transition state mimics for the chemical oxidation of binuclear Pd$_2$(II,II) complexes (Scheme 31b). Alternatively, complex **80** can be viewed as a snap shot along the reaction coordinate of oxidation of Pd$_2$(II) to Pd$_2$(III) in which a shorter Pd–Pd distance has been stabilized by reducing the distance between electron donor and oxidant.

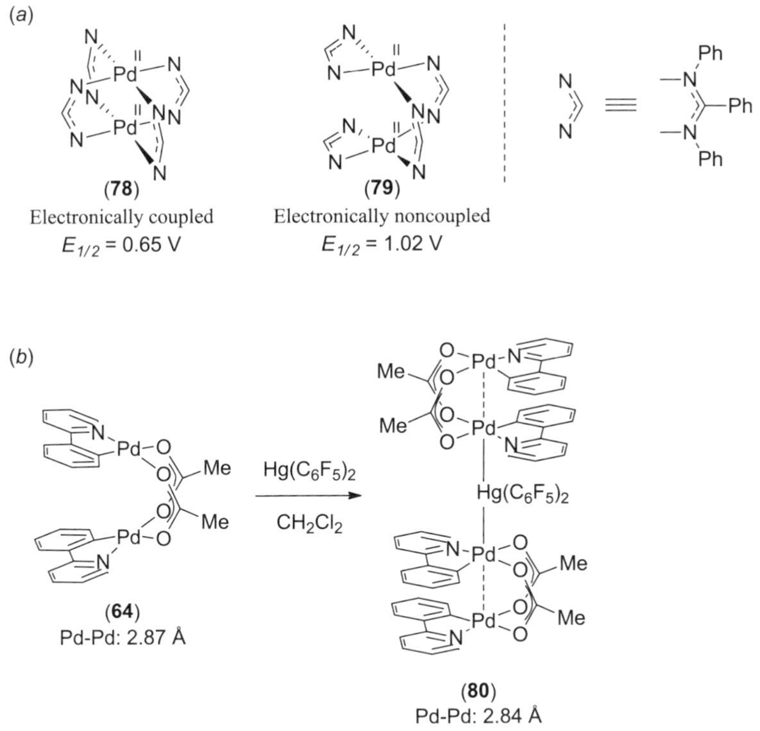

(a)

(78)
Electronically coupled
$E_{1/2}$ = 0.65 V

(79)
Electronically noncoupled
$E_{1/2}$ = 1.02 V

(b)

(64)
Pd-Pd: 2.87 Å

(80)
Pd-Pd: 2.84 Å

Scheme 31. (a) The HOMO of **78** is 370 mV higher in energy than the HOMO of **79**, consistent with M–M interaction facilitating oxidation. Potentials are relative to SCE in CH_2Cl_2. (b) Treatment of complex **64** with $Hg(C_6F_5)_2$ affords pentanuclear complex **80** with metallophilic interactions between the binuclear Pd units and the Hg center. Contraction of the Pd–Pd distance within the binuclear units is consistent with partial Pd–Pd bond formation upon electron transfer from the binuclear unit to the electron deficient Hg(II).

E. C–X Bond-Forming Reductive Elimination from Binuclear Pd(III) Complexes

In both C–H chlorination (301) and arylation (289) reactions discussed above (Schemes 28 and 30, respectively), oxidation is turnover limiting, and thus the proposed high-valent Pd_2 intermediates are not observed during catalysis; consumption of the oxidized species during catalysis is faster than oxidation to generate proposed high-valent intermediates. By employing stronger oxidants than are typically employed in catalysis (i.e., $PhICl_2$ in lieu of NCS, where NCS = N-chlorosuccinimide) and cryogenic temperature, a growing family of binuclear

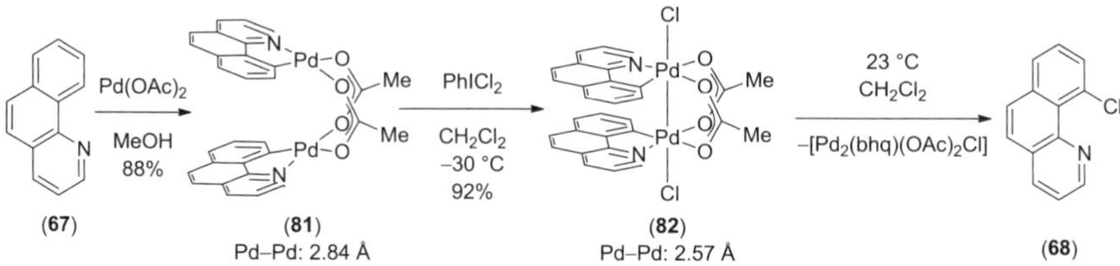

Scheme 32. The C–H palladation of benzo[*h*]quinoline with Pd(OAc)₂ affords complex **81**. Oxidation of **81** affords Pd₂(III,III) complex **82** with attendant contraction of the Pd–Pd distance. Warming the binuclear Pd₂(III,III) complex **82** to 23 °C affords **68** and Pd(II) complexes, consistent with C–Cl reductive elimination from **82**.

organopalladium(III) complexes has been characterized, allowing direct interrogation of the C−X bond-forming chemistry from these species (288).

The first organometallic chemistry from Pd(III) was reported during study of the redox chemistry of binuclear Pd_2(II,II) complex **81** (Scheme 32) (288). The C−H palladation of benzo[*h*]quinoline with Pd(OAc)₂ affords complex **81** in which the Pd centers are held in proximity by bridging acetate ligands. Oxidation of complex **81** with $PhICl_2$ afforded binuclear Pd(III) complex **82**. The Pd−Pd distance in **82** is 0.27 Å shorter than the corresponding distance in **81**, which is consistent with the formation of a Pd−Pd single bond, as predicted from the molecular orbital scheme in Fig. 1. Warming complex **82** results in the formation of compound **68** and Pd(II) containing complexes, the products of C−Cl reductive elimination. Rapid ligand exchange following reductive elimination prevents assignment of the Pd(II) structure following reductive elimination, but the empirical formula [Pd₂(bhq)(OAc)₂Cl] has been established by both combustion analysis as well as chemical methods (bhq = benzo[*h*]quinolinyl) (288, 303).

Experimental investigation of the reaction kinetics of C−Cl bond-forming reductive elimination from Pd₂(III,III) complex **82** suggests that reductive elimination proceeds without cleavage of the Pd–Pd core (288, 303). A DFT study of the mechanism of C−Cl reductive elimination supported this contention and showed that the transition state for C−Cl reductive elimination is early with respect to cleavage of the Pd−Pd bond (Fig. 11).

The role of the binuclear Pd core of **82** in C−Cl reductive elimination was investigated by examination of the barrier to reductive elimination as a function of Pd−Pd distance (303). Computationally forced elongation of the Pd−Pd bond attenuates the overlap of the $4d_{z^2}$ orbitals and thus decreases electronic communication between the metals as compared to low-energy structure **A** (Fig. 11). As can be seen by the computed activation barrier to reductive elimination as a function of Pd−Pd separation in **81**, elongation of the Pd−Pd bond results in a monotonic increase in the reductive elimination activation barrier (Scheme 33). Comparison

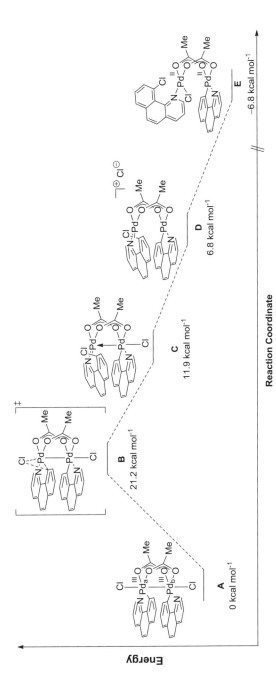

Reaction Coordinate

Figure 11. Summary of experimental and theoretical investigations of the mechanism of C–Cl reductive elimination from Pd$_2$(III,III) complex **82**. The C–Cl bond formation proceeds without precleavage of the Pd–Pd bond. The identity of the Pd(II) species formed after reductive elimination is not known; computationally **D** evolves to **E** and thus **E** is drawn to illustrate that C–Cl reductive elimination from **B** is exothermic.

277

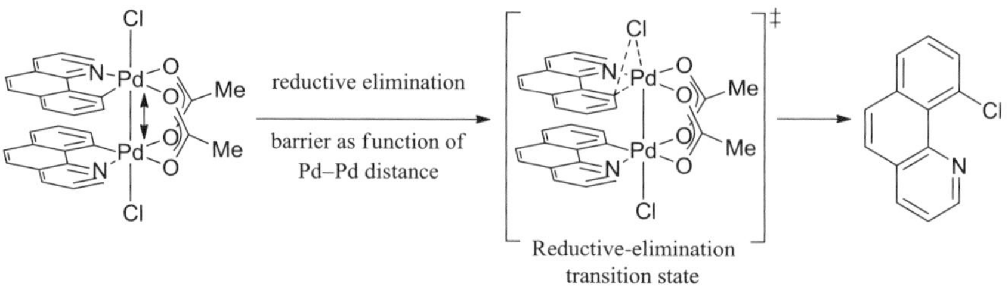

Scheme 33. The activation barrier to C–Cl reductive elimination from **82** increases as the Pd–Pd distance is elongated.

of the activation barriers across the series indicates that M−M cooperation during reductive elimination lowers the energy barrier by ~30 kcal mol^{-1}.

Pd−Pd Distance (Å)	Ground-State Energy (kcal mol^{-1})	Activation Energy (kcal mol^{-1}) (relative to **A**)
2.62	0.0 (**A**)	17.1
2.95	6.2	21.8
3.30	20.5	33.3
3.65	35.3	46.0

Two mechanisms have been put forward to explain the apparent role of the Pd−Pd bond in facilitating C−Cl reductive elimination (287). First, because the Pd−Pd bond is not significantly cleaved in the reductive-elimination transition state, the apical chloride ligand involved in reductive elimination is still subject to the strong trans influence of the Pd−Pd bond. The M−M bonds typically exhibit strong trans influences (304–306), and during the reductive elimination from **82**, the trans influence of the Pd−Pd bond might facilitate reductive elimination. Second, because both metal centers participate in the reductive elimination reaction, as the transition metal fragment is reduced by two electrons, electron density is redistributed across both Pd centers. Delocalization, or sharing, of the additional electron density between two metal centers avoids reductive elimination intermediates in which a single metal center develops significant electronic charge. Consistent with the Pauling electroneutrality principle (307, 308), the sharing of the electronic reorganization during bimetallic reductive elimination between the two metals should lower the activation barrier to the redox process by avoiding charge-separated intermediates.

F. Unbridged Pd(III)−Pd(III) Bonds

In the studies that have implicated binuclear intermediates in catalysis, acetate ligands, which can bridge metal centers, have been present. A question regarding

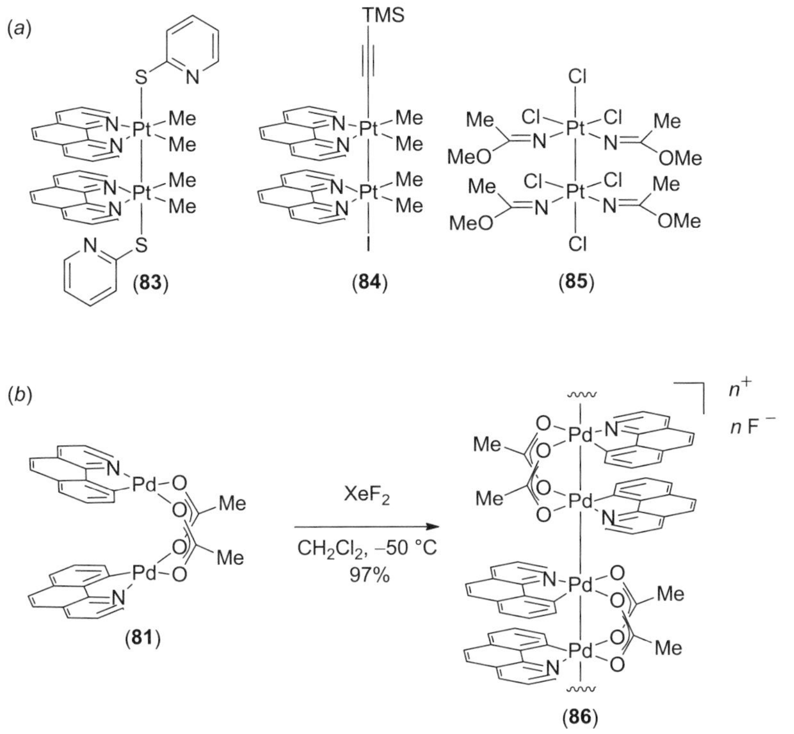

Scheme 34. (*a*) Diplatinum (III,III) complexes, featuring Pt–Pt bonds that are not supported by bridging ligands, have been prepared by oxidation of mononuclear Pt(II) complexes. (*b*) Oxidation of binuclear Pd$_2$(II,II) complex **81** with XeF$_2$ affords polynuclear Pd(III) chains in which unbridged Pd(III)–Pd(III) bonds bridge discrete binuclear units. Structure **86** is the first characterized example of Pd(III)–Pd(III) bonds unsupported by bridging ligands. The unbridged Pd–Pd bond lengths are 2.9718(6) and 2.9823(2) Å, while the distance between acetate-bridged Pd atoms is 2.7206(4) Å.

the potential generality of binuclear intermediates in catalysis is the requirement for bridging ligands to prearrange the binuclear core. If bridging ligands are not present, putative binuclear intermediates would contain unbridged Pd–Pd bonds. Unbridged Pt(III)–Pt(III) bonds, generated by oxidation of mononuclear Pt(II) complexes have been observed; Pt$_2$(III,III) complexes **83–85** have been characterized by X-ray crystallography (Scheme 13) (309–311). The first example of unbridged Pd(III)–Pd(III) bonds, obtained by treatment of binuclear Pd$_2$(II,II) complex **81** with XeF$_2$ to afford solution-stable Pd(III) chain **86**, was recently reported (Scheme 34) (312). X-ray crystallographic analysis revealed an infinite chain of Pd nuclei with alternating Pd–Pd bond lengths; the unbridged Pd–Pd distances are 2.9718(6) and 2.9823(2) Å, while the distance between acetate-bridged Pd atoms is 2.7206(4) Å. The stability of unbridged Pd(III)−Pd(III) bonds between the acetate-bridged binuclear units in complex **86** suggests that bridging ligands may not be required to preorganize polynuclear intermediates in Pd catalyzed oxidation reactions.

G. Palladium–Palladium Bond-Cleavage Pathways

In 2010, Sanford and co-workers (313) reported that treatment of binuclear complex **81** with Togni's I(III) reagent **87** affords mononuclear Pd(IV) complex **89** (Scheme 35*a*). Examination of the reaction kinetics of oxidation of binuclear Pd$_2$(II,II) complex **81** to afford mononuclear Pd(IV) complex **89** revealed that oxidation is rate determining, and that initial oxidation proceeds from a binuclear Pd$_2$(II,II) complex without fragmentation (143). The experimentally determined rate law suggests binuclear cation **88**, which could be formualted as either a Pd$_2$(III,III) or mixed-valent Pd$_2$(II,IV) complex (see below), as the initial product

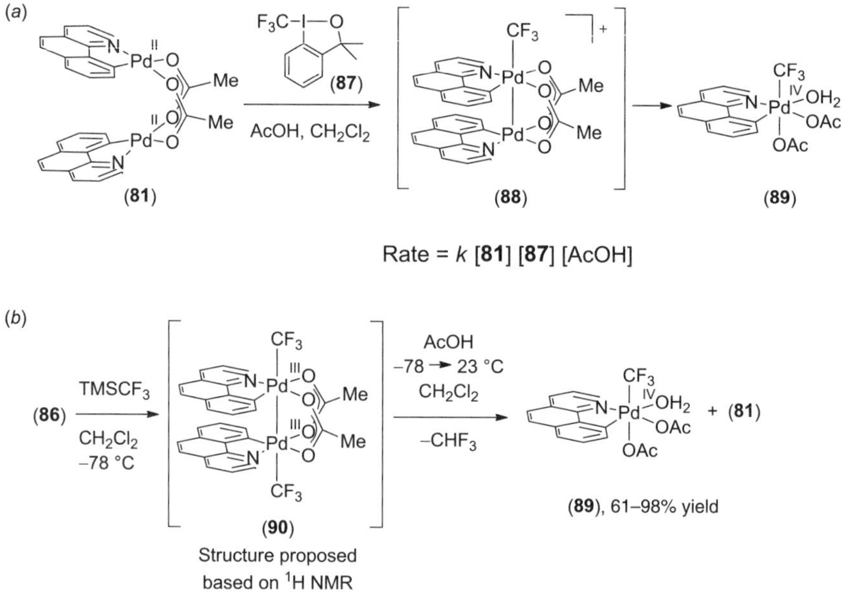

Rate = k [**81**] [**87**] [AcOH]

Scheme 35. (*a*) Oxidation of binuclear Pd$_2$(II,II) complex **81** to mononuclear Pd(IV) compelx **89** proceeds via initial oxidation to binuclear cation **88**, which could be formulated as either a Pd$_2$(III,III) or mixed-valent Pd$_2$(II,IV) complex, followed by Pd–Pd heterolysis. (*b*) Treatment of Pd(III) wire **86** with TMSCF$_3$ affords a complex assigned as trifluoromethyl Pd$_2$(III,III) complex **90**. Warming **90** in the presence of AcOH affords mononuclear Pd(IV) complex **89** and binuclear Pd$_2$(II,II) complex **81**, confirming the viability of Pd(III)–Pd(III) heterolysis.

of oxidation and suggests that the mononuclear Pd(IV) complex is formed by formal Pd–Pd heterolysis from **88** (Scheme 14*a*). Heterolysis affords mononuclear Pd(IV) complex **89** and one-half an equivalent of binuclear $Pd_2(II,II)$ complex **81**, which can undergo oxidation with additional **87** to afford additional mononuclear Pd(IV) complex **89**. Density functional theory investigation of potential fragmentation pathways from **88** suggests that fragmentation faces a lower activation barrier than does potential $C–CF_3$ reductive elimination (314). With the use of modified reaction conditions, Pd–Pd fragmentation to generate mononuclear Pd (IV) complex **89** has been directly observed (Scheme 14*b*) (314). Treatment of Pd (III) fluoride wire **86** with $TMSCF_3$ (TMS = trimethylsilyl) affords a new compound, assigned as **90** based on ^1H NMR data. Warming complex **90** in the presence of AcOH affords mononuclear Pd(IV) complex **89**, as well as binuclear $Pd_2(II,II)$ complex **81**. The transformation of **86** to **89** confirms the viability of the proposed Pd(III)–Pd(III) heterolysis reaction. Prolonged heating of mononuclear Pd(IV) complex **89** does afford the $C–CF_3$ product of reductive elimination, but the available experimental data suggests that reductive elimination does not proceed directly from complex **89** (i.e., production of 10-trifluoromethylbenzo [*h*]quinoline displays an induction period during the thermolysis of **89**) (143). The identity of the complex from which $C–CF_3$ bond formation proceeds is not currently known.

The reactivity of oxidized binuclear Pd complexes is subject to significant ligand effects; the identity of the axial X ligand can determine whether disproportionative fragmentation to Pd(IV) and Pd(II) (e.g., Scheme 35) or direct reductive elimination without fragmentation of the binuclear core (e.g., Scheme 30) is preferred. Canty, Yates and co-workers (315, 316) reported computational studies of ligand effects in binuclear Pd complexes in which the $4d_{z^2}$ orbital population was examined as a function of apical ligand identity (Fig. 12). Based on the computed $4d_{z^2}$ orbital population, and the effect of apical ligand identity on the heterolytic Pd–Cl dissociation energy, it was proposed that $Pd_2(III,III)$ complexes should be formulated as mixed-valent $Pd_2(II,IV)$ dimers if their apical ligands have significantly different trans-influences. Further work will be required to understand the potential role of Pd–Pd cleavage in Pd catalyzed arylation reactions (e.g., Scheme 32), as well as other Pd catalyzed oxidation reactions.

A report by Mirica and co-workers (317) also proposed the synthesis of a mononuclear Pd(IV) complex by initial oxidation to a $Pd_2(III,III)$ intermediate that undergoes subsequent Pd–Pd cleavage chemistry to afford the observed mononuclear Pd(IV) complex (**93**) (Scheme 36). Binuclear $Pd_2(II,II)$ complex **91** displays an unusually close Pd–Pd interaction for unbridged Pd(II) centers (3.11 Å), which has been suggested to result from significant mixing of the $5p_z$ and 5s orbitals with the $4d_z^2$ orbital. Density functional theory calculations of a derivative of **91**, in which the methyl ligands are replaced with acetonitrile ligands, afforded a computed Pd–Pd bond order of 0.19. Based on the close proximity of

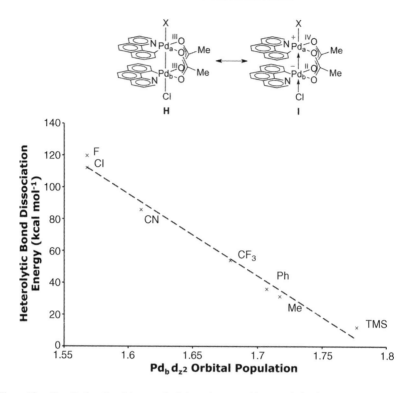

Figure 12.   Density functional theory calculations show a positive correlation between sigma-donating ability of apical ligands on Pd_a and population of the d_z^2 orbital of Pd_b. The d_z^2 population is negatively correlated with heterolytic bond cleavage energy. Taken together, these data suggest that increasingly sigma-donating ligands favor a mixed-valent $Pd_2(II,IV)$ electronic structure over valence symmetric $Pd_2(III,III)$ formulations (TMS = trimethylsilyl). [Adapted from (315)].

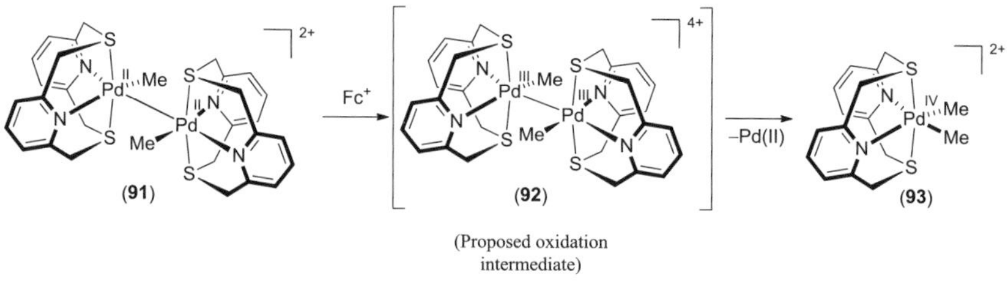

Scheme 36.   Oxidation of binuclear $Pd_2(II,II)$ complex **91** to mononuclear Pd(IV) complex **93** has been proposed to proceed via $Pd_2(III,III)$ complex **92**. A close Pd–Pd contact, which is proposed to result from mixing of the $4d_z^2$ orbital with the 5 s and $5p_z$ orbitals, is observed in binuclear $Pd_2(II,II)$ complex **91**.

the Pd centers in **91**, oxidation was proposed to proceed via a $Pd_2(III,III)$ intermediate prior to eventual formation of the observed mononuclear Pd(IV) complex **93**. In other cases, oxidation of Pd(II) to Pd(IV) has been shown to proceed through $Pd_2(III,III)$ complexes in which there is a chloride bridge between the two Pd centers and no direct Pd–Pd contact (318).

H. Dipalladium Catalysts in Aerobic Oxidation

Based on the observed two-electron redox chemistry displayed in the oxidation of **96** (Pd_2hpp_4) (hppH = hexahydro-2*H*-pyrimido[1,2-*a*]pyrimidine) to binuclear Pd(III) complex $Pd_2hpp_4Cl_2$, as well as the proposed advantages of redox synergy between two Pd centers during redox catalysis, complex **96** was investigated as a C–H oxidation catalyst (Scheme 37) (142). Complex **96** was found to be a competent catalyst for the α-hydroxylation of carbonyl compounds using O_2 as oxidant. A variety of tertiary alcohols are accessible by this protocol. Preliminary investigations suggest that free radicals are not involved in the reaction mechanism, as radical clock substrate **99** underwent oxidation without formation of side-products of radical chemistry being observed. Complex **96** has been found to be a dioxygenase-like catalyst for the α-hydroxylation of carbonyl compounds; based

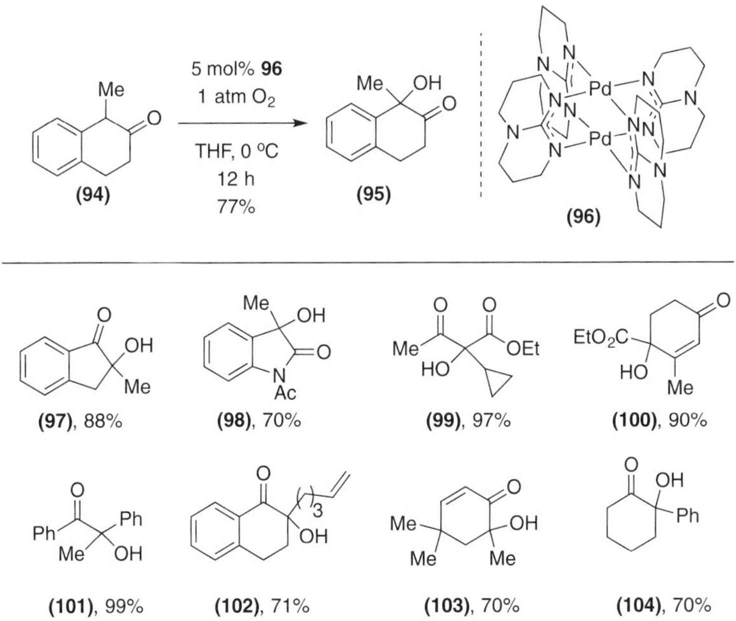

Scheme 37. Complex **96** displays dioxygenase activity in the α-hydroxylation of carbonyl compounds.

on oxygen uptake experiments, it was determined that one-half of an equivalent of O_2 was consumed in the hydroxylation reaction and thus both oxygen atoms of O_2 are incorporated in substrate oxidation. The role of the binuclear core of **96** and the mechanism of oxygen activation are currently unresolved questions.

I. Outlook

The examination of binuclear Pd complexes in redox catalysis is still in its infancy and information regarding the potential generality of mechanisms involving cooperative redox reactions in multinuclear intermediates is not available. Binuclear oxidation intermediates may be more prevalent than generally appreciated: two cases [directed, $Pd(OAc)_2$ catalyzed chlorination and arylation] have thus far been investigated and both involve binuclear oxidation intermediates. Binuclear intermediates have been detected in the formation of mononuclear Pd(IV) complex **89** (314), and unbridged Pd(III)−Pd(III) bond-containing complexes have been characterized (e.g., **91**) (312), suggesting the bridging ligands are not required for productive M−M interactions.

While hypotheses have been advanced to rationalize the productive effect of M−M redox chemistry in catalysis, significant additional work needs to be pursued before the generality and the role of binuclear redox chemistry in catalysis can be appreciated. As exemplified by the chemistry of Pd trifluoromethyl complex **90**, ligand structure can exert a powerful influence on the course of reactivity, and the nuclearity of the complex from which product formation proceeds is not necessarily the initially formed oxidation product.

IV. PARALLELS BETWEEN DIRHODIUM AND DIPALLADIUM SYSTEMS

A. Metallicity

In 2010, Powers et al. (303) proposed a two-tiered nomenclature scheme for discussing redox chemistry of transition metal complexes. In this scheme, *nuclearity* was used to describe the number of transition metal centers that were present in the complex undergoing a given transformation; both Rh_2 and Pd_2 complexes discussed herein are *binuclear* by this definition. *Metallicity* was defined to be the number of transition metal centers that participate in redox chemistry during a redox transformation. For example, consider the addition of MeI to the $Au_2(I,I)$ complex **105** (Scheme 38) (319–321). *Monometallic* oxidative addition would proceed at a single Au center, to afford the Au(I)/Au(III) mixed-valent species **106**, before comproportionative isomerization to complex **107** (Scheme 38*a*). In contrast, *bimetallic* oxidative addition would proceed with

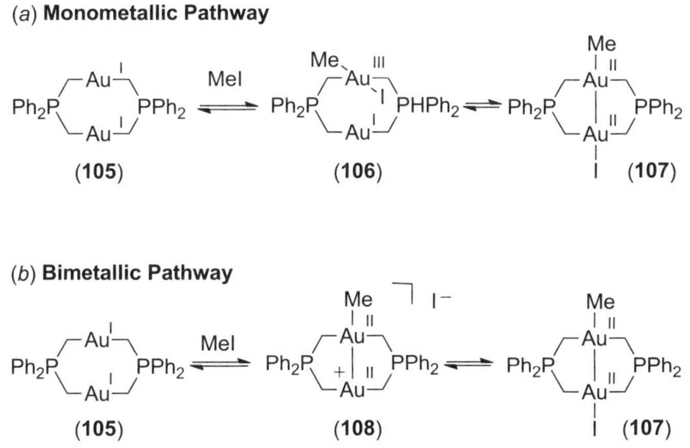

(a) **Monometallic Pathway**

Ph$_2$P〈Au—Au〉PPh$_2$ (105) ⇌ (MeI) ⇌ Ph$_2$P〈Me–Au(III)···I / Au〉PHPh$_2$ (106) ⇌ Ph$_2$P〈Me–Au(II) / Au(II)〉PPh$_2$ (107)

(b) **Bimetallic Pathway**

Ph$_2$P〈Au—Au〉PPh$_2$ (105) ⇌ (MeI) ⇌ [Ph$_2$P〈Me–Au(II) / Au(II)〉PPh$_2$]$^+$ I$^-$ (108) ⇌ Ph$_2$P〈Me–Au(II) / Au(II)〉PPh$_2$ (107)

Scheme 38. Mono- vs bimetallic oxidation of a digold(I) complex.

concurrent formation of an Au–Au bond (**108**); both Au centers participate in redox chemistry in the bimetallic case (Scheme 38b). The examples from both Rh$_2$ and Pd$_2$ chemistry that have been presented suggest that bimetallic redox cooperativity can be involved in both group transfer and well as organometallic C–H functionalization reactions and that this cooperativity can be mediated by the M–M bond in these complexes. Mechanistically, the Rh$_2$ and Pd$_2$ systems described herein diverge. Dirhodium systems, operating via an ATM pathway, accomplish C–H functionalization via group transfer. The majority of Pd$_2$ systems presented accomplish C–H functionalization via oxidation of organometallic intermediates. Despite these critical differences in mechanism, the M–M bonds in both Rh$_2$ and Pd$_2$ systems play an important role in moderating important reaction barriers. During carbene-transfer reactions, delocalization of electron density over the binuclear Rh$_2$ core is proposed to allow for both increased electrophilicity at the carbene carbon, as well as increased facility of Rh–C cleavage during the C–H functionalization. For these Rh$_2$ systems, a 3c–4e bonding manifold in the putative carbene and nitrene intermediates engenders these species with strongly electrophilic character, lowering the barriers to reaction with C–H bonds. In analogy, for C–Cl reductive elimination from **82**, which proceeds without fragmentation of the binuclear Pd$_2$ core, it has been proposed that the delocalization of electron density over a binuclear unit lowers the activation barrier to reductive elimination.

Both binuclear Rh(II) and Pd(III) complexes appear to exhibit enhanced reactivity of the ligands in the apical position, which, in both cases has been proposed to be a result of the strong trans influence of the M–M bond (304–306). The M–M bond in both the Rh$_2$ and Pd$_2$ systems presented herein also exhibit strong kinetic trans effects. An example of this is the group-transfer reactivity of

Rh_2 species: since the Rh_2-L bonds are longer and weaker for axial ligands than they would be in a mononuclear complex, an axial carbene or nitrene ligand is significantly activated when bound to the Rh_2 core. As yet, there is no good measure of either the structural or kinetic trans effect of $Pd(III)-Pd(III)$ bonds, but the closely related $Pt(III)-Pt(III)$ bond has been shown to exert a strong trans effect on the axial ligands of paddlewheel complexes (304–306).

Important ligand effects on the nature of $Pd_2(III,III)$ complexes have been uncovered: for binuclear $Pd(III)$ chloride complexes, such as **82**, the $C-Cl$ reductive elimination proceeds from a binuclear core without fragmentation of the $Pd-Pd$ bond while oxidation of **82** affords a mononuclear $Pd(IV)$ complex, which has been proposed to arise from heterolysis of a $Pd(III)-Pd(III)$ in preference to $C-CF_3$ reductive elimination (314). In principle, it may be possible to alter $Rh(II)-Rh(II)$ bond polarity in an analogous manner. However, introduction of two *different* neutral axial donors is a synthetic challenge. Padwa and coworkers (322) explored computationally the electronic structure of variously substituted $X-Rh-Rh=CH_2$ carbene complexes. They find subtle changes in the polarity of the $Rh-Rh=CH_2$ unit as X is varied from weaker to stronger σ donors. An alternative method for tuning $Rh-Rh$ bond polarity is by using N,O chelating equatorial ligands. The [4,0] isomer of such a complex should have a substantially polarized $Rh-Rh$ bond. To our knowledge, the importance of bond polarity derived from this ligand arrangement to catalysis is not known.

B.   Stabilization of Unusual Oxidation States Using M−M Bonds

Mononuclear $Rh(II)$ and $Pd(III)$ complexes are fairly rare. Although mononuclear complexes containing $Rh(II)$ and $Pd(III)$ have been prepared (156, 323–325), stabilization of these oxidation states generally requires careful synthetic engineering; neither $Rh(II)$ nor $Pd(III)$ are easily accessible oxidation states for simple Werner-type complexes (3). In contrast, a significant number of binuclear $Rh(II)$ complexes and a growing number of binuclear $Pd(III)$ complexes are known. Singly bonded $Rh_2(II,II)$ and $Pd_2(III,III)$ species are formally isoelectronic, and the nature of the $M-M$ bonds in these species is quite similar. These compounds are similarly stabilized by bridging bidentate ligands (e.g., carboxylate ions). The frontier orbitals of Rh_2 and Pd_2 species are all $M-M$ antibonding in character. Thus, removal of electrons from this orbital manifold via oxidation is expected to enhance $M-M$ bonding. This effect allows for the stability of the $Pd(III)$ oxidation state, which is ordinarily difficult to achieve.

Both Rh_2 and Pd_2 species can undergo disproportionation under certain conditions. For example, phosphine ligands can induce disproportionation of $Rh_2(TFA)_4$ into mononuclear $Rh(III)$ and $Rh(I)$ species (Scheme 39*a*) (326, 327). Disproportionation of $Rh_2(II,III)$ species has been suggested to be relevant to $C-H$ amination chemistry (Scheme 39*b*) (133). Du Bois and co-worker (238) proposed

(a) Rh^{II}——Rh^{II} ⟶ Rh^{III} + Rh^{I}

(b) Rh^{II}——Rh^{III} ⟶ Rh^{III} + $0.5\,Rh_2^{II,II}$

(c) Pd^{III}——Pd^{III} ⟶ Pd^{IV} + $0.5\,Pd_2^{II,II}$

Scheme 39. Heterolysis of mixed-valent Rh_2 and Pd_2 species. (a) From $Rh_2(II,II)$ species. (b) From mixed-valent $Rh_2(II,III)$: this fragmentation can potentially reduce the concentration of active catalyst over time. (c) From $Pd_2(III,III)$ resulting in a mononuclear Pd(IV) species that can be reintroduced into bimetallic catalysis following reductive elimination to Pd(II).

that the disproportionation of $Rh_2(II,III)$ reduces the concentration of active catalyst over time, ultimately leading to catalyst deactivation in C−H amination reactions. Evidence for the heterolysis of $Pd_2(III,III)$ dimers with CF_3^- ligands was shown via the direct observation of mononuclear Pd(IV) species (Scheme 35) (313).

C. Outlook

While electronically similar, dirhodium and dipalladium complexes participate in C−H functionalization reactions via disparate mechanisms. Most well-defined dipalladium complexes discovered to date are thought to make carbon–heteroatom bonds by the OM pathway, whereas many dirhodium complexes function via an ATM. However, it is important to emphasize that Pd_2 systems are not all relegated to an OM reaction pathway. The α-hydroxylation of carbonyls using binuclear $Pd_2(II,II)$ (Scheme 37) is a potential example of a group-transfer reaction that is mediated by binuclear Pd complexes, and stands in contrast to the aromatic C−X bond-forming reactions via organometallic mechanisms that are more commonly observed from $Pd_2(III,III)$ (e.g., Scheme 32) (142). While most well-characterized $Pd_2(III,III)$ complexes have axial halide or other X^- type ligands, the activation of O_2 comprises a unique example of a Pd_2 intermediate that potentially crosses over into the ATM paradigm. These results raise the question of whether Rh_2 systems can be engineered that operate via an OM pathway, or, perhaps more excitingly, whether more general Pd_2 group-transfer catalysts can be discovered.

It is worth contemplating whether lessons learned from Rh_2 chemistry may be of relevance to Pd_2 systems, and vice versa. For example, the importance of one-electron chemistry is now better understood in Rh_2 chemistry, especially with respect to C−H amination. One-electron redox processes are essentially absent from discussion in Pd_2 chemistry. Given that mixed-valent $Pd_2(II,III)$ species are known and can be stable, single-electron transfer mechanisms involving such species as intermediates may merit future investigation. Likewise, while much of the discussion of Pd_2 chemistry centered on the catalyst structure and nuclearity, Rh_2 catalysts or intermediates with nuclearity other than two are not generally considered. This discrepancy is due partly to the fact that the $Rh^{(II)}-Rh^{(II)}$ single bond is a strong defining structural characteristic [in contrast to the ~0.1 bond order between Pd(II) atoms]. However, further aggregation of multiple Rh_2 dimers into polynuclear structures may be worth considering.

V. SUMMARY

Both Rh_2 and Pd_2 complexes participate in C−H functionalization reactions. In this chapter, we discussed in detail the nature of the M−M bonds in these compounds, outlined the current scope of the chemistry available to these bimetallic species, and documented the current level of understanding of the mechanism of action of these catalysts. For Rh_2 catalysts, the key mechanistic features are an ATM that involves intermediates having axial carbene (carbenoid) or nitrene (nitrenoid) ligands. We have seen that in Rh_2 carbenoid chemistry, the mechanism can be highly substrate controlled: the electronic structures of diazo compounds heavily influence reaction coordinate thermodynamics. In Rh_2 nitrenoid chemistry, catalyst structure plays an important role, and single-electron transfer steps are a recurring theme. In conjunction with new results in the field of Ru_2 nitrenoid chemistry, the idea of oxidation-state control over mechanism and selectivity was presented. Mechanistic studies have shown that $Pd_2(III,III)$ dimers, which contain a Pd–Pd single bond, are competent for the C−X bond-forming steps of the catalytic reaction. Future research will be focused on understanding the role of discrete binuclear complexes and the potential participation of disproportionative Pd–Pd cleavage reactions, which have been demonstrated to be potential pathways to mononuclear Pd(IV) complexes. A unified blueprint for how the M−M bond impacts the reactivity of Pd_2 and Rh_2 complexes does not yet exist, but we hope that this chapter conveys the current understanding of these systems and will enable the design of new, highly efficient, and perhaps less expensive catalysts for C−H functionalization in the future.

NOTE ADDED IN PROOF

The first spectroscopic characterization of a metastable Rh_2 carbenoid intermediate has now been reported (328).

ACKNOWLEDGMENTS

KPK and JFB thank the Chemical Sciences, Geosciences, and Biosciences Division, Office of Basic Energy Sciences, Office of Science, U.S. Department of Energy for support

(DE-FG02-10ER16204), and the CCI Center for Selective C–H Functionalization supported by NSF (CHE-1205646). DCP and TR thank the National Science Foundation (CHE-0952753) and the Air Force (FA9550-10-1-0170) for financial support.

ABBREVIATIONS

ATM	Atom-transfer mechanism
bhq	Benzo[*h*]quinoline
bpy	2,2'-Bipyridine
cap	Caprolactamate
cat	Catalytic
CMD	Concerted metalation–deprotonation
CV	Cyclic voltammetry
DAniF	*N,N'*-Di-*p*-methoxyphenylformamidinate
DESI–MS	Desorption electrospray ionization mass spectrometry
DFT	Density functional theory
dpb	*N,N'*-Diphenylbenzamidinate
DTolF	*N, N'*-Di-*p*-tolylformamidinate
EMAC	Extended metal-atom chain
EPR	Electron paramagnetic resonance
esp	$\alpha,\alpha,\alpha',\alpha'$-Tetramethyl-1,3-benzenedipropanoate
espn	$\alpha,\alpha,\alpha',\alpha'$-Tetramethyl-1,3-benzenedipropanamidate
EXAFS	Extended X-ray absorption fine structure
Fc	Ferrocene
^{1}H NMR	Proton nuclear magnetic resonance
HOMO	Highest occupied molecular orbital
hp	Hydroxypyridinate
hppH	Hexahydro-2*H*-pyrimido[1,2-α]pyrimidine
KIE	Kinetic isotope effect
Lm	*m* supporting mechanism
LUMO	Lowest unoccupied molecular orbital
Mes	Mesityl
MOF	Metal–organic framework
MS	Molecular sieves
nb	Nonbonding
NCS	*N*-Chlorosuccinimide
OAc	Acetate
oct	Octanoate
OM	Organometallic mechanism
PC	Orthometalated phosphine

PCET	Proton-coupled electron transfer
pfb	Perfluorobutyrate
phpy	2-Phenylyridyl
piv	Pivalate
py	Pyrolidinate
rec	2-[3-(1-Carboxy-1-methylethoxy)phenoxy]-2-methylpropanoate
RT	Room temperature
S-biDOSP	(5R,5'R)-5,5'-(1,3-phenylene)bis[1-[(4-dodecylphenyl)sulfonyl]-L-prolinate
S-biTISP	(5R,5'R)-5,5'-(1,3-phenylene)bis[1-[[2,4,6-tris(1-methylethyl)phenyl]sulfonyl]-L-prolinate
S-BSB	(S)-(-)-(Phenylsulfonyl)prolinate
SCE	Saturated calomel electrode
SCF-Xα-SW	Self-consistent field-Xα-scattered wave
S-DOSP	(S)-(-)-(p-Dodecylphenylsulfonyl)prolinate
S-PTAD	Adamantan-1-yl-(1,3-dioxo-1,3-dihydro-isoindol-2-yl)-acetate
S-PTA	N-Phthaloyl-(s)-alaninate
Tces	2,2,2-Trichloroethoxysulfonamide
T-HYDRO	70% tert-Butylhydroperoxide in water
tfacam	Trifluoroacetamide, F_3CCONH
TFA	Trifluoroacetate
TMS	Trimethylsilyl
TPA	Triphenylaceate
Troc	2,2,2-Trichloroethoxycarbamate
Ts	Tosyl
UV	Ultraviolet

REFERENCES

1. J. A. Bertrand, F. A. Cotton, and W. A. Dollase, *J. Am. Chem. Soc.*, 85, 1349 (1963).

2. F. A. Cotton, N. F. Curtis, C. B. Harris, B. F. G. Johnson, S. J. Lippard, J. T. Mague, W. R. Robinson, and J. S. Wood, *Science*, 145, 1305 (1964).

3. F. A. Cotton, C. A. Murillo, and R. A. Walton, *Multiple Bonds between Metal Atoms*, Oxford University Press, 2005.

4. D. Lawton and R. Mason, *J. Am. Chem. Soc.*, 87, 921 (1965).

5. J. V. Brencic and F. A. Cotton, *Inorg. Chem.*, 8, 7 (1969).

6. F. A. Cotton and W. K. Bratton, *J. Am. Chem. Soc.*, 87, 921 (1965).

7. M. J. Bennett, K. G. Caulton, and F. A. Cotton, *Inorg. Chem.*, 8, 1 (1969).

8. F. A. Cotton, B. G. Deboer, M. D. Laprade, J. R. Pipal, and D. A. Ucko, *J. Am. Chem. Soc.*, 92, 2926 (1970).

9. G. S. Muraveiskaya, G. A. Kukina, V. S. Orlova, O. N. Evstafeva, and M. A. Poraikoshits, *Doklady Akademii Nauk Sssr*, 226, 596 (1976).

10. D. M. Collins, F. A. Cotton, S. Koch, M. Millar, and C. A. Murillo, *J. Am. Chem. Soc.*, 99, 1259 (1977).

11. F. A. Cotton and J. L. Thompson, *J. Am. Chem. Soc.*, 102, 6437 (1980).

12. P. G. Rasmussen, J. E. Anderson, O. H. Bailey, M. Tamres, and J. C. Bayon, *J. Am. Chem. Soc.*, 107, 279 (1985).

13. F. A. Cotton, J. H. Matonic, and C. A. Murillo, *J. Am. Chem. Soc.*, 119, 7889 (1997).

14. K. Umakoshi and Y. Sasaki, *Adv. Inorg. Chem.*, 40, 187 (1993).

15. F. A. Cotton, J. D. Gu, C. A. Murillo, and D. J. Timmons, *J. Am. Chem. Soc.*, 120, 13280 (1998).

16. E. Peligot, *C. R. Acad. Sci.*, 19, 609 (1844).

17. E. Peligot, *Ann. Chim. Phys.*, 12, 528 (1844).

18. F. A. Cotton and R. Poli, *Inorg. Chem.*, 26, 3652 (1987).

19. F. A. Cotton, L. M. Daniels, and C. A. Murillo, *Angew. Chem. Int. Ed.*, 31, 737 (1992).

20. F. A. Cotton, L. M. Daniels, L. R. Falvello, and C. A. Murillo, *Inorg. Chim. Acta*, 219, 7 (1994).

21. F. A. Cotton, L. M. Daniels, and C. A. Murillo, *Inorg. Chim. Acta*, 224, 5 (1994).

22. T. Nguyen, A. D. Sutton, M. Brynda, J. C. Fettinger, G. J. Long, and P. P. Power, *Science*, 310, 844 (2005).

23. Y. C. Tsai, H. Z. Chen, C. C. Chang, J. S. K. Yu, G. H. Lee, Y. Wang, and T. S. Kuo, *J. Am. Chem. Soc.*, 131, 12534 (2009).

24. M. H. Chisholm, in *Metal–Metal Bonding*, G. Parkin, Ed. Springer-Verlag, Berlin, 2010, pp. 29–57.

25. M. H. Chisholm and B. J. Lear, *Chem. Soc. Rev.*, 40, 5254 (2011).

26. M. H. Chisholm and A. M. Macintosh, *Chem. Rev.*, 105, 2949 (2005).

27. M. H. Chisholm and N. J. Patmore, *Acc. Chem. Res.*, 40, 19 (2007).

28. J. F. Berry, in *Metal–Metal Bonding*, G. Parkin, Ed. Springer-Verlag, Berlin, 2010, pp. 1–28.

29. B. O. Roos and L. Gagliardi, *Inorg. Chem.*, 45, 803 (2005).

30. G. L. Manni, A. L. Dzubak, A. Mulla, D. W. Brogden, J. F. Berry, and L. Gagliardi, *Chem. Eur. J.*, 18, 1737 (2012).

31. A. Corma, H. Garcia, and F. X. L. Xamena, *Chem. Rev.*, 110, 4606 (2010).

32. L. Q. Ma and W. B. Lin, *Top. Curr. Chem.*, 293, 175 (2010).

33. R. J. Kuppler, D. J. Timmons, Q. R. Fang, J. R. Li, T. A. Makal, M. D. Young, D. Q. Yuan, D. Zhao, W. J. Zhuang, and H. C. Zhou, *Coord. Chem. Rev.*, 253, 3042 (2009).

34. D. Farrusseng, S. Aguado, and C. Pinel, *Angew. Chem. Int. Ed.*, 48, 7502 (2009).

35. L. Q. Ma, C. Abney, and W. B. Lin, *Chem. Soc. Rev.*, 38, 1248 (2009).

36. A. U. Czaja, N. Trukhan, and U. Muller, *Chem. Soc. Rev.*, 38, 1284 (2009).

37. J. Lee, O. K. Farha, J. Roberts, K. A. Scheidt, S. T. Nguyen, and J. T. Hupp, *Chem. Soc. Rev.*, *38*, 1450 (2009).

38. Paulisse. R, Reimling. H, E. Hayez, A. J. Hubert, and P. Teyssie, *Tetrahedron Lett.* 2233 (1973).

39. A. J. Hubert, A. F. Noels, A. J. Anciaux, and P. Teyssie, *Synthesis-Stuttgart*, 600 (1976).

40. R. R. Schrock, M. L. Listemann, and L. G. Sturgeoff, *J. Am. Chem. Soc.*, *104*, 4291 (1982).

41. M. H. Chisholm and F. A. Cotton, *Acc. Chem. Res.*, *11*, 356 (1978).

42. M. H. Chisholm, *Polyhedron*, *2*, 681 (1983).

43. M. H. Chisholm, *J. Chem. Soc., Dalton Trans.*, 1781 (1996).

44. W. Zhang and J. S. Moore, *Adv. Synth. Catal.*, *349*, 93 (2007).

45. O. Coutelier, G. Nowogrocki, J. F. Paul, and A. Mortreux, *Adv. Synth. Catal.*, *349*, 2259 (2007).

46. N. Saragas, G. Floros, P. Paraskevopoulou, N. Psaroudakis, S. Koinis, M. Pitsikalis, and K. Mertis, *J. Mol. Catal. A*, *303*, 124 (2009).

47. M. H. Chisholm, D. M. Hoffman, J. M. Northius, and J. C. Huffman, *Polyhedron*, *16*, 839 (1997).

48. G. Floros, N. Saragas, P. Paraskevopoulou, L. Choinopoulos, S. Koinis, N. Psaroudakis, M. Pitsikalis, and K. Mertis, *J. Mol. Catal. A*, *289*, 76 (2008).

49. M. H. Chisholm, E. E. Delbridge, A. R. Kidwell, and K. B. Quinlan, *Chem. Commun.*, 126 (2003).

50. B. A. Burroughs, B. E. Bursten, S. Chen, M. H. Chisholm, and A. R. Kidwell, *Inorg. Chem.*, *47*, 5377 (2008).

51. S. Fantacci, N. Re, M. Rosi, A. Sgamellotti, M. F. Guest, P. Sherwood, and C. Floriani, *J. Chem. Soc., Dalton Trans.*, 3845 (1997).

52. M. H. Chisholm, E. R. Davidson, and K. B. Quinlan, *J. Am. Chem. Soc.*, *124*, 15351 (2002).

53. S. T. Chen and M. H. Chisholm, *Inorg. Chem.*, *48*, 10358 (2009).

54. K. R. Mann, N. S. Lewis, V. M. Miskowski, D. K. Erwin, G. S. Hammond, and H. B. Gray, *J. Am. Chem. Soc.*, *99*, 5525 (1977).

55. A. F. Heyduk and D. G. Nocera, *Science*, *293*, 1639 (2001).

56. J. Rosenthal, J. Bachman, J. L. Dempsey, A. J. Esswein, T. G. Gray, J. M. Hodgkiss, D. R. Manke, T. D. Luckett, B. J. Pistorio, A. S. Veige, and D. G. Nocera, *Coord. Chem. Rev.*, *249*, 1316 (2005).

57. A. J. Esswein, A. S. Veige, and D. G. Nocera, *J. Am. Chem. Soc.*, *127*, 16641 (2005).

58. T. S. Teets, T. R. Cook, B. D. McCarthy, and D. G. Nocera, *J. Am. Chem. Soc.*, *133*, 8114 (2011).

59. K. A. Hofmann and G. Bugge, *Chem. Ber.*, *41*, 312 (1908).

60. J. K. Barton, H. N. Rabinowitz, D. J. Szalda, and S. J. Lippard, *J. Am. Chem. Soc.*, *99*, 2827 (1977).

61. K. Sakai, Y. Kizaki, T. Tsubomura, and K. Matsumoto, *J. Mol. Catal.*, *79*, 141 (1993).

62. K. Sakai and K. Matsumoto, *J. Mol. Catal.*, *62*, 1 (1990).

63. K. Sakai and K. Matsumoto, *J. Coord. Chem.*, *18*, 169 (1988).

64. K. Matsumoto and T. Watanabe, *J. Am. Chem. Soc.*, *108*, 1308 (1986).

65. K. Matsumoto and M. Ochiai, *Coord. Chem. Rev.*, *231*, 229 (2002).

66. M. Ochiai and K. Matsumoto, *Chem. Lett.*, 270 (2002).

67. W. Z. Chen, J. Yamada, and K. Matsumoto, *Synth. Commun.*, *32*, 17 (2002).

68. Y. S. Lin, S. Takeda, and K. Matsumoto, *Organometallics*, *18*, 4897 (1999).

69. K. Matsumoto, Y. Nagai, J. Matsunami, K. Mizuno, T. Abe, R. Somazawa, J. Kinoshita, and H. Shimura, *J. Am. Chem. Soc.*, *120*, 2900 (1998).

70. K. Sakai, T. Tsubomura, and K. Matsumoto, *Inorg. Chim. Acta*, *234*, 157 (1995).

71. K. Matsumoto, K. Mizuno, T. Abe, J. Kinoshita, and H. Shimura, *Chem. Lett.*, 1325 (1994).

72. K. Sakai and K. Matsumoto, *J. Mol. Catal.*, *67*, 7 (1991).

73. L. Ackermann, *Chem. Rev.*, *111*, 1315 (2011).

74. A. S. Borovik, *Chem. Soc. Rev.*, *40*, 1870 (2011).

75. H. M. L. Davies and D. Morton, *Chem. Soc. Rev.*, *40*, 1857 (2011).

76. R. I. McDonald, G. S. Liu, and S. S. Stahl, *Chem. Rev.*, *111*, 2981 (2011).

77. T. C. Boorman and I. Larrosa, *Chem. Soc. Rev.*, *40*, 1910 (2011).

78. C. M. Che, V. K. Y. Lo, C. Y. Zhou, and J. S. Huang, *Chem. Soc. Rev.*, *40*, 1950 (2011).

79. J. F. Hartwig, *Chem. Soc. Rev.*, *40*, 1992 (2011).

80. H. J. Lu and X. P. Zhang, *Chem. Soc. Rev.*, *40*, 1899 (2011).

81. J. Wencel-Delord, T. Droge, F. Liu, and F. Glorius, *Chem. Soc. Rev.*, *40*, 4740 (2011).

82. S. Y. Zhang, F. M. Zhang, and Y. Q. Tu, *Chem. Soc. Rev.*, *40*, 1937 (2011).

83. M. Zhou and R. H. Crabtree, *Chem. Soc. Rev.*, *40*, 1875 (2011).

84. O. Baudoin, *Chem. Soc. Rev.*, *40*, 4902 (2011).

85. S. H. Cho, J. Y. Kim, J. Kwak, and S. Chang, *Chem. Soc. Rev.*, *40*, 5068 (2011).

86. T. W. Lyons and M. S. Sanford, *Chem. Rev.*, *110*, 1147 (2010).

87. H. M. L. Davies and J. R. Manning, *Nature (London)*, *451*, 417 (2008).

88. B. Meunier, S. P. de Visser, and S. Shaik, *Chem. Rev.*, *104*, 3947 (2004).

89. R. G. Bergman, *Nature (London)*, *446*, 391 (2007).

90. A. J. Canty and G. Vankoten, *Acc. Chem. Res.*, *28*, 406 (1995).

91. A. D. Ryabov, I. K. Sakodinskaya, and A. K. Yatsimirsky, *J. Chem. Soc., Dalton Trans.*, 2629 (1985).

92. M. Gomez, J. Granell, and M. Martinez, *J. Chem. Soc., Dalton Trans.*, 37 (1998).

93. B. Martin-Matute, C. Mateo, D. J. Cardenas, and A. M. Echavarren, *Chem. Eur. J.*, *7*, 2341 (2001).

94. D. L. Davies, S. M. A. Donald, and S. A. Macgregor, *J. Am. Chem. Soc.*, *127*, 13754 (2005).

95. M. Lafrance, C. N. Rowley, T. K. Woo, and K. Fagnou, *J. Am. Chem. Soc.*, *128*, 8754 (2006).

96. D. Garcia-Cuadrado, A. A. C. Braga, F. Maseras, and A. M. Echavarren, *J. Am. Chem. Soc.*, *128*, 1066 (2006).

97. M. Lafrance and K. Fagnou, *J. Am. Chem. Soc.*, *128*, 16496 (2006).

98. M. Lafrance, S. I. Gorelsky, and K. Fagnou, *J. Am. Chem. Soc.*, *129*, 14570 (2007).

99. O. Daugulis, H. Do, and D. Shabashov, *Acc. Chem. Res.*, *42*, 1074 (2009).

100. C. Jia, T. Kitamura, and Y. Fujiwara, *Acc. Chem. Res.*, *34*, 633 (2001).

101. H. M. L. Davies and R. E. J. Beckwith, *Chem. Rev.*, *103*, 2861 (2003).

102. M. P. Doyle, R. Duffy, M. Ratnikov, and L. Zhou, *Chem. Rev.*, *110*, 704 (2009).

103. M. P. Doyle and D. C. Forbes, *Chem. Rev.*, *98*, 911 (1998).

104. J. R. Hansen and H. M. L. Davies, *Coord. Chem. Rev.*, *252*, 545 (2008).

105. J. Du Bois, *Org. Process Res. Dev.*, *15*, 758 (2011).

106. H.-K. Yip, T.-F. Lai, and C.-M. Che, *J. Chem. Soc., Dalton Trans.*, 1639 (1991).

107. B.-H. Xia, C.-M. Che, and Z.-Y. Zhou, *Chem. Eur. J.*, *9*, 3055 (2003).

108. F. A. Cotton, M. Matusz, R. Poli, and X. J. Feng, *J. Am. Chem. Soc.*, *110*, 1144 (1988).

109. Q.-J. Pan, H.-X. Zhang, X. Zhou, H.-G. Fu, and H.-T. Yu, *J. Phys. Chem. A*, *111*, 287 (2006).

110. J. E. Bercaw, A. C. Durrell, H. B. Gray, J. C. Green, N. Hazari, J. A. Labinger, and J. R. Winkler, *Inorg. Chem.*, *49*, 1801 (2010).

111. T. Kawamura, M. Maeda, M. Miyamoto, H. Usami, K. Imaeda, and M. Ebihara, *J. Am. Chem. Soc.*, *120*, 8136 (1998).

112. M. P. Doyle and T. Ren, *Prog. Inorg. Chem.*, *49*, 113 (2001).

113. J. E. McGrady, R. Stranger, and T. Lovell, *Inorg. Chem.*, *37*, 3802 (1998).

114. F. A. Cotton, B. G. Deboer, M. D. Laprade, J. R. Pipal, and D. A. Ucko, *Acta Crystallogr., Sect. B: Struct. Sci.*, B 27, 1664 (1971).

115. J. J. Ziolkowski, M. Moszner, and T. Glowiak, *J. Chem. Soc., Chem. Commun.*, 760 (1977).

116. J. F. Berry, E. Bill, E. Bothe, F. A. Cotton, N. S. Dalal, S. A. Ibragimov, N. Kaur, C. Y. Liu, C. A. Murillo, S. Nellutla, J. M. North, and D. Villagran, *J. Am. Chem. Soc.*, *129*, 1393 (2007).

117. F. A. Cotton, E. A. Hillard, and C. A. Murillo, *J. Am. Chem. Soc.*, *124*, 5658 (2002).

118. R. G. Pearson, *J. Chem. Ed.*, *45*, 581 (1968).

119. R. G. Pearson, *J. Chem. Ed.*, *45*, 643 (1968).

120. F. A. Cotton, E. V. Dikarev, M. A. Petrukhina, and S.-E. Stiriba, *Inorg. Chem.*, *39*, 1748 (2000).

121. F. A. Cotton and T. R. Felthouse, *Inorg. Chem.*, *19*, 2347 (1980).

122. M. B. Hall, *Polyhedron*, 6, 679 (1987).

123. B. O. Roos, A. C. Borin, and L. Gagliardi, *Angew. Chem. Int. Ed.*, *46*, 1469 (2007).

124. B. O. Roos, *Collect. Czech. Chem. Commun.*, *68*, 265 (2003).

125. L. Dubicki and R. L. Martin, *Inorg. Chem.*, *9*, 673 (1970).

126. J. G. Norman and H. J. Kolari, *J. Am. Chem. Soc.*, *100*, 791 (1978).

127. D. S. Martin, T. R. Webb, G. A. Robbins, and P. E. Fanwick, *Inorg. Chem.*, *18*, 475 (1979).

128. G. Bienek, W. Tuszynski, and G. Gliemann, *Z. Naturforsch B Chem. Sci.*, *33*, 1095 (1978).

129. V. M. Miskowski, W. P. Schaefer, B. Sadeghi, B. D. Santarsiero, and H. B. Gray, *Inorg. Chem.*, *23*, 1154 (1984).

130. J. W. Trexler, A. F. Schreiner, and F. A. Cotton, *Inorg. Chem.*, *27*, 3265 (1988).

131. M. Y. Chavan, T. P. Zhu, X. Q. Lin, M. Q. Ahsan, J. L. Bear, and K. M. Kadish, *Inorg. Chem.*, *23*, 4538 (1984).

132. J. G. Norman, G. E. Renzoni, and D. A. Case, *J. Am. Chem. Soc.*, *101*, 5256 (1979).

133. D. N. Zalatan and J. Du Bois, *J. Am. Chem. Soc.*, *131*, 7558 (2009).

134. K. P. Kornecki and J. F. Berry, *Chem. Eur. J.*, *17*, 5827 (2011).

135. D. L. Lichtenberger, J. R. Pollard, M. A. Lynn, F. A. Cotton, and X. J. Feng, *J. Am. Chem. Soc.*, *122*, 3182 (2000).

136. D. S. Martin, R. A. Newman, and L. M. Vlasnik, *Inorg. Chem.*, *19*, 3404 (1980).

137. V. M. Miskowski and H. B. Gray, *Inorg. Chem.*, *27*, 2501 (1988).

138. C. Lin, T. Ren, E. J. Valente, J. D. Zubkowski, and E. T. Smith, *Chem. Lett.*, 753 (1997).

139. V. M. Miskowski, T. M. Loehr, and H. B. Gray, *Inorg. Chem.*, *26*, 1098 (1987).

140. D. Penno, V. Lillo, I. O. Koshevoy, M. Sanaú, M. A. Ubeda, P. Lahuerta, and E. Fernández, *Chem. Eur. J.*, *14*, 10648 (2008).

141. D. Penno, F. Estevan, E. Fernández, P. Hirva, P. Lahuerta, M. Sanaú and M. A. Úbeda, *Organometallics*, *30*, 2083 (2011).

142. G. J. Chuang, W. K. Wang, E. Lee, and T. Ritter, *J. Am. Chem. Soc.*, *133*, 1760 (2011).

143. D. C. Powers and T. Ritter, *Organometallics*, *32*, 2042 (2013).

144. F. A. Cotton, I. O. Koshevoy, P. Lahuerta, C. A. Murillo, M. Sanaú, M. A. Ubeda, and Q. Zhao, *J. Am. Chem. Soc.*, *128*, 13674 (2006).

145. T. Kawamura, H. Katayama, and T. Yamabe, *Chem. Phys. Lett.*, *130*, 20 (1986).

146. R. Boca, *Coord. Chem. Rev.*, *248*, 757 (2004).

147. J. Telser and R. S. Drago, *Inorg. Chem.*, *23*, 3114 (1984).

148. M. Mikuriya, D. Yoshioka, and M. Handa, *Coord. Chem. Rev.*, *250*, 2194 (2006).

149. M. A. S. Aquino, *Coord. Chem. Rev.*, *170*, 141 (1998).

150. W. R. Hagen, *Adv. Inorg. Chem.*, *38*, 165 (1992).

151. C. L. Yao, L. P. He, J. D. Korp, and J. L. Bear, *Inorg. Chem.*, *27*, 4389 (1988).

152. T. Kawamura, H. Katayama, H. Nishikawa, and T. Yamabe, *J. Am. Chem. Soc.*, *111*, 8156 (1989).

153. K. Das, K. M. Kadish, and J. L. Bear, *Inorg. Chem.*, *17*, 930 (1978).

154. J. L. Bear, B. Han, Z. Wu, E. Van Caemelbecke, and K. M. Kadish, *Inorg. Chem.*, *40*, 2275 (2001).

155. J. L. Bear, E. Van Caemelbecke, S. Ngubane, V. Da-Riz, and K. M. Kadish, *Dalton Trans.*, *40*, 2486 (2011).

156. L. M. Mirica and J. R. Khusnutdinova, *Coord. Chem. Rev.*, *257*, 299 (2013).

157. J. F. Berry, F. A. Cotton, S. A. Ibragimov, C. A. Murillo, and X. Wang, *Inorg. Chem.*, *44*, 6129 (2005).

158. P. A. Pittet, L. Dadci, P. Zbinden, A. Abouhamdan, and A. E. Merbach, *Inorg. Chim. Acta*, *206*, 135 (1993).

159. N. R. Candeias, C. A. M. Afonso, and P. M. P. Gois, *Org. Biomol. Chem.*, *10*, 3357 (2012).

160. M. P. Doyle, M. Y. Eismont, D. E. Bergbreiter, and H. N. Gray, *J. Org. Chem.*, *57*, 6103 (1992).

161. J. Lloret, F. Estevan, K. Bieger, C. Villanueva, and M. A. Ubeda, *Organometallics*, *26*, 4145 (2007).

162. M. P. Doyle, M. Yan, H. M. Gau, and E. C. Blossey, *Org. Lett.*, *5*, 561 (2003).

163. K. Takeda, T. Oohara, M. Anada, H. Nambu, and S. Hashimoto, *Angew. Chem. Int. Ed.*, *49*, 6979 (2010).

164. K. Takeda, T. Oohara, N. Shimada, H. Nambu, and S. Hashimoto, *Chem. Eur. J.*, *17*, 13992 (2011).

165. D. K. Kumar, A. S. Filatov, M. Napier, J. Sun, E. V. Dikarev, and M. A. Petrukhina, *Inorg. Chem.*, *51*, 4855 (2012).

166. T. Nagashima and H. M. L. Davies, *Org. Lett.*, *4*, 1989 (2002).

167. E. V. Dikarev, D. K. Kumar, A. S. Filatov, A. Anan, Y. W. Xie, T. Asefa, and M. A. Petrukhina, *ChemCatChem*, *2*, 1461 (2010).

168. G. Bertrand, *Carbene Chemistry*, Marcel Dekker, New York, 2002.

169. E. Buchner and L. Feldmann, *Chem. Ber.*, *36*, 3509 (1903).

170. H. Staudinger and O. Kupfer, *Chem. Ber.*, *45*, 501 (1912).

171. G. Maas, in *Organic Synthesis, Reactions and Mechanisms*, Springer, Berlin/Heidelberg, 1987, pp. 75–253.

172. M. P. Doyle, *Chem. Rev.*, *86*, 919 (1986).

173. M. P. Doyle, M. A. McKervey, and T. Ye, *Modern Catalytic Methods for Organic Synthesis with Diazo Compounds: From Cyclopropanes to Ylides*, John Wiley & Sons, Inc., New York, 1998.

174. P. Yates, *J. Am. Chem. Soc.*, *74*, 5376 (1952).

175. F. Greuter, J. Kalvoda, and O. Jeger, *Proc. Chem. Soc., London*, 349 (1958).

176. W. von E. Doering and W. R. Roth, *Tetrahedron*, *19*, 715 (1963).

177. M. M. Diaz-Requejo, T. R. Belderrain, M. C. Nicasio, S. Trofimenko, and P. J. Perez, *J. Am. Chem. Soc.*, *124*, 896 (2002).

178. A. Demonceau, A. F. Noels, A. J. Hubert, and P. Teyssie, *J. Chem. Soc., Chem. Commun.*, 688 (1981).

179. M. Terada and Y. Toda, *Angew. Chem. Int. Ed.*, *51*, 2093 (2012).

180. H. M. L. Davies and Y. Lian, *Acc. Chem. Res.*, *45*, 923 (2012).

181. X. C. Wang, X. F. Xu, P. Y. Zavalij, and M. P. Doyle, *J. Am. Chem. Soc.*, *133*, 16402 (2011).

182. N. Selander and V. V. Fokin, *J. Am. Chem. Soc.*, *134*, 2477 (2012).

183. B. T. Parr, S. A. Green, and H. M. L. Davies, *J. Am. Chem. Soc.*, *135*, 4716 (2013).

184. E. Nakamura, N. Yoshikai, and M. Yamanaka, *J. Am. Chem. Soc.*, *124*, 7181 (2002).

185. P. Pelphrey, J. Hansen, and H. M. L. Davies, *Chem. Sci.*, *1*, 254 (2010).

186. R. Cohen, B. Rybtchinski, M. Gandelman, H. Rozenberg, J. M. L. Martin, and D. Milstein, *J. Am. Chem. Soc.*, *125*, 6532 (2003).

187. J. F. Berry, *Dalton Trans.*, *41*, 700 (2012).

188. M. P. Doyle, *J. Am. Chem. Soc.*, *115*, 958 (1993).

189. A. Padwa, D. J. Austin, A. T. Price, M. A. Semones, M. P. Doyle, M. N. Protopopova, W. R. Winchester, and A. Tran, *J. Am. Chem. Soc.*, *115*, 8669 (1993).

190. E. Nadeau, Z. Li, D. Morton, and H. M. L. Davies, *Synlett*, *2009*, 151 (2009).

191. H. M. L. Davies and C. Venkataramani, *Angew. Chem. Int. Ed.*, *41*, 2197 (2002).

192. J. Hansen, J. Autschbach, and H. M. L. Davies, *J. Org. Chem.*, *74*, 6555 (2009).

193. H. M. L. Davies and S. A. Panaro, *Tetrahedron*, *56*, 4871 (2000).

194. M. P. Doyle, J. H. Griffin, V. Bagheri, and R. L. Dorow, *Organometallics*, *3*, 53 (1984).

195. Z. Qu, W. Shi, and J. Wang, *J. Org. Chem.*, *66*, 8139 (2001).

196. D. F. Taber and S. C. Malcolm, *J. Org. Chem.*, *63*, 3717 (1998).

197. H. M. L. Davies and A. R. Dick, in *Top. Curr. Chem.*, J.-Q. Yu and Z. Shi, Ed. Springer Verlag, Berlin/Heidelberg, Berlin, 2010, pp. 303–345.

198. G. L. Rempel, P. Legzdins, H. Smith, and G. Wilkinson, *Inorg. Synth.*, *13*, 90 (1972).

199. G. Winkhaus and P. Ziegler, *Z. Anorg. Allgem. Chem.*, *350*, 51 (1967).

200. J. L. Bear, J. Kitchens, and M. R. Willcott, *J. Inorg. Nucl. Chem.*, *33*, 3479 (1971).

201. F. A. Cotton and J. G. Norman, *J. Am. Chem. Soc.*, *94*, 5697 (1972).

202. C. R. Wilson and H. Taube, *Inorg. Chem.*, *14*, 405 (1975).

203. G. H. P. Roos and M. A. McKervey, *Synth. Commun.*, *22*, 1751 (1992).

204. M. A. Golubnichaya, I. B. Baranovskii, G. Y. Mazo, and R. N. Shchelokov, *Russ. J. Inorg. Chem.*, *26*, 2868 (1981).

205. P. R. Bontcev, M. Miteva, E. Zhecheva, D. Mechandjiev, G. Pneumatikakis, and C. Angelopoulos, *Inorg. Chim. Acta*, *152*, 107 (1988).

206. M. Koralewicz, F. P. Pruchnik, A. Szymaszek, K. Wajda-Hermanowicz, and K. Wrona-Grzegorek, *Transition Met. Chem. (London)*, *23*, 523 (1998).

207. H. M. L. Davies, P. R. Bruzinski, D. H. Lake, N. Kong, and M. J. Fall, *J. Am. Chem. Soc.*, *118*, 6897 (1996).

208. J. F. Berry, F. A. Cotton, P. L. Huang, C. A. Murillo, and X. P. Wang, *Dalton Trans.*, 3713 (2005).

209. H. M. L. Davies and N. Kong, *Tetrahedron Lett.*, *38*, 4203 (1997).

210. H. M. L. Davies and S. A. Panaro, *Tetrahedron Lett.*, *40*, 5287 (1999).

211. C. G. Espino, K. W. Fiori, M. Kim, and J. Du Bois, *J. Am. Chem. Soc.*, *126*, 15378 (2004).

212. K. P. Kornecki and J. F. Berry, *Eur. J. Inorg. Chem.*, 562 (2012).

213. J. Bickley, R. Bonar-Law, T. McGrath, N. Singh, and A. Steiner, *New J. Chem.*, *28*, 425 (2004).

214. M. P. Doyle, *J. Org. Chem.*, *71*, 9253 (2006).

215. A. J. Catino, J. M. Nichols, H. Choi, S. Gottipamula, and M. P. Doyle, *Org. Lett.*, *7*, 5167 (2005).

216. K. P. Kornecki and J. F. Berry, *Chem. Commun.*, *48*, 12097 (2012).

217. A. J. Catino, E. F. Raymon, and M. P. Doyle, *J. Am. Chem. Soc.*, *126*, 13622 (2004).

218. M. O. Ratnikov, L. E. Farkas, E. C. McLaughlin, G. Chiou, H. Choi, S. H. El-Khalafy, and M. P. Doyle, *J. Org. Chem.*, *76*, 2585 (2011).

219. A. J. Catino, J. M. Nichols, R. E. Forslund, and M. P. Doyle, *Org. Lett.*, *7*, 2787 (2005).

220. K. Guthikonda and J. Du Bois, *J. Am. Chem. Soc.*, *124*, 13672 (2002).

221. F. Estevan, P. Lahuerta, J. Lloret, D. Penno, M. Sanau, and M. A. Ubeda, *J. Organomet. Chem.*, *690*, 4424 (2005).

222. P. Hirva, P. Lahuerta, and J. Perez-Prieto, *Theor. Chem. Acc.*, *113*, 63 (2005).

223. J. Lloret, K. Bieger, F. Estevan, P. Lahuerta, P. Hirva, J. Perez-Prieto, and M. Sanau, *Organometallics*, *25*, 5113 (2006).

224. J. Lloret, F. Estevan, P. Lahuerta, P. Hirva, J. Perez-Prieto, and M. Sanau, *Chem. Eur. J.*, *15*, 7706 (2009).

225. E. V. Dikarev, B. Li, and H. T. Zhang, *J. Am. Chem. Soc.*, *128*, 2814 (2006).

226. J. Hansen, B. Li, E. Dikarev, J. Autschbach, and H. M. L. Davies, *J. Org. Chem.*, *74*, 6564 (2009).

227. G. Timmer and J. F. Berry, *Chem. Sci.*, *3*, 3038 (2012).

228. H. Nishiyama, Y. Itoh, H. Matsumoto, S.-B. Park, and K. Itoh, *J. Am. Chem. Soc.*, *116*, 2223 (1994).

229. J. L. Maxwell, K. C. Brown, D. W. Bartley, and T. Kodadek, *Science*, *256*, 1544 (1992).

230. F. Collet, R. H. Dodd, and P. Dauban, *Chem. Commun.*, 5061 (2009).

231. G. Dequirez, V. Pons, and P. Dauban, *Angew. Chem. Int. Ed.*, *51*, 2 (2012).

232. D. Zalatan and J. Bois, in *Top. Curr. Chem.*, J.-Q. Yu and Z. Shi, Ed. Springer Berlin/ Heidelberg, 2010, pp. 347–378.

233. H. Kwart and A. A. Khan, *J. Am. Chem. Soc.*, *89*, 1951 (1967).

234. R. A. Abramovitch, T. D. Bailey, T. Takaya, and V. Uma, *J. Org. Chem.*, *39*, 340 (1974).

235. Y. Yamada, T. Yamamoto, and M. Okawara, *Chem. Lett.*, *4*, 361 (1975).

236. R. Breslow and S. H. Gellman, *J. Am. Chem. Soc.*, *105*, 6728 (1983).

237. C. G. Espino, P. M. Wehn, J. Chow, and J. Du Bois, *J. Am. Chem. Soc.*, *123*, 6935 (2001).

238. K. W. Fiori and J. Du Bois, *J. Am. Chem. Soc.*, *129*, 562 (2007).

239. D. N. Zalatan and J. Du Bois, *J. Am. Chem. Soc.*, *130*, 9220 (2008).

240. K. W. Fiori, C. G. Espino, B. H. Brodsky, and J. Du Bois, *Tetrahedron*, *65*, 3042 (2009).

241. W. Lwowski, *Nitrenes*, Wiley, New York, 1970.

242. E. F. V. Scriven, *Azides and nitrenes*, Academic Press, New York, 1984.

243. R. D. Grigg, J. W. Rigoli, S. D. Pearce, and J. M. Schomaker, *Org. Lett.*, *14*, 280 (2011).

244. J. V. Mulcahy and J. Du Bois, *J. Am. Chem. Soc.*, *130*, 12630 (2008).

245. L. A. Boralsky, D. Marston, R. D. Grigg, J. C. Hershberger, and J. M. Schomaker, *Org. Lett.*, *13*, 1924 (2011).

246. A. R. Thornton and S. B. Blakey, *J. Am. Chem. Soc.*, *130*, 5020 (2008).

247. A. R. Thornton, V. I. Martin, and S. B. Blakey, *J. Am. Chem. Soc.*, *131*, 2434 (2009).

248. S. Beaumont, V. Pons, P. Retailleau, R. H. Dodd, and P. Dauban, *Angew. Chem. Int. Ed.*, *49*, 1634 (2010).

249. H. Lebel, K. Huard, and S. Lectard, *J. Am. Chem. Soc.*, *127*, 14198 (2005).

250. H. Lebel and K. Huard, *Org. Lett.*, *9*, 639 (2007).

251. Q. Nguyen, K. Sun, and T. G. Driver, *J. Am. Chem. Soc.*, *134*, 7262 (2012).

252. R. H. Perry, T. J. Cahill, J. L. Roizen, J. Du Bois, and R. N. Zare, *Proc. Natl. Acad. Sci. USA*, *109*, 18295 (2012).

253. A. Norder, S. A. Warren, E. Herdtweck, S. M. Huber, and T. Bach, *J. Am. Chem. Soc.*, *134*, 13524 (2012).

254. X. Lin, C. Zhao, C.-M. Che, Z. Ke, and D. L. Phillips, *Chem. Asian J.*, *2*, 1101 (2007).

255. S. Uemura and S. R. Patil, *Chem. Lett.*, *11*, 1743 (1982).

256. E. C. McLaughlin, H. Choi, K. Wang, G. Chiou, and M. P. Doyle, *J. Org. Chem.*, *74*, 730 (2008).

257. M. E. Harvey, D. G. Musaev, and J. Du Bois, *J. Am. Chem. Soc.*, *133*, 17207 (2011).

258. J. S. Pap, S. DeBeer George, and J. F. Berry, *Angew. Chem. Int. Ed.*, *47*, 10102 (2008).

259. A. K. Musch Long, G. H. Timmer, J. S. Pap, J. L. Snyder, R. P. Yu, and J. F. Berry, *J. Am. Chem. Soc.*, *133*, 13138 (2011).

260. N. R. Deprez and M. S. Sanford, *Inorg. Chem.*, *46*, 1924 (2007).

261. K. Muñiz, *Angew. Chem. Int. Ed.*, *48*, 9412 (2009).

262. X. Chen, K. M. Engle, D. H. Wang, and J.-Q. Yu, *Angew. Chem. Int. Ed.*, *48*, 5094 (2009).

263. L.-M. Xu, B.-J. Li, Z. Yang, and Z.-J. Shi, *Chem. Soc. Rev.*, 712 (2010).

264. P. Sehnal, R. J. K. Taylor, and I. J. S. Fairlamb, *Chem. Rev.*, *110*, 824 (2010).

265. J. M. Davidson and C. Triggs, *Chem. Ind. (London)*, 457 (1966).

266. J. M. Davidson and C. Triggs, *J. Chem. Soc. A*, 1324 (1968).

267. J. M. Davidson and C. Triggs, *J. Chem. Soc. A*, 1331 (1968).

268. M. O. Unger and R. A. Fouty, *J. Org. Chem.*, *34*, 18 (1969).

269. L. Eberson and Gomezgon. L, *J. Chem. Soc., Chem. Commun.*, 263 (1971).

270. F. R. S. Clark, C. B. Thomas, J. S. Willson, and R. O. C. Norman, *J. Chem. Soc., Perkin Trans.*, 1289 (1974).

271. D. R. Bryant, J. E. McKeon, and B. C. Ream, *Tetrahedron Lett.* 3371 (1968).

272. Vanhelde. R and G. Verberg, *Recl. Trav. Chim. Pay B.*, *84* (1965).

273. P. M. Henry, *J. Org. Chem.*, *36*, 1886 (1971).

274. L. Eberson and L. Jonsson, *J. Chem. Soc., Chem. Commun.*, 885 (1974).

275. R. O. C. Norman, W. J. E. Parr, and C. B. Thomas, *J. Chem. Soc., Perkin Trans.*, 369 (1974).

276. L. M. Stock, K. Tse, L. J. Vorvick, and S. A. Walstrum, *J. Org. Chem.*, *46*, 1757 (1981).

277. A. R. Dick, M. S. Remy, J. W. Kampf, and M. S. Sanford, *Organometallics*, *26*, 1365 (2007).

278. C. K. Pal, S. Chattopadhyay, C. Sinha, and A. Chakravorty, *J. Organomet. Chem.*, *439*, 91 (1992).

279. A. L. Seligson and W. C. Trogler, *J. Am. Chem. Soc.*, *114*, 7085 (1992).

280. L. Boisvert, M. C. Denney, S. K. Hanson, and K. I. Goldberg, *J. Am. Chem. Soc.*, *131*, 15802 (2009).

281. M. P. Lanci, M. S. Remy, W. Kaminsky, J. M. Mayer, and M. S. Sanford, *J. Am. Chem. Soc.*, *131*, 15618 (2009).

282. P. K. Byers, A. J. Canty, B. W. Skelton, and A. H. White, *J. Chem. Soc., Chem. Commun.*, 1722 (1986).

283. A. R. Dick, J. W. Kampf, and M. S. Sanford, *J. Am. Chem. Soc.*, *127*, 12790 (2005).

284. J. M. Racowski, A. R. Dick, and M. S. Sanford, *J. Am. Chem. Soc.*, *131*, 10974 (2009).

285. D. C. Powers and T. Ritter, *Top. Organomet. Chem.*, *35*, 129 (2011).

286. T. Yoneyama and R. H. Crabtree, *J. Mol. Catal. A*, *108*, 35 (1996).

287. D. C. Powers and T. Ritter, *Acc. Chem. Res.*, *45*, 840 (2012).

288. D. C. Powers and T. Ritter, *Nat. Chem.*, *1*, 302 (2009).

289. N. R. Deprez and M. S. Sanford, *J. Am. Chem. Soc.*, *131*, 11234 (2009).

290. D. Kalyani, A. R. Dick, W. Q. Anani, and M. S. Sanford, *Tetrahedron*, *62*, 11483 (2006).

291. V. I. Bakhmutov, J. F. Berry, F. A. Cotton, S. Ibragimov, and C. A. Murillo, *Dalton Trans.* 1989 (2005).

292. R. N. Pandey and P. M. Henry, *Can. J. Chem.*, *52*, 1241 (1974).

293. R. N. Pandey and P. M. Henry, *Can. J. Chem.*, *53*, 1833 (1975).

294. S. Winstein, J. McCaskie, H. B. Lee, and P. M. Henry, *J. Am. Chem. Soc.*, *98*, 6913 (1976).

295. A. D. Ryabov, *Synthesis-Stuttgart*, 233 (1985).

296. J. A. Jordan-Hore, C. C. C. Johansson, M. Gulias, E. M. Beck, and M. J. Gaunt, *J. Am. Chem. Soc.*, *130*, 16184 (2008).

297. R. Giri, J. Liang, J. G. Lei, J. J. Li, D. H. Wang, X. Chen, I. C. Naggar, C. Y. Guo, B. M. Foxman, and J.-Q. Yu, *Angew. Chem. Int. Ed.*, *44*, 7420 (2005).

298. Y. Fuchita, M. Kawakami, and K. Shimoke, *Polyhedron*, *10*, 2037 (1991).

299. K. J. Stowers and M. S. Sanford, *Org. Lett.*, *11*, 4584 (2009).

300. D. C. Powers, M. A. L. Geibel, J. Klein, and T. Ritter, *J. Am. Chem. Soc.*, *131*, 17050 (2009).

301. D. C. Powers, D. Y. Xiao, M. A. L. Geibel, and T. Ritter, *J. Am. Chem. Soc.*, *132*, 14530 (2010).

302. M. Kim, T. J. Taylor, and F. P. Gabbaï, *J. Am. Chem. Soc.*, *130*, 6332 (2008).

303. D. C. Powers, D. Benitez, E. Tkatchouk, W. A. Goddard, III, and T. Ritter, *J. Am. Chem. Soc.*, *132*, 14092 (2010).

304. L. S. Hollis, M. M. Roberts, and S. J. Lippard, *Inorg. Chem.*, *22*, 3637 (1983).

305. K. A. Alexander, S. A. Bryan, F. R. Fronczek, W. C. Fultz, A. L. Rheingold, D. M. Roundhill, P. Stein, and S. F. Watkins, *Inorg. Chem.*, *24*, 2803 (1985).

306. C. M. Che, W. M. Lee, T. C. W. Mak, and H. B. Gray, *J. Am. Chem. Soc.*, *108*, 4446 (1986).

307. L. Pauling, *The Nature of the Chemical Bond*, Cornell University Press, New York, 1960, pp. 1–664.

308. S. J. Lippard and J. M. Berg, *Principles of Bioinorganic Chemistry*, University Science Books, Mill Valley, CA, 1994.

309. K. J. Bormington, M. C. Jennings, and R. J. Puddephatt, *Organometallics*, *27*, 6521 (2008).

310. G. Bandoli, P. A. Caputo, F. P. Intini, M. F. Sivo, and G. Natile, *J. Am. Chem. Soc.*, *119*, 10370 (1997).

311. A. J. Canty, M. G. Gardiner, R. C. Jones, T. Rodemann, and M. Sharma, *J. Am. Chem. Soc.*, *131*, 7236 (2009).

312. M. G. Campbell, D. C. Powers, J. Raynaud, M. J. Graham, P. Xie, E. Lee, and T. Ritter, *Nat. Chem.*, *3*, 949 (2011).

313. Y. D. Ye, N. D. Ball, J. W. Kampf, and M. S. Sanford, *J. Am. Chem. Soc.*, *132*, 14682 (2010).

314. D. C. Powers, E. Lee, A. Ariafard, M. S. Sanford, B. F. Yates, A. J. Canty, and T. Ritter, *J. Am. Chem. Soc.*, *134*, 12002 (2012).

315. A. Ariafard, C. J. T. Hyland, A. J. Canty, M. Sharma, N. J. Brookes, and B. F. Yates, *Inorg. Chem.*, *49*, 11249 (2010).

316. A. Ariafard, C. J. T. Hyland, A. J. Canty, M. Sharma, and B. F. Yates, *Inorg. Chem.*, *50*, 6449 (2011).

317. J. Luo, J. R. Khusnutdinova, N. P. Rath, and L. M. Mirica, *Chem. Commun.*, *48*, 1532 (2012).

318. J. R. Khusnutdinova, N. P. Rath, and L. M. Mirica, *Angew. Chem. Int. Ed.*, *50*, 5532 (2011).

319. K. J. Bonnington, M. C. Jennings, and R. J. Puddephatt, *Organometallics*, *27*, 6521 (2008).

320. S. Jamali, S. M. Nabavizadeh, and M. Rashidi, *Inorg. Chem.*, *44*, 8594 (2005).

321. J. D. Basil, H. H. Murray, J. P. Fackler, J. Tocher, A. M. Mazany, B. Trzcinska-Bancroft, H. Knachel, D. Dudis, T. J. Delord, and D. Marler, *J. Am. Chem. Soc.*, *107*, 6908 (1985).

322. S. M. Sheehan, A. Padwa, and J. P. Snyder, *Tetrahedron Lett.*, *39*, 949 (1998).

323. J. R. Khusnutdinova, N. P. Rath, and L. M. Mirica, *J. Am. Chem. Soc.*, *132*, 7303 (2010).

324. K. R. Dunbar, and S. C. Haefner, *Organometallics*, *11*, 1431 (1992).

325. S. C. Haefner, K. R. Dunbar, and C. Bender, *J. Am. Chem. Soc.*, *113*, 9540 (1991).

326. J. Telser and R. S. Drago, *Inorg. Chem.*, *23*, 2599 (1984).

327. J. Telser and R. S. Drago, *Inorg. Chem.*, *25*, 2989 (1986).

328. K. P. Kornecki, J. F. Briones, V. Boyarskikh, F. Fullilove, J. Autschbach, K. E. Schrote, K. M. Lancaster, H. M. L. Davies, and J. F. Berry, *Science*, *342*, 351 (2013).

Activation of Small Molecules by Molecular Uranium Complexes

HENRY S. LA PIERRE AND KARSTEN MEYER

Friedrich-Alexander-University Erlangen-Nuremberg, Department of Chemistry and Pharmacy; Inorganic Chemistry, Erlangen, Bavaria, DE

CONTENTS

Progress in Inorganic Chemistry, Volume 58, First Edition. Edited by Kenneth D. Karlin.
© 2014 John Wiley & Sons, Inc. Published 2014 by John Wiley & Sons, Inc.

2. Complex [{(Cp*)(η^8-C$_8$H$_4$(1,4-Si(iPr)$_3$)$_2$)U}$_2$(μ,η^2:η^2-N$_2$)]
3. [(Cp*)$_3$U(η^1-N$_2$)]
4. Complex [(Ar(tBu)N)$_3$U(μ,η^1:η^1-N$_2$)Mo(N(tBu)Ph)$_3$]
5. Complex [{(OAr)$_3$U}$_2$(μ-η^2:η^2-N$_2$)]
6. Complex [K(dme)$_4$][K(dme){(Et$_8$-calix[4]tetrapyrrole)U}$_2$(μ-NK)$_2$]
7. Studying N$_2$ Cleavage Via Azide Reduction

VI. DIOXYGEN

    A. Uranium(III) Complexes
       1. Ligand Redistribution Reactions
       2. Bridging Oxo Synthesis
    B. Uranium(V) Complexes
       1. Formation of μ-Oxo and -Peroxide Ligands

VII. CARBON DIOXIDE

    A. Uranium(III) Complexes
       1. Reductive C=O Bond Cleavage to CO and μ-Oxo
       2. η^1-CO$_2$ Coordination
       3. Reductive Disproportionation to CO and CO$_3{}^{2-}$
       4. Insertion into a U−C Bond
       5. Insertion into a U−N Bond
       6. Insertion into a U−S Bond
    B. Uranium(IV) Complexes
       1. Insertion into a U−N Bond
       2. Insertion into a U−C Bond
       3. Insertion into a U−O Bond
       4. Insertion into a U−S Bond
    C. Uranium(V) Complexes
       1. Imido Multiple-Bond Metathesis

VIII. NITROUS OXIDE

    A. Uranium(III) Complexes
       1. Bridging Oxo Synthesis
    B. Uranium(IV) Complexes
       1. Terminal Oxo Synthesis

IX. WATER

    A. Uranium(III) Complexes
       1. Hydroxide Synthesis
       2. Uranyl Polymer and Cluster Synthesis
    B. Uranium(IV) Complexes
       1. Hydroxo and Cluster Synthesis
       2. Alkoxide Hydrolysis and Oxidation
    C. Uranium(VI) Complexes
       1. Imido Hydrolysis
    D. Uranyl(VI/V) Complexes

I. INTRODUCTION

This chapter reviews the activation of small molecules by molecular uranium complexes. The phrase "small molecule" has a variety of meanings depending on the field of chemistry. In the case of organometallic and bioinorganic chemistry, small molecule activation typically refers to the reaction chemistry of small, ubiquitous molecules, such as dihydrogen (H_2), dinitrogen (N_2), carbon monoxide (CO), or methane (CH_4). These thermodynamically and relatively (to very) kinetically stable molecules are of societal import because they are central to both industrial and biological chemistry (1). The explicit exploration of the small molecule activation chemistry of molecular uranium complexes is a relatively recent phenomenon in academic chemistry. The pursuit of molecular uranium chemistry, in particular uranium(III) chemistry, has been dependent on the development of new synthetic methods to readily prepare low-valent and low-coordinate uranium complexes (2–10). It is also dependent on the advent of modern spectroscopy and crystallography.

Uranium is a particularly intriguing element with which to study small molecule activation. The metal ion has four readily accessible oxidation states (III, IV, V, and VI), and typically participates in 1 or 2 electron reductions employing the III/IV and III/V redox couples (11, 12). However, uranium complexes can, as part of bi- or trimetallic systems or with the participation of a redox active ligand, be employed in 3, 4, 5, 6, or 8 electron reductions (13). Additionally, due to the polarized nature of uranium ligand multiple and single bonds, it can participate in redox neutral metathesis reactions or insertion reactions with small molecules. The strongly oxidizing nature of uranium(VI) complexes can even lead to the oxidative activation of small molecules (Section X.C). From an electronic structure perspective, these reactions are complex, since the 5f, 5d, and 6p orbitals can mix and participate in σ, π, and δ bonding with supporting ligands and small molecules. In the case of low-valent uranium complexes, the role of the 5f orbitals is prominent due to contraction of the 5d orbitals across the actinide series and, as result, the 5f orbitals can effectively overlap with ligand orbitals. This behavior contrasts strongly with that of the lanthanides, in which the 4f orbitals are much more contracted and participate to a lesser extent in covalent bonding (14–17). The pseudo-core 6p orbitals are significant in high-valent uranium-ligand multiple bond complexes, of which uranyl and its analogs are the most prominent examples (18–36). Hence, the study of small molecule activation by uranium complexes can play an important role in understanding the bonding and electronic structure of uranium complexes. With the continued international nuclear endeavor, such knowledge is imperative and can aid in the design of new nuclear fuels, including uranium nitride or carbide (37–41), the development of nuclear waste separation and chemical remediation schemes (42–52), and refinement of theoretical models of the bonding in the more radioactive late actinides (53–56).

While the interest in small-molecule activation chemistry of molecular uranium is recent, and builds on the long-standing pursuit of transition metal activation of small molecules (1), uranium has been known to play a key role as part of a solid-state catalyst in the Haber–Bosch process (conversion of N_2 and H_2 to ammonia) since 1909 (57). Modern, industrial catalysts for the Haber–Bosch process do not employ uranium, but the possibility of developing significantly improved industrial catalysts with uranium has helped to spark further interest in the field (58–67). Such an argument for the development of uranium catalysts is advanced by claiming the dramatically low cost of molecular uranium precursors that are orders of magnitude cheaper than second- or third-row late-transition metals and about the same as iron (65). This argument, however, fails to take account of the significant waste disposal costs even of depleted uranium and, most significantly, fails to account for the public's potentially strong negative reaction to any product made with a uranium process (cf. the similar difficulties with genetically modified crops). Small-molecule activation studies with uranium, however, do allow for the correlation of molecular–electronic structure/reactivity relationships. Because the reactivity observed is often dramatically different from analogous transition metal and lanthanide systems, the f-orbital participation in chemical bonding can be interrogated. These fundamental insights into structure and reactivity are the main goal of this chapter. It is hoped that the analysis of the divergent reactivity patterns of uranium will inspire the development of unique and valuable chemical processes that meet the stringent criteria for a commercially and culturally acceptable process.

II. SCOPE AND ORGANIZATION

This chapter is designed to review the reaction chemistry of uranium complexes with small molecules of industrial and biological importance. Specifically, carbon monoxide (CO), nitrogen monoxide (NO), dinitrogen (N_2), dioxygen (O_2), carbon dioxide (CO_2), nitrous oxide (N_2O), dihydrogen (H_2), saturated hydrocarbons, unsaturated hydrocarbons (alkenes, alkynes, and arenes), and water (H_2O) are covered. This chapter is limited to molecular systems, and, where appropriate, comparisons with lanthanide or transition metal systems will be made, but are by no means exhaustive.

There are several previous reviews that cover aspects of the chemistry presented here. Evans and Kozimor (13) presented the reductive transformations of uranium complexes, including discussion of reductive small molecule activation. This chapter also contains an excellent review of the relevant electrochemical potentials of uranium(III) and (IV) complexes. Ephritikhine (68) provided a summary of recent developments in coordination chemistry that touched on small-molecule activation. Similarly, Arnold also provided a broad overview of recent developments in the field, as well as an excellent explication of the importance of uranium small-molecule activation (69). The field has developed rapidly, and a number of recent reviews have covered different aspects of uranium and actinide chemistry (52, 70–88). As such, there will be some overlap, but none of these articles have comprehensively discussed the small-molecule chemistry of molecular uranium complexes. The growing body of actinide catalysis literature is incorporated, but it is by no means complete. Readers are referred to several reviews (58–65). Note that the use of bond notation in text (i.e., $U-C$ bond) is meant to denote valency only, not bond order. Where appropriate, bond order may be discussed separately. No literature search is perfect. We apologize in advance for the omission of any relevant literature meeting the delineated scope.

III.   CARBON MONOXIDE

Interest in CO complexes of uranium initially stemmed from the potential application of $U(CO)_6$ in isotope separation methodologies during the Manhattan Project (89–91). Homoleptic, transition metal carbonyl complexes are often volatile and potentially less corrosive than UF_6. While many attempts employing high-temperature reductive methods to synthesize $U(CO)_6$ were made, none were successful (90). Subsequent argon matrix studies indicated that $U(CO)_6$ could be produced with a v_{CO} of 1961 cm^{-1} (cf. with free CO $v_{CO} = 2145$ cm^{-1}), but it is only stable < 20 K (91, 92). The U(IV) complex, $UF_4(CO)$, was similarly investigated at 20 K and was shown to have a v_{CO} of 2182 cm^{-1}, thus indicating the absence of π back-bonding (93). Similarly, UO_2 was shown to absorb CO gas < 20 K (94). This section describes the direct evidence for the formation of molecular uranium carbonyl complexes and the varied reactivity of CO with well-defined uranium(III, IV, and V) complexes.

A.   Uranium(III) Complexes

1.   Complex [(Cp′)₃U(CO)]

a. **Synthesis and Characterization.**   In 1986, Andersen and co-workers (95) reported the first molecular uranium carbonyl complex stable at room temperature and pressure. While there were previous indications that uranium complexes could bind CO from the CO insertion chemistry of uranium(IV) alkyls and hydrides (Sections III.B.1 and III.B.2), no uranium carbonyl had been isolated. The exposure of a green solution of [(TMSC₅H₄)₃U] (TMS = trimethylsilyl; TMSC₅H₄ = trimethylsilylcyclopentadienyl anion) to 1 atm of CO resulted in a rapid color change of the solution to a deep burgundy to yield [(TMSC₅H₄)₃U(CO)] (Scheme 1). Exposing the solution to vacuum or purging with argon results in the loss of CO and the return of the green colored solution, indicative of [(TMSC₅H₄)₃U]. Infared spectrscopy (IR) studies revealed a v_{CO} of 1976 cm^{-1}. Isotopic labeling studies match reasonably well with those predicted by the simple harmonic oscillator model (for $^{13}CO v_{CO}$ is 1935 cm^{-1} (predicted is 1931 cm^{-1}) and for $C^{18}O v_{CO}$ is 1932 cm^{-1}, (predicted is 1930 cm^{-1})) and suggest the carbonyl ligand is C bound and the $U-C-O$ moiety is linear. While the binding of CO by [(TMSC₅H₄)₃U] is reversible,

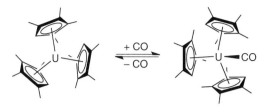

Scheme 1.   Reversible CO binding by [(C₅HMe₄)₃U].

the authors note that $[(TMSC_5H_4)_3U(CO)]$ is stable for 2 years at $-80\,°C$. A bond dissociaton energy (BDE) has been measured for the $U-C(CO)$ bond in $[(TMSC_5H_4)_3U(CO)]$ of $10\,kcal\,mol^{-1}$ (96). This value is close to that predicted by DFT of $14\,kcal\,mol^{-1}$ (97). This bond strength is substantially less than that for similar transition metal complexes, such as $[Cp^*_2Zr(CO)_2]$ ($Cp^* =$ pentamethylcyclopentadienyl anion), which has a BDE of $40\,kcal\,mol^{-1}$ (98).

The first crystallographically characterized uranium carbonyl complex was prepared by Carmona and co-workers in 1995 (99). Treatment of a cherry-red solution of $[(C_5HMe_4)_3U]$ ($C_5HMe_4 =$ tetramethylcyclopentadienyl anion) with an atmosphere of CO affords a purple solution (Scheme 1). Cooling this solution under an atmosphere of CO to $-20\,°C$ leads to the formation of purple–black crystals, which were identified by nuclear magnetic resonance (NMR) and IR spectroscopy, as well as by X-ray crystallography (XRD) (Fig. 1), to be

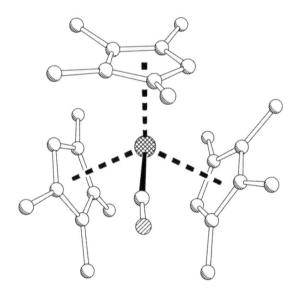

Figure 1.   Crystal structure of $[(C_5HMe_4)_3U(CO)]$.

$[(C_5HMe_4)_3U(CO)]$. Solution IR reveals a strong resonance at $1900\,cm^{-1}$, while solid-state IR in KBr gives a CO stretch at $1880\,cm^{-1}$. Like the CO complexation by $[(TMSC_5H_4)_3U]$, CO binding by $[(C_5HMe_4)_3U(CO)]$ is reversible and exposure to vacuum slowly leads to the isolation of the starting complex. The X-ray crystal structure reveals that the CO ligand is C bound, as expected from the IR vibrational band, and that the U$-$C$-$O angle is nearly linear, $175.2(6)°$, and the C$-$O bond is $1.42(7)\,\text{Å}$ (cf. with $1.28\,\text{Å}$ for free CO). The three cyclopentadienyl anion (Cp) ligands are planar and arranged with the hydrogen substituents of the Cp ligands in the same direction about the complex in order to minimize steric clash between the ligands. In a subsequent study, Carmona and co-workers (100) also synthesized and characterized in solution $[(1,3\text{-}(TMS)_2C_5H_3)_3U(CO)]$ $(1,3\text{-}(TMS)_2C_5H_3 =$ 1,3-trimethylsilylcyclopentadienyl anion) and $[(tBuC_5H_4)_3U(CO)]$ $(tBuC_5H_4 =$ tert-butylcyclopentadienyl anion) complexes. Although they were not able to obtain crystal structures of these complexes, they found that the π donating ability of the $[(Cp')_3U]$ (Cp$' =$ substituted cyclopentadienyl anion) complex was dependent on the nature of the Cp ligand substituents. Accordingly, they reported for $[(1,3\text{-}TMSC_5H_3)_3U(CO)]$ a ν_{CO} of $1988\,cm^{-1}$ and for $[(tBuC_5H_4)_3U(CO)]$ a ν_{CO} of $1960\,cm^{-1}$. In other words, the π donating ability of the $[(Cp)_3U]$ complexes decreases in the order $C_5HMe_4 > Me_3CC_5H_4 > TMSC_5H_4 > 1,3\text{-}TMSC_5H_3$.

In 2003, Evans et al. (101) furnished a further crystallographic example of a uranium(III) CO complex with the isolation of $[(Cp^*)_3U(CO)]$. Treatment of a benzene solution of $[(Cp^*)_3U]$ with 1 atm of CO results in a color change from brown to black (Scheme 1). While the complex can be isolated by removing volatiles under reduced pressure, $[(Cp^*)_3U(CO)]$ is unstable to vacuum as a solid. The proton NMR (^{1}H NMR) spectrum reveals a single resonance at δ 0.25 ppm (parts per million) consistent with η^1-CO (rather than insertion chemistry). Vibrational IR studies on $[(Cp^*)_3U(CO)]$ as a Nujol mull gives a resonance at $1917\,cm^{-1}$ ($1922\,cm^{-1}$ as KBr pellet). Assignment as a CO stretch was confirmed by isotope-labeling studies with ^{13}CO, which gave an observed stretch at $1877\,cm^{-1}$ (predicted $1879\,cm^{-1}$). Crystals suitable for X-ray diffraction studies can be obtained from a room temperature saturated solution (Fig. 2). As expected, based on the case of $[(C_5HMe_4)_3U(CO)]$ and theoretical studies, the CO ligand in $[(Cp^*)_3U(CO)]$ is C bound to the uranium center with a U$-$C distance of 2.383 (6) Å and C$-$O distance of $1.13(1)\,\text{Å}$. The reactivity of $[(Cp^*)_3U]$ with CO contrasts strongly with that of $[(Cp^*)_3Sm]$ and $[(Cp^*)_3Nd]$ with CO (101, 102). The lanthanide complexes undergo CO insertion chemistry with Cp* ligands to afford carbonium ion complexes (Scheme 2). In discussing the contrasting reactivity of the 4f and 5f systems, Evans et al. (101, 102) suggest that the reactivity is consistent with the hypothesis that lanthanide organometallic bonding is more ionic than in actinide complexes and, to the extent of their involvement in the U$-$CO bond, that the 5f orbitals have a greater radial extension than the 4f orbitals.

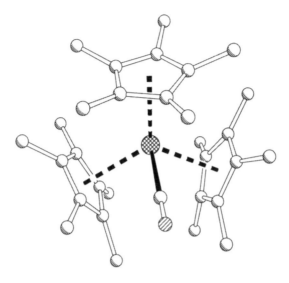

Figure 2.   Crystal structure of [(Cp*)$_3$U(CO)].

b. Electronic Structure.   In a series of papers between 1987–1989, Bursten et al. (103–105) advanced a model of the electronic structure of [(Cp)$_3$U] and [(Cp)$_3$U (CO)] complexes. These Xα-SW molecular orbital (MO) investigations built on the initial work by Tatsumi and Hoffmann (106) that predicted the stability of uranium carbonyl complexes based on the hypothetical complex [(Cp)$_3$U(CO)]$^{3+}$. The salient feature of the Bursten et al. (103–105) model is that the 5f^3 electrons are engaged in π back-bonding with the 2π orbitals of CO (Scheme 3, left). They propose that this donation of electron density from the 5f^3 electrons is the basis of the reduced CO stretching frequencies of [(Cp′)$_3$U(CO)] complexes. Furthermore, these studies predict that the isocarbonyl is unstable despite the oxophilicity of uranium. Additionally, Bursten et al. (105) predict the stability of a uranium(III) nitrosyl complex, [(Cp′)$_3$U(NO)] (Section IV.A.3), based on the increased π acceptance of NO and participation in electron-transfer reactions in comparison to CO.

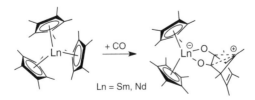

Scheme 2.   Carbon monoxide insertion chemistry of [(Cp*)$_3$Ln]; where Ln = Sm and Nd.

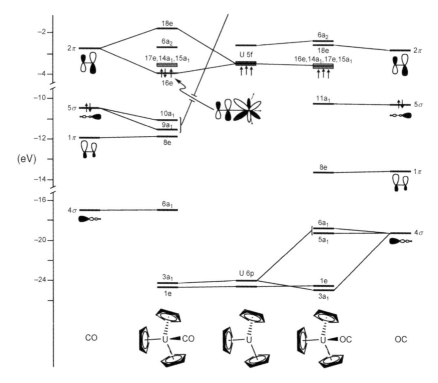

Scheme 3.   Bursten's model of carbonyl (left) and isocarbonyl (right) bonding in [(Cp)₃U(CO)].
[Adapted from (103)].

Eisenstein and co-workers (97) reexamined this bonding model in light of the series of differentially substituted [(Cp′)₃U(CO)] that have been synthesized subsequent to [(CpTMS)₃U(CO)]. In light of this data, the authors note that the CO stretching frequency is highly dependent on the nature of the Cp′ substituent. This observation is in contrast to reduced early metal metallocene complexes, in which the CO stretching frequency is independent of the Cp′ substituent (97). The CO stretching frequencies increase in the order C₅HMe₄ < Cp* < C₅H₄TMS. To account for this experimental observation, the authors' propose that back-bonding to the bound CO is derived from the filled π-symmetry orbitals of the [(Cp′)₃U] fragment, which are largely ligand-based orbitals, as developed in Scheme 4(*a*), and the interaction with CO (*b*). This model for bonding is supported by density functional theory (DFT) calculations that correctly match the magnitude and trend of the observed CO stretching frequencies. In support of this argument, the d⁰ vanadium bis(imido) complex [(PEt₃)₂(N(*t*-Bu))₂V(CO)] [Al(PFTB)₄] (PFTB = *per*-fluoro-*tert*-butoxide) is proposed to bind CO via an

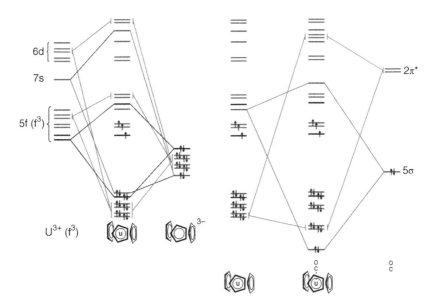

Scheme 4.   Eisenstein's model of carbonyl bonding in [(Cp)₃U(CO)]: left – symmetry adapted orbitals of [Cp₃]³⁻ and right – overlap with 2π orbitals of CO. [Adapted from (97)].

analogous filled π-symmetry ligand orbital to π* CO orbital donation (107). The divergent reactivity of aryloxide and amide complexes of U(III) with CO (Sections III.A.2 and III.A.4) may be due to the absence in these systems of π-symmetry supporting ligand orbitals of appropriate energy to interact with the bound CO. The bonding of the [(Cp′)₃U] complexes with CO analogs, (e.g., isocyanides and [Cp*Al] and [Cp*Ga]), also has been explored in detail (95, 100, 108, 109). Note also that a number of papers have covered theoretical treatments of uranium halide carbonyl complexes (110–112).

2.   *Complex [{((^{tBu}ArO)₃tacn)U}₂(μ:η¹,η¹-CO)]*

While initial theoretical studies by Bursten et al. (105) suggested that isocarbonyl monouranium(III) complexes are unlikely to be isolable, Castro-Rodriguez and Meyer (113) recently isolated a unique bridging carbonyl–isocarbonyl complex. Treatment of a degassed pentane solution of [((^{tBu}ArO)₃tacn)U], ((^{tBu}ArOH)₃tacn = 1,4,7-tris(3,5-di-*tert*-butyl-2-hydroxybenzylate)-1,4,7-triazacyclononane), with 1 atm of CO results in a slight color change from red–brown to brown. Work-up and crystallization from benzene results in the isolation of [{((^{tBu}ArO)₃tacn)U}₂(μ:η¹,η¹-CO)]. The IR spectrum of [{((^{tBu}ArO)₃tacn)U}₂(μ:η¹,η¹-CO)] as Nujol mull reveals a vibrational band at 2092 cm⁻¹, indicative of bound CO. The X-ray crystal structure, obtained by recrystallization from benzene at room temperature,

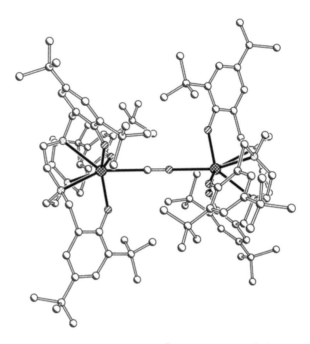

Figure 3.   Crystal structure of [{((^{tBu}ArO)$_3$tacn)U}$_2$(μ:η^1,η^1-CO)].

reveals a dinuclear structure with the bound CO ligand centered on an inversion center (Fig. 3). Based on the average displacement of the U ion below the plane defined by the aryloxide oxygen atoms, the authors propose that the complex is mixed-valent U(IV/III), with a bridging carbonyl radical anion ligand. It is proposed that this complex forms by the capping of a charge-separated [((^{tBu}ArO)$_3$tacn) U$^+$(IV)$-$CO$^{•-}$] by another equivalent of the trivalent starting complex. A similar bonding situation has been observed in a cobalt–ytterbium bridging isocarbonyl complex, [(CO)$_3$Co$-$CO$-$Yb(Cp*)$_2$(thf)], where thf = tetrahydrofuran (ligand), described by Tilley and Andersen (114).

3.   Complex [(η^8-C$_8$H$_6$(1,4-Si(iPr)$_3$)$_2$)(η^5-C$_5$Me$_{5-n}$H$_n$)U(thf)], (n = 0, 1)

With the appropriate steric and electronic structure, uranium(III) complexes can reductively homologate CO. In a number of ground-breaking publications, Cloke and co-workers (115) first reported the homologation of CO to the deltate dianion (C$_3$O$_3$$^{2-}$) by 2 equiv of [($\eta^8$-C$_8H_6$(1,4-Si(iPr)$_3$)$_2$)($\eta^5$-Cp*)U(thf)]. This process is selective (in the presence of excess CO at 1 atm and room temperature, rt) and [{(η^8-C$_8$H$_6$(1,4-Si(iPr)$_3$)$_2$)(η^5-Cp*)U}$_2$((μ-η^1:η^2-C$_3$O$_3$)] is the sole product (Scheme 5). This reactivity of molecular uranium complexes with CO is quite

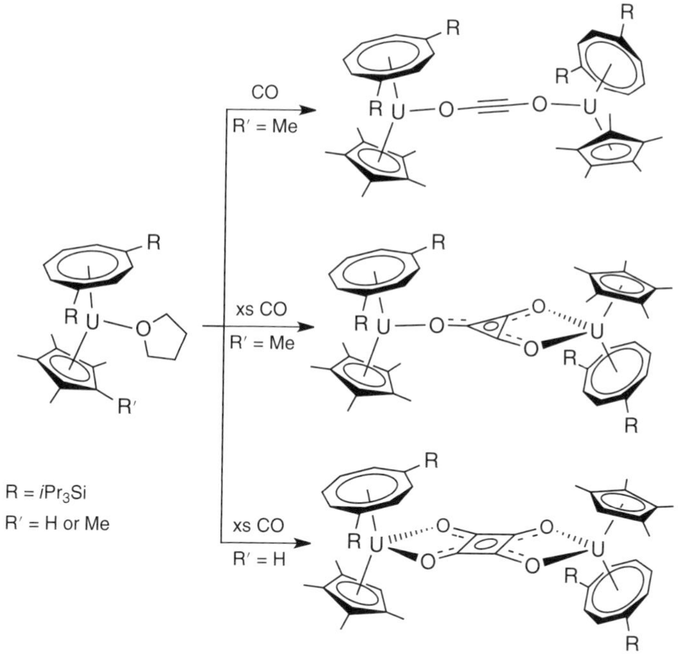

Scheme 5.   Reductive homologation of CO by [(η^8-C$_8$H$_6$(1,4-Si(iPr)$_3$)$_2$)(Cp')U(thf)] complexes.

unique as the synthesis of homologated CO products (Fig. 4) was previously limited to reduction by molten alkali metals, electrochemical methods, and surface catalysis methods (116–120). The reaction proceeds in pentane at $-78\,°C$ and the product is recrystallized from diethyl ether. The X-ray crystal structure reveals an asymmetric binding of the C$_3$O$_3^{2-}$ dianion as μ-η^1:η^2-C$_3$O$_3$, (Fig. 5). The oxidation state of each uranium center is assigned as U(IV) on the basis of DFT studies. Notably, the C–C bonds of the deltate dianion are not equivalent. Further DFT studies suggested that the basis of this structural feature was an agostic C–C bond interaction with the uranium center. In solution, ^{1}H NMR studies do not reflect an asymmetric binding of the deltate dianion. The ^{13}C NMR studies employing isotopically labeled ^{13}CO confirmed this assignment as a broad single resonance for the deltate dianion is observed at room temperature at δ 225 ppm. Variable temperature (VT) NMR studies did not reveal a freezing out of the isomers at low temperature, which prevented a determination of the mechanism for the interconversion of the asymmetric environments.

Subsequent studies investigated the role of sterics on the selectivity of the CO homologation reaction. Cloke and co-workers (121) found the use C$_5$Me$_4$H instead of Cp* led to the selective formation of squarate dianion, (C$_4$O$_4^{2-}$) under analogous

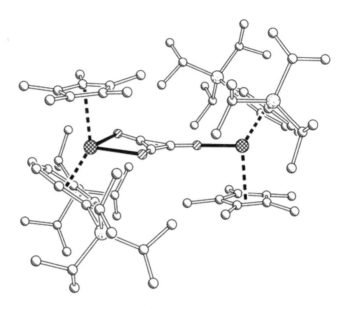

Figure 4.   Crystal structure of $[\{(\eta^8\text{-}C_8H_6(1,4\text{-}(i\text{Pr})_3)_2)(\eta^5\text{-}Cp^*)U\}_2(\mu\text{-}\eta^1{:}\eta^2\text{-}C_3O_3)]$.

conditions. Exposure of $[(\eta^8\text{-}C_8H_6(1,4\text{-}Si(i\text{Pr})_3)_2)(\eta^5\text{-}C_5Me_4H)U(thf)]$ to excess CO at 1 atm and $-30\,^{\circ}C$ in toluene leads to the selective formation of $[\{(\eta^8\text{-}C_8H_6(1,4\text{-}Si(i\text{Pr})_3)_2)(\eta^5\text{-}C_5Me_4H)U\}_2((\mu\text{-}\eta^2{:}\eta^2\text{-}C_4O_4)]$. This slight change in the sterics of the mixed COT/Cp (COT = cyclooctatetraenyl dianion) metallocene sandwich had precedent in the observed change in N_2 coordination by group IV (4) metallocenes in switching to C_5Me_4H from Cp^* (122, 123). Recrystallization from Et_2O/THF (THF = Tetrahydrofuran solvent) reveals, by X-ray diffraction, a C_i symmetric dimer (Fig. 6). In contrast to the structure of $[\{(\eta^8\text{-}C_8H_6(1,4\text{-}Sii\text{Pr}_3)_2)(\eta^5\text{-}Cp^*)U\}_2((\mu\text{-}\eta^1{:}\eta^2\text{-}C_3O_3)]$, the analysis of $[\{(\eta^8\text{-}C_8H_6(1,4\text{-}Si(i\text{Pr})_3)_2)(\eta^5\text{-}C_5Me_4H)U\}_2((\mu\text{-}\eta^2{:}\eta^2\text{-}C_4O_4)]$ does not reveal any agostic $C-C$ interactions between the U centers and the squarate dianion, which is probably due to the increased

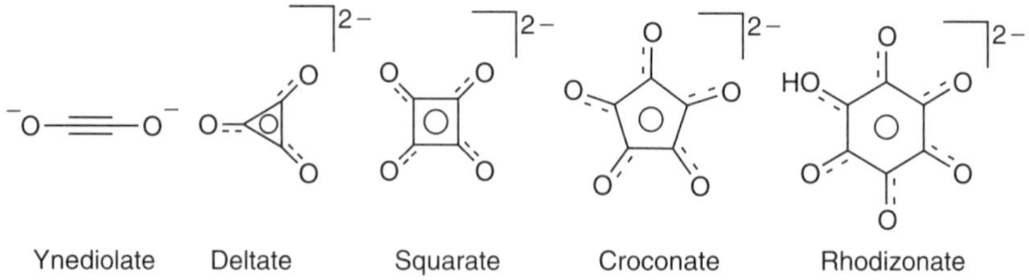

Ynediolate   Deltate   Squarate   Croconate   Rhodizonate

Figure 5.   Cyclic, aromatic, oxocarbon anions $C_nO_n^{2-}$, $n = 2$–6.

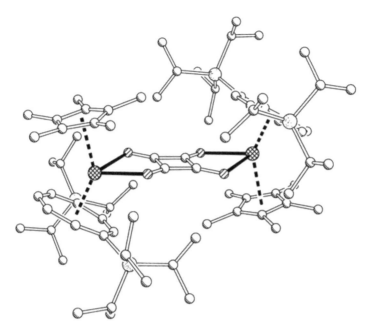

Figure 6.   Crystal structure of [{(η^8-C$_8$H$_6$(1,4-Si(*i*Pr)$_3$)$_2$)(η^5-C$_5$Me$_4$H)U}$_2$(μ-η^2:η^2-C$_4$O$_4$)].

steric requirements for including four CO equivalents. Recently, it has been shown that the functionalization of the squarate anion can be achieved by treating the complex [{(η^8-C$_8$H$_6$(1,4-Si(*i*Pr)$_3$)$_2$)(η^5-C$_5$Me$_4$H)U}$_2$((μ-η^2:η^2-C$_4$O$_4$)] with TMSCl, which furnishes the TMS protected squarate dianion and the uranium(IV) chloride complex (124). Unfortunately, this transformation is not general, and analogous reactions for the cleavage of ynediolate and deltate complexes lead to intractable mixtures.

In order to further probe the role of sterics in the formation of the squarate and deltate dianions by the mixed COT/Cp sandwich complexes, and potentially reveal the mechanism of CO homologation, the ynediolate complex resulting from the dimerization of CO was sought. It was found, by NMR, that treating [(η^8-C$_8$H$_6$(1,4-Si(*i*Pr)$_3$)$_2$)(η^5-Cp*)U(thf)] with 0.95 equiv of CO in toluene at −78 °C provided a mixture of the ynediolate [(U(η^8-C$_8$H$_6$(1,4-Si(*i*Pr)$_3$)$_2$)(η^5-Cp*))$_2$((μ-η^1:η^1-C$_2$O$_2$)] and the known deltate complex [(U(η^8-C$_8$H$_6$(1,4-Si(*i*Pr)$_3$)$_2$)(η^5-Cp*))$_2$((μ-η^1:η^2-C$_3$O$_3$)] (125). These complexes can eventually be separated by fractional crystallization, and an X-ray diffraction study of [{(η^8-C$_8$H$_6$(1,4-Si(*i*Pr)$_3$)$_2$)(η^5-Cp*)U}$_2$(μ-η^1:η^1-C$_2$O$_2$)] shows a nearly linear ynediolate linkage (Fig. 7). Significantly, treatment of this ynediolate complex with additional equivalents of ^{13}CO, even under elevated temperatures, results in no

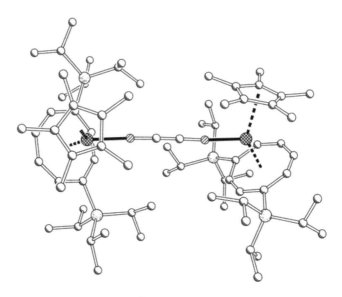

Figure 7. Crystal structure of $[\{(\eta^8\text{-}C_8H_6(1,4\text{-}Si(iPr)_3)_2)(\eta^5\text{-}Cp^*)U\}_2((\mu\text{-}\eta^1\text{:}\eta^1\text{-}C_2O_2)]$.

further reaction, and none of the deltate complex is formed. The lack of reactivity with excess CO implies that the ynediolate is not an intermediate in the formation of the deltate dianion; and therefore, the production of both products proceeds through different intermediates (Scheme 6). The tuning of the steric and electronic properties of the mixed COT–Cp uranium sandwich complexes eventually led to the development of a ligand system that selectively forms the ynediolate in the presence of excess CO. Namely, treating $[(\eta^8\text{-}C_8H_6(1,4\text{-}TMS)_2)(\eta^5\text{-}C_5Me_4TMS)U]$ (reduced sterics at the COT ligand and increased strerics at the Cp′ derivative) with 1 atm CO at −78 °C, and allowing to warm to room temperature, leads to the exclusive formation of the ynediolate complex, $[\{(\eta^8\text{-}C_8H_6(1,4\text{-}TMS)_2)(\eta^5\text{-}C_5Me_4TMS)U\}_2((\mu\text{-}\eta^1\text{:}\eta^1\text{-}C_2O_2)]$ (124).

Both Green and co-workers (126) and Maron and co-workers (127) developed computational models of the CO homologation mechanism in the formation of the ynediolate and deltate dianions that account for the experimental observation that the ynediolate complex, $[\{(\eta^8\text{-}C_8H_6(1,4\text{-}Si(iPr)_3)_2)(\eta^5\text{-}Cp^*)U\}_2((\mu\text{-}\eta^1\text{:}\eta^1\text{-}C_2O_2)]$, does not convert to the deltate complex, $[\{(\eta^8\text{-}C_8H_6(1,4\text{-}Si(iPr)_3)_2)(\eta^5\text{-}Cp^*)U\}_2((\mu\text{-}\eta^1\text{:}\eta^2\text{-}C_3O_3)]$. Green and co-workers (126), using DFT methods, were able to identify a "zigzag" intermediate formed by the coupling of two mono-CO complexes of the parent U(III) mixed COT/Cp sandwich complex (Scheme 6). Maron and co-workers (127) identified an alternate reactive intermediate that leads to the formation of the deltate anion via the addition of a third equivalent of CO. In

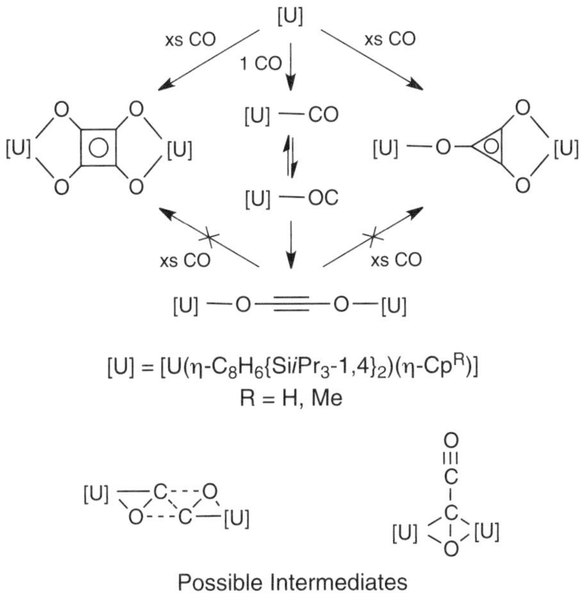

Scheme 6.   Mechanistic hypotheses for CO homologation.

this intermediate, the $C_2O_2^{2-}$ intermediate is bound asymmetrically to one of the two CO linkages in a μ^2-η^2:η^2-C_2O_2 fashion (Scheme 6). Further details on these systems can be found in the original papers and a subsequent review article (128). The chemistry of these systems in mixed-gas reactions CO–NO and CO–H$_2$ are described in the context of the additional small molecule reactant later in this chapter (Sections IV.A.2 and X.A).

4.   Ynediolate $(OCCO)^{2-}$ Systems

a.   Complexes [(N(TMS)$_2$)$_3$U] and [(OAr)$_3$U].   The selective homologation CO to ynediolate ligands is not limited to U(III) mixed COT–Cp complexes. A recent reexamination of the reactivity of the classic [(N(TMS)$_2$)$_3$U] complex (129) by Arnold et al. (130) revealed that a CO homologation reaction proceeds under 1 atm of CO at room temperature. The purple color of [(N(TMS)$_2$)$_3$U] gradually changes to amber in either alkane or arene solvents to give the ynediolate complex, [{(N(TMS)$_2$)$_3$U}$_2$(μ-η^1:η^1-C$_2$O$_2$)] in 82% yield (the reaction does not proceed in either pyridine (Py) solvent or THF). The formation of the ynediolate was confirmed by X-ray crystallography and by an isotopic-labeling experiment with ^{13}CO, which gives a single ^{13}C NMR resonance for the ynediolate product at δ 171 ppm. Functionalization of the ynediolate with a variety of silanes and boranes

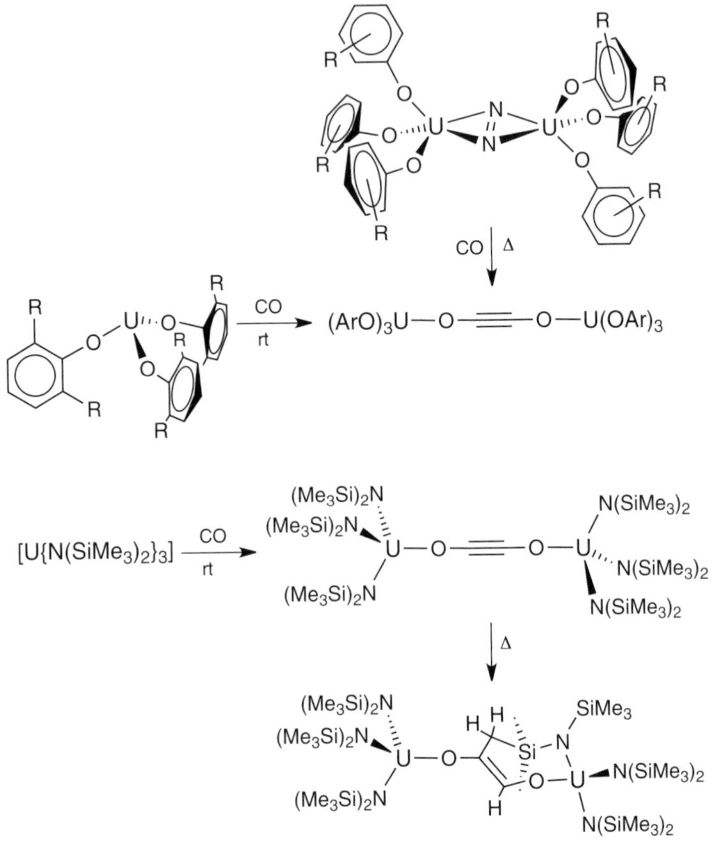

Scheme 7.   Reductive coupling of CO by [(N(TMS)$_2$)$_3$U], [(OAr)$_3$U], and [{(OAr)$_3$U}$_2$(μ-η^2:η^2N$_2$)] and the internal functionalization to form a chelated enediolate.

was attempted, however, neither functionalization of the C≡C triple bond nor cleavage of the U−O bonds was observed. Warming a solution of [{(N(TMS)$_2$)$_3$U}$_2$(μ-η^1:η^1-C$_2$O$_2$)] to 75 °C does result in the intramolecular C−H addition of a silyl methyl C−H bond across the C−C triple bond to give a uranium coordinated enediolate (Scheme 7).

Arnold and co-workers (131) also demonstrated that simple U(III) tris(aryl-oxide) complexes can homologate CO. Both ODtbp (O-2,6-tBuC$_6$H$_3$ anion) and OTtbp (O-2,4,6-tBuC$_6$H$_2$ anion) supported uranium complexes lead to the homologation of CO and the formation of an ynediolate under 1 atm of CO in noncoordinating solvents (Scheme 7). Intriguingly, the N$_2$ complex [(U(OTtbp)$_3$)$_2$(μ-η^2:η^2-N$_2$)] also results in the homologation of CO if briefly heated to 80 °C in an atmosphere of CO. However, other simple U(III) complexes fail to couple CO under analogous conditions. It seems that the η^6–arene–U interaction in [(ODipp)$_3$U]$_2$ (ODipp = O-2,6-iPr-C$_6$H$_3$ anion) and η^1–arene–U interaction in [(N(SiPhMe$_2$)$_2$)$_3$U] prevent CO coupling (131); variable pressure *in situ* IR vibrational studies are necessary to determine whether or not they are capable of transient CO binding. Based on these studies, the CO homologation by uranium(III) complexes is sensitive to steric environment, which must both enforce an unsaturated coordination sphere and also be sufficiently soft to allow intermolecular, bimetallic coupling. The variable coordination environment of the tris(aryloxide) complexes may be the basis of their more kinetically favorable coupling of CO than the tris(amide) [(N(TMS)$_2$)$_3$U], however, differences in their U(III/IV) redox couple cannot be ruled out.

b. Complex [(TREN^{DMSB})U]. Liddle and co-workers (132), building on the results by Cloke and Arnold, employed the silyl-capped TREN ligand developed by Scott and co-workers (133, 134) to effect a closed synthetic cycle for the selective homo-coupling of CO to silyl protected ynediolates. Specifically, treating the uranium(III) complex [(TREN^{DMSB})U] (TREN^{DMSB} = N(CH$_2$CH$_2$NSiMe$_2$$t$Bu)$_3$ trianion) with an excess of CO at room temperature (rt) and 1 atm results in the formation of the ynediolate complex, [{(TREN^{DMSB})U}$_2$(μ-η^1:η^1-C$_2$O$_2$)], in 62% isolated yield (Scheme 8) (134). Nuclear magnetic resonance studies indicate that the conversion is quantitative and that the yield is limited due to solubility. Warming a solution of [{(TREN^{DMSB})U}$_2$(μ-η^1:η^1-C$_2$O$_2$)] in benzene results in the insertion of the ynediolate unit into a N−Si bond of the ligand and in the

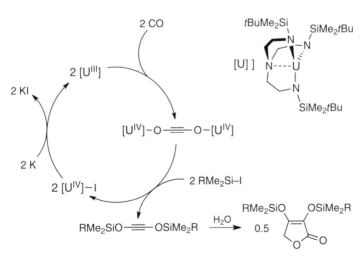

Scheme 8.   Synthetic cycle for reductive coupling of CO by [(TREN^{DMSB})U].

abstraction of a proton from solvent to give a dimeric chelated enediolate (Scheme 8). The ynediolate can be selectively removed from $[\{(TREN^{DMSB})U\}_2(\mu\text{-}\eta^1\text{:}\eta^1\text{-}C_2O_2)]$ by treating the dimer with Me_2PhSiI or Me_3SiI to give 2 equiv of the U(IV) iodide, $[(TREN^{DMSB})UI]$, and free bis(organosiloxy)acetylene, which undergoes subsequent chemistry dependent on the nature of the silyl group. The U(IV) iodide complex, $[(TREN^{DMSB})UI]$, can be reduced by excess potassium to furnish the trivalent starting complex $[(TREN^{DMSB})U]$ (Scheme 8). Thus, this system affords access to a closed synthetic cycle for the synthesis of complex organic species from CO.

B. Uranium(IV) Complexes

In contrast to the limited reactivity of uranium(III) complexes with CO, the chemistry of uranium(IV) with CO is quite rich. While there are no known monomeric uranium(IV) carbonyl complexes, uranium(IV) complexes have a diverse CO insertion chemistry. Studies in this area began in the late 1970's and built on comparative chemistry to group IV (4) metallocene hydride and alkyl CO insertion chemistry (135). Continued recent interest in these reactions stems from the utility of using CO as a readily available C_1 carbon source for industrial synthesis.

1. Insertion into U–C Bonds

In 1978, Marks and co-workers (136) presented the first example of CO insertion into a uranium(IV) U–C bond. Treatment of $[(Cp^*)_2U(Me)_2]$ at $-80\,°C$ in toluene with 1-atm CO leads to the consumption of 2 equiv of CO (Scheme 9). Warming to room temperature leads to the isolation of a dinuclear product, namely, $[((Cp^*)_2U)_2(OCMe=CMeO)_2]$, bridged by 2 equiv of the enediolate. The connectivity was indicated by IR, which revealed vibrational bands at 1655 $(v_{c=c})$, 1252 and 1220 (v_{CO}) cm^{-1}, and by cryoscopy molecular measurements, which further suggested the formation of a dimer. An X-ray diffraction study of the thorium analogue confirmed this assignment. If bulkier alkyl substituents are employed, a monomeric uranium enediolate is formed. Exposing $[(Cp^*)_2U(CH_2TMS)_2]$ at $-80\,°C$ to CO gives $[(Cp^*)_2U(OC(CH_2TMS)=C(CH_2TMS)O)]$ (Scheme 9). However, treating the monoalkyl $[(Cp^*)_2UCl(CH_2TMS)]$ with CO results in the isolation of an O-bound enolate, $[(Cp^*)_2UCl(O(TMS)C=CH_2)]$. Marks and co-workers (136) hypothesize that the mechanism for the formation of this enolate complex proceeds via silyl migration at a nucleophilic O-bound acyl–carbene (Scheme 9). This mechanistic proposal is substantiated by subsequent studies (Section IV.B.1).

Fischer and co-workers (137) later studied the insertion of CO into uranium(IV) alkyls supported by a tris(Cp) ligand systems, $[(Cp)_3UR]$ (R = Me, Et, iPr, n-C_4H_9, $t$$C_4H_9$). These complexes insert CO into the σ-bonded alkyl U–C bond. All

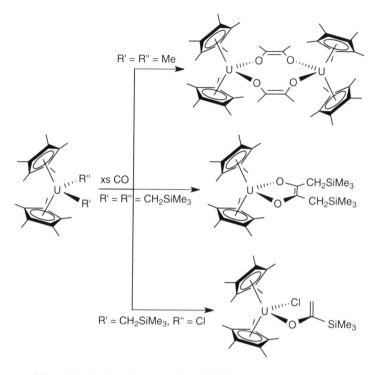

Scheme 9.   Insertion chemistry of $[(Cp^*)_2U(R)_2]$ and $[(Cp^*)_2URCl]$ CO.

compounds were isolated on a preparative scale; however, crystallographic characterization remained elusive. The formation of an η^2-acyl was determined on the basis of IR vibrational characterization. Cryoscopic molecular weight determination of the insertion products of the smaller alkyls (R = Me and Et) indicate that a dinuclear complex is formed, likely a bridging enediolate (Scheme 10). However, this product could not be confirmed spectroscopically. Ephritikhine later examined these systems and found that heating the η^2-acyl complexes $[(Cp)_3U(\eta^2\text{-OC}(R))]$ results in the precipitation of an insoluble material and the formation of alkyl benzene (Scheme 10) (138). Presumably, the alkyl benzenes are formed via carbene insertion into the Cp ligand with C–O bond cleavage. Such a mechanism was supported by ^{13}C NMR studies of the products resulting from the incorporation of ^{13}CO. Andersen and co-workers (139) also revisited these systems with the mono-methyl Cp derivatives (C_5MeH_4 = methylcyclopentadienyl anion). Treating $[(C_5MeH_4)_3U(t\text{-}C_4H_9)]$ with CO gives the η^2-acyl $[(C_5MeH_4)_3U(\eta^2\text{-OC}(t\text{-}C_4H_9))]$ (Scheme 10). The formation of the η^2-acyl was confirmed by the observation of an IR stretch at $1490\,\mathrm{cm^{-1}}$. In an attempt

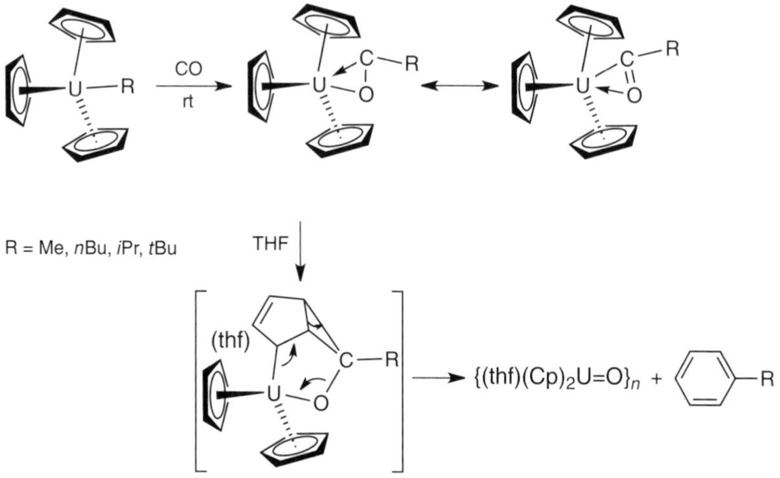

Scheme 10.   Reactivity of [(Cp)$_3$UR] with CO.

to observe the formation of an η^1 species, [(C$_5$MeH$_4$)$_3$U(η^1-OC(t-C$_4$H$_9$))] was found to decompose with the formation of *meta*- and *para-tert*-butyl toluene.

Simpson and Andersen (140) also examined the reactivity of CO with uranium(IV) carbon bonds in an amide coordination environment. The cyclo-metalated complex, [(N(TMS)$_2$)$_2$(κ^2-(N,C)-CH$_2$SiMe$_2$N(SiMe$_3$))U], which can be prepared by a variety of methods (141), reacts with CO (18 atm) selectively at the U–C σ bond (Scheme 11). Similar to the formation of the O-bound enolate in [(Cp*)$_2$UCl(O(TMS)C=CH$_2$)], a silyl group migration occurs to form a chelated O-bound enolate ligand with an exocyclic methylene moiety to the complex, [(N (TMS)$_2$)$_2$(κ^2-(N,O)-OC(=CH$_2$)SiMe$_2$N(SiMe$_3$))U] (Scheme 11). Although this product was not crystallographically characterized, the formation of the chelated O-bound enolate was established by comparison with the thorium analog. In recent years, Ephritikhine and co-workers (142, 143) expanded this chemistry by examining the CO insertion chemistry of the bis(C-metalated) U(IV) complex,

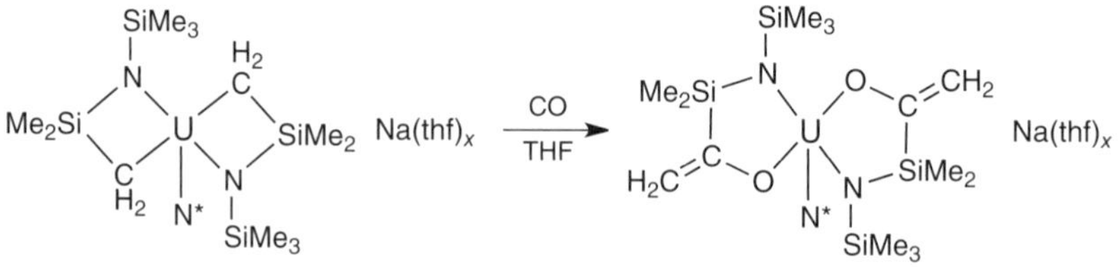

Scheme 11.   Carbon monoxide insertion chemistry of U(IV) with the ((κ^2-(N,C)-CH$_2$SiMe$_2$N (SiMe$_3$))$^{2-}$ ligand.

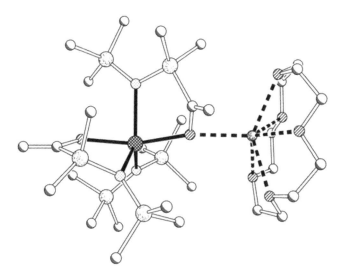

Figure 8.   Crystal structure of [Na(15-C-5)][U(N(TMS)$_2$)(κ^2-(N,O)-OC(=CH$_2$)SiMe$_2$N(SiMe$_3$))$_2$].

[(N(TMS)$_2$)(κ^2-(N,C)$-$CH$_2$SiMe$_2$N(SiMe$_3$))$_2$U]$^{1-}$, with CO. This bis(C-metalated) U(IV) complex, [(N(TMS)$_2$)(κ^2-(N,C)$-$CH$_2$SiMe$_2$N(SiMe$_3$))$_2$U]$^{1-}$, reacts analogously to the neutral, mono-cyclometaled complex and yields a product of two CO insertion reactions, [(N(TMS)$_2$)(κ^2-(N,O)-OC(=CH$_2$)SiMe$_2$N(SiMe$_3$))$_2$U]$^{1-}$ (Scheme 11). The solid-state structure is dependent on the nature of the supporting cation and crown ether. The structure of [Na(15-C-5)][(N(TMS)$_2$)(κ^2-(N,O)-OC (=CH$_2$)SiMe$_2$N(SiMe$_3$))$_2$U] (15$-$C$-$5 = 1,4,7,10,13-pentaoxacyclopentadecane) is depicted in Fig. 8 and clearly confirms earlier spectroscopic work supporting the formation of an exocyclic methylene. This ligand system in turn supports a variety of redox chemistry (142). The insertion of CO into the non-chelated U$-$C σ bond of [(N(TMS)$_2$)$_3$(Me)U] also has been described (144).

Arnold et al. (145) attempted to extend the uranium(III) CO homologation chemistry to a uranium(III) complex supported by a bidentate alkoxy-tethered NHC (N-heterocyclic carbene) ligand. However, the reaction of [L(N(TMS)$_2$)$_2$U] (L = OCMe$_2$CH$_2$(1-C{NCH$_2$CH$_2$NR}); R = 2,6-*i*Pr$_2$$-C_6H_3$) with CO at 80 °C results in CO insertion chemistry identical to that observed by Andersen and Ephritikhine (140–144). Presumably [L(N(TMS)$_2$)$_2$U] undergoes net bimetallic "oxidative elimination" of H$_2$ with formation of a [(κ^2-(N,C)-CH$_2$SiMe$_2$N (SiMe$_3$))]$^{2-}$ ligand, which, in turn, undergoes CO insertion and silyl migration to give the exocyclic methylene (Scheme 12).

Evans et al. (146) developed a variety of CO insertion chemistry based on the bis(Cp) "tuck-in" complexes. The silyl, alkyl "tuck-in" ligands in the complex

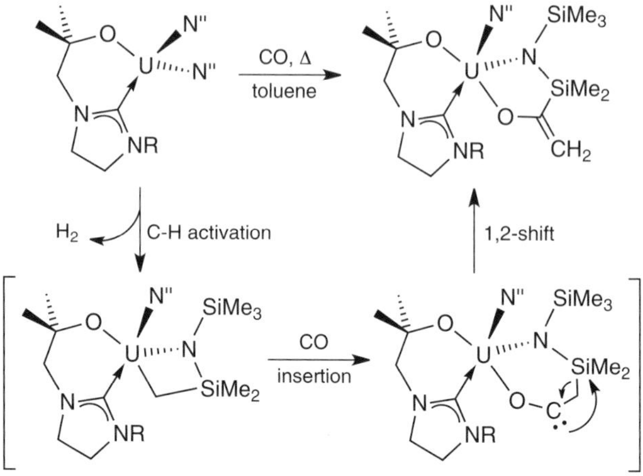

Scheme 12.   CO insertion chemistry of $[L(N(TMS)_2)_2U]$ $(L = OCMe_2CH_2(1-C\{NCH_2CH_2NR\})$; $R = 2,6\text{-}iPr_2\text{-}C_6H_3)$, $(N'' = N(TMS)_2)$.

$[(\eta^5:\eta^1\text{-}C_5Me_4SiMe_2CH_2)_2U]$ are prepared by methane elimination from $[(\eta^5\text{-}C_5Me_4SiMe_3)_2U(Me)_2]$. Treating a red solution of $[(\eta^5:\eta^1\text{-}C_5Me_4SiMe_2CH_2)_2U]$ with CO in benzene gives the product of double insertion, $[(\eta^5:\eta^1\text{-}C_5Me_4SiMe_2C-(=CH_2)O)_2U]$ (Scheme 13 and Fig. 9). As it has been observed in other silyl supported alkyl complexes, the observed product is an O-bound enolate. In this case, no η^2-acyl intermediate is observed. However, if the steric pressure at the uranium center is increased, an unusual product is observed. The complex $[(\eta^5:\eta^1\text{-}C_5Me_4SiMe_2CH_2)_2U]$ undergoes a single insertion reaction with 1,3-diisopropyl-carbodiimide to give $[(\eta^5:\eta^1\text{-}C_5Me_4SiMe_2CH_2)U(\eta^5\text{-}C_5Me_4SiMe_2CH_2)C(iPrN)_2-\kappa^2N,N']$ (Scheme 13) (147). The subsequent reaction with CO induces drastic changes in the coordination environment, including the breaking of a $U-CH_2$ bond and a $C-N$ linkage in the amidinate ligand. One η^5-cyclopentadienyl ligand has been removed per metal, and one substituent on that ring, either a methyl or a silyl, has migrated. Newly formed bonds include $U-O$, two $C-C$, and $C-N$.

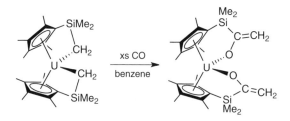

Scheme 13.   Carbon monoxide insertion chemistry of $[(\eta^5:\eta^1\text{-}C_5Me_4SiMe_2CH_2)_2U]$.

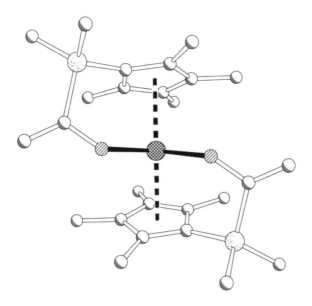

Figure 9.   Crystal structure of $[(\eta^5:\eta^1\text{-}C_5Me_4SiMe_2C\text{-}(=CH_2)O)_2U]$.

Despite the complexity of the transformation, the sole product observed is $[\{\mu\text{-}[\eta^5\text{-}$ $C_5Me_4SiMe_2CH_2C(=NiPr)O\text{-}\kappa^2O,N]U[OC(C_5Me_4Si\text{-}Me_2CH_2)CN(iPr)\text{-}\kappa^2O,N]\}_2]$ (Scheme 13 and Fig. 10). This transformation suggests that there remains a wealth of chemistry to be found in CO insertion reactions. Less sterically pressured single insertion derivatives of $[(\eta^5:\eta^1\text{-}C_5Me_4SiMe_2CH_2)_2U]$ give the expected O-bound enolate products (148). Note, however, that many authors have pursued the analogous insertion of isocyanides, which differ primarily in the increased stability of the η^2-iminoacyl and in the propensity of isocyanides to form adducts (rather than undergo insertion reactions) due to their stronger σ donating characteristics (140, 144, 149, 150).

2.   Insertion into U=C Bonds

In 1981, Gilje and co-workers (151) reported the synthesis of $[(Cp)_3U$ $(=CHPMe_2Ph)]$ and $[(Cp)_3U(=CHPMePh_2)]$, in which the U—C bond has multiple–bond character to some extent. These green complexes readily insert CO into the U=C bond to give η^2-acyl complexes of the type $[(Cp)_3U(\eta^2\text{-}$ $OCCHPMePh_2)]$. The addition of CO to a green toluene solution of $[(Cp)_3U$ $(=CHPMePh_2)]$ leads to a rapid (<30 min) change in color to deep red. The crystal structure of the product complex $[(Cp)_3U(\eta^2\text{-}OC_{(1)}C_{(2)}HPMePh_2)]$ reveals an η^2-acyl coordination mode (Fig. 11). Equation 1 shows the resonance structures of

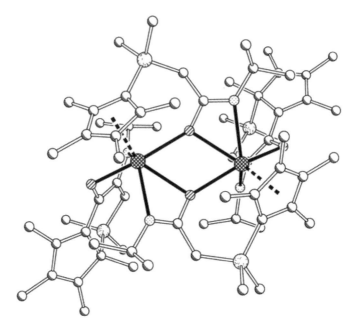

Figure 10.   Crystal structure of {μ-[η^5-C$_5$Me$_4$SiMe$_2$CH$_2$C(=NiPr)O−κ^{2}O,N]U[OC(C$_5$Me$_4$Si−Me$_2$ CH$_2$)CN(iPr)-κ^{2}O,N]}$_2$.

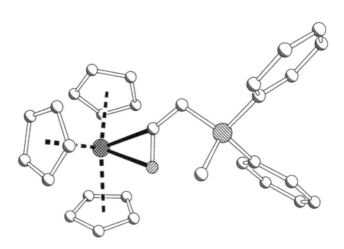

Figure 11.   Crystal structure of [(Cp)$_3$U(η^2-OCCHPMePh$_2$)].

$[(Cp)_3U(\eta^2\text{-OCCHPMePh}_2)]$. The Cl–C2 bond length is 1.37(3) Å, which suggests olefinic character, but the C–O bond length is 1.27(3) Å, not quite as long as expected for a single C–O bond, which suggests that two limiting resonance structures contribute to the ground state (Eq. 1). Insertion into the U=C bond in $[(Cp)_3U(=CHPMe_2Ph)]$ is a general reaction. A variety of transition metal carbonyl complexes, including $[CpCo(CO)_2]$, $[CpMn(CO)_2]$, $[CpFe(CO)_2]_2$, and $[W(CO)_6]$, will insert the complexed CO ligand into the U=C bond to give bridging heterobimetallic complexes with a variety of coordination modes (152–157).

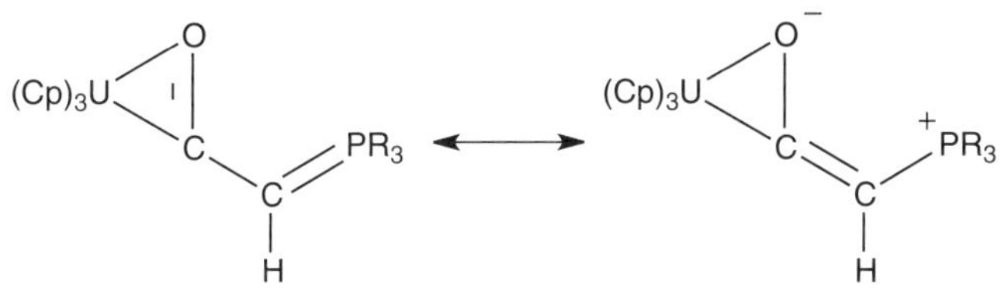

3. Insertion into U–N and U–H Bonds

Carbon monoxide will also insert into uranium(IV) U–N bonds. Marks and co-workers (158) found that CO inserts, under forcing conditions (95–100 °C and < 1 atm of CO), into the U–N bond of $[(Cp^*)_2UCl(NEt_2)]$ in toluene to give the η^2-carbamoyl product, $[(Cp^*)_2UCl(\eta^2\text{-OC(NEt}_2))]$ (Scheme 14). The η^2-carbamoyl complex has a CO stretch at 1559 cm^{-1} and shows a temperature sensitive ^{1}H NMR spectrum. A solid-state structure of the thorium analog reveals that the

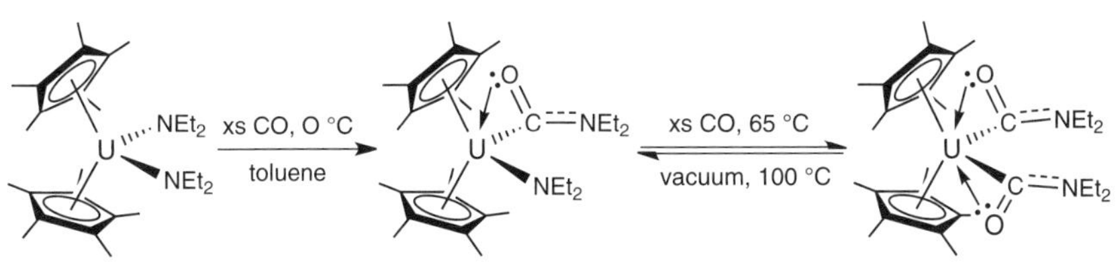

Scheme 14.   Carbon monoxide insertion chemistry of $[(Cp^*)_2UCl(NEt_2)]$ and $[(Cp^*)_2U(NEt_2)_2]$.

η^2-carbamoyl is disordered over two orientations within the metallocene wedge. A VT NMR study demonstrated that for [(Cp*)$_2$UCl(η^2-OC(NEt$_2$))] the η^2-carbamoyl moiety has an inversion barrier of $\Delta G^* = 8.9 \pm 0.5$ kcal mol^{-1} at $-70\,°C$. The bis(dialkylamide) complex, [(Cp*)$_2$U(NEt$_2$)$_2$], also undergoes CO insertion into the U$-$N bonds. The first insertion to form the mono-carbamoyl is quite rapid in comparison to the mono-chloride complex. Exposure of [(Cp*)$_2$U(NEt$_2$)$_2$] to excess CO at 1 atm for 2 h at $0\,°C$ gives selectively [(Cp*)$_2$U(NEt$_2$)(η^2-OC (NEt$_2$))]. The CO stretching frequency in this complex is 1491 cm^{-1}. The insertion of a second equivalent of CO requires more forcing conditions ($65\,°C$ and 1-atm CO) to yield [(Cp*)$_2$U(η^2-OC(NEt$_2$))$_2$]; with a CO stretching frequency in this complex of 1499 cm^{-1}. The molecular connectivity was confirmed by an X-ray crystallographic study, which revealed a single isomer in the solid state. Remarkably, the insertion of the second equivalent of CO is reversible, and exposing the complex as a solid to vacuum at $100\,°C$ gives selectively the mono-carbamoyl complex [(Cp*)$_2$U(NEt$_2$)(η^2-OC(NEt$_2$))]. Note that Zanella and co-workers (137) observed similar results for their tris(Cp) system described above. Carbon monoxide inserts into both U$-$N and U$-$P bonds in complexes of the type [(Cp)$_3$UR] (R $=$ NEt$_2$ or P(C$_6$H$_5$)$_2$). Berthet and co-workers (159, 160) also reexamined the bis(metallocene) uranium(IV) amide as a cation. In this example, CO inserts into the U$-$N bond of [(Cp*)$_2$U(NMe$_2$)(thf)]$^+$ to give [(Cp*)$_2$U(η^2-OC (NMe$_2$))(thf)]$^+$. In contrast to [(Cp*)$_2$UCl(NEt$_2$)], this system proceeds rapidly at room temperature even in a coordinating solvent (e.g., THF). The insertion of CO into U$-$H bonds has been mentioned in the case of the U(IV) complex [{(Cp*)$_2$UH$_2$}$_2$] and the U(III) complex [(Cp*)$_2$UH(DMPE)] [DMPE $= 1,2$-bis (dimethylphosphino)ethane], but it has not been described (161–163). However, the related work with thorium(IV) dihydrides and monohydrides has been reported in detail (164, 165).

C. Uranium(V) Complexes

1. Complex [((^{Ad}ArO)$_3$tacn)U(=NTMS)]

There is only one example of a high-valent, uranium(V) complex reacting with CO. Treating trivalent [((^{Ad}ArO)$_3$tacn)U] ((^{Ad}ArOH)$_3$tacn $= 1,4,7$-tris-(3-adamantyl-5-*tert*-butyl-2-hydroxybenzylate)-1,4,7-triazacyclononane) with TMSN$_3$ affords [((^{Ad}ArO)$_3$tacn)U(=NTMS)] (166). The U$-$N bond in this imido complex [2.1219(18) Å] is longer and significantly more bent than in the analogous imido complex supported by the *tert*-butyl ligand, [((^{tBu}ArO)$_3$tacn)U(=NTMS)] – [1.989(5) Å]. Accordingly, the adamantyl complex, [((^{Ad}ArO)$_3$tacn)U(=NTMS)], but not the *tert*-butyl complex [((^{tBu}ArO)$_3$tacn)U(=NTMS)], reacts with π acids, including CO. Treating a solution of [((^{Ad}ArO)$_3$tacn)U(=NTMS)] with 1 atm of CO results in the formation of the uranium(IV) isocyanate complex,

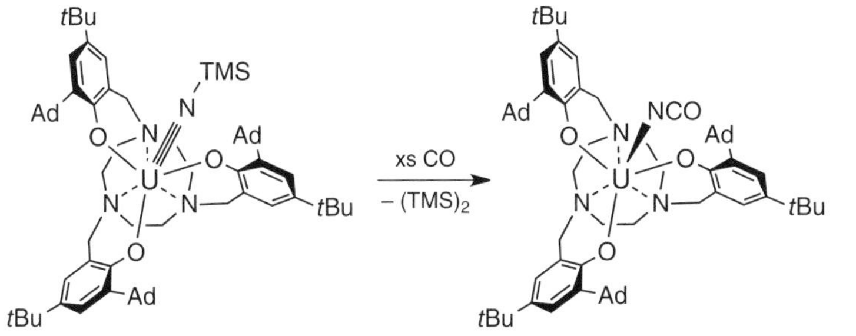

Scheme 15.   Reaction of $[((^{Ad}ArO)_3tacn)U(=NTMS)]$ with CO.

$[((^{Ad}ArO)_3tacn)U(\eta^1-N^{12/13}CO)]$, and hexamethyldisilane $(Me_3Si-SiMe_3)$ (Scheme 15). The formation of the isocyanate from CO was confirmed by an X-ray diffraction study and ^{13}CO isotopic labeling. An intense absorption at 2185 shifts to 2122 cm^{-1} in the labeled reaction. This reaction is similar to the nitrene-transfer reaction of $[(Cp'')_2Ti(=NTMS)]$ (Cp$'' = $ 1,3-TMS-C_5H_3 anion) with excess CO to give $[(Cp'')_2Ti(CO)_2]$ and TMSNCO, except in this case the metal center undergoes only a one-electron reduction along with the concomitant formation of the silyl radical Me$_3$Si$^{\bullet}$ (167). There may be an intermediate nitride in this reaction; whether N$-$Si bond cleavage precedes or follows N$-$C bond formation is unknown. In the case of the reaction the π acid MeNC, a closed synthetic cycle for the production of MeN=C=NCH$_2$Cl can be established.

IV.   NITROGEN MONOXIDE

A.   Uranium(III) Complexes

1.   Complex $[(ODtbp)_3U]$

Nitrogen monoxide has been used as an oxygen-atom transfer reagent. Burns and co-workers (168) found that the careful addition of 0.5 equiv of NO to a solution of $[(ODtbp)_3U]$ affords $[\{U(ODtbp)_3\}_2(\mu\text{-}O)]$. If excess NO is used, the reaction is not clean. Also, the fate of the nitrogen atom of NO is not reported. This oxygen-atom transfer chemistry is similar to that described for CO$_2$ and N$_2$O (Sections VII.A.1, VII.A.3 and VIII).

2.   Complex $[(\eta^8\text{-}C_8H_6(1,4\text{-}Si(iPr)_3)_2)(\eta^5\text{-}Cp^*)U]$

Cloke and co-workers (169) studied the reactivity of mixed COT–Cp uranium (III) complexes with 1 equiv of NO. These reactions produced a large number of

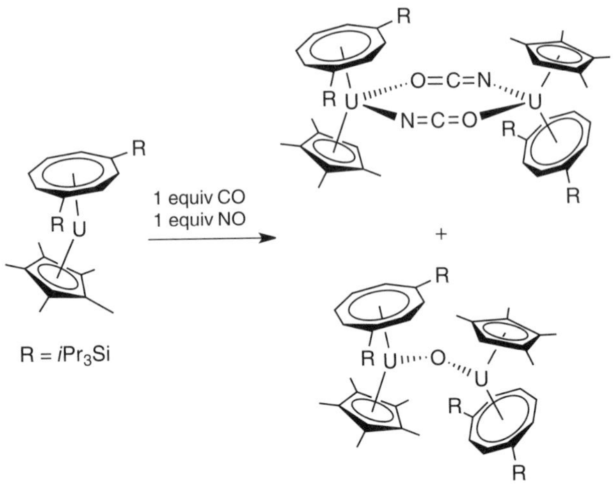

Scheme 16.   Reaction of $[(\eta^8\text{-}C_8H_6(1,4\text{-}Si(iPr)_3)_2)(\eta^5\text{-}Cp^*)U]$ with CO and NO.

uncharacterized products. Based on the working hypothesis that these reactions may be producing an unstable hyponitrite analogue of the ynediolate complexes (Scheme 16), the reactivity of these systems in the presence of 0.5 equiv of ^{13}CO was examined in an effort to stabilize a $N_2CO_3{}^{2-}$ analog of the deltate dianion $C_3O_3{}^{2-}$. The reaction of $[(\eta^8\text{-}C_8H_6(1,4\text{-}Si(iPr)_3)_2)(\eta^5\text{-}Cp^*)U]$ with 1 equiv of NO and 0.5 equiv of ^{13}CO at $-80\,^\circ C$ afforded a reaction mixture with two primary products as observed by ^1H NMR spectroscopy and a single new enriched carbon resonance at δ 249 ppm in the ^{13}C NMR spectrum. The products were separated by fractional crystallization and identified by standard spectroscopic techniques as the bridging μ-oxo complex $[\{(\eta^8\text{-}C_8H_6(1,4\text{-}Si(iPr)_3)_2)(\eta^5\text{-}Cp^*)U\}_2(\mu\text{-}O)]$ and the bridging bis(cyanate) complex, $[\{(\eta^8\text{-}C_8H_6(1,4\text{-}Si(iPr)_3)_2)(\eta^5\text{-}Cp^*)U\}_2(\mu\text{:}\eta^1,\eta^1\text{-}NCO)_2]$. The bridging fulminate CNO^{1-} could not be ruled out entirely due to disorder in the bridging light atoms in the X-ray structure of $[\{U(\eta^8\text{-}C_8H_6(1,4\text{-}Si(iPr)_3)_2)(\eta^5\text{-}Cp^*)\}_2(\mu\text{:}\eta^1,\eta^1\text{-}NCO)_2]$, but that seems unlikely as it would require the cleavage of the stronger $C-O$ bond rather than the weaker $N-O$ bond. In addition to being the first well-defined example of uranium NO reactivity, one mechanistic hypothesis for the cleavage of the $N-O$ bond may imply the formation of a uranium nitride (a complex that, until recently, had remained elusive (170), Section IV.A.3).

3.   Complex $[(C_5Me_4H)_3U]$

Despite the prediction of stable uranium nitrosyls by Bursten et al. in 1989 (105), it was not until 2012 that Evans and co-workers (171) reported the

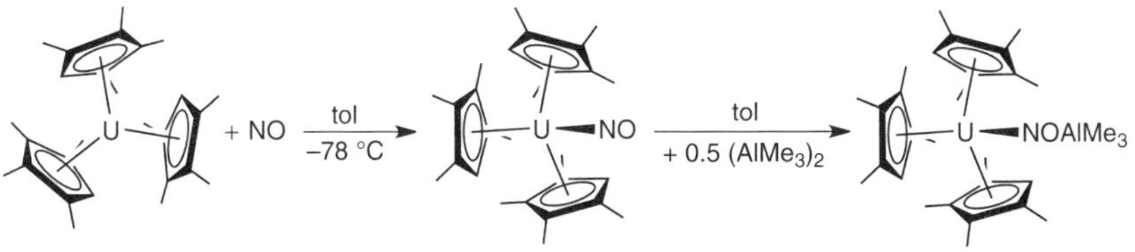

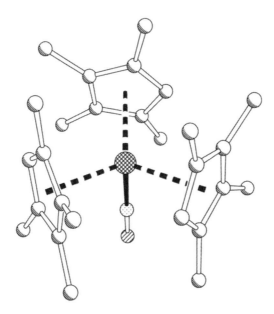

Scheme 17.   Reaction of [(C$_5$Me$_4$H)$_3$U] with NO and subsequent reactivity with Me$_6$Al$_2$.

first uranium nitrosyl complex. The reaction of 1 equiv of NO gas with [(Cp′)$_3$U] complexes at −78 °C was pursued. While the reaction of NO with [(Cp*)$_3$U] and its derivative [(C$_5$Me$_4$SiMe$_3$)$_3$U] resulted in intractable mixtures, the reaction with [(C$_5$Me$_4$H)$_3$U] in toluene led to the isolation of [(C$_5$Me$_4$H)$_3$U(NO)] in 62% after crystallization from toluene (tol) (Scheme 17). Similar approaches to the NO chemistry of [(TMSC$_5$H$_4$)U(thf)] afforded no identifiable products (172). Crystals of [(C$_5$Me$_4$H)$_3$U(NO)] suitable for an X-ray diffraction study were obtained from toluene. The molecular structure reveals a distorted tetrahedral geometry (Fig. 12).

Figure 12.   Crystal structure of (C$_5$Me$_4$H)$_3$U(NO).

The U$-$N$-$O angle is linear at 180.0(4)°, which is indicative of a NO$^+$ formal oxidation state. However, all other metric parameters, spectroscopy, and magnetism imply an NO$^-$ and U(IV) formulation, which suggest the linear U$-$N$-$O angle is due to the steric constraints of the system. The N$-$O bond length is 1.231(5) Å, which is close to that observed in NO^{1-} complexes of 1.26 Å. Similarly, the U$-$N bond length is 2.013(4) Å, which is expected for a formal U$-$N double bond, as in U(IV) imides, 1.952(12) to 2.097(5) Å. Isotopic labeling with ^{15}NO allowed the identification of the N$-$O IR stretch at 1439 and 1373 cm^{-1} (1416 and 1366 cm^{-1} with the ^{15}N isotopomer), which is also consistent with a NO^{1-} formulation.

The electronic structure of [(C$_5$Me$_4$H)$_3$U(NO)] is unique, but similar to that predicted by Bursten et al. (105) for [Cp$_3$U(NO)] (Section IV.A.3). The DFT studies suggest that the NO is reduced to NO^{1-} and the singlet and triplet states should be nearly isoenergetic. However, superconducting quantum interference device (SQUID) magnetization studies clearly indicate that the singlet state is at least 700 cm^{-1} lower in energy than the triplet. This insulation of the singlet state from the excited triplet state is hypothesized to be due to the π back-bonding of the f^2 electrons with the NO π* orbitals. Consistent with the complex's proposed electronic structure, and the known reactivity of transition metal NO^{1-} complexes, [(C$_5$Me$_4$H)$_3$U(NO)] reacts with Me$_6$Al$_2$ to give [(C$_5$Me$_4$H)$_3$U(=N$-$OAlMe$_3$)] (Scheme 17). The identity of this complex was confirmed by NMR and IR spectroscopy, as well as elemental analysis (171).

B.   Uranium (IV and V) Complexes

1.   Complex [(N(TMS)$_2$)$_3$UCl]

While not a reaction with NO, Schelter and co-workers (173) recently demonstrated that uranium(IV and V) complexes can undergo one-electron oxidation by nitrite [ONO]$^{1-}$ to give NO and uranium(V) and (VI) complexes selectively (173). Specifically, treating [(N(TMS)$_2$)$_3$UCl] with Na[ONO] leads to the known uranium(V) oxo [(N(TMS)$_2$)$_3$UO] and NO. Alternatively, treating [(N(TMS)$_2$)$_3$UClX] (X = F, Cl, Br) with Na[ONO] gives the uranium(VI) oxo complexes [(N(TMS)$_2$)$_3$UOX], with the halide trans to the oxo ligand. The trans arrangement of the halide and oxo ligands may be a manifestation of the inverse trans-influence (19–21). Recently, Cantat and co-workers (174) also reported the oxidation of [UCl$_4$] with 6 equiv of [PPh$_4$][ONO] to give the nitrite uranyl complex [PPh$_4$][UO$_2$(κ^1-(O)-NO$_2$)$_2$(κ^2-(O,O)$-$NO$_2$)$_2$] with the concomitant production of NO.

V. DINITROGEN

A. Uranium(III) Complexes

1. Complex [{(TREN^{DMSB})U}$_2$(μ,η^2:η^2-N$_2$)]

Although there was indirect evidence for N_2 complexes of uranium(III) since 1982 (163), the first isolated uranium dinitrogen complex was not reported until 1998. In a series of two papers, Scott and co-workers (175, 176) reported the synthesis of [{(TREN^{DMSB})U}$_2$(μ,η^2:η^2-N$_2$)]. The reduction of [(TREN^{DMSB}) UCl] with a potassium mirror in pentane afforded the mixed-valence U(III/IV) chloride bridged dimer, [{(TREN^{DMSB})U}$_2$(μ-Cl)] (Scheme 18). Careful sublimation of this complex allowed the cleavage of the Lewis acid–base pair and the isolation of the uranium(III) complex [(TREN^{DMSB})U]. Dissolving [(TREN^{DMSB})U] in benzene under 1 atm of N_2 leads to a rapid color change from purple to red. Based on changes in its ^{1}H NMR spectrum, slightly higher pressure of N_2 leads to clean conversion to a new species. This new complex, identified as the bridging N_2 complex [{(TREN^{DMSB})U}$_2$(μ,η^2:η^2-N$_2$)], can be isolated by cooling the red

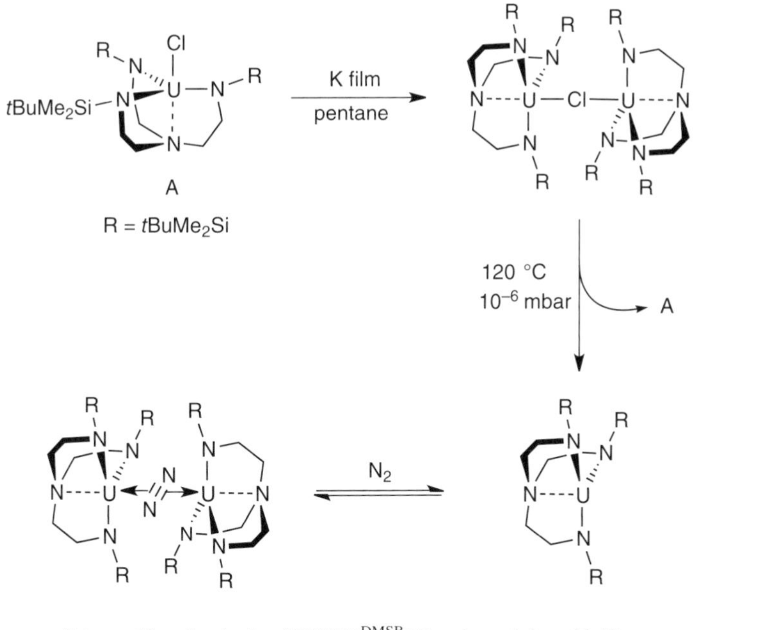

Scheme 18.   Synthesis of [(TREN^{DMSB})U] and reactivity with N_2.

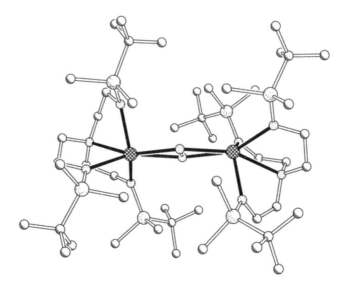

Figure 13.   Crystal structure of $[\{U(TREN^{DMSB})\}_2(\mu,\eta^2:\eta^2\text{-}N_2)]$.

pentane solution to $-20\,^{\circ}C$. The binding of N_2 is reversible and freeze–pump cycles result in the formation of the purple starting complex. The molecular structure, as determined by X-ray diffraction (Fig. 13), reveals a bridging side-on N_2 ligand with the TREN ligands inter-digitated. The uranium atoms are slightly out of the plane of the N atoms by 0.84 and 0.85 Å. Most saliently, the N—N distance in N_2 is 1.109(7) Å, essentially the same as that found in free N_2, 1.0975 Å. The IR spectra of $^{14}N_2$ and the isotopically enriched $^{15}N_2$ are super-impossible, as expected for this high-symmetry complex with the nonactivated N_2 ligand, and no further conclusion can be drawn concerning the bonding of the N_2 ligand.

The authors originally proposed that the bonding between the uranium centers and the N_2 ligand is essentially $N_2 \rightarrow U$ σ bonding. This hypothesis is based on the bond metrics and the complexes' observed room temperature magnetic moment of $3.22\ \mu_B$ per uranium center (Evans' method in solution), which is very similar to that observed for the free uranium(III) precursor, $3.06\ \mu_B$. Subsequent DFT studies suggested that the uranium centers participate in π back-bonding into the N_2 ligand via π symmetry 5f orbitals (177, 178). Such bonding interaction, however, is expected to weaken and thus lengthen the N_2 N—N bond. One possible explanation for the lack of N_2 activation (as determined by IR vibrational spectroscopy and X-ray crystallography) might be the high nodality of the 5f orbitals. Lengthening the N_2 N—N bond may lead to a decreased effective overlap between the 5f orbitals and the N_2 π^* orbitals. Alternatively, steric clash between the TREN ligands on both U centers may prevent bringing the U ions closer together and increasing the reduction of the N_2 ligand. Either way, the activation of N_2 in this system is limited.

2. *Complex [{(Cp*)(η^8-C$_8$H$_4$(1,4-Si(iPr)$_3$)$_2$)U}$_2$(μ,η^2:η^2-N$_2$)]*

Cloke and co-workers (6, 179) also synthesized a side-on bound bridging N_2 complex, employing their pentalene–Cp mixed sandwich complexes. The pentalene ligand, $[\eta^8$-C$_8$H$_4$(1,4-Si(iPr)$_3$)$_2]^{2-}$ is a ring-closed version of the COT ligand described previously and provides a lower symmetry and slightly smaller steric profile for uranium than in the [(Cp′)$_3$U] complexes. The "base-free" starting material is made in two steps from solvent-free UI$_3$. The black complex, [(Cp*)(η^8-C$_8$H$_4$(1,4-Si(iPr)$_3$)$_2$)U], reacts with N_2 in pentane at room temperature to give a green–black solution. The binding of N_2 in this system is also reversible, and the application of freeze–pump–thaw cycles regenerates the starting complex as evidenced by the ^{1}H NMR spectrum, which is indicative of the C_2 symmetric starting material. The conversion to the N_2 complex is not complete even under 50 psi of N_2. However, the N_2 adduct can be isolated and crystallographically characterized (Fig. 14) by fractional crystallization in pentane at $-20\,°$C under a slight overpressure of N_2 (5 psi). The N_2 complex is binuclear with a μ-η^2:η^2-N$_2$ ligand. Unlike in [{(TREN^{DMSB})U}$_2$(μ,η^2:η^2-N$_2$)], the N—N bond of N_2 in [{(Cp*)(η^8-C$_8$H$_4$(1,4-SiiPr$_3$)$_2$)U}$_2$(μ,η^2:η^2-N$_2$)] is significantly lengthened to

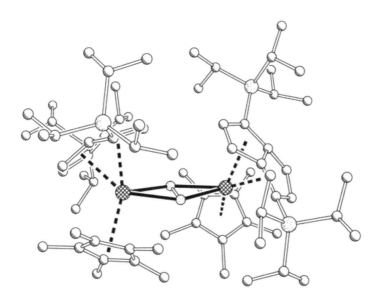

Figure 14.   Crystal structure of [{(Cp*)(η^8-C$_8$H$_4$(1,4-Si(iPr)$_3$)$_2$)U}$_2$(μ,η^2:η^2-N$_2$)].

1.232(10) Å, and, thus, indicates a significant activation resulting in a reduced formal bond order of 2. Unfortunately the N−N stretch could not be identified in the IR spectrum, but clearly the N_2 ligand in $[\{(Cp^*)(\eta^8$-$C_8H_4(1,4$-$Si(iPr)_3)_2)$ $U\}_2(\mu,\eta^2{:}\eta^2$-$N_2)]$ undergoes reduction to N_2^{2-}. This interpretation is further supported by a DFT study (179). Note that O'Hare and co-workers (180) also presented electrochemical evidence for the formation of uranium dinitrogen complex in the reduction of the related bis(pentalene) complex $[(Pn^*)_2U]$ $(Pn^* =$ permethylpentalene) under dinitrogen (180).

3. $[(Cp^*)_3U(\eta^1\text{-}N_2)]$

The sole example of a terminal, end-on monometallic dinitrogen complex of uranium was prepared using the $[(Cp^*)_3U]$ complex, which Evans has employed to make the isostructural CO complex (Section III.A.1) (181). In contrast to the binding of CO in $[(Cp^*)_3U(CO)]$, dinitrogen binding in $[(Cp^*)_3U(\eta^1$-$N_2)]$ is quite weak. There is no observable interaction between $[(Cp^*)_3U]$ and N_2 at room temperature and 1 atm. However, increasing the pressure to 80 psi leads to a darkening of the solution and the crystallization of $[(Cp^*)_3U(\eta^1$-$N_2)]$. Note that, in solution, reducing pressure to 1 atm results in loss of N_2. An X-ray structure shows that in the solid state $[(Cp^*)_3U(\eta^1$-$N_2)]$ is very similar to all other $[(Cp^*)_3UL]$ complexes. All centroid–U–centroid angles are 120° and the centroid–U–N(N_2) angle is 90°. The N−N distance of the N_2 ligand is 1.120(4) Å, the same as in free N_2. The IR stretch for N_2 was found at 2207 cm^{-1} (2132 cm^{-1} for $^{15}N_2$). This value is 124 cm^{-1} less than in free N_2 (2331 cm^{-1} by Raman), which, as a whole, is indicative of only weak N_2 activation.

4. Complex $[(Ar(tBu)N)_3U(\mu,\eta^1{:}\eta^1\text{-}N_2)Mo(N(tBu)Ph)_3]$

An end-on coordination mode also has been observed in a Mo−U hetero-bimetallic complex (182). Cummins and co-workers (183, 184) prepared the tris(anilide) complex of uranium(III), namely, $[(Ar(tBu)N)_3U(thf)]$ (Ar = 3,5-dimethyl-C_6H_3). This complex does not react with N_2. However, 1:1 mixtures of $[(Ar(tBu)N)_3U(thf)]$ and either $[(N(tBu)Ph)_3Mo]$ or $[(N(Ad)Ph)_3Mo]$ under 1-atm N_2 at room temperature afford μ-$\eta^1{:}\eta^1$-N_2 bridging complexes. In contrast to the thoroughly detailed chemistry of $[(N(tBu)Ph)_3Mo]$ with N_2, which leads to N_2 cleavge, the addition of 1 equiv of the uranium complex leads to the selective formation of the orange heterobimetallic complex, $[(Ar(tBu)N)_3U(\mu,\eta^1{:}\eta^1$-$N_2)Mo$ $(N(tBu)Ph)_3]$. This N_2 adduct is thermally stable and does not undergo N_2 cleavage to afford uranium and molybdenum nitrido complexes. An XRD structure of $[(Ar(tBu)N)_3U(\mu,\eta^1{:}\eta^1$-$N_2)Mo(N(tBu)Ph)_3]$ (Fig. 15) confirms the μ-$\eta^1{:}\eta^1$-N_2 connectivity and gives an N−N bond length of 1.232(11) Å, which suggests reduction to N_2^{2-} and a bond order of 2. The Mo−N(amide) and U−N(amide)

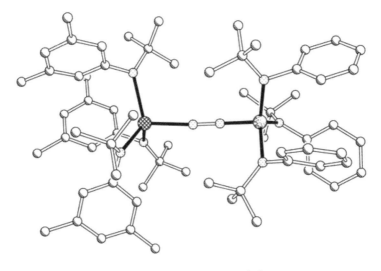

Figure 15.   Crystal structure of $[(Ar(tBu)N)_3U(\mu,\eta^1:\eta^1-N_2)Mo(N(tBu)Ph)_3]$.

bond lengths are consistent with a Mo(IV) and U(IV) oxidation assignment. Furthermore, in the case of $[(Ar(tBu)N)_3U(\mu,\eta^1:\eta^1-N_2)Mo(N(Ad)Ph)_3]$ the N_2 stretch could be identified in the IR spectrum via isotopic labeling at $1568\ cm^{-1}$.

5.   Complex $[\{(OAr)_3U\}_2(\mu-\eta^2:\eta^2-N_2)]$

The most recent example of N_2 coordination and activation comes from a reexamination of uranium tris(aryloxide) chemistry (131). Arnold and co-workers (131) prepared the $[(OAr)_3U]$ (OAr = ODtbp and OTtbp) complexes from $[(N(TMS)_2)_3U]$ in hexane under argon. In contrast to the long established chemistry of $[(ODipp)_3U]$, which forms a bimetallic η^6-bridging aryloxide complex in noncoordinating solvents (Section XIII.A), the $[(OAr)_3U]$ (OAr = ODtbp and OTtbp) complexes are monomers (Scheme 19) (168, 185). In the case that $[(ODtbp)_3U]$ is prepared under N_2, two complexes are isolated: The expected monomer, as was described in the original report (186), and a small amount of a red bimetallic bridging N_2 complex, $[\{(ODtbp)_3U\}_2(\mu-\eta^2:\eta^2-N_2)]$. The analogous complex $[(OTtbp)_3U]$ can be prepared under argon. Exposure of hexane solutions of $[(OTtbp)_3U]$ to N_2, in turn, affords the yellow complex $[\{(OTtbp)_3U\}_2$ $(\mu-\eta^2:\eta^2-N_2)]$ in 80% yield. The solid-state N−N bond lengths in $[\{(OAr)_3U\}_2$ $(\mu-\eta^2:\eta^2-N_2)]$ range from 1.163(19) to 1.236(5) Å (multiple independent molecules in asymmetric units). The range of observed bond lengths and their cocrystallized solvent dependent nature suggests that N_2 is bound in a soft potential well and that the N_2 ligand is reduced to a certain degree.

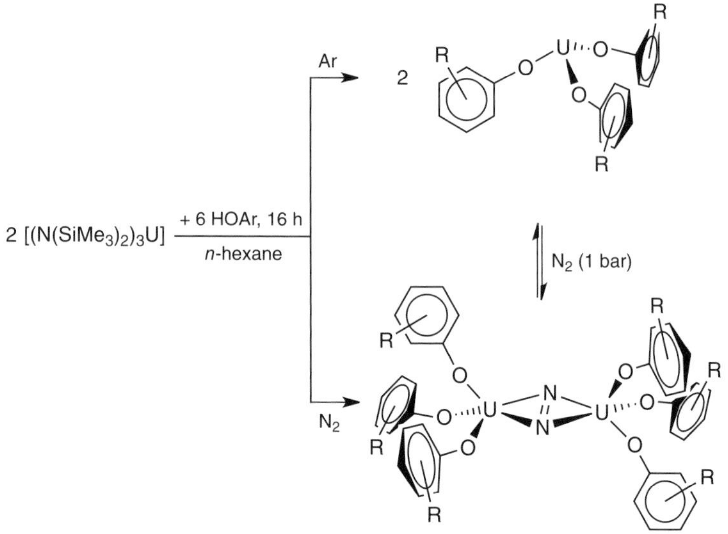

Scheme 19.   Synthesis of [{(OAr)$_3$U}$_2$(μ-η^2:η^2-N$_2$)].

6.   Complex [K(dme)$_4$][K(dme){(Et$_8$-calix[4]tetrapyrrole)U}$_2$(μ-NK)$_2$]

Notably, an example of N$_2$ cleavage by uranium(III) in the presence of an excess of an external strong reductant has also been presented. Gambarotta and co-workers (187) prepared the uranium(III) complex [Et$_8$-calix[4]tetrapyrrole) U-(dme)][K(dme)] from [UI$_3$(dme)$_2$] where dme = 1,2-dimethoxyethane (ligand). Exposure of the tetrapyrrole complex to N$_2$ with potassium napthalide results in the formation of the mixed-valent [U(IV/V)] bridging nitride complex [K(dme)$_4$] [K(dme){(Et$_8$-calix[4]tetrapyrrole)U}$_2$(μ-NK)$_2$]. In the absence of N$_2$ degradative reactions proceed, presumably with solvent and/or adventitious grease. The presence of U(V) was confirmed from the near-IR electronic absorption spectrum with an absorption at 1247 nm. Further evidence for bridging nitrides (as opposed to oxos) includes the observation of hyperfine coupling in the electron paramagnetic resonance (EPR) spectrum of the complex prepared under [15]N$_2$ (but not of the complex prepared under argon). Note that this complex can be isolated reproducibly only under N$_2$ (not Ar).

7.   Studying N$_2$ Cleavage Via Azide Reduction

One of the goals of these studies of N$_2$ activation is the development of synthetic methods for the reductive functionalization of N$_2$. In the absence of direct methods for the cleavage of N_2 by molecular uranium complexes, a number of groups have sought to study a possible intermediate in the reductive functionalization of dinitrogen (a uranium nitride). This moiety, as mentioned elsewhere, has until recently remained elusive. Based on analogous transition metal chemistry, the principal synthetic approach has been the reductive cleavage of azide (N_3^{1-}) (188–191). The azide anion, as well as alkyl and aryl azides, is isoelectronic with CO_2 [Section VII.A.1 for the reductive cleavage of CO_2 and readers are referred to two excellent reviews on metal–ligand multiple bonding in actinide complexes for details of imide synthesis from the reduction of alkyl and aryl azides (79, 80)]. Uranium azide complexes have been prepared by a number of oxidative transfer and salt metathesis methods. Stewart (192) found that trityl azide (trityl = triphenylmethane) will oxidize $[(N(TMS)_2)_3U]$ to give $[(N(TMS)_2)_3U$ $(N_3)]$ and Gomberg's dimer. Similarly, Meyer and co-workers (193) observed the formation of the uranium azide, $[((^{Ad}ArO)_3tacn)U(N_3)]$ from the respective uranium(III) precursor and $TMSN_3$ (193). Kiplinger and co-workers (194) developed an unusual, but very effective oxidative azide-transfer reagent for actinide chemistry, $[(PPh_3)Au(N_3)]$ (PPh_3 = triphenylphosphine). Treatment of a variety of uranium(III) metallocenes gives uranium(IV) azide complexes, gold metal, and free PPh_3. Sodium azide has been employed as an azide-transfer reagent by a number of groups. Ephritikhine used it in 1991 to synthesize the uranium(IV) azide $[(TMSC_5H_4)_3U(N_3)]$ and the mixed (III/IV) dinuclear complex $[\{(TMSC_5H_4)_3U\}_2(\mu\text{-}N_3)]$ (195). Similar methodology has been used by Evans et al. (196), Hayton and co-workers (197), Schelter and co-workers (198), and Ephritikhine and co-workers (142, 199) in the synthesis of uranium(IV/V) azide metallocene, aryloxide, and amide complexes. Most significantly Crawford et al. (200) prepared the uranium(IV) homoleptic heptaazide complex $[(nBu)_4N]_2[U(N_3)_7]$ from $[(nBu)_4N]_2[UCl_6]$, $[(nBu)_4N]Br$, and AgN_3 in acetonitrile. Mazzanti and co-workers (201) found that this remarkable complex could be prepared in situ with the less dangerous azide-transfer reagent CsN_3 (AgN_3 is a shock-sensitive explosive).

The reductive cleavage of azide by U(III) and photochemically by U(IV) has been observed. Evans et al. (202) found that treating the complex $[(Cp^*)_2U((\mu\text{-}\eta^2\text{:}\eta^1\text{-}Ph)_2BPh_2)]$ with NaN_3 affords and octanuclear ring complex, $[(Cp^*)_2U(\mu\text{-}N)U(\mu\text{-}N_3(Cp^*)_2]_4$, comprised of repeating bridging nitrides and azides. Evans et al. (196) also prepared a nitrido-centered cluster by treating $[Cp^*UI_2(thf)_3]$ with 3 equiv of NaN_3 to give $[Cp^*U(\mu\text{-}I)_2(\mu^3\text{-}N)]$. The identity of the central bridging nitrido ligand (as opposed to O^{2-} or OH^{1-}) was confirmed by mass spectrometry. Nocton and co-workers (201) prepared a similar bridging nitrido complex by the reduction of the in situ prepared $[U(N_3)_7]^{2-}$ by $[UI_3(thf)_4]$. The product of this reaction, $\{[(Cs(acn)_3][U_4(\mu^4\text{-}N)(\mu\text{-}N_3)_8\text{-}(acn)_8I_6]\}_\infty$, where acn = acetonitrile (ligand), is a one-dimensional (1D) polymer in the solid state and contains a nitrido ligand bridging four uranium centers. Hayton and co-workers (203) also were able to obtain a bridging nitride ligand by the reduction of NaN$_3$ by [(N(TMS)$_2$)$_3$U] followed by subsequent oxidation by Me$_3$NO (Me$_3$NO = trimethylamine N-oxide) to afford a linear NUO uranyl-like motif capped by U(IV) in the complex [Na(dme)$_2$][(N(TMS)$_2$)$_2$(O)U(μ-N)(CH$_2$SiMe$_2$NTMS)U(N(TMS)$_2$)$_2$]. Cummins and co-workers (204) also has found that the tris(anilide) complex, [(N(tBu)Ar)$_3$U(thf)] (Ar = 3,5-Me$_2$C$_6$H$_3$), reduces NaN$_3$ to give a bridging nitride, Na[{(N(tBu)Ar)$_3$}U]$_2$(μ-N)]. This complex undergoes sequential one-electron oxidation to give U(IV/V) and (V/V) complexes. Oxidizing [(N(tBu)Ar)$_3$U(thf)] with the borane-protected azide [NMe$_4$][(C$_6$F$_5$)$_3$B(N$_3$)] gives the imido–borane complex, [NMe$_4$][(N(tBu)Ar)$_3$U=NB(C$_6$F$_5$)$_3$], which might be considered a borane-capped nitride complex (205). However, neither this uranium(V) complex nor the uranium(VI) imido–borane complex [(N(tBu)Ar)$_3$U=NB(C$_6$F$_5$)$_3$] achieved by oxidation of [NMe$_4$][(N(tBu)Ar)$_3$U=NB(C$_6$F$_5$)$_3$] by I$_2$ can be deprotected to afford a free, terminal nitride. As discussed further in Section XI.C, Kiplinger and co-workers (206) tried a photochemical route to a terminal nitride by the photolysis of [(Cp*)$_2$UN$_3$(N(TMS)$_2$)], which results in the 1,1-C−H bond addition to a putative uranium(VI) nitride [(Cp*)$_2$UN(N(TMS)$_2$)] (to give [(Cp*)$_2$U(η^5:η^1-C$_5$Me$_4$CH$_2$NH)(N(TMS)$_2$)]. Only recently has a terminal, monomeric uranium nitride been prepared. Liddle and co-workers (170) found that the oxidation of the bulky TREN uranium complex, [(TREN^{TIPSi})U] (TREN^{TIPSi} = N(CH$_2$CH$_2$NSi(iPr)$_3$)$_3$), by NaN$_3$ followed by chelation of the coordinated Na ion with 2 equivalents of 12-C-4 results in the isolation a uranium(V) nitrido complex, [Na(12-C-4)$_2$][(TREN^{TIPSi})U≡N]. This complex has been shown to undergo oxidation with I$_2$ to give a uranium(VI) nitrido complex, [(TREN^{TIPSi})U≡N] (207).

VI. DIOXYGEN

A. Uranium(III) Complexes

1. Ligand Redistribution Reactions

The controlled reactivity of O$_2$ with uranium is largely unexplored. Particularly in the case of low-valent [U(III/IV)] complexes, most practitioners rigorously avoid exposure to O$_2$. However, Brennan (172) in his thesis with Andersen, reported the oxidation of [(TMSC$_5$H$_4$)$_3$U] by 1 equiv of O$_2$. The only isolated product of the reaction is [(TMSC$_5$H$_4$)$_4$U] as confirmed by mass spectrometry, NMR, and X-ray crystallography. Notably, in the solid state, all four [TMSC$_5$H$_4$]$^{1-}$ ligands are bound η^5. The oxidation of [(N(TMS)$_2$)$_3$U] with O$_2$ was also studied (172). The sole characterized product is the uranium(IV) cyclometalated complex, [U(N(TMS)$_2$)$_2$(κ^2-(N,C)-CH$_2$SiMe$_2$N(SiMe$_3$))]. The authors hypothesize that an aminyl radical is generated that abstracts the H atom from the γ-position of the ligand. The oxidation of tris(aryloxide) complexes gave similar results. Burns and co-workers (168) found that treating $[(ODtbp)_3U]$ with O_2 affords the uranium(IV) ligand redistribution product $[(ODtbp)_4U]$. Mazzanti and co-workers (49) also studied the oxidation of UI_3 with O_2, however, the optimized reaction employs H_2O and pyridine N-oxide (Section IX.A).

2. Bridging Oxo Synthesis

Kanellakopulos and co-workers (208) reported a well-defined and characterized example of oxidation and oxygen-atom transfer from O_2. Exposing a THF solution of the sterically unconstrained uranium(III) tris(Cp) complex, $[(Cp)_3U(thf)]$, with 1 equiv of dry O_2 affords the bridging oxo complex $[\{(Cp)_3U\}_2(\mu\text{-}O)]$. The molecular structure, as determined by X-ray crystallography, reveals a linear $U-O-U$ linkage (180°) with a crystallographically enforced inversion center at the oxygen atom. The $U-O$ bond is short at 2.0881(4) Å and similar to that observed in $[\{(TMSC_5H_4)_3U\}_2(\mu\text{-}O)]$ (Section IX.A) of 2.1052(2) Å (209). Andersen (129) also has reported that the oxidation of $[(N(TMS)_2)_3U]$ with O_2 gives the uranium(V) terminal oxo complex, $[(N(TMS)_2)_3U{=}O]$. The reaction product was subsequently shown to be the bridging oxo complex, $[\{(N(TMS)_2)_3U\}_2(\mu\text{-}O)]$ (210, 211).

B.   Uranium(V) Complexes

1.   Formation of μ-Oxo and -Peroxide Ligands

Recently, two well-defined examples of O_2 reactivity were reported for uranium(V) uranyl complexes. Mazzanti and co-workers (212) prepared the thermally stable uranium(V) uranyl trimer complex supported by an aza β-diketiminate ligand [L = 2-(4-tolyl)-1,3-bis(quinolyl)malondiiminate], $[UO_2L]_3$. The oxidation of this complex by dry O_2 in acetonitrile gives a μ-oxo dimer of hexavalent uranyl, $[(UO_2L)_2(\mu\text{-}O)]$ (Scheme 20). Arnold and co-workers (213) observed similar reactivity for two uranyl(V) cations supported in the cleft of a Pacman-like polypyrollic ligand (L). Specifically, the two-electron reduction of O_2 to peroxide is accomplished by the binuclear uranium(V) complex $K_2[(OUO)_2(L)]$ in pyridine (Scheme 20). The product binuclear uranium(VI) complex, $K_2[(\mu{:}\eta^2,\eta^2\text{-}O_2)(UO_2)_2(L)]$, features a bridging peroxide derived from O_2, in which the $O-O$ bond is 1.433(7) Å. In order to include the two uranyl groups, the Pacman ligand opens to prevent intramolecular clash between the *endo* oxo groups. This feature is best described by the inter-cleft bite angle of 90.18° (in comparison to the bis(μ-O) precursors, which have an angle of 61.38° and 65.18°.)

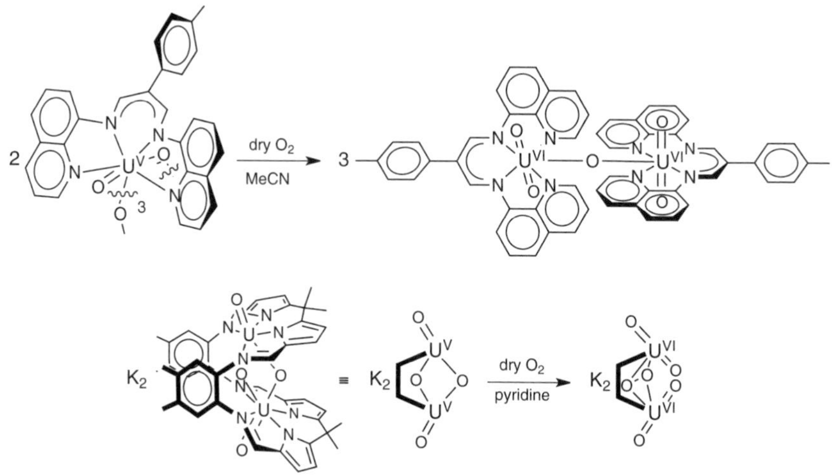

Scheme 20.   Dioxygen reactions of pentavalent uranyl complexes.

VII.   CARBON DIOXIDE

A.   Uranium(III) Complexes

1.   Reductive C=O Bond Cleavage to CO and μ-Oxo

The reactivity of uranium(III) complexes with CO_2 was initially explored with the $[(TMSC_5H_4)_3U]$ complex by Berthet et al. (209). It was found that exposure of a green solution of $[(TMSC_5H_4)_3U]$ to 1 atm of CO_2, in excess, immediately led the color of the solution to change to deep red. The resultant red complex was isolated and characterized by ^1H NMR and X-ray crystallography, revealing a bridging μ-oxo moiety in $[\{(TMSC_5H_4)_3U\}_2(\mu\text{-}O)]$ (Scheme 21). The authors suggest that the mechanism proceeds through a bimetallic μ-CO_2 complex. This proposition is based on the related chemistry of CS_2, which under similar conditions and with identical supporting ligands, forms the complex $[\{(TMSC_5H_4)_3U\}_2(\mu:\eta^2,\eta^2\text{-}CS_2)]$. This complex has been isolated and fully characterized (214). In such a mechanism, CO is extruded from the intermediate μ-CO_2 complex. Recent theoretical studies by Meyer and co-workers (215) suggest that the stability of the $\mu:\eta^2,\eta^2\text{-}CS_2$ complex is due to the thermodynamically unfavorable formation of CS gas in comparison to CO.

Castro-Rodriguez and Meyer (113) expanded on these initial studies by employing the U(III) complex $[((^{tBu}ArO)_3tacn)U]$, which doubly reduces CO_2 to form the U(IV) oxo-bridged complex $[\{(^{tBu}ArO)_3tacnU\}_2(\mu\text{-}O)]$ with concomitant evolution of CO (Scheme 21). The complex $[\{(^{tBu}ArO)_3tacnU\}_2(\mu\text{-}O)]$ was

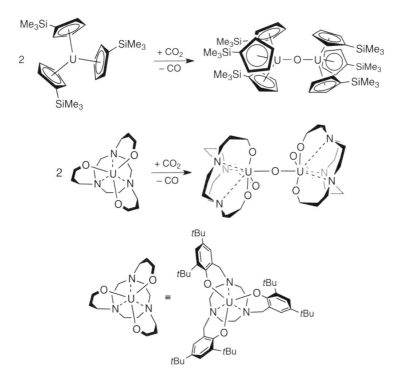

Scheme 21.   Reductive cleavage reactions of CO_2 by uranium(III) complexes.

characterized by ^{1}H NMR, ultraviolet–visible (UV–vis), and X-ray crystallography. The CO produced in the cleavage reaction was detected by IR spectroscopy. The authors report the observation of a colorless intermediate, which may be the unstable dinuclear uranium CO_2 bridged complex, similar to the one proposed in the [(TMSC$_5$H$_4$)$_3$U] reaction with CO_2. The bridging O^{2-} ligand in [{(^{tBu}ArO)$_3$tacnU}$_2$(μ-O)] is buried deep inside the complex core, encased by interdigitating *tert*-butyl groups as revealed by its solid-state molecular structure, X-ray diffraction. As result of this kinetic protection, no further reactivity with CO_2 or other reagents was observed. However, modification of the ligand arms or the anchor afford complexes susceptible to subsequent reactivity with CO_2, CS_2, and COS (Section VII.B.3).

2. η^1-CO_2 Coordination

Starkly different reactivity with CO_2 is observed when the sterically encumbered, adamantyl functionalized derivative, [((^{Ad}ArO)$_3$tacn)U] is employed (193).

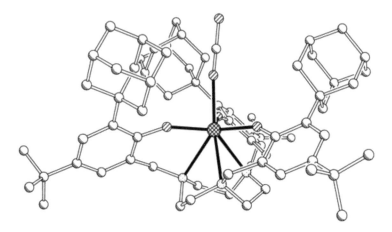

Figure 16.   Crystal structure of [((^{Ad}ArO)$_3$tacn)U(η^1-CO$_2$)].

The complex [((^{Ad}ArO)$_3$tacn)U] reacts with excess CO$_2$ at room temperature, either in solution or in the solid state, to give a monometallic, charge separated [η^1-CO$_2$]$^{\bullet 1-}$ complex, namely, [((^{Ad}ArO)$_3$tacn)U(η^1-CO$_2$)], in which the uranium has undergone a one-electron oxidation. The solid-state structure (Fig. 16) reveals that the CO$_2$ ligand is bound in a linear η^1-O$_\alpha$CO$_\beta$ fashion with U–O–C and O–C–O angles of 171.1(2)° and 178.0(3)°, respectively. Interestingly, the O$_\alpha$–C bond distance of 1.122(4) Å is significantly shorter than the C–O$_\beta$ distance of 1.277(4) Å, suggesting that [((^{Ad}ArO)$_3$tacn)U(η^1-CO$_2$)] possesses charge-separated resonance structures as depicted in Eq. 2. Figure 16 shows the molecular structure of [((^{Ad}ArO)tacn)U(η^1-CO$_2$)].

$$U^{IV} \overset{+}{\longleftarrow} O = \overset{\bullet}{C} - O^- \longleftrightarrow U^{III} \longleftarrow \overset{+}{O} \equiv C - O^-$$

3.   Reductive Disproportionation to CO and CO$_3{}^{2-}$

The Meyer lab has expanded the scope of reductive CO$_2$ transformations at U(III) by both varying the ortho substituents and the nature of the chelating anchor of the aryloxide arms. These modifications in turn protect varying portions of the coordination sphere. By employing N- and mesitylene-anchored ligands, the trivalent uranium complexes, [((^{Ad}ArO)$_3$N)U] ((^{Ad}ArOH)$_3$N) = tris(2-hydroxy-3-adamantyl-5-methylbenzyl)amine)) and [((^{tBu}ArO)$_3$mes)U] ((^{tBu}ArOH)$_3$mes) = 1,3,5-trimethyl-2,4,6-tris(2,4-di-*tert*-butyl-hydroxybenzyl)-methylbenzene)), can be

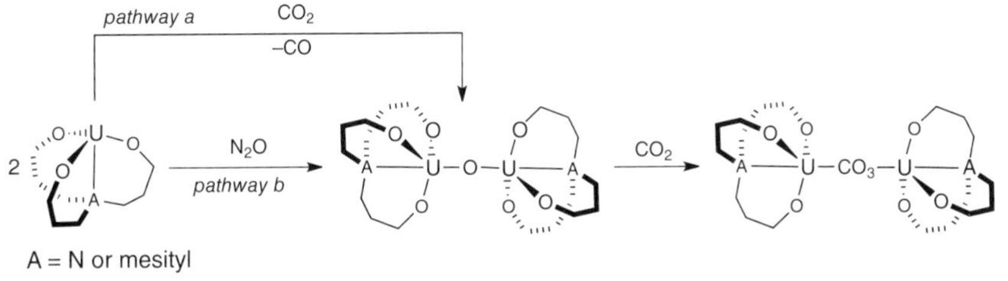

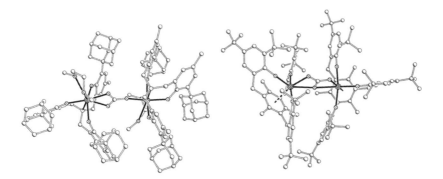

Scheme 22.   Formation of carbonate complexes.

prepared (216, 217). These complexes have markedly different reactivity compared to $[(({}^{tBu}ArO)_3tacn)U]$ and $[(({}^{Ad}ArO)_3tacn)U]$. In the presence of excess CO_2, both complexes $[(({}^{Ad}ArO)_3N)U]$ and $[(({}^{tBu}ArO)_3mes)U]$ undergo reductive disproportionation of CO_2 to form uranium carbonate bridged complexes, $[\{(({}^{Ad}ArO)_3N)U\}_2(\mu\text{-}\eta^1\text{:}\kappa^2\text{-}CO_3)]$ and $[\{(({}^{tBu}ArO)_3mes)U\}_2(\mu\text{-}\kappa^2\text{:}\kappa^2\text{-}CO_3)]$, with concomitant evolution of CO (Scheme 22, Pathway a) (217).

The molecular structures of $[\{(({}^{Ad}ArO)_3N)U\}_2(\mu\text{-}\eta^1\text{:}\kappa^2\text{-}CO_3)]$ and $[\{(({}^{tBu}ArO)_3mes)U\}_2(\mu\text{-}\kappa^2\text{:}\kappa^2\text{-}CO_3)]$ feature two different binding modes of the CO_3^{2-} ligand (Fig. 17) (217). In $[\{(({}^{Ad}ArO)_3N)U\}_2(\mu\text{-}\eta^1\text{:}\kappa^2\text{-}CO_3)]$, the $U-\eta^1(O)$ bond is the shortest uranium–carbonate bond, measuring 2.210(6) Å, while the two longer $U-O$ carbonate bonds are nearly equivalent at 2.414(6) and 2.439(6) Å. Within the carbonate unit of $[\{(({}^{Ad}ArO)_3N)U\}_2(\mu\text{-}\eta^1\text{:}\kappa^2\text{-}CO_3)]$, there are two shorter $C-O$ distances measuring 1.263(10) and 1.281(10) Å and one longer $C-O$ distance at 1.305(11) Å. The complex $[\{(({}^{tBu}ArO)_3mes)U\}_2(\mu\text{-}\kappa^2\text{:}\kappa^2\text{-}CO_3)]$ has two short $U-O$ carbonate bonds [2.333(4) and 2.332(3) Å] and two long $U-O$ bonds [2.659(4) and 2.603(4) Å]. The carbonate unit in $[\{(({}^{tBu}ArO)_3mes)U\}_2(\mu\text{-}\kappa^2\text{:}\kappa^2\text{-}CO_3)]$ exhibits two shorter $C-O$ distances [1.279(7) and 1.285(6) Å] and

Figure 17.   Crystal structures of $[\{(({}^{Ad}ArO)_3N)U\}_2(\mu\text{-}\eta^1\text{:}\kappa^2\text{-}CO_3)]$ (left) and $[\{(({}^{tBu}ArO)_3mes)U\}_2(\mu\text{-}\kappa^2\text{:}\kappa^2\text{-}CO_3)]$ (right).

one longer C$-$O distance of 1.305(6) Å. These distances are comparable to Gardiner's dimeric samarium(III) bridging carbonate complex that features the same binding mode (218).

The oxo-bridged species $[\{((^{Ad}ArO)_3N)U\}_2(\mu\text{-O})]$ and $[\{((^{tBu}ArO)_3\text{mes})U\}_2(\mu\text{-O})]$ can be independently synthesized by treating $[((^{Ad}ArO)_3N)U]$ and $[((^{tBu}ArO)_3\text{mes})U]$ with excess N_2O (Section VIII.A) (217). Both $[\{((^{Ad}ArO)_3N)U\}_2(\mu\text{-O})]$ and $[\{((^{tBu}ArO)_3\text{mes})U\}_2(\mu\text{-O})]$ reacted with CO_2 to form $[\{((^{Ad}ArO)_3N)U\}_2(\mu\text{-}\eta^1\text{:}\kappa^2\text{-}CO_3)]$ and $[\{((^{tBu}ArO)_3\text{mes})U\}_2(\mu\text{-}\kappa^2\text{:}\kappa^2\text{-}CO_3)]$, respectively, verifying them as intermediates in the reductive disproportionation pathways (Scheme 22, Pathway b). Detailed DFT studies on the pathway leading to complex $[\{((^{tBu}ArO)_3\text{mes})U\}_2(\mu\text{-}\kappa^2\text{:}\kappa^2\text{-}CO_3)]$ also support the formation of $[\{((^{tBu}ArO)_3\text{mes})U\}_2(\mu\text{-O})]$ as an intermediate (219).

In comparison to kinetic inertness of $[\{((^{tBu}ArO)_3\text{tacn}U\}_2(\mu\text{-O})]$, the reactivity of uranium oxo-bridged species $[\{((^{Ad}ArO)_3N)U\}_2(\mu\text{-O})]$ and $[\{((^{tBu}ArO)_3\text{mes})U\}_2(\mu\text{-O})]$ is noteworthy. This divergent reactivity can be attributed to the increased accessibility of the nucleophilic bridging O^{2-} ligand in $[\{((^{Ad}ArO)_3N)U\}_2(\mu\text{-O})]$ and in $[\{((^{tBu}ArO)_3\text{mes})U\}_2(\mu\text{-O})]$. This reactivity pattern has been employed to synthesize bridging mixed-chalcogenide carbonates, $[\{((^{Ad}ArO)_3N)U\}_2(\mu\text{-}CE_3)]$ (CE_3^{2-}, $E = O$, S, Se), by the reaction of bridging chalcogenide $[\{((^{Ad}ArO)_3N)U\}_2(\mu\text{-E})]$ ($E = O$, S, and Se) complexes with CO_2, COS, and CS_2 (220, 221). The addition of CO_2 to the μ-S and μ-Se complexes produces $[\{((^{Ad}ArO)_3N)U\}_2(\mu\text{-}\eta^1\text{:}\kappa^2\text{-}CO_2S)]$ and $[\{((^{Ad}ArO)_3N)U\}_2(\mu\text{-}\eta^1\text{:}\kappa^2\text{-}CO_2Se)]$, respectively (220).

Further modification of the ortho substituent of the tris(aryloxide) tacn ligand complexes from *tert*-butyl (i.e., $[(^{tBu}ArO)_3\text{tacn})U]$ to neopentyl, $[((^{Neop,Me}ArO)_3\text{tacn})U]$, $[(^{Neop,Me}ArOH)_3\text{tacn} = 1,4,7\text{-tris}(2\text{-hydroxy-5-methyl-3-neopentylbenzyl})\text{-}1,4,7\text{-triazacyclononane}]$ opens up the ligand frame work significantly and allows the development of catalytic reductive disproportionation of CO_2 (Scheme 23) (222). Like $[(^{tBu}ArO)_3\text{tacn})U]$, $[(^{Neop,Me}ArO)_3\text{tacn})U]$ can be

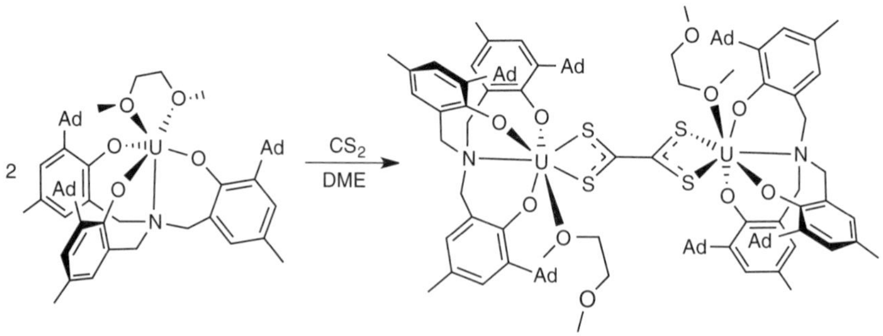

Scheme 23.   Synthesis of $[\{((^{Ad}ArO)_3\text{tacn})U\}_2(\mu\text{-}\kappa^2\text{:}\kappa^2\text{-}C_2S_4)]$.

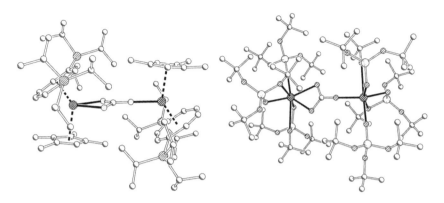

Figure 18. Crystal structures of $[\{(\eta^8\text{-}C_8H_6(1,4\text{-}Si(iPr)_3)_2)(Cp^*)U\}_2(\mu\text{-}\eta^1:\kappa^2\text{-}CO_3)]$ (left) and $[\{(OSi(O(tBu))_3)_3U\}_2(\mu\text{-}\eta^1:\kappa^2\text{-}CO_3)]$ (right).

oxidized with N_2O to $[\{((^{\text{Neop,Me}}ArO)_3tacn)U\}_2(\mu\text{-}O)]$ (Section VIII.A). This bridging oxo complex can, in turn, be converted into the carbonate $[\{((^{\text{Neop,Me}}ArO)_3tacn)U\}_2(\mu\text{-}\kappa^2:\kappa^2\text{-}CO_3)]$ with excess CO_2. Both $[\{((^{\text{Neop,Me}}ArO)_3tacn)U\}_2(\mu\text{-}O)]$ and $[\{((^{\text{Neop,Me}}ArO)_3tacn)U\}_2(\mu\text{-}\kappa^2:\kappa^2\text{-}CO_3)]$ can be reduced back to $[((^{\text{Neop,Me}}ArO)_3$ tacn)U] with a potassium graphite and the production of K_2O and K_2CO_3. Under 1 atm of CO_2, a solution of $[((^{\text{Neop,Me}}ArO)_3tacn)U]$ over potassium graphite can repeat this cycle several times, until the reaction is terminated by formation of an insoluble polynuclear U(IV) complex $\{K[((^{\text{Neop,Me}}ArO)_3tacn)U)(\kappa^2\text{-}CO_3)]\}_4$ that precipitates.

Treating Cloke's $[(\eta^8\text{-}C_8H_6(1,4\text{-}Si(iPr)_3)_2)(\eta^5\text{-}C_5Me_{5-n}H_n)U(thf)]$ $(n = 0, 1)$ with an excess of CO_2 at $-30\,°C$ also gives the carbonate-bridged complexes $[\{(\eta^8\text{-}C_8H_6(1,4\text{-}SiiPr_3)_2)(\eta^5\text{-}C_5Me_{5-n}H_n)U\}_2(\mu\text{-}\eta^1:\kappa^2\text{-}CO_3)]$ $(n = 0, 1)$ by the reductive disproportionation of CO_2 (223). The molecular structure of $[\{(\eta^8\text{-}C_8H_6(1,4\text{-}SiiPr_3)_2)(Cp^*)U\}_2(\mu\text{-}\eta^1:\kappa^2\text{-}CO_3)]$ features the carbonate ligand bound in $\mu\text{-}\eta^1:\kappa^2\text{-}CO_3$ fashion similar to $[\{((^{\text{Ad}}ArO)_3N)U\}_2(\mu\text{-}\eta^1:\kappa^2\text{-}CO_3)]$ (Fig. 18). Recently, Mazzanti and co-workers (224) also have reported similar chemistry. The dimeric uranium(III) complex, $[U(OSi(O(tBu))_3)_2(\mu\text{-}OSi(O(tBu))_3)]_2$, prepared from $[(N(TMS)_2)_3U]$ and $HOSi(O(tBu))_3$, reacts with excess CO_2 to give $[\{(OSi(O(tBu))_3)_3U\}_2(\mu\text{-}\eta^1:\kappa^2\text{-}CO_3)]$ (Fig. 18). Carbon monoxide is formed concomitantly in this reaction (as confirmed by ^{13}C NMR and by binding to vanadocene). The complex $[U(OSi(O(tBu))_3)_2(\mu\text{-}OSi(O(tBu))_3)]_2$ also reacts with CS_2 to give the second example of a U(IV/IV) bridged CS_2 complex.

Given the isolation of $[((^{\text{Ad}}ArO)_3tacn)U(\eta^1\text{-}CO_2)]$ and the reduction cleavage and disproportionation chemistry demonstrated by uranium(III) complexes, it is quite surprising that the reductive coupling of CO_2 to oxalate $C_2O_4^{2-}$ has not yet been reported. However, the U(III) complex $[((^{\text{Ad}}ArO)_3N)U]$ was found to

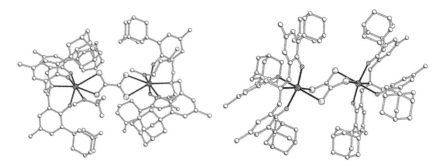

Figure 19. Crystal structure of [{((^{Ad}ArO)$_3$tacn)U}$_2$(μ-κ^2:κ^2-C$_2$S$_4$)] (left) and [{((^{Ad}ArO)$_3$tacn)U}$_2$ (μ-κ^2:κ^2-C$_2$S$_4$)]$^{2-}$ (right).

reductively couple CS$_2$, to form the tetra–thiooxalate complex [{((^{Ad}ArO)$_3$tacn)U}$_2$ (μ-κ^2:κ^2-C$_2$S$_4$)] (Scheme 23) (225). The molecular structure of [{((^{Ad}ArO)$_3$tacn)U}$_2$ (μ-κ^2:κ^2-C$_2$S$_4$)] features the C$_2$S$_4^{2-}$ ligand in an unprecedented coordination mode, (μ-κ^2(S,S'):κ^2(S'',S''')-C$_2$S$_4$ (Fig. 19). Further reduction of [{((^{Ad}ArO)$_3$tacn)U}$_2$ (μ-κ^2:κ^2-C$_2$S$_4$)] by two electrons is possible and generates the uranium ethylenete-trathiolate C$_2$S$_4^{4-}$ complex, namely, [Na(dme)$_3$]$_2$[{((^{Ad}ArO)$_3$tacn)U}$_2$(μ-κ^2: κ^2-C$_2$S$_4$)] (225).

In a reaction related to the reductive coupling of CO$_2$ to oxalate, Meyer and co-workers (226) recently demonstrated C—C bond formation between CO$_2$ and a dinuclear uranium(IV) enediolate. The enediolate complex can be prepared by treating [((^{tBu}ArO)$_3$tacn)U] with 0.5 equiv of the diketones: benzil or di-*tert*-butyl diketone. These reactions give the yellow–brown complexes [{((^{tBu}ArO)$_3$tacn)U}$_2$ (μ-η^1:η^1-(O(Ph)C=C(Ph)O)] and [{((^{tBu}ArO)$_3$tacn)U}$_2$(μ-η^1:η^1-(O(*t*Bu)C=C(*t*Bu) O)]. Exposure of these complexes to excess CO$_2$ results in C—C bond formation between CO$_2$ and the enolate, thus giving an asymmetric complex as characterized by XRD in the case of the reaction of [{((^{tBu}ArO)$_3$tacn)U}$_2$(μ-η^1:η^1-(O(*t*Bu)C= C(*t*Bu)O)] with CO$_2$ (Scheme 24 and Fig. 20). The newly introduced CO$_2$ fragment bridges both uranium centers. Such reactivity holds promise for developing syntheti-cally useful C—C bond-forming reactions with uranium(III) complexes and the abundant C$_1$ carbon source, CO$_2$.

4. Insertion into a U—C Bond

Bart and co-workers (227) reported a well-defined reaction of a CO$_2$ insertion into the uranium(III) benzyl ligand U—C bond. Exposing trivalent [Tp*$_2$U(CH$_2$Ph)] [Tp* = hydro-tris(3,5-dimethylpyrazolyl)borate] to 1 atm of CO$_2$ resulted in the generation of the uranium carboxylate complex [Tp*$_2$U(η^2-O$_2$CCH$_2$Ph)] (Scheme 25). The [Tp*$_2$U(η^2-O$_2$CCH$_2$Ph)] complex has been crystallographically

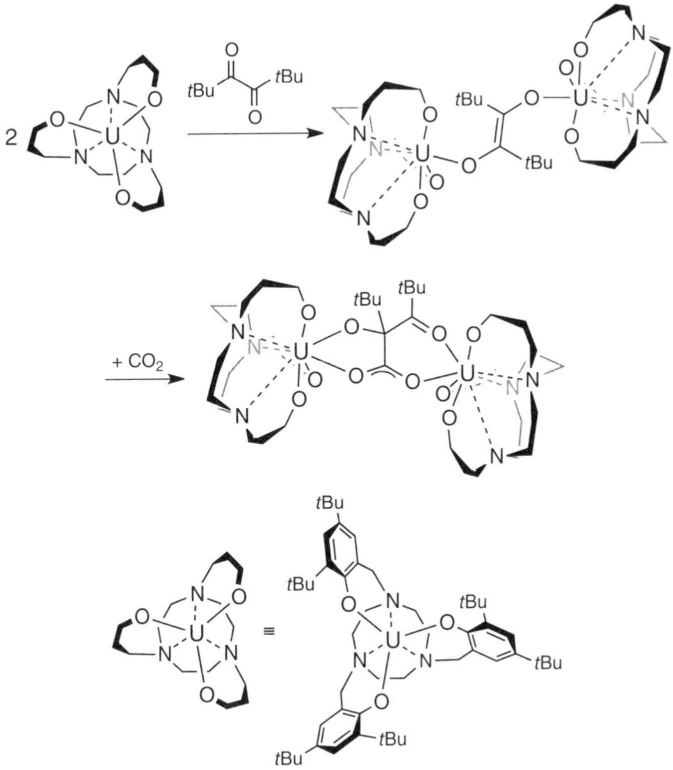

Scheme 24. The C–C bond forming with CO_2 and $[\{(((^{tBu}ArO)_3tacn)U\}_2(\mu\text{-}\eta^1:\eta^1\text{-}(O(tBu)C=C(tBu)O)]$.

characterized and features the carboxylate ligand bound in a η^2 (O,O′)-fashion. The O–C bond lengths are 1.228(12) and 1.269(13) Å and indicate that there is electron delocalization over the O–C–O unit. The carboxylate unit can be liberated from the U(III) center by treating $[Tp^*_2U(\eta^2\text{-}O_2CCH_2Ph)]$ with iodotrimethylsilane to form the silyl ester, $Me_3SiOC(O)CH_2Ph$, and the uranium mono-iodide species, $[Tp^*_2UI]$, which can react with 1 equiv of KCH_2Ph to regenerate $[Tp^*_2U(CH_2Ph)]$ and establish a closed synthetic cycle. Treating $[Tp^*_2U(CH_2Ph)]$ with acidic terminal alkynes gives the complexes $[Tp^*_2U(CCTMS)]$ and $[Tp^*_2U(CCPh)]$ (228). Exposure of either of these complexes to CO_2 also results in CO_2 insertion into the U–C bond to give $[Tp^*_2U(\eta^2\text{-}O_2CCTMS)]$ and $[Tp^*_2U(\eta^2\text{-}O_2CCPh)]$, respectively, as forest green complexes. The metrical parameters of the Tp^* ligands in the solid-state structures of $[Tp^*_2U(\eta^2\text{-}O_2CCH_2Ph)]$ and $[Tp^*_2U(\eta^2\text{-}O_2CCPh)]$ are nearly identical, suggesting that the carboxylate substitution has little influence on the structure.

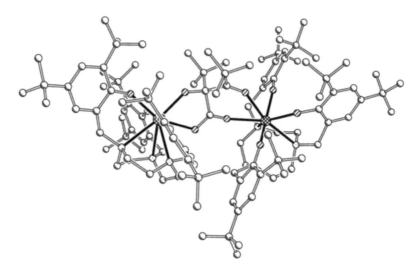

Figure 20.   Crystal structure of $[\{((({}^{tBu}ArO)_3tacn)U\}_2(\mu\text{-}O({t}Bu)C(CO_2)C({t}Bu)=O)]$.

5.   Insertion into a U−N Bond

The protonolysis reaction of $[Tp^*_2U(CH_2Ph)]$ developed by Bart and co-workers (228) was extended to amines. Hence, addition of a solution of aniline or benzylamine to $[Tp^*_2U(CH_2Ph)]$ resulted in the formation of $[Tp^*_2U(NHPh)]$ and $[Tp^*_2U(NHCH_2Ph)]$, respectively. Exposure of the complexes to CO_2 also results in rapid insertion of CO_2 and the formation of the carbamates, $[Tp^*_2U (\eta^2\text{-}O_2CNHPh)]$ and $[Tp^*_2U(\eta^2\text{-}O_2CNHCH_2Ph)]$. The insertion of CO_2 was confirmed by the shift in the N−H IR stretch.

In reexamining the small molecule chemistry of $[(N(TMS)_2)_3U]$, Arnold and co-workers (131) found that under 1 atm of carbon dioxide, CO_2 inserts into the U−N amide bond of $[U(N(TMS)_2]_3$, followed by reductive splitting, resulting in elimination of $O=C=NTMS$ and formation of tetravalent $[U(OTMS)_4]$. Analysis of the volatiles from the reaction revealed an IR stretch at $2185\ cm^{-1}$, corresponding to isocyanate formation. Similar reactivity was observed with a uranium(III) bis(amide), supported by a bidentate alkoxy-tethered NHC ligand (145). The reaction of $[(N(TMS)_2)_2UL]$ $(L = OCMe_2CH_2(1\text{-}C\{NCH_2CH_2NR\});\ R = 2,6\text{-}iPr_2\text{-}C_6H_3)$ with CO_2 gives an insoluble material, tentatively deemed $[U(L)N(TMS)_2(OTMS)(OCNTMS)]_n$ on the basis of the IR spectrum, which includes an IR stretch at $2183\ cm^{-1}$, corresponding to isocyanate incorporation. The use of a bulky dialkyl amide may result in redox neutral conversion to the uranium(III) tris(carbamate) because $[(N(Et)_2)_4U]$ undergoes a similar reaction (Section VII.B.1).

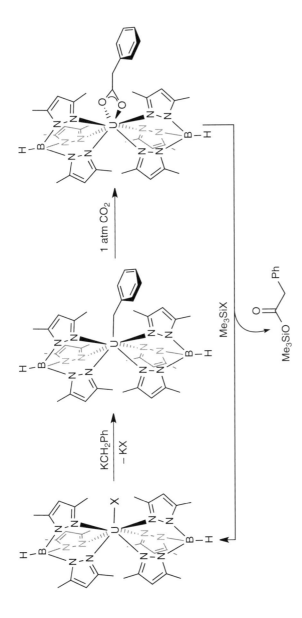

Scheme 25. Uranium(III) synthetic cycle for the synthesis of Me₃SiOC(O)CH₂Ph.

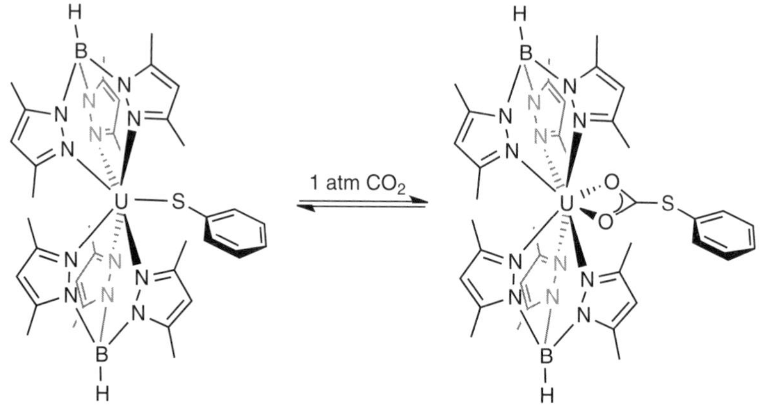

Scheme 26.   Reversible insertion of CO_2 into $U-S$ bond in $[(Tp^*)_2U(SPh)]$.

6.   Insertion into a $U-S$ Bond

The protonolysis reaction of $[Tp^*_2U(CH_2Ph)]$, developed by Bart (Section VII.A.4), was also extended to thiols. Exposure of the complex $[Tp^*_2U(SPh)]$ to CO_2 results in a rapid change in the color of the solution from deep blue to green (228). Removing the volatiles under vacuum resulted in the isolation of a blue residue, which was confirmed by NMR to be the starting material $[Tp^*_2U(SPh)]$. This observation implies reversible CO_2 insertion into the $U-S$ σ bond. The insertion reaction was confirmed by monitoring the reaction *in situ* by NMR (Scheme 26). Similar reactivity has been observed for the uranium(IV) thiol $[(Cp)_3USiPr]$ (Section VII.B.4) (229).

B.   Uranium(IV) Complexes

1.   Insertion into a $U-N$ Bond

One of the first reports on the chemistry of uranium with CO_2, was the preparation of a uranium tetra(carbamate) complex $[U(O_2CNEt_2)_4]$. In 1974, Bagnall and Yanir (230) reported the insertion of CO_2 into the $U-N$ bonds of $[U(NEt_2)_4]$. On exposing a green solution of $[U(NEt_2)_4]$ in hexane to an atmosphere of CO_2, a violent, exothermic reaction occurred and was only controllable at $0-5\,^{\circ}C$. The formation of the carbamates was confirmed by IR spectroscopy. Note that the analogous reactions with CS_2, COS, and CSe_2 were successful as well. The complex $[U(O_2CNEt_2)_4]$ was also prepared several years later by adding UCl_4

to a pretreated solution of $NHEt_2$ and CO_2 (231). The authors claim that formation of $[U(O_2CNEt_2)_4]$ proceeds through a salt metathesis of pre-formed diethylammonium carbamate. Berthet and co-workers (159, 160) also demonstrated the ready insertion of CO_2 into a U−N amide bond. The cationic metallocene complex, $[(Cp^*)_2U(N(Me)_2)(thf)][B(Ph)_4]$, reacts with CO_2 to give the carbamate $[(Cp^*)_2U(O_2CN(Me)_2)(thf)][B(Ph)_4]$ in THF within 2 h. The carbamate complex was characterized by NMR and IR spectroscopy, as well as elemental analysis.

Carbon dioxide insertion into the U−N bond of U(IV) amide complexes $[(({}^{tBu}ArO)_3tacn)U(NHMes)]$ and $[(({}^{Ad}ArO)_3tacn)U(NHMes)]$ also has been studied by Meyer and co-workers (232). The reaction of these compounds with CO_2 produced the carbamate complexes $[(({}^{tBu}ArO)_3tacn)U(\eta^1\text{-}OC(O)N(H)Mes)]$ and $[(({}^{Ad}ArO)_3tacn)U(\eta^2\text{-}OC(O)N(H)Mes)]$, respectively. The molecular structure of $[(({}^{tBu}ArO)_3tacn)U(\eta^1\text{-}OC(O)N(H)Mes)]$ features an η^1 binding mode of the carbamate ligand, while $[(({}^{Ad}ArO)_3tacn)U(\eta^2\text{-}OC(O)N (H)Mes)]$ has the carbamate ligand bound in a $\eta^2\text{-}(O,O')$-fashion. The discrepancy in binding modes can be attributed to steric differences in the two ligand systems. Due to the less sterically hindering *tert*-butyl substituents in $[(({}^{tBu}ArO)_3tacn)U(\eta^1\text{-}OC(O)N(H)Mes)]$, the carbamate is capable of forming one strong U−O interaction [2.227(3) Å]. The adamantyl groups in $[(({}^{Ad}ArO)_3tacn)U(\eta^2\text{-}OC(O)N(H)Mes)]$, however, prevent the carbamate ligand from approaching the U ion. To compensate for this, two weaker U−O bonds are formed, measuring 2.434(4) and 2.527(4) Å.

2. Insertion into a U−C Bond

The insertion of CO_2 into U−C bonds was first documented in 1985 by Moloy and Marks (233). Exposure of $[(Cp^*)_2U(Me)_2]$ to excess CO_2 resulted in double insertion of CO_2 into U−C bonds to form the bis(acetate) complex. Confirmation of the bidentate acetate ligands and its monomeric formulation as $[(Cp^*)_2U(OAc)_2]$ (as opposed to a dimeric bridging structure as seen in $[\{Cp_2U\}_2(\mu\text{-}\eta^1\text{:}\eta^1\text{-}OAc)_2])$ was shown by IR (C−O stretch $< 1590\,cm^{-1}$) and cryoscopic molecular weight determination in benzene. Addition of 1 equiv of CO_2 resulted in a single insertion of CO_2 into one U−C bond to form the uranium methyl acetate complex $[(Cp^*)_2U(Me)(OAc)]$.

Evans et al. (234) presented four examples of CO_2 insertion into a U−C bond. In the presence of 80 psi carbon dioxide, CO_2 will undergo a double insertion reaction into the U−C bonds of the uranium(IV) bis(phenylacetylide) complex $[(Cp^*)_2U(C\equiv CPh)_2]$. This reaction results in formation of $[(Cp^*)_2U (O_2C\equiv CCPh)_2]$, which was characterized by ${}^1H$ NMR, and IR spectroscopy, since single crystals for an XRD study were not obtainable. In recent chemistry, this reaction was extended to the insertion of CO_2 into uranium allyl U−C

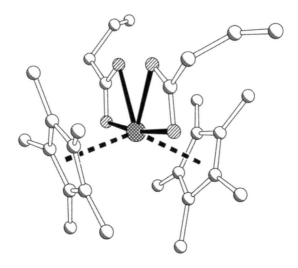

Figure 21.   Crystal structure of $[(Cp^*)_2U(\eta^2\text{-OOCCH}_2C(H)CH_2)_2]$.

bonds (235). $[(Cp^*)_2UCl(\eta^3\text{-CH}_2C(R)CH_2)]$ (R = H, Me) reacts with CO_2 to give the monoinsertion products, $[(Cp^*)_2UCl(\eta^2\text{-OOCCH}_2C(R)CH_2)]$ (R = H, Me), which, like the acetylide reactions, were characterized by NMR and IR. However, the reaction with bis(allyl) complex $[(Cp^*)_2U(\eta^3\text{-CH}_2C(H)CH_2)(\eta^1\text{-CH}_2C(H)$ $CH_2)]$ led to a unique crystallographically characterizable bis(acetate) complex $[(Cp^*)_2U(\eta^2\text{-OOCCH}_2C(H)CH_2)_2]$ (Fig. 21) (235). It was found that treating this bis(acetate) with TMSCl at slightly elevated temperatures leads to the formation of $[(Cp^*)_2UCl_2]$ and TMS protected carboxylate, thereby meeting the requirements for creating a closed synthetic cycle.

Similarly, CO_2 insertion into a uranium cyclopentadienyl methyl amidinate complex $[(Cp^*)_2\{(iPr)NC(Me)N(iPr)\text{-}\eta^2\text{-N,N}'\}U(Me)]$ generates the acetate complex $[(Cp^*)_2\{(iPr)NC(Me)N(iPr)\text{-}\eta^2\text{-N,N}'\}U(\eta^1\text{-O}_2CMe)]$ (236). The molecular structure of $[(Cp^*)_2\{(iPr)NC(Me)N(iPr)\text{-}\eta^2\text{-N,N}'\}U(\eta^1\text{-O}_2CMe)]$ demonstrates the unusual η^1 acetate binding (Fig. 22). The uranium tethered alkyl complex, $[(\eta^5\text{:}\eta^1\text{-C}_5Me_4SiMe_2CH_2)_2U]$, inserts CO_2 to form the tethered carboxylate complex, $[(\eta^5\text{:}\eta^1\text{-C}_5Me_4SiMe_2CH_2CO_2)_2U]$ (Scheme 27) (237).

Recently, Mazzanti and co-workers (238) also presented the insertion of CO_2 into U–C in a salan (diamine bis(phenolate)) supported complex. Carbon dioxide readily reacts with $[U(\text{salan-}(tBu)_2)(CH_2SiMe_3)_2]$ to give $[U(\text{salan-}(tBu)_2)$ $(O_2CCH_2SiMe_3)_2]$. The precise coordination of the carboxylate ligands is unknown, but it is likely that both coordinate in the same manner because the ^1H NMR implies C_2 symmetry.

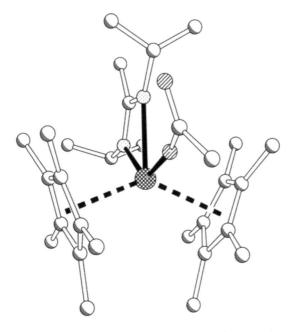

Figure 22.   Crystal structure of $[(Cp^*)_2(i\text{PrNC(Me)N}i\text{Pr-}\eta^2\text{-N,N}')U(\eta^1\text{-O}_2\text{CMe})]$.

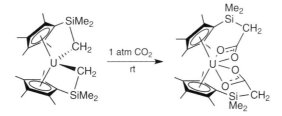

Scheme 27.   Double insertion of CO_2 into U−C bonds of $[(\eta^5{:}\eta^1\text{-C}_5\text{Me}_4\text{SiMe}_2\text{CH}_2)_2U]$.

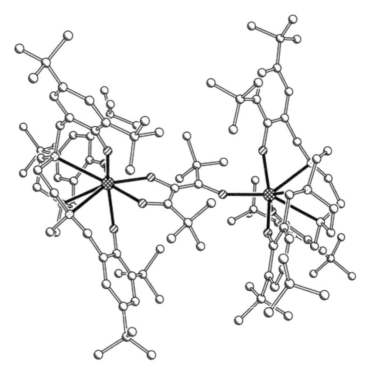

Figure 23.   Crystal structure of [{(((^{tBu}ArO)$_3$tacn)U}$_2$(μ-η^1:η^2-(tBu)COCOCO(tBu))].

3.   Insertion into a U−O Bond

The reactions of the tetravalent uranium oxo-bridged species [{(((^{Ad}ArO)$_3$N)U}$_2$(μ-O)] and [{(((^{tBu}ArO)$_3$mes)U}$_2$(μ-O)] with CO_2 to form bridging carbonate complexes, which was described previously (Section VII.A.3), represent prime examples of CO_2 insertion into U−O bonds (217). This chemistry has been extended to bridging triketone enolates (226). Treating [(((^{tBu}ArO)$_3$tacn)U] with 0.5 equiv of the triketone tBuCOCOCOtBu gives the bridging dinuclear uranium(IV) triketone enolate [{(((^{tBu}ArO)$_3$tacn)U}$_2$(μ-η^1:η^2-(tBuCOCOCOtBu)] (Fig. 23). The addition of CO_2 to this dinuclear species leads to insertion into the η^1-U−O bond of the complex as evidenced by preliminary XRD data (Scheme 28).

Arnold and co-workers (131) also observed CO_2 insertion into U−O bonds during the reaction of CO_2 with the uranium dinitrogen complex [(U(OTtbp)$_3$)$_2$(μ-η^2:η^2-N$_2$)]. The complex [(OTtbp)$_2$U(μ-O)(μ-O$_2$COTtbp)$_2$U(OTtbp)$_2$] is

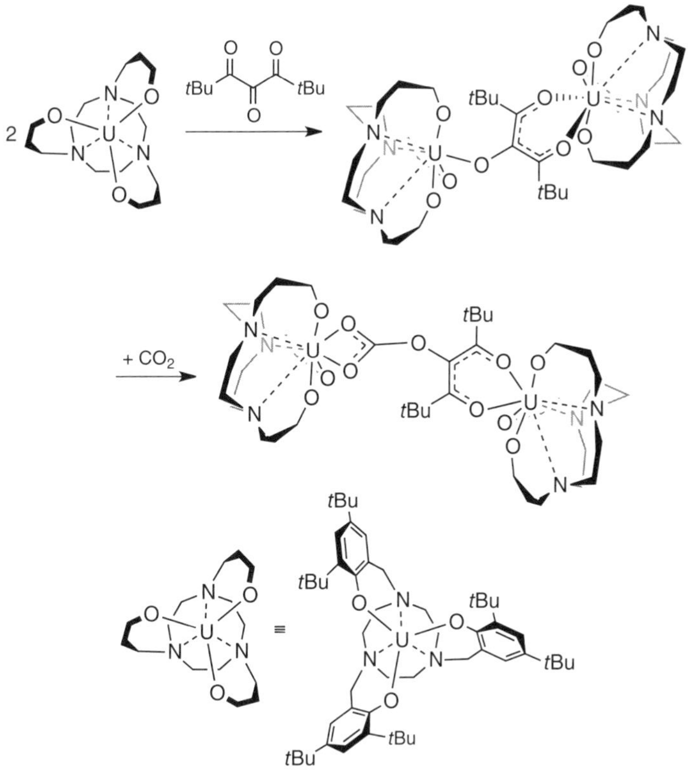

Scheme 28.   Insertion of CO_2 into U−O bond of [{((^{tBu}ArO)$_3$tacn)U}$_2$(μ-η^1:η^2-(tBu)COCOCO(tBu))].

formed at elevated temperatures by the simultaneous insertion of CO_2 and reductive cleavage of CO_2 (Scheme 29). The molecular structure of [(OTtbp)$_2$ U(μ-O)(μ-O$_2$COTtbp)$_2$U(OTtbp)$_2$] exhibits two bridging tri-$tert$-butylphenyl carbonate ligands coordinated in a η^2-(O′,O″)-fashion with each oxygen bound to a different uranium center (Fig. 24).

4.   Insertion into a U−S Bond

The first example of CO_2 insertion into a U−S bond was reported by Ephritikhine and co-workers in 1996 (229). Carbon dioxide inserts reversibly into the U−S bond of [(Cp)$_3$U(S(iPr))] to form [(Cp)$_3$U(O$_2$CS(iPr))]. Few details of this reaction are reported. This study was followed up by examining CO_2

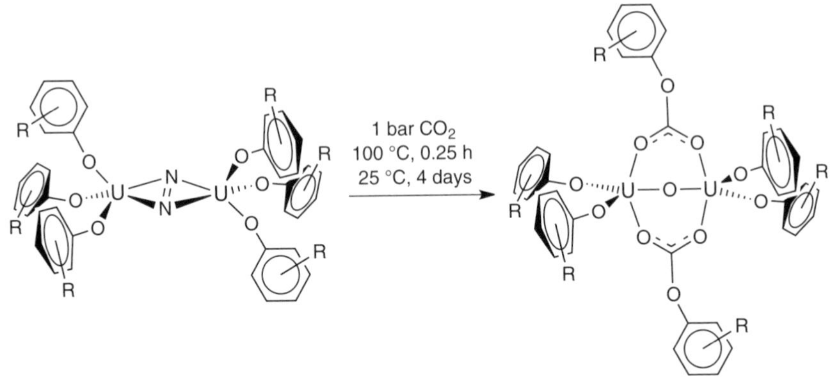

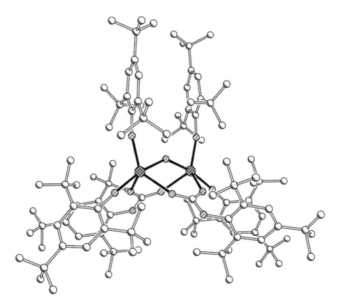

Scheme 29.   Reaction of CO_2 with $[(U(OTtbp)_3)_2(\mu\text{-}\eta^2\text{:}\eta^2\text{-}N_2)]$.

insertion into U−S bonds of $[(Cp^*)_2U(S(tBu))_2]$ (239). The uranium thiocarbonate complex $[(Cp^*)_2U(O_2CS(tBu))_2]$ was produced in the reaction of CO_2 with $[(Cp^*)_2U(S(tBu))_2]$ (Scheme 30). A single-crystal X-ray diffraction study of $[(Cp^*)_2U(O_2CS(tBu))_2]$ is not reported.

Figure 24.   Crystal structure of $[\{(OTtbp)_2U\}^2(\mu\text{-}O)(\mu\text{-}O_2COTtbp)_2]$.

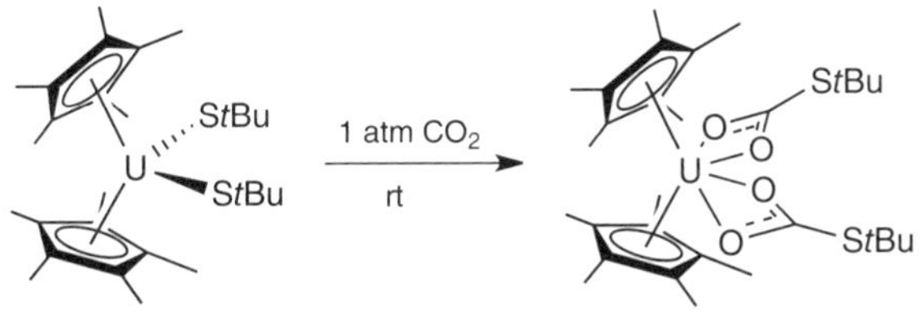

Scheme 30.   Insertion of CO_2 into U−S bonds of $[(Cp^*)_2U(StBu)_2]$.

C.   Uranium(V) Complexes

1.   Imido Multiple-Bond Metathesis

The uranium imido complex $[((^{Ad}ArO)_3tacn)U(NTMS)]$, described previously (Section III.C.1), has been shown to be highly reactive toward π-acids, such as CO and methyl isocyanide (166). Similar reactivity has been observed for uranium mesitylimido complexes $[((^RArO)_3tacn)U(NMes)]$ (R = tBu and Ad) and CO_2. Treatment of $[((^RArO)_3tacn)U(NMes)]$ with CO_2 results in multiple-bond metathesis to form isocyanates and uranium terminal oxo species $[((^RArO)_3tacn)U(O)]$ (R = tBu and Ad) (Scheme 31) (232). Although the U(V) oxo complexes $[((^RArO)_3tacn)U(O)]$ are stable, note that the synthesis of U(V) complexes with terminal oxo ligands cannot be accomplished with standard methods (e.g., by treating the trivalent U(III) complexes with oxygen-atom transfer reagents). The molecular structures of the $[((^RArO)_3tacn)(O)]$ complexes reveal terminal oxo ligands coordinated at the axial position of an idealized C_3 symmetrical complex

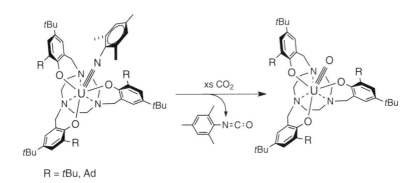

R = tBu, Ad

Scheme 31.   Multiple-bond metathesis of uranium(V) mesityl imido complexes with CO_2.

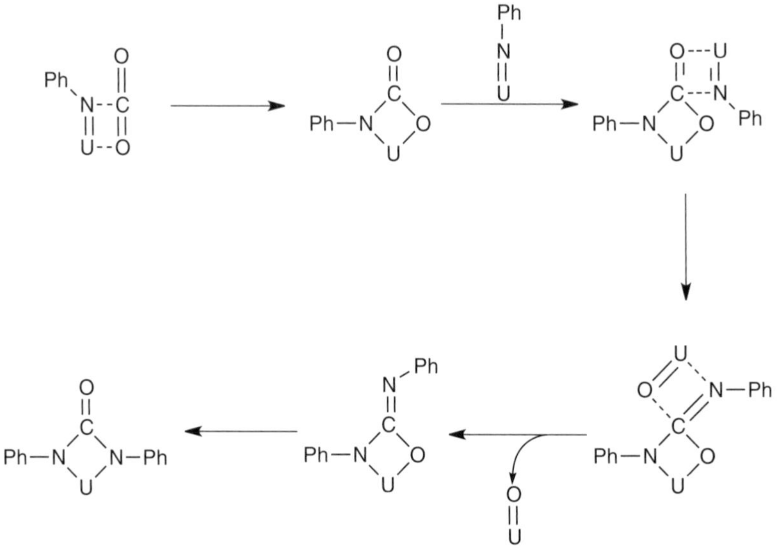

Scheme 32.   Possible mechanism of carbimate formation from a uranium(V) phenyl imido and CO_2.

with U=O bond distances of 1.848(8) and 1.848(4) Å for tBu and Ad, respectively. The formation of $[((^R ArO)_3 tacn)U(O)]$ is proposed to occur through a [2+2]-cycloaddition, with the formation of the stable U=O moiety and mesityl isocyanate as the driving force.

The less sterically hindered U(V) phenylimido complex $[((^{tBu} ArO)_3 tacn)U(NPh)]$ reacts with CO_2 to yield $[((^{tBu} ArO)_3 tacn)U(O)]$ and the uranium diphenyl ureate complex $[((^{tBu} ArO)_3 tacn)U(\eta^2 \text{-}N,N' \text{-}(NPh)_2 CO)]$ (232). The complex $[((^{tBu} ArO)_3 tacn)U(\eta^2 \text{-}N,N' \text{-}(NPh)_2 CO)]$ could be formed by two possible mechanisms. Two successive [2+2] cycloadditions could occur. The first addition generates a uranium carbimate intermediate, which undergoes a second cyclo-addition of $[((^{tBu} ArO)_3 tacn)U(NPh)]$ to form a dinuclear species that ultimately liberates uranium oxo complex $[((^{tBu} ArO)_3 tacn)U(O)]$ and $[((^{tBu} ArO)_3 tacn) U(\eta^2 \text{-}N,N' \text{-}(NPh)_2 CO)]$ (Scheme 32). Alternatively, complex $[((^{tBu} ArO)_3 tacn) U(\eta^2 \text{-}N,N' \text{-}(NPh)_2 CO)]$ could have formed from the reaction of the U(V) imido $[((^{tBu} ArO)_3 tacn)U(NPh)]$ with the phenyl-isocyanate that is formed during the reaction. An XRD structure of $[((^{tBu} ArO)_3 tacn)U(\eta^2 \text{-}N,N' \text{-}(NPh)_2 CO)]$ reveals the ureate ligand coordinated to the uranium center in a η^2-(N,N')-fashion (Fig. 25).

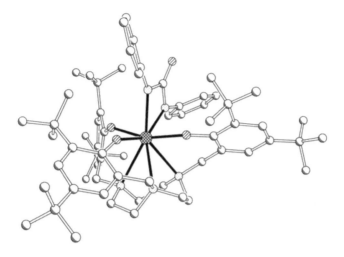

Figure 25.   Crystal structure of $[((^{tBu}ArO)_3tacn)U(\eta^2\text{-}N,N'\text{-}(NPh)_2CO)]$.

VIII.   NITROUS OXIDE

A.   Uranium(III) Complexes

1.   Bridging Oxo Synthesis

Despite the utility of N_2O as an oxgen-atom transfer reagent, it has been employed relatively infrequently in f-element chemistry (1, 240). With the recent isolation of a transition metal complex of N_2O (240) and the importance of N_2O in climate change (300 times more potent greenhouse gas than CO_2 on a per molecule basis (241)), the reactivity of this small molecule with actinides may receive increased attention. Uranium(III) chemistry with N_2O has so far been limited to the construction of dinuclear μ-O complexes. Brennan (172) studied the oxidation of $[(C_5H_4Me)_3U(thf)]$ with N_2O. The addition of 1 equiv of an ethereal solution of N_2O via syringe to a THF solution of $[(C_5H_4Me)_3U(thf)]$ resulted in the immediate precipitation of a green microcrystalline material, which was completely insoluble in toluene and precluded definitive NMR analysis. The product was assigned on the basis of IR data as $[\{(C_5H_4Me)_3U\}_2(\mu\text{-}O)]$ in comparison to the thoroughly characterized, isostructural bridging chlacogenide complexes $[\{(C_5H_4Me)_3U\}_2(\mu\text{-}E)]$, where $E = S$, Se, or Te. If the reaction of $[(C_5H_4Me)_3U(thf)]$ with N_2O was performed in toluene, an unidentified, insoluble red compound precipitated. Definitive crystallographic analysis of the product of N_2O oxidation of a $[(Cp')_3U]$ complex was obtained by Ephritikhine and co-workers (209) in

1991. Oxidation of the complex [(TMSC$_5$H$_4$)$_3$U] with N$_2$O also gives [{(TMSC$_5$H$_4$)$_3$U}$_2$(μ-O)] as described previously in the section on CO$_2$. It was found that exposure of a green solution of [(TMSC$_5$H$_4$)$_3$U] or of a powder of [(TMSC$_5$H$_4$)$_3$U] to 1 atm of N$_2$O in excess gives [{(TMSC$_5$H$_4$)$_3$U}$_2$(μ-O)] as confirmed by spectroscopic analysis (Scheme 21).

This reactivity pattern was extended to tris(aryloxide) complexes by Burns and co-workers (168). In contrast to the reactivity, Burns observed between [(ODtbp)$_3$U] and O$_2$, treating [(ODtbp)$_3$U] with N$_2$O affords [{U(ODtbp)$_3$}$_2$ (μ-O)] as was observed with NO (Section IV.A.1). Meyer and co-workers (217), in turn, has used this reactivity in order to explore the mechanism of reductive CO$_2$ disproportionation to CO and CO$_3^{2-}$ as described in the CO$_2$ section. Namely, the oxidation of [((^{Ad}ArO)$_3$N)U], [((^{tBu}ArO)$_3$mes)U], and [((^{Neop,Me}ArO)$_3$tacnU)] with excess N$_2$O gives, respectively, [{((^{Ad}ArO)$_3$N)U}$_2$(μ-O)], [{((^{tBu}ArO)$_3$mes)U}$_2$ (μ-O)], and [{(^{Neop,Me}ArO)$_3$tacnU}$_2$(μ-O)] (Schemes 22 and 23) (217, 222).

B. Uranium(IV) Complexes

1. Terminal Oxo Synthesis

In one case, the oxidation of a uranium(IV) complex with N$_2$O results in the synthesis of a terminal oxo complex. Arney and Burns (243) reported the unique, base-free uranium(IV) metallocene imido complexes [(Cp*)$_2$U(N(2,6-iPr$_2$C$_6$H$_3$)] and [(Cp*)$_2$U(N(2,4,6-tBu$_3$C$_6$H$_3$)]. The oxidation of these complexes by N$_2$O (or amine N-oxides) produces the uranium(VI) oxo, imido complexes [(Cp*)$_2$ U(N(2,6-iPr$_2$C$_6$H$_3$)(O)] and [(Cp*)$_2$U(N(2,4,6-tBu$_3$C$_6$H$_3$)(O)] (Fig. 26) (243). At

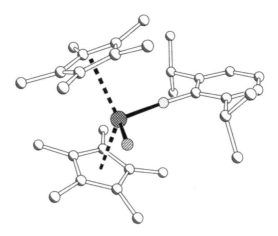

Figure 26. Crystal structure of [(Cp*)$_2$U(N(2,6-iPr$_2$C$_6$H$_3$)(O)].

the time, these were among the first examples of organometallic uranium(VI) complexes and are still among relatively few examples of uranium(VI) monooxo complexes (20, 173, 244, 245). The use of a bulky aryl imide is necessary: oxidizing [(Cp*)$_2$U(NPh)(py)] (py = pyridine, ligand) with pyridine N-oxide results in the net oxo-imido exchange and the isolation of [(Cp*)$_2$U(NPh)$_2$], "UO$_2$," and Cp*$_2$ (242). The oxidation state of [(Cp*)$_2$U(N(2,6-iPr$_2$C$_6$H$_3$)(O)] and [(Cp*)$_2$U(N(2,4,6-tBu$_3$C$_6$H$_3$)(O)] was confirmed by UV–vis–near-IR spectroscopy (the complexes lacked metal-based transitions). The ^{1}H NMR revealed the complexes to be temperature-independent paramagnets, as has been observed in [UO$_2$]$^{2+}$ complexes (246).

IX. WATER

A. Uranium(III) Complexes

1. Hydroxide Synthesis

Lukens et al. (247) synthesized the only uranium(III) hydroxide complexes in 1996. Careful hydrolysis of either [(Cp$''$)$_3$U] (Cp$''$ = 1,3-(TMS)$_2$C$_5$H$_3$) or [(Cp‡)$_3$U] (Cp‡ = 1,3-(tBu)$_2$C$_5$H$_3$) gives the dimeric hydroxide complexes [{(Cp$''$)$_2$U}$_2$(μ-OH)$_2$] and [{(Cp‡)$_2$U}$_2$(μ-OH)$_2$]. Since [(Cp‡)$_3$U] is difficult to prepare, [{(Cp‡)$_2$U}$_2$(μ-OH)$_2$] is best synthesized by the hydrolysis of the dimeric hydride complex [{(Cp‡)$_2$U}$_2$(μ-H)$_2$]. The complexes undergo a formal "oxidative elimination" at elevated temperatures to give H$_2$ and the uranium(IV) complexes [{(Cp$''$)$_2$U}$_2$(μ-O)$_2$] or [{(Cp‡)$_2$U}$_2$(μ-O)$_2$]. This nonradical process is likely intramolecular based on isotopic labeling in cross-over experiments. The observed kinetic isotope effect (KIE) and entropy of activation (from Eyring analysis) are consistent with an α-OH-[1,2]-elimination from a transiently formed hydride, [{(Cp$''$)$_2$U}$_2$(μ-OH)(μ-O)(H)] (Scheme 33).

The use of less sterically demanding Cp ligands results in oxidation of uranium(III) to uranium(IV). Kanellakopulos and co-workers (208) also were able to synthesize the bridging oxo complex [{(Cp)$_3$U}$_2$(μ-O)] (Section VI.A.2 also prepared from O$_2$) by oxidation with H$_2$O. This reaction likely proceeds via [(Cp)$_3$U(OH)], since the rapid sublimation of [(Cp)$_3$U(OH)], prepared indirectly from NaOH and [(Cp)$_3$U(Br)], also yields [{(Cp)$_3$U}$_2$(μ-O)].

2. Uranyl Polymer and Cluster Synthesis

The oxidation of uranium(III) halides by H$_2$O has also been explored. Such controlled hydrolysis reactions for the formation of non-uranyl core clusters model uranium speciation during bioremedition (248). As mentioned in the Section VI.A.1

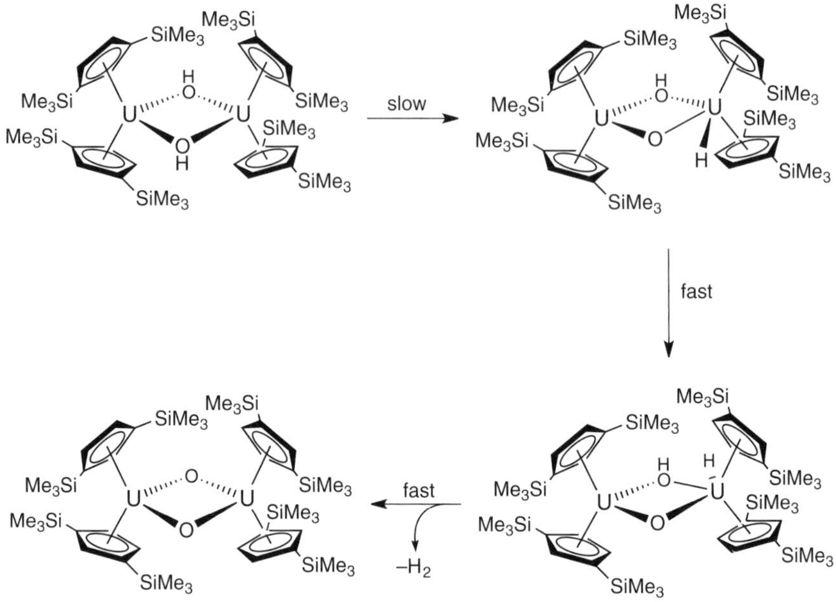

Scheme 33.   Mechanism of H_2 elimination from $[\{(Cp'')_2U\}_2(\mu\text{-OH})_2]$.

on O_2 reactivity, Natrajan et al. (49) found that the oxidation of $UI_3(thf)_4$ with O_2 affords the uranyl(V) polymer $\{[UO_2py_5][KI_2py_2]\}_n$ in 28% yield. In order to achieve a more reliable synthesis, controlled hydrolysis with an equivalent of H_2O and 1 equiv of pyridine N-oxide in pyridine (py) ligand gives $\{[UO_2py_5][KI_2py_2]\}_n$ in 54% yield. The molecular structure reveals a polymeric, cation–cation structure with potassium bridging $[O=U=O]^+$ subunits (Fig. 27). The use of excess O_2 or pyridine N-oxide gives the uranyl(VI) complex $[UO_2I_2Py_3]$. If this hydrolysis is carried out with 2 equiv of H_2O in ACN (acetonitrile, solvent) in the presence of KOTf (OTf = triflate), a brown solution forms, which over 15 h forms an emerald green solution. Crystallization by diffusion of diisopropyl ether gives the dodec-anuclear cluster $[U_{12}(\mu^3\text{-OH})_8(\mu^3\text{-O})_{12}I_2(\mu^2\text{-OTf})_{16}\text{-(acn)}_8]\cdot2ACN\cdot2H_2O$ with a $U_{12}O_{20}$ core (249). The diffusion of diisopropyl ether into the reaction mixture before completion of the reaction leads to the isolation of a mixture of different intermediate products. X-ray diffraction analysis revealed the presence of three additional species with different structures, which all contain the same U_6O_8 core. The composition of the cluster is highly dependent on the nature of the supporting ligands in the reaction mixture. If the triflate starting material $U(OTf)_3(acn)_3$ is used instead of $UI_3(thf)_4$, the large $U_{12}O_{20}$ core structure does not form and only the U_6O_8

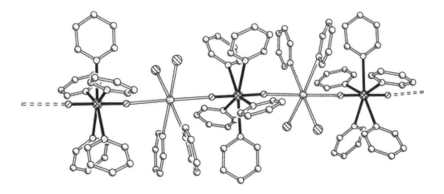

Figure 27.   Crystal structure of $\{[UO_2py_5][KI_2py_2]\}_n$.

core clusters are observed. The use of potassium benzoate (instead of triflate) as a model of humic acid (a necessary cofactor of uranyl bioreduction (250)) results in a U_6O_8 cluster, $[U_6O_4(OH)_4(PhCOO)_{12}(py)_3]$ (Fig. 28) (251). There are a few U(IV) clusters presenting the U_6O_8 or the $U_6O_4(OH)_4$ core, and their formation by oxidation and/or hydrolysis of uranium(III) or (IV) complexes in the presences of bidentate, bridging ligands appears to be a general reaction (50, 249, 252, 253). Notably, $[U_6O_4(OH)_4(PhCOO)_{12}(py)_3]$ can also be prepared by reaction of the uranyl(V) polymer, $\{[UO_2py_5][KI_2py_2]\}_n$ and benzoic acid (254). The inclusion of bidentate amines, such as TMEDA (TMEDA = tetramethylethylenediamine), can

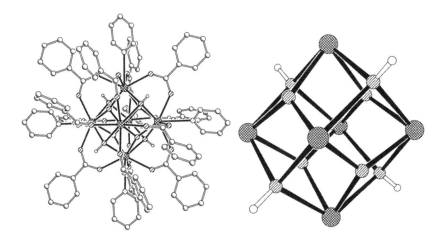

Figure 28.   Crystal structure of $[U_6O_4(OH)_4(C_6H_5COO)_{12}(py)_3]$ (left) and core (right).

lead to the construction of larger clusters. The hydrolysis of 16 equiv of $UI_3(thf)_4$ with 24 equiv of H_2O and potassium benzoate leads to selective synthesis of a $\{U_{16}O_{22}(OH)_2\}$ core cluster complex (251).

Ephritikhine and co-workers (255) explored the controlled hydrolysis of terpyridine (terpy) complexes of uranium(III). The oxidation of $[UI_2(terpy)_2(py)][I]$ or $[U(OTf)_2(terpy)_2(py)][OTf]$ with H_2O in acetonitrile solutions gives tri- and tetrameric μ-O complexes, $[\{UI(terpy)_2(\mu\text{-}O)\}_2\{UI_2(terpy)\}]I_4$ and $[\{U(OTf)(terpy)_2(\mu\text{-}O))(\mu\text{-}OTf)U(terpy)\}_2(\mu\text{-}OTf)_2(\mu\text{-}O)][OTf]_4\cdot py\cdot ACN$, respectively. Mazzanti and co-workers (256) also reported the hydrolysis of uranium(III) complexes supported by the neutral tripodal N-donor ligand tpa (tris((2-pyridine)methyl) amine). The oxidation of $[(tpa)_2U]I_3$ by adventitious water in ACN during crystallization gives the uranium(IV) bis(hydroxide) $[U(tpa)_2(OH)_2]I_2\cdot 3ACN$. The synthesis of this complex is not reproducible on scale. The controlled hydrolysis of $[(tpa)_2U]I_3$ by 1 or 2 equiv of H_2O gives $\{[U(tpa)(\mu\text{-}O)I]_3(\mu^3\text{-}I)\}I_2\cdot 3ACN$. Similar results were reported by Scott and co-workers (134) in the serendipitous isolation of the monohydroxide complex $[(TREN^{DMSB})U(OH)(CH_2PMe_3)]$ during the attempted crystallization of $[U(TREN^{DMSB})U(CH_2PMe_3)]$, presumably due to adventitious H_2O. Kozimor and co-workers (257) also reported a uranium(III) bis(μ-O) complex, $\{(Cp^*)_2U\}_2(\mu\text{-}O)$, and a uranium(IV) terminal oxo complex, $(Cp^*)_2U(O)(C(NMeCMe)_2)$, by reduction of $[(Cp^*)_3U]$ with potassium graphite (in the presence of the carbene $C(NMeCMe)_2$ in the latter case). The source of the oxygen atom in both cases is unknown, but is speculated to be either O_2 or H_2O.

B.   Uranium(IV) Complexes

1.   Hydroxo and Cluster Synthesis

There are several examples of redox neutral hydrolysis of uranium(IV) complexes to form bridging oxo and terminal hydroxide complexes. Andersen and co-workers (247) prepared dinuclear bis(μ-O) complexes by the hydrolysis of $[(Cp'')_2U(Me)_2]$ and $[(Cp^\ddagger)_2U(Me)_2]$ to give $[\{(Cp'')_2U\}_2(\mu\text{-}O)_2]$ and $[\{(Cp^\ddagger)_2U\}_2(\mu\text{-}O)_2]$. These complexes are the products of oxidative elimination of H_2 from $[\{(Cp'')_2U\}_2(\mu\text{-}OH)_2]$ and $[\{(Cp^\ddagger)_2U\}_2(\mu\text{-}OH)_2]$ as described (Section IX.A.1). Gilje and co-workers (258) reported the hydrolysis of $[(Cp^*)_2UCl_2(HNSPh_2)]$ with 1 equiv of $H_2O\cdot HNSPH_2$. This reaction gives the terminal uranium(IV) hydroxide complex, $([Cp^*)_2UCl(OH)(HNSPh_2)]$. The authors propose that it is an intermediate in the formation of the tetra uranium oxo cluster $[Cp(Cl)(HNSPh_2)U(\mu\text{-}O)(\mu\text{-}O)_2U(Cl)(HNSPh_2)_2]_2$ from $[(Cp^*)_2UCl_2(HNSPh_2)]$ and excess $H_2O\cdot HNSPh_2$. Similar reactivity has been observed by Ephritikhine and co-workers (259) in the uranium(IV) tris(Cp) complexes. Controlled hydrolysis of the complexes $[(C_5H_4R)_3UH]$ (R = tBu or

TMS) gives the hydroxides $[(C_5H_4R)_3U(OH)]$. Notably, in contrast to the reactivity of the less bulky $[(Cp)_3U(OH)]$ described in Section IX.A.1 (208), heating $[(C_5H_4TMS)_3U(OH)]$ does not give $[(C_5H_4TMS)_3U(\mu\text{-O})]$ and H_2O. Instead, hydrolysis of $[(C_5H_4TMS)_3U(\mu\text{-O})]$ with 1 equiv of H_2O gives 2 equiv of the terminal hydroxide $[(C_5H_4TMS)_3U(OH)]$. Refluxing $[(C_5H_4TMS)_3U(OH)]$ in alkanes gives the trimer $[\{(C_5H_4TMS)_2U(\mu\text{-O})\}_3]$ and free C_5H_5TMS. Recently, Girolami and co-workers (260) noted that refluxing $[UCl_4]$ and sodium N,N-dimethylaminodiboranate, $Na(H_3BNMe_2BH_3)$, in DME gives bridging oxo complex $[U_2(\mu\text{-O})(BH_4)_6(dme)_2]$ in 7% yield (not the bridging hydride, $[U_2(\mu\text{-H})(BH_4)_6(dme)_2]$). The oxygen atom is most likely derived from adventitious H_2O.

2. Alkoxide Hydrolysis and Oxidation

Duval and co-workers (261) have reported that the photolysis of the uranyl dication supported triphenylphosphine oxide (OPPh$_3$), $[UO_2(OPPh_3)_4][OTf]_2$, in either methanol (MeOH) or Et_2O/ACN gives the U(IV) trans-alkoxide dication complexes $[U(OR)_2(OPPh_3)_4][OTf]_2$ (R = Me or Et) (Scheme 34). Hydrolysis of $[U(OR)_2(OPPh_3)_4][OTf]_2$ with excess water gives the starting uranyl complex, $[UO_2(OPPh_3)_4][OTf]_2$, and free alcohol. The fate of H atoms necessary to balance this hydrolysis reaction is not reported, but presumably leads to the oxidative elimination of dihydrogen.

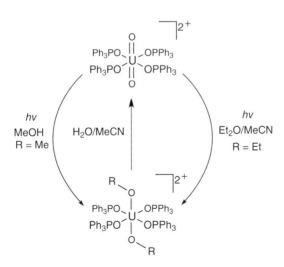

Scheme 34.   Hydrolysis and oxidation of $[U(OR)_2(OPPh_3)_4][OTf]_2$ (R = Me or Et) derived from $[U(O)_2(OPPh_3)_4][OTf]_2$.

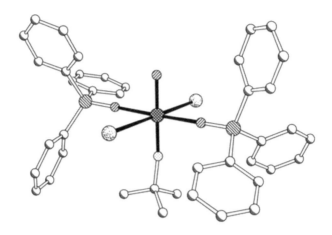

Figure 29. Crystal structure of [UO(N(tBu))I$_2$(OPPh$_3$)$_2$].

C. Uranium(VI) Complexes

1. Imido Hydrolysis

Hayton et al. (28) reported the isolation of a uranium(VI) oxo imido complex, which represents a structural link between uranyl and trans bis(imido) uranium complexes. The oxo imido complex, [UO(N(tBu))I$_2$(OPPh$_3$)$_2$], was prepared by treating [U(N(tBu))$_2$I$_2$(thf)$_2$] with B(C$_6$F$_5$)$_3$·H$_2$O in THF followed by 2 equiv of OPPh$_3$ in toluene. The molecular structure of [UO(N(tBu))I$_2$(OPPh$_3$)$_2$] reveals a linear trans oxo–imido [178.4(3)°] arrangement with a U−O$_{yl}$ (O$_{yl}$ = terminal oxygen atom) bond length of 1.764(5) Å and a U−N$_{imido}$ bond length of 1.821 (7) Å (Fig. 29).

D. Uranyl(VI/V) Complexes

The aqueous chemistry of uranyl [UO$_2$]$^{2+}$ and [UO$_2$]$^+$ has been extensively studied because of its central role in understanding the speciation of uranium in the environment, nuclear fuel processing, and waste remediation (78). In particular, [UO$_2$]$^{2+}$ O$_{yl}$ exchange with water was originally studied in 1949 in order to disambiguate the possible composition of uranyl as either [UO$_2$]$^{2+}$ or [U(OH)$_4$]$^{2+}$ (262). In 1961, Gordon and Taube (263) established that [UO$_2$]$^+$ exchanges the O$_{yl}$ ligands with water at least 3×10^9 times faster than [UO$_2$]$^{2+}$. The breadth and depth of subsequent studies is beyond the scope of this chapter, and interested readers are referred to several recent reviews on aqueous uranyl coordination chemistry (71, 76, 82–84). In particular, readers should see Fortier and Hayton's (78) review on uranyl functionalization, which details the mechanism and kinetics of O_{yl} exchange with water and the recent article by Gibson and co-workers (264) on oxo exchange in $[UO_2]^+$.

X.   DIHYDROGEN

A.   Uranium(III) Complexes

A significant body of work has explored the synthesis and reactivity of molecular and polymeric (i.e., UH_3) uranium hydrides (9, 265). However, relatively few of these complexes are prepared directly from H_2. While uranium(III) hydride complexes are known, they are generally prepared from uranium(IV) precursors via hydrogenolysis (Section X.B). A mixture of the uranium(III) hydride, $[\{(Cp^*)_2U(H)\}_2]$, and the uranium(IV) hydride complex, $[\{(Cp^*)_2U(H_2)\}_2]$, can be directly prepared from $[(Cp^*)_2U(CH(TMS)_2)]$ with excess H_2 (266). The details are discussed in the subsequent uranium(IV) section.

The direct reaction of a non-hydride, non-alkyl, uranium(III) complex with H_2 is limited to a recent example from Cloke and co-workers (267), in which the reactivity of $[\eta^8-C_8H_6(1,4-Si(iPr)_3)_2)(\eta^5-Cp^*)U]$ with mixtures of CO and H_2 was explored. Adding 1 equiv of ^{13}CO, followed by 2 equiv of H_2 to a toluene solution of $[\eta^8-C_8H_6(1,4-Si(iPr)_3)_2)(\eta^5-Cp^*)U]$ at $-78\,°C$, followed by warming to room temperature, revealed a single new compound by ^{13}C NMR (single quartet resonance at $\delta = 319\,ppm$, with $J_{CH} = 137\,Hz$). Examination of the molecular structure determined by XRD confirmed the formation of a uranium(IV) methoxide complex $[\eta^8-C_8H_6(1,4-Si(iPr)_3)_2)(\eta^5-Cp^*)U(O^{13}Me)]$, resulting from the reduction of CO (Scheme 35). The methoxide ligand can be liberated with TMSOTf to give TMSOMe and $[\eta^8-C_8H_6(1,4-Si(iPr)_3)_2)(\eta^5-Cp^*)U(OTf)]$ to close a synthetic cycle. The mechanism of CO hydrogenation by $[\eta^8-C_8H_6(1,4-SiiPr_3)_2)(\eta^5-Cp^*)U]$ is not known, but the absence of hydride formation (i.e., $[\eta^5-C_8H_6(1,4-Si(iPr)_3)_2)(\eta^5-Cp^*)U(H)])$ under pure H_2 and the lack of reactivity of the ynediolate complex, $[(U(\eta^8-C_8H_6(1,4-Si(iPr)_3)_2)(\eta^5-Cp^*))_2((\mu-\eta^1:\eta^1-C_2O_2)]$, with H_2, suggest that the reaction may proceed via direct H_2 reduction of the "zigzag" intermediate proposed for the formation of the above-mentioned ynediolate complex, $[\{(\eta^8-C_8H_6(1,4-Si(iPr)_3)_2)(\eta^5-Cp^*)U\}_2((\mu-\eta^1:\eta^1-C_2O_2)]$. Alternatively, the formation of the putative hydride $[\eta^8-C_8H_6(1,4-Si(iPr)_3)_2)(\eta^5-Cp^*)U(H)]$ in low concentration may lead to the reduction of CO via a formyl ligand, as has been observed in Zr and Ce hydride complexes, as well as in U acyl complexes (135, 268–271). In particular, Maatta and Marks (271) observed that the uranium(IV) acyl $[(Cp^*)_2UCl(\eta^2-OC(Ph))]$ can be catalytically hydrogenated to the alkoxide $[(Cp^*)_2UCl(OCH_2Ph)]$ by $[\{(Cp^*)_2Th(H)_2\}_2]$ under an atmosphere of H_2. More recently, Andersen and co-workers (270) reported that the complex $[(Cp''')_2CeH]$

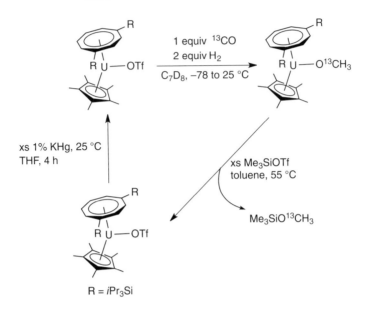

Scheme 35.   Hydrogenation of ^{13}CO by [η^8-C$_8$H$_6$(1,4-Si(iPr)$_3$)$_2$)(η^5-Cp*)U] and closed synthetic cycle for the formation of TMSOMe and [η^8-C$_8$H$_6$(1,4-Si(iPr)$_3$)$_2$)(η^5-Cp*)U(OTf)].

(Cp''' = [1,2,4-(tBu)$_3$C$_5$H$_2$]$^-$ anion) will also hydrogenate CO in a 10:1 mixture H$_2$:CO to the terminal methoxide [(Cp''')$_2$Ce(OMe)] via a dinuclear cerium formaldehyde dianion intermediate, [{(Cp''')$_2$Ce}$_2$(μ-CH$_2$O)].

B.   Uranium(IV) Complexes

Uranium(IV) alkyl complexes undergo ready hydrogenolysis. Marks and co-workers (161) first reported that the hydrogenolysis of [(Cp*)$_2$U(Me)$_2$] gives [{(Cp*)$_2$U(H)$_2$}$_2$] in 1978 (Scheme 36). The structure [{(Cp*)$_2$U(H)$_2$}$_2$] was originally inferred by comparison of its IR spectrum with that of the thorium analogue, the solid-state structure of which was eventually determined by neutron diffraction (272). The complex [{(Cp*)$_2$U(H)$_2$}$_2$] is in equilibrium with the dimeric uranium(III) hydride [{(Cp*)$_2$U(H)}$_2$] and free H$_2$ via a binuclear reductive elimination. In a subsequent study, the production of the uranium(IV) hydride complex, [{(Cp*)$_2$U(H)$_2$}$_2$], by σ bond metathesis was confirmed by use of D$_2$ gas, which produced exclusively CH$_3$D (162). A monomeric uranium(III) hydride can be prepared by the hydrogenolysis of [(Cp*)$_2$U(Me)$_2$] or [(Cp*)$_2$U(CH$_2$TMS)$_2$] with 1 atm H$_2$ in the presence of DMPE. This reaction gives the DMPE stabilized mononuclear complex [(Cp*)$_2$U(H)DMPE] (Scheme 36), which was confirmed by XRD analysis. However, the hydride could not be located in the difference

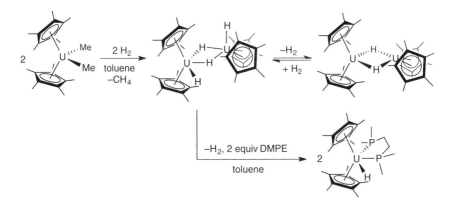

Scheme 36.   Hydrogenolysis of $[(Cp^*)_2U(Me)_2]$ to give $[\{(Cp^*)_2U(H)_2\}_2]$, $[\{(Cp^*)_2U(H)\}_2]$, and $[(Cp^*)_2U(H)(DMPE)]$.

map (163), nor could a hydride resonance be observed by ^{1}H NMR spectroscopy. Regardless, IR spectroscopy revealed a stretch at $1219 \, cm^{-1}$, which shifts to $870 \, cm^{-1}$ in the complex prepared from D_2 and is assigned as the U–H stretch (163). Note that the complex $[(Cp^*)_2U(H)DMPE]$ reacts with N_2, CO, and THF to give mixtures of uranium(III) and (IV) complexes. Hydrogenolysis of the uranium(IV) monoalkyl complexes $[(Cp^*)_2U(R)(Cl)]$ (R = Me, CH_2TMS) gives the trimer $[\{(Cp^*)_2U(Cl)\}_3]$ cleanly (266). This reaction is in contrast to the thorium analogue, which gives the chlorido hydrido complex $[(Cp^*)_2Th(H)(Cl)]$ under identical conditions (161).

The mechanism of alkyl hydrogenolysis has been studied by Lin and Marks (273) for the complexes $[(Cp^*)_2U(CH_2TMS)(O\text{-}tBu)]$ and $[(Cp^*)_2Th(CH_2TMS)(O\text{-}tBu)]$ among several others. These reactions are first order in both the H_2 and actinide complex. In the case of $[(Cp^*)_2Th(CH_2TMS)$ $(O\text{-}tBu)]$, the KIE is $k_H/k_D = 2.5(4)$, which implies that H–H bond cleavage is rate determining. Furthermore, there was no evidence for a preequilibrium formation of an η^2-H_2 complex. This procedure to synthesize dinuclear uranium(III) bridging hydride complexes has been employed by Andersen and co-workers (247) to prepare $[\{(Cp^\ddagger)_2U\}_2(\mu\text{-}H)_2]$ from $[(Cp^\ddagger)_2U(Me)_2]$, as necessary for the preparation of his uranium(III) hydroxide complex, $[\{(Cp^\ddagger)_2U\}_2(\mu\text{-}OH)_2]$ (Section IX.A.1).

Evans et al. (274) revisited Marks' studies in 2007 and obtained X-ray structural data for the uranium(III) complex, $[\{(Cp^*)_2U(\mu\text{-}H)\}_2]$, and the uranium(IV) complex, $[\{(Cp^*)_2U(H)_2\}_2]$. In the case of $[\{(Cp^*)_2U(\mu\text{-}H)\}_2]$, the bridging hydrides could not be located, but it possesses a U $\cdots$ U distance of 3.8530(7) and 3.8651(7) Å (two unique molecules). The bimetallic core is similar to that

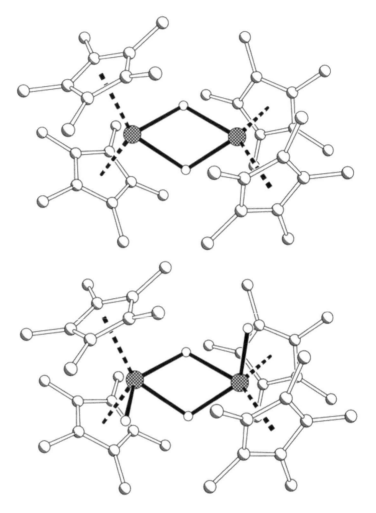

Figure 30.   Crystal structures of [{(Cp*)$_2$U(μ-H)}$_2$] (XRD) and [{(Cp*)$_2$U(H)$_2$}$_2$] (neutron diffraction).

observed in [{(Cp*)$_2$Sm(μ-H)}$_2$]. In the case of [{(Cp*)$_2$U(H)$_2$}$_2$], the bridging hydrides could be located in the difference map giving U−H distance of 1.94(9) Å and a U···U distance of 3.606(6) Å, as expected for the smaller radius of uranium(IV) in comparison to uranium(III). Subsequently, Evans and co-workers (275) developed a method to grow large crystals of [{(Cp*)$_2$U(H)$_2$}$_2$] and sealed them under H$_2$ for analysis by neutron diffraction. The molecular structure of [{(Cp*)$_2$U(H)$_2$}$_2$] determined by neutron diffraction allows the location of the terminal hydrides as well (Fig. 30). The bridging hydrides were observed with U−H bond lengths of 2.148(7)

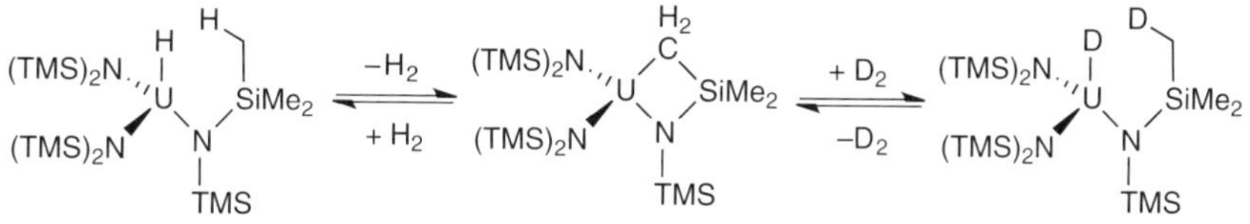

Scheme 37.   Hydrogen–deuterium exchange in $[(N(TMS)_2)_3UH]$.

and 2.134(9) Å, while the terminal hydride has a U−H bond length of 2.052(15) Å. Evans et al. (274) explored the reactivity of these hydride complexes with unsaturated substrates, such as azobenzene and COT. The hydrides act as masked forms of U(II), releasing H_2 and providing up to 4 reducing equiv per metal center to give, for example, the uranium(VI) bis(imide) $[(Cp^*)_2U(NPh)_2]$ (prepared by the addition of 2 equiv of azobenzene to $[\{(Cp^*)_2U(\mu\text{-}H)\}_2]$) (274).

Andersen and co-workers (141, 276, 277) have also explored the reactivity of uranium(IV) hydrides. Under an atmosphere of D_2, complete deuterium exchange of all 55 hydrogens in $[(N(TMS)_2)_3UH]$ is observed. This reaction is completely reversible, exposing perdeutero-$[(N(TMS)_2)_3UD]$ to H_2 results in clean conversion to the perprotio-$[(N(TMS)_2)_3UH]$ (Scheme 37). The activation of H_2 proceeds via σ-bond metathesis. The deuteration of the ligand requires the σ activation of C−H bonds. The complex $[U(N(TMS)_2)_2(\kappa^2\text{-}(N,C)\text{-}CH_2SiMe_2N(SiMe_3))]$ was shown to be an intermediate in the reaction because pyrolysis of $[(N(TMS)_2)_3UH]$ results in the elimination of H_2 and the formation of $[U(N(TMS)_2)_2(\kappa^2\text{-}(N,C)\text{-}CH_2SiMe_2N(SiMe_3))]$ (141). Treating the cyclo-metalated complex $[U(N(TMS)_2)_2(\kappa^2\text{-}(N,C)\text{-}CH_2SiMe_2N(SiMe_3))]$ with D_2 gives the perdeuterio-$[(N(TMS)_2)_3UD]$. Surprisingly, the analogous reaction of $[(Cp')_3UR]$ with H_2 to give $[(Cp')_3UH]$ has not been reported (265). While this lack of reactivity may be due to steric constraints, the U(III) anions $[Cp_3UR]Li$ (R = Me, nBu) readily undergo σ bond metathesis with dihydrogen in THF to give the corresponding hydride $[Cp_3UH]Li$ complex and free alkane (278).

C. Uranium(VI) Complexes

Uranium(VI) bis(imido) complexes react with molecular hydrogen via a formal [3+2]-cycloaddition. This reaction has some precedent in transition metal chemistry: [3+2]-cycloaddition pathways for the oxidation of H_2 by mid- to late metals have been known for a century (279). For example, the reduction of permanganate by H_2 is believed to proceed through a Mn(V) intermediate and a [3+2]-cycloaddition pathway (280, 281). In the case of uranium, Burns and co-workers (242, 282) observed that the uranium(VI) complexes $[(Cp^*)_2U(NPh)_2]$ and $[(Cp^*)_2U(NAd)_2]$ both react with H_2 to give uranium(IV) bis(amide)

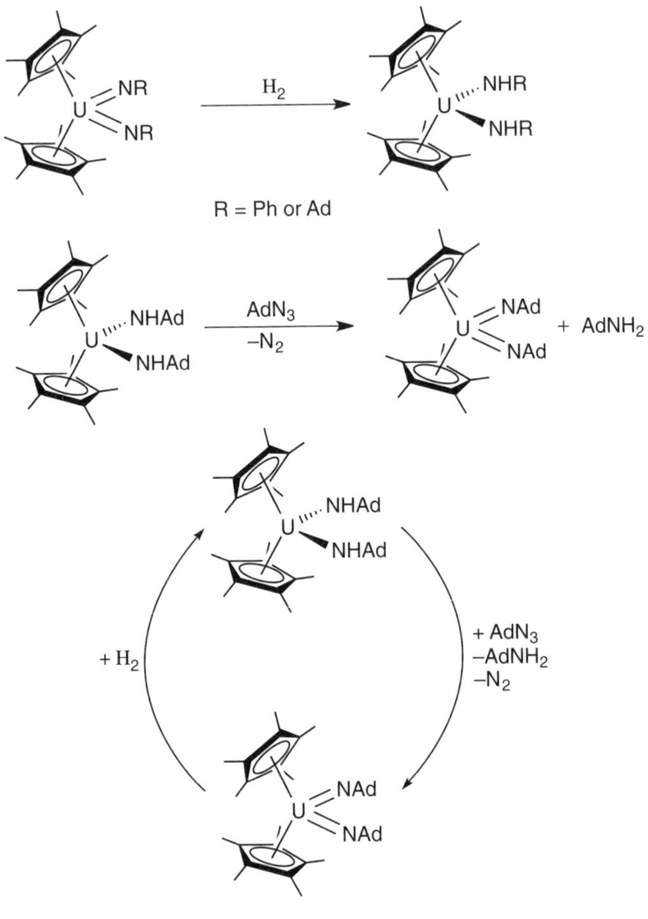

Scheme 38. Dihydrogen homolytic cleavage by $[(Cp^*)_2U(NPh)_2]$ and $[(Cp^*)_2U(NAd)_2]$ and the application of this reaction in the catalytic reduction of AdN_3.

complexes, $[(Cp^*)_2U(NHPh)_2]$ and $[(Cp^*)_2U(NHAd)_2]$. The admantyl bis (imido) reacts substantially faster ($t_{1/2} = 4\,h$ vs $t_{1/2} = 21\,h$). It was found that the addition of AdN_3 to $[(Cp^*)_2U(NHAd)_2]$ results in the reduction of AdN_3 to $AdNH_2$ and the oxidation of $[(Cp^*)_2U(NHAd)_2]$ to $[(Cp^*)_2U(NAd)_2]$. This reaction can in turn be rendered catalytic (Scheme 38): at 55 °C under 1-atm H_2, AdN_3 is catalytically reduced to $AdNH_2$ by 20 mol. % $[(Cp^*)_2U(NAd)_2]$. Preliminary mechanistic studies suggest that the U$-$N bonds are not broken during catalysis and the amides serve as hydrogen-atom donors facilitated by the U(IV/VI) couple.

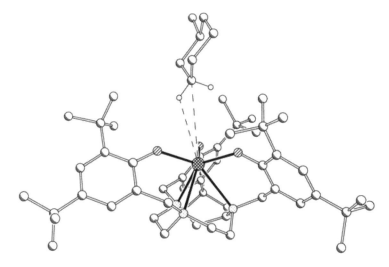

Figure 31.   Crystal structure of $[(({}^{tBu}ArO)_3tacn)U(C_7H_{14})]$ measured at 100 K.

XI.   SATURATED HYDROCARBONS

A.   Uranium(III) Complexes

Meyer and co-workers (283) reported a series of uranium(III) complexes with exceedingly rare examples of intermolecular alkane adducts. The crystallization of $[(({}^{tBu}ArO)_3tacn)U]$ from pentane with 50 equiv of cycloalkane (cyclopentane, hexane, and methyl cyclopentane, and cyclohexane) leads to the isolation of $[(({}^{tBu}ArO)_3tacn)U(C_7H_{14})]$ (C_7H_{14} = methylcyclohexane) in the case methylcyclohexane (Fig. 31). The complex $[(({}^{tBu}ArO)_3tacn)U(C_7H_{14})]$ has $U \cdots C$ distance of 3.864 Å and short contacts between the peripheral t-butyl groups and the axial alkane ligand (from 2.12–2.71 Å). The uranium–carbon distances are less than the sum of the van der Waals radii for a $U-CH_2$ or $U-CH_3$ contact (3.9 Å) and suggest a bonding interaction between the alkane and the U center. X-ray diffraction analysis at 5 K allowed the refinement of all hydrogen atoms and established an η^2-C,H coordination of the alkane ligand.

B.   Uranium(IV) Complexes

While lanthanum and thorium complexes have been shown to activate methane and saturated alkanes by σ-bond metathesis (284, 285), comparable intermolecular reactivity with uranium and $C-H$ bonds of sp^3 or sp^2 hydridized carbon atoms has not been observed. However, intramolecular σ-bond metathesis reactions have been demonstrated recently. Evans et al. (286) has shown that the uranium hydride complexes $[\{(Cp^*)_2U(\mu\text{-}H)\}_2]$ and $[\{(Cp^*)_2U(H)_2\}_2]$ eliminate H_2 thermally and form a dinuclear "tuck-in–tuck-over" complex, $[\{(\mu\text{-}\eta^5:\eta^1:\eta^1\text{-}C_5Me_3(CH_2)_2) Cp^*U\}(\mu\text{-}H)_2\{U(Cp^*)_2\}]$ (Scheme 39). This complex can undergo subsequent reactivity via σ bond metathesis or serve as a latent form of U(II) (287). As mentioned previously, the silyl, alkyl "tuck-in" ligands in the complex $[(\eta^5:\eta^1\text{-}C_5Me_4SiMe_2CH_2)_2U]$ are prepared by methane elimination from $[(\eta^5\text{-}C_5Me_4 SiMe_3)_2U(Me)_2]$ (146). Note that the ligand metallation observed by Andersen and co-workers (141) during $H-D$ exchange by $[(N(TMS)_2)_3UH]$ also constitutes an example of intramolecular sp^3 $C-H$ bond activation (Section X.B) (141). Evans et al. (288) also employed the $[(Cp^*)_2U(Me)_2]$ in the chelation-assisted $C-H$ activation of sp^2 $C-H$ bonds of arenes. Treating $[(Cp^*)_2U(Me)_2]$ with PhTe$-$TePh results in the elimination of MeTePh and CH_4 and the net activation of the phenyl ring via σ bond metathesis to give $[(Cp^*)_2U(\eta^2\text{-}(Te,C)\text{-}TeC_6H_4)]$ (Scheme 39) (288). An analogous reaction for the ortho metallation of pyridine N-oxide has been developed by Kiplinger and co-workers (289–291). Pyridine N-oxide reacts with $[(Cp^*)_2U(R)_2]$ (R = Me, CH_2Ph) to give $[(Cp^*)_2UR(\eta^2\text{-}(O,C)\text{-}ONC_5H_4)]$ (R = Me, CH_2Ph) (Scheme 39). In this case, one alkyl group is retained. Diaconescu (292–296) has pushed this reactivity pattern, employing a ferrocene diamide derivative (297, 298). The complex, $[(NN^{fc})U(CH_2Ph)_2]$ (NN^{fc} = ferrocene($NSitBuMe_2$)$_2$), can ortho-metalate N-heterocycles that, in turn, undergo ring-opening and $C-C$ bond-forming reactions within the coordination sphere (292–296). In the only example of its kind, Meyer and co-workers (299) showed that heating the sterically pressured uranium(III) diazoalkane complex $[((^{Ad}ArO)_3tacn)U(\eta^1\text{-}N_2CPh_2)]$ results in cleavage of the ortho-$C-H$ bond of one of the diphenyldiazomethane phenyl substitutents, followed by nitrogen insertion into the new $U-C$ bond, producing the phenyl-substituted indazole uranium complex, $[((^{Ad}ArO)_3tacn)U(\eta^2\text{-}3\text{-}Ph\text{-}Ind)]$ (Ind = $C_7H_5N_2$, indazolate anion) and the concomitant production of dihydrogen.

C. Uranium(VI) Complexes

High-valent uranium complexes with metal-ligand multiple bonds have also activated sp^3 hybridized $C-H$ bonds intramolecularly. Burns and co-workers (300) observed that the complex $[(Cp^*)_2U(NAd)_2]$ will add a Cp^* methyl $C-H$ bond across the bis(imido) moeity in a formal [3+2]-cycloaddition in benzene at reflux (Scheme 40). Burns and co-workers (301) subsequently found that thermolysis of the organometallic uranium mixed imido, hydrazinado complex $[(Cp^*)_2U(=N\text{-}2,4,6\text{-}tBu_3C_6H_2)(=N-N=CPh_2)]$ results in net $C-H$ bond addition of one of the tBu groups across the imido, hydrazinado moiety to give the uranium(IV) bis(amide) complex (Scheme 40). This reactivity parallels the previously described H_2, formal [3+2] cycloaddition. Kiplinger and co-workers (206) subsequently

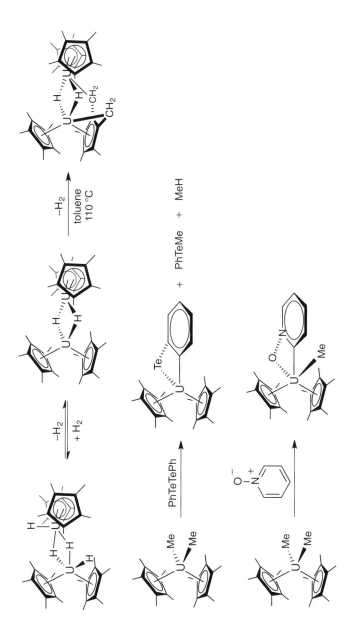

Scheme 39. Synthesis of "tuck-in–tuck-over" complex, $[\{(\mu\text{-}\eta^5:\eta^1:\eta^1\text{-}C_5Me_3(CH_2)_2)Cp^*U\}(\mu\text{-}H)_2\{U(Cp^*)_2\}]$ and arene sp^2 C–H bond activation.

379

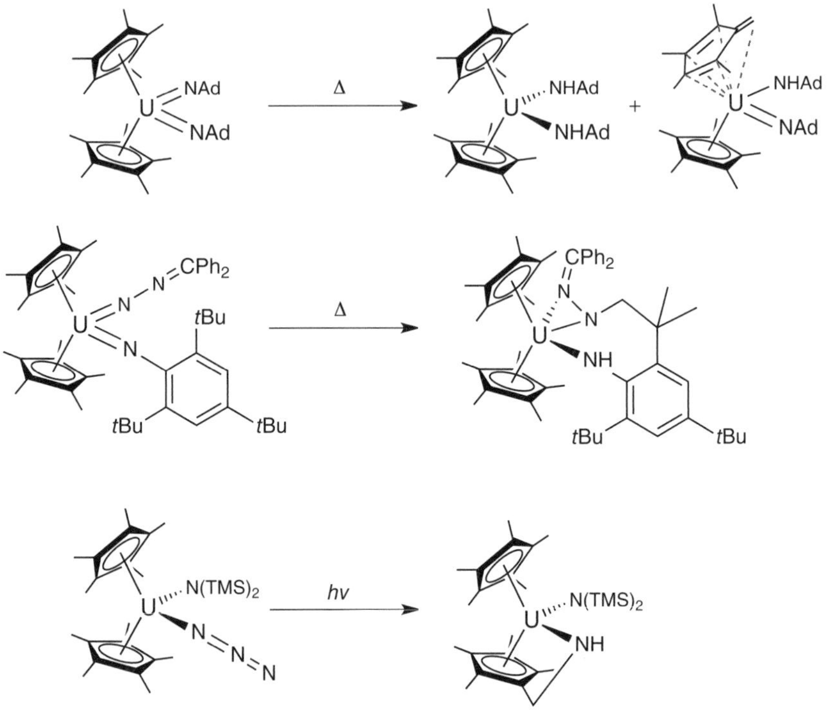

Scheme 40. The C–H bond addition reactions of [(Cp*)₂U(=N-2,4,6-*t*Bu₃C₆H₂)(=N–N=CPh₂)] and [(Cp*)₂UN(N(TMS)₂)].

reported the intramolecular, 1,1-C–H bond addition to a putative uranium(VI) nitride [(Cp*)₂UN(N(TMS)₂)] (derived from the azide [(Cp*)₂UN₃(N(TMS)₂)] by photolysis) to give [(Cp*)U(η^5:η^1-C₅Me₄CH₂NH)(N(TMS)₂)] (Scheme 40).

XII. ALKENES AND ALKYNES

A. Uranium(III) Complexes

1. Reductive Coupling of Alkynes

The reductive coupling of internal alkynes by uranium(III) complexes and an external reductant has been known since 1982, and a variety of uranium complexes are now known to give C–C coupled products (266). Marks and co-workers (266) originally reported that the uranium(III) trimer [{(Cp*)₂UCl}₃] or the monomeric,

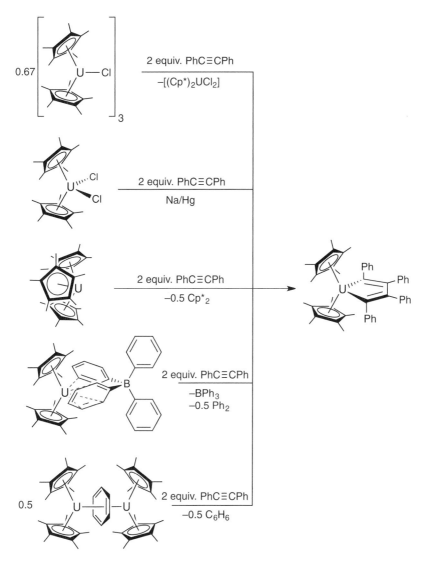

Scheme 41.   Reductive coupling of diphenylacetylene.

$[(Cp^*)_2UCl_2]Na \cdot thf_x$ will disproportionate in the presence of diphenylacetylene to give $[(Cp^*)_2U(C_4Ph_4)]$ and $[(Cp^*)_2UCl_2]$ (Scheme 41). Quantitative conversion to $[(Cp^*)_2U(C_4Ph_4)]$ can be achieved employing excess Na/Hg amalgam. More recently, Evans et al. (302) showed that the uranium(III) complexes $[(Cp^*)_3U]$, $[(Cp^*)_2U((\mu\text{-Ph})_2BPh_2)]$, and $[\{(Cp^*)_2U\}_2(C_6H_6)]$ (Section III.B.3), all serve as

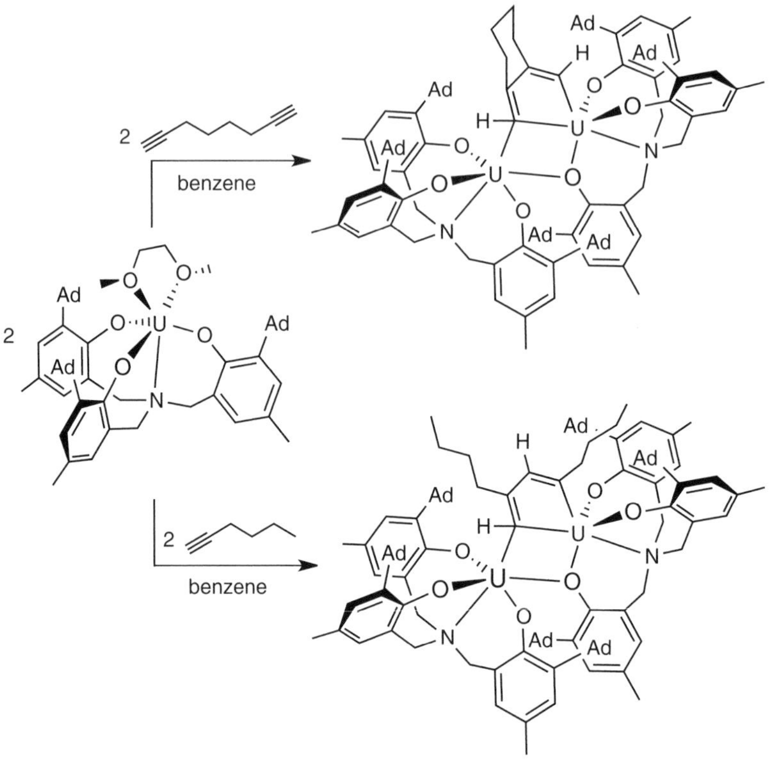

Scheme 42.   Reductive coupling of mono- and dialkynes by $[((^{Ad}ArO)_3N)U(dme)]$.

synthetic equivalents to monomeric "uranium(II)." Treating these complexes with 2 equiv of diphenylacetylene gives the previously characterized $[(Cp^*)_2U(C_4Ph_4)]$ (Scheme 41).

Meyer and co-workers (303) explored the chemistry of mono- and dialkyne substrates with $[((^{Ad}ArO)_3N)U(dme)]$. This complex reacts with 1 equiv of the terminal bis(alkynes), 1,7-octadiyne or 1,6-heptadiyne, undergoing intramolecular C–C coupling and cyclization to form the uranium(IV) complexes $[((^{Ad}ArO)_3N)U\}_2$ $(\mu-\eta^2:\eta^1-1,2-(CH)_2$-cyclohexane)] and $[((^{Ad}ArO)_3N)U\}_2(\mu-\eta^2:\eta^2-1,2-(CH)_2$-cyclopentane)] (Scheme 42). With 2 equiv of the terminal monoalkynes, 1-hexyne or 4-$tert$-butyl-phenylacetylene, $[((^{Ad}ArO)_3N)U(dme)]$ in benzene gives the complexes, $[\{[((^{Ad}ArO)_3N)U\}_2(\mu-\eta^2:\eta^1-2-nBu-1,3$-octadiene)] and $[\{[((^{Ad}ArO)_3N)U\}_2(\mu-\eta^2: \eta^1-1,3$-di-$(p-tBu$-phenyl)butadiene))], in an intermolecular C–C coupling reaction. These U(IV) vinyl complexes could be considered models of cycloisomerization reactions catalyzed by gold complexes (304–306).

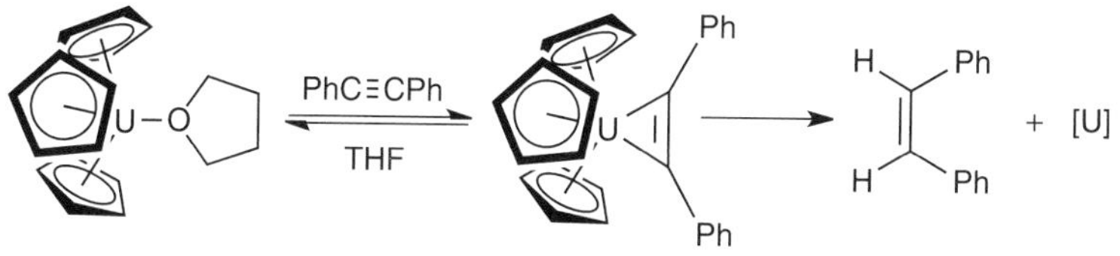

Scheme 43.   Formation of η_2-alkyne complex of uranium(III).

2. η^2-Alkyne Adduct

Ephritikhine and co-workers (278) spectroscopically observed, with ^{1}H NMR and ^{13}C NMR, a potential intermediate in these U(III) mediated C–C coupling reactions of alkynes, namely, a η^2-alkyne complex. The complex [(Cp)$_3$U(thf)] reacts with excess (50-fold) diphenylacetylene in THF to give [(Cp)$_3$ U(η^2-PhC≡CPh)] (PhC≡CPh = diphenylacetylene) (Scheme 43). This complex decomposes relatively rapidly to cis-stilbene (0.3 equiv after 3 h) and an unidentified uranium-containing product. Labeling experiments indicate that the H atoms do not come from the solvent. The authors suggest that hydrogen-atom abstraction from the Cp rings is the operative mechanism. More definitive spectroscopic data for a uranium(IV) η^2-alkyne complex has also been obtained (Section XII.B.3).

3. Terminal Alkyne Protonolysis

Bart and co-workers (228) and Takats and co-workers (307) have prepared unusual examples of uranium(III) acetylide complexes. As mentioned previously, treating [Tp*$_2$U(CH$_2$Ph)] with acidic terminal alkynes gives the complexes [Tp*$_2$U(C≡CTMS)] and [Tp*$_2$U(C≡CPh)] and concomitantly produces toluene. Infrared absorption bands for the C≡C triple bonds in [Tp*$_2$U(C≡CTMS)] and [Tp*$_2$U(C≡CPh)] appear at 1997 and 2049 cm^{-1}, respectively, indicating that this bond has been slightly reduced upon coordination to the uranium center as compared to free (trimethylsilyl)acetylene (2034 cm^{-1}) and phenylacetylene (2110 cm^{-1}). Takats and co-workers (307) also prepared [Tp*$_2$U(C≡CPh)] by an alternative method. Reducing the complex [Tp*$_2$UI] in the presence of phenyl-acetylene with Na/Hg amalgam gives [Tp*$_2$U(C≡CPh)] in 41% yield. A depro-tonation route employing [Tp*$_2$U(N(TMS)$_2$)] proved unreliable.

B. Uranium(IV) Complexes

Alkyne protonolysis reactions and insertion reactions of alkynes and alkenes with uranium(IV) complexes have become very sophisticated, including applica-tions in catalysis (65). Uranium complexes readily catalyze transformations (e.g., hydroamination, hydrogenation, and hydrosilylation). These reactions provide synthetically useful routes to enynes, imines, alkynylimines, and vinylsilanes. The kinetic, thermodynamic, and mechanistic factors that govern these useful transformations are beyond the scope of this chapter and readers are referred to several recent reviews and articles (58–64, 295, 308, 309). This section focuses on the stoichiometric transformation of ethylene and small alkynes.

1. Terminal Alkyne Protonolysis

The synthesis of uranium(IV) acetylides by the protonolysis of terminal alkynes was first explored in the $[U(N(TMS)_2)_2(\kappa^2\text{-}(N,C)\text{-}CH_2SiMe_2N(SiMe_3))]$ complex by Dormond and co-workers (310). The synthesis of alkyl complexes larger than methyl or aryl complexes of the type $[(N(TMS)_2)_3U(R)]$ all result in γ-deprotonation of the ligand to give $[U(N(TMS)_2)_2(\kappa^2\text{-}(N,C)\text{-}CH_2SiMe_2N(SiMe_3))]$. However, treatment of this complex with a variety of terminal alkyl alkynes results in the ready synthesis of complexes of the type $[(N(TMS)_2)_3U(C{\equiv}CR)]$ (where R = n-propyl, n-butyl, n-pentyl, n-hexyl, or t-butyl). These acetylide complexes were shown to selectively insert into ketones. In order to study ferromagnetic coupling between U(IV) centers, Shores and co-workers (311) used this synthetic approach to build ethynylbenzene bridged complexes based on Scott's bulky TREN^{DMSB} ligand. The $[(TREN^{DMSB})UI]$ complex also undergoes ligand cyclometalation with transmetalation of benzyl potassium to give the cyclometalated complex $[(N(CH_2CH_2NSiMe_2tBu)_2(\kappa^2\text{-}(N,C)\text{-}CH_2CH_2NSiMe_2CMe_2(CH_2)U]$ (312). Protonation of this cyclometalated complex with terminal alkynes allows for the construction of mono-, di-, or trimetallic uranium(IV) systems, bridged by ethynyl benzene complexes that possess weak ferromagnetic coupling, such as $[\{(TREN^{DMSB})U\}_2(1,4\text{-}(C{\equiv}C-)-C_6H_4)]$ (Scheme 44) (311). Cummins and co-workers (313) also observed similar reactivity in the construction of a bimetallic μ-acetylide complex. The reaction between $[(N(tBu)Ar)_3UI]$ (Ar = 3,5-Me$_2$C$_6$H$_3$) and sodium acetylide produces $[\{(N(tBu)Ar)_3U\}_2(\mu\text{-}C_2)]$ and acetylene. The symmetric C≡C stretching mode was identified in the Raman spectrum at

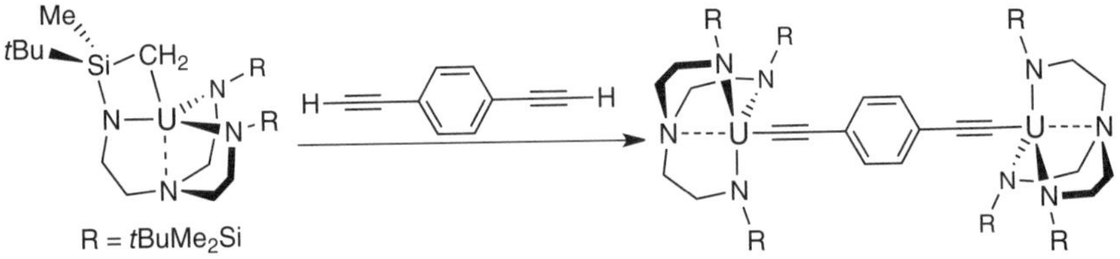

Scheme 44. Synthesis of dinuclear bridging ethynyl benzene complex $[\{(TREN^{DMSB})U\}_2$ $(1,4\text{-}(C{\equiv}C)-C_6H_4)]$.

1904 cm^{-1}. The connectivity was confirmed by an XRD analysis, which revealed a $\mu\text{-}\eta^1\text{:}\eta^1$ coordination mode for the bridging acetylide (or "dicarbide").

Eisen and co-workers (314) applied the protonolysis of terminal alkynes in the reaction of $[(Cp^*)_2U(Me)_2]$ with 2 equiv of phenyl acetylene or *tert*-butyl acetylene to give $[(Cp^*)_2U(CCR)_2]$ (R = phenyl or *tert*-butyl). Alternatively, $[(Cp^*)_2U(C(CR)_2]$ (R = phenyl or *tert*-butyl) can be prepared from the bis(amide) complex $[(Cp^*)_2U(NHR)_2]$ (R = 2,6-dimethyl-C_6H_3). These procedures are typically performed *in situ* to prepare precatalysts for alkyne oligomerization and hydroamination. Crystallographic characterization of $[(Cp^*)_2U(C{\equiv}CPh)_2]$ was obtained by Evans et al. (234) for his studies of the insertion reactions of $[(Cp^*)_2 U(C{\equiv}CR)_2]$, as described previously (Section VII.B.2). Note that Kiplinger and co-workers (315) developed an alternative method for the synthesis of uranium(V) acetylide complexes via the oxidation of uranium(IV) imido complexes $[(Cp^*)_2U (N\text{-}2,6\text{-}i\text{Pr}{-}C_6H_3)]$ with copper acetylide to give copper metal and the uranium(V) complex, $[(Cp^*)_2U(N\text{-}2,6\text{-}i\text{Pr}{-}C_6H_3)(C{\equiv}CPh)]$. This method can also be applied to uranium(III) amides to give uranium(IV) acetylides. The complex $[((Et)_2N)_3U]$ $[BPh_4]$ also undergoes protonolysis with terminal alkynes as evidenced by NMR spectroscopy (Section XII.B.3) (316).

2. Alkyne and Alkene Insertion Chemistry

Two systems have been demonstrated to selectively insert 1 equiv of ethylene into U−H or U−C bonds. The thorium(IV) hydride, $[\{(Cp^*)_2Th(H)_2\}_2]$, reacts with an atmosphere of ethylene to give the mononuclear bis(ethyl) complex, $[(Cp^*)_2Th(Et)_2]$ (Scheme 45) (162). This ready insertion of alkenes has been coupled to the σ bond metathesis reactions of uranium(IV) alkyls in these systems with H_2 to give the parent complex $[\{(Cp^*)_2U(H)_2\}_2]$ and to develop a catalytic

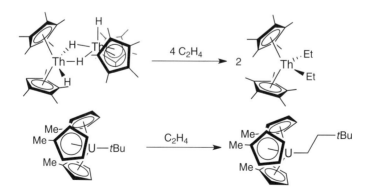

Scheme 45.   Ethylene insertion reactions.

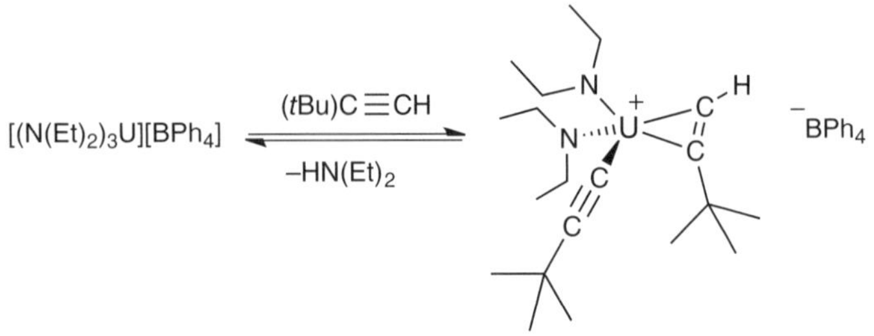

Scheme 46.   Observation of $[((Et)_2N)_2U(C{\equiv}C(tBu))(\eta^2\text{-HC}{\equiv}C(tBu))][BPh_4]$.

reduction of alkenes. $[\{(Cp^*)_2U(H)_2\}_2]$ rapidly reduces 1-hexene to n-hexane under an atmosphere of H_2. Similarly, Andersen and co-workers (139) observed the insertion of ethylene into a U–C σ bond in $[(C_5H_4Me)_3U(tBu)]$ to give $[(C_5H_4Me)_3U(CH_2CH_2\text{-}tBu)]$. For reactions of alkenes in catalytic intramolecular hydroamination reactions, see the work of Marks and co-workers (63, 308, 309). For the insertion of alkynes into U–C, U–N, and U–H bonds, which are key elementary steps in a variety of catalytic transformations, see the aforementioned reviews (58–62).

3. η^2-Alkyne Adduct

Eisen and co-workers (316) also presented spectroscopic evidence for a uranium(IV) η^2-alkyne complex $[((Et)_2N)_3U][BPh_4]$ as a precatalyst for alkyne dimerization. In the presence of a large excess of *tert*-butylalkyne, the complex $[((Et)_2N)_3U][BPh_4]$ undergoes protonolysis to give $[((Et)_2N)_2U(C{\equiv}C(tBu))]$ $[BPh_4]$, which is then trapped as the η^2-alkyne complex by another equivalent of *tert*-butylalkyne and thus yielding $[((Et)_2N)_2U(C{\equiv}C(tBu))(\eta^2\text{-HC}{\equiv}C(tBu))]$ $[BPh_4]$ (Scheme 46). This complex has a half-life of 6 h. The ^{1}H NMR spectrum shows one acetylide signal (C–H) at −2.14 ppm, which correlated in the DEPT (distortionless enhancement by polarization transfer) NMR and two-dimensional (2D) C–H correlation to a carbon at −19.85 ppm with a coupling constant of $J =$ 250 Hz. In addition, two tBu group signals have been found in the ^{1}H and ^{13}C DEPT, and C–H correlation NMR spectra. These results clearly indicate that a second equivalent of acetylene is bound to the cationic complex. The C≡C stretching of the free alkyne ($2108 \, cm^{-1}$) is not visible, and two signals at lower frequencies are observed at 2032 and $2059 \, cm^{-1}$, as expected for η^2-alkyne complexes.

4.   [2+2] Addition and Metallacycle Formation

Like low-coordinate transition metal, metal–ligand mutiple-bond complexes, uranium(IV) imido and hydrazinado complexes undergo [2+2]-cycloaddition reactions with alkynes to give metallacyclobutenes. Andersen and co-workers (317, 318) prepared "base-free" uranium(IV) oxo and imido complexes supported by the bulky Cp derivative, Cp'''. These complexes [(Cp''')$_2$UNMe] and [(Cp''')$_2$UO] have divergent chemistry with internal alkynes. The imido complex, [(Cp''')$_2$UNMe], reacts with dimethyl- or diphenyl acetylene to give the azametallocylcobutene complexes [(Cp''')$_2$U(N(Me)C(R)=C(R)] (R = Me or Ph) (Scheme 47). In the presence of excess MeNH$_2$, the imine MeN=C(R)CH$_2$R is released, and the uranium(IV) bis(amide) complex is formed, [(Cp''')$_2$U(NHMe)$_2$]. These stoichiometric reactions can be coupled for the catalytic hydroamination of internal alkynes. In contrast to [(Cp''')$_2$UNMe], [(Cp''')$_2$UO] will not form metallacycles with internal alkynes, but will undergo nucleophilic addition with TMSCl, a reaction in which [(Cp''')$_2$UNMe] will not participate. The DFT studies suggest that the dichotomous reactivity of the two apparently multiply bonded ligands in an isostructural environment is derived from the essentially σ only bonding in [(Cp''')$_2$UO] (319). This bonding structure leads to a stronger U−O bond. This result is counterintuitive because increased bond order and/or covalency has been conflated with increased bond strength. Since, on the one hand, there is little to no π bonding (or multiple-bond character) in the U−O bond, it does not participate in [2+2]-cycloaddition reactions. However, due to the charge localization at the oxygen atom, [(Cp''')$_2$UO] is a potent nucleophile. On the other hand, the complex [(Cp''')$_2$U(NMe)] has a weaker U=N bond, but multiple-bond character. As a result, it undergoes in [2+2]-cycloaddition with internal alkynes. Bart and co-workers (320) recently presented another example of this reaction pattern. The addition of trimethylsilyldiazomethane (TMSCHN$_2$) to the previously discussed [Tp*$_2$U(CH$_2$Ph)] results in the production of 1,2-diphenyl-ethane and a transiently stable η^1-hydrazonido complex, [Tp*$_2$U(η^1-TMSCHN$_2$)]. This product can be trapped by the addition of phenylacetylene to give the azametallocy-clobutene complex, [Tp*$_2$U((N(N=CHTMS)C(H)=C(Ph))], as definitively identified by XRD and IR spectroscopy (Scheme 47).

XIII. ARENES

A. Half-Sandwich Complexes

Metal–arene complexes are well known for the transition metals and lanthanides, but remain rare for the 5f elements (321). Remarkably, an η^6-benzene complex of uranium was reported in 1971 by a modified version of Fischer's Friedel–Craft reaction (322, 323). Refluxing a mixture of UCl$_4$, AlCl$_3$, and powdered aluminum metal in benzene for 7 h, followed by filtration and standing for 2–3 days, led to the isolation of black crystals. X-ray diffraction studies revealed the molecular structure of the η^6-benzene complex [(η^6-C$_6$H$_6$)U(AlCl$_4$)$_3$]

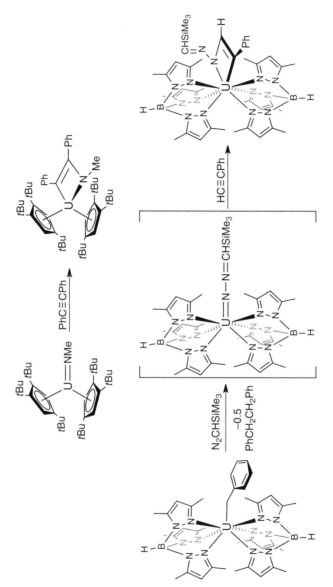

Scheme 47. The [2+2]-cycloaddition of internal and terminal alkynes with uranium(IV) imido and hydrazonido complexes.

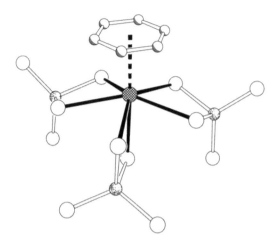

Figure 32.   Crystal structure of $[(\mu^6\text{-}C_6H_6)U(AlCl_4)_3]$.

(Fig. 32), with a mean $U-C$ distance of 2.91(1) Å. Modification of these methods and the use of hexamethylbenzene (C_6Me_6) results in the synthesis of a variety mono-, di-, and trinuclear formally uranium(III) and (IV) η^6-arene complexes including $[(\eta^6\text{-}C_6Me_6)U(AlCl_4)_3]$, $[(U_2(\eta^6\text{-}C_6Me_6)_2Cl_4(\mu\text{-}Cl)_3][AlCl_4]$, $[(U(\eta^6\text{-}C_6Me_6)Cl_2(\mu\text{-}Cl)_3UCl_2(\eta^6\text{-}C_6Me_6)][AlCl_4]$, $[(U(\eta^6\text{-}C_6Me_6)Cl_2(\mu\text{-}Cl)_3UCl_2(\mu\text{-}Cl)_3UCl_2(\eta^6\text{-}C_6Me_6))]$, and $[(U_3(\mu^3\text{-}Cl)_3(\mu^2\text{-}Cl)_3(\mu^1,\eta^2\text{-}AlCl_4)_3(\eta^6\text{-}C_6Me_6)_3)]$ $[AlCl_4]$ (324–326). These methods also have been extended to a mono-toluene (PhMe) complex, $[(\eta^6\text{-}PhMe)U(AlCl_4)_3]$ (327). Ephritikine and co-workers (328, 329) developed a more direct method for the synthesis of mononuclear uranium η^6-arene complexes by refluxing $[U(BH_4)_4]$ in mesitylene ($C_6H_3Me_3$) to get $[(\eta^6\text{-}C_6H_3Me_3)U(BH_4)_3]$. This complex readily exchanges the arene ligand in THF to give $[U(BH_4)_3(thf)]$ or, in benzene with hexamethylbenzene, to give $[(\eta^6\text{-}C_6Me_6)U(BH_4)_3]$. The complex $[(\eta^6\text{-}C_6H_3Me_3)U(BH_4)_3]$ also serves as a precursor to Cp complexes: treating $[(\eta^6\text{-}C_6H_3Me_3)U(BH_4)_3]$ with CpH results in the production of H_2 and the isolation of $[CpU(BH_4)_3]$ (328).

Sattelberger and co-workers (330) found that this arene-bonding motif could be extended to tris(aryloxide) complexes. Protonolysis of $[((TMS)_2N)_3U]$ with HODipp yields the homoleptic tris(aryloxide) $[(ODiip)_3U]$, which, in solution, appears to be monomeric by ^1H NMR because there is only one resonance observed corresponding to the isopropyl methyl protons. In the solid state, however, when crystallized from hexane, $[(ODiip)_3U]$ has a dimeric centro-symmetric structure, in which the aryl ligands bridge and coordinate in an η^6 fashion. The two η^6-arene bridges $[U-C(av) = 2.92(2)$ Å] hold the dinuclear unit together with a $U\cdots U$ separation of 5.34 Å. This propensity of uranium(III) tris(aryloxide) complexes to form η^6-arene interactions has been employed by Meyer and co-workers (216) to template arene and aryloxide coordination in the complex $[((^{tBu}ArO)_3mes)U]$, described previously. Evans et al. (331) also observed the formation of an η^6-arene complex in the reaction of $[(C_5Me_4H)_2U(Me)_2]K$ with 2 equiv of $[HNEt_3][BPh_4]$ to give H_2, triethylamine (NEt_3), $K[BPh_4]$, and $[(C_5Me_4H)_2 U(\mu\text{-}\eta^6:\eta^1\text{-Ph})(\mu\text{-}\eta^1:\eta^1\text{-Ph})BPh_2)]$. This connectivity is in contrast to the Cp* systems in which two phenyl rings are coordinated in an η^2-fashion, $[(Cp^*)_2 U((\mu\text{-}\eta^2:\eta^1\text{-Ph})_2BPh_2)]$ (332). In $[(C_5Me_4H)_2U(\mu\text{-}\eta^6:\eta^1\text{-Ph})(\mu\text{-}\eta^1:\eta^1\text{-Ph})BPh_2)]$, the U—C distances in the η^6-phenyl are between 2.902(4) and 3.066(4) Å. Uranium bis(arene) sandwiches remain unknown, but have been studied by computational methods (112, 333–335).

B. Dinuclear, Inverted Sandwich Complexes

1. $[\{(Ar(R)N)_2U\}_2(\mu^2\text{-}\eta^6:\eta^6\text{-Arene})]$

Dinuclear inverted sandwich complexes, in which an arene is bound in a $\mu^2\text{-}\eta^6:\eta^6$ fashion, are relatively uncommon, and there are only a few examples throughout the transition metals (336, 337). The first report of a uranium dinuclear inverted sandwich complex was in 2000 by Cummins and co-workers (338). Although several methods were found to make $[\{(Ar(R)N)_2U\}_2(\mu^2\text{-}\eta^6:\eta^6\text{-arene})]$ (R = t-Bu or Ad, Ar = 3,5-$C_6H_3Me_2$), depending on the arene (benzene or toluene) and on the R group, the most straightforward method was the reduction of $[(Ar(tBu)N)_2UI]$ in benzene or toluene with KC_8 (potassium graphite), which, in the case of the toluene adduct, gave the product $[\{(Ar(tBu)N)_2U\}_2(\mu^2\text{-}\eta^6:\eta^6\text{-}C_7H_8)]$ in 40% yield. The chemical shifts of the bound toluene were assigned by deuterium labeling: the four resonances are found at +18.7, −65.0, −83.6, and −88.8 ppm in the ^2H NMR spectrum. Structural data for the adamantyl derivative $[\{(Ar(Ad)N)_2U\}_2(\mu^2\text{-}\eta^6:\eta^6\text{-}C_7H_8)]$ revealed that the U—C average distance is 2.594(9) Å and C—C (μ-toluene, average) is 1.438(13) Å (Fig. 33). Thus, the U—C average distance is much shorter than in the mono-arene complexes (~2.91 Section XIII.A), and the C—C bond distance is slightly longer than in free toluene (0.04 Å). Such structural parameters suggest that the uranium valency falls between two possible formulations: two divalent uranium centers, and a neutral arene, or two tetravalent uranium centers and an arene tetraanion. A recent combined density functional and CASSCF/CASPT2 theoretical and uranium L_3 edge X-ray absorption near-edge spectroscopy (XANES) study clearly establishes that the f orbitals host the unpaired electrons, followed energetically by two δ bonds formed by filled uranium f orbitals and lowest unoccupied molecular orbitals (LUMOs) of toluene, in agreement with latter formal valency assignment (339). The chemical reactivity reveals $[\{(Ar(R)N)_2U\}_2(\mu^2\text{-}\eta^6:\eta^6\text{-arene})]$ to behave as a latent source of divalent uranium and to react as a 4 e$^-$ reductant. Accordingly,

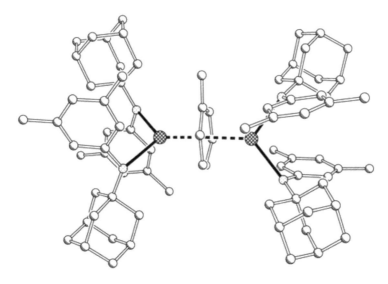

Figure 33.   Crystal structure of [{(Ar(Ad)N)$_2$U}$_2$(μ^2-η^6:η^6-C$_7$H$_8$)].

reactions with Ph$_2$S$_2$ and azobenzene result in the elimination of free arene and the isolation of dimeric U(IV) complexes [{(Ar(R)N)$_2$U(SPh)}$_2$(μ^2-SPh)$_2$] and [{(Ar(R)N)$_2$U}$_2$(μ^2-NPh)$_2$]. Preliminary DFT studies suggest a highly covalent-bonding model in which 4 e$^-$ are stabilized in δ back-bonding with the arene LUMO orbitals. Such δ back-bonding interactions have previously been proposed by Bursten and co-workers (340) to describe the bonding in [U(C$_7$H$_7$)$_2$]$^{1-}$.

2.   Complex [{(Mes(tBu)CN)$_3$U}$_2$(μ^2-η^6:η^6-Arene)]$^{1-/2-}$

Modification of the supporting ligand set to the bulky, and more π-accepting ketimide ligand, [{(Mes(*t*Bu)CN)]$^{1-}$ (Mes = mesityl), allowed for the construction of inverted arene sandwich complexes in two charge states and for the synthesis of these complexes in DME so that a variety arenes could be studied (341). The dianionic arene complexes are prepared by the reduction of [(Mes(*t*Bu)CN)$_3$UI] with 4 equiv of KC$_8$ or a Na mirror in DME with 0.5 equiv of arene. This reaction leads to the isolation of complexes of the type [{(Mes(*t*Bu)CN)$_3$U}$_2$(μ^2-η^6:η^6-arene)]$^{2-}$, where the arene can be benzene, toluene, naphthalene, biphenyl, *trans*-stilbene, or *p*-terphenyl. In all cases, both uranium centers are bound to the same aryl ring. This connectivity is in contrast to d block dimetal complexes of naphthalene (napth) that bind to adjacent rings in a "slipped-inverted sandwich" structure (342, 343). In the XRD determined molecular structure of K$_2$[{(Mes(*t*Bu)CN)$_3$U}$_2$(μ^2-η^6:η^6-arene)], the 12 U−C distances are quite short (from 2.565(11) to 2.749(10) Å) (344). These monoanion complexes can also be prepared by reduction of [(Mes(*t*Bu)CN)$_3$UI] with 2 equiv of KC_8 in DME and 0.5 equiv of arene. The two charge states can, in turn, be interconverted by reduction with K/anthracene or oxidation with [Cp_2Fe] [OTf], respectively. Notably, mono- and dianion complexes react in a convergent manner with Ph_2S_2. The reaction of [{(Mes(tBu)CN)$_3$U}$_2$(μ^2-η^6:η^6-napth)]$^{2-}$ with Ph_2S_2 (2 equiv) yields the dinuclear trithiolate-bridged uranium(IV) derivative, [{(Mes(tBu)CN)$_3$U}$_2$((μ-SPh)$_3$)]$^{1-}$ and, presumably, MSPh (M = Na, K) in a net 4 e^- oxidation. The monoanion [{(Mes(tBu)CN)$_3$U}$_2$(μ^2-η^6:η^6-napth)]$^{1-}$ complex was found to be oxidized by 3 e^- with 1.5 equiv of Ph_2S_2 to give the same product, [{(Mes(tBu)CN)$_3$U}$_2$((μ-SPh)$_3$)]$^{1-}$ (Scheme 48)(344). As a part of these studies, the authors advance a more detailed bonding model, which emphasizes the covalent nature of the δ back-bonding in these complexes and the isolobal relationship of [{(Mes(tBu)CN)$_3$U}$_2$(μ^2-η^6:η^6-C_6H_6)]$^{2-}$ with the triple-decker transition metal complex [{CpV}$_2$(μ^2-η^6:η^6-C_6H_6)] (336).

3. Complex [{(Cp*$_2$U}$_2$(μ^2-η^6:η^6-Arene)] and Analogs

Subsequent to the initial reports by Cummins, Evans and co-workers (345) reported the synthesis of [{(Cp*$_2$U}$_2$(μ^2-η^6:η^6-C_6H_6)] by the reduction of [(Cp*)$_3$U] in benzene with KC_8 or potassium mirror with 18-C-6 (1,4,7,10,13,16-hexaoxacyclooctadecane). The coproduct in these reactions is KCp*. The authors propose that this reaction proceeds by ionic metathesis with a benzene anion. The solid-state molecular structure of [{(Cp*$_2$U}$_2$(μ^2-η^6:η^6-C_6H_6)] reveals that the bridging ligand is not planar, but the 1.42(2)–1.462(18) Å C–C bond distances and 117.5(12)°–121.4(12)° C–C–C angles are indistinguishable within the error limits from those in free benzene (Fig. 34). A single Cp* ligand per uranium can be exchanged with a variety of supporting ligands (X), including KOAr (Ar = 2,6-di-tBu-4-MeC_6H_2), LiCH(TMS)$_2$, and KN(TMS)$_2$ to give complexes of the type [{(Cp*) (X)U}$_2$(μ^2-η^6:η^6-C_6H_6)] (345, 346). The solid-state structures of these complexes indicate that the bonding in the [U_2(μ^2-η^6:η^6-C_6H_6)] core is largely unperturbed by substitution, although in some complexes the buckling of the benzene to a distorted chair conformation is more pronounced. The authors note that the magnetic and spectroscopic data are most consistent with the formulation of the complex as formally two trivalent uranium centers and a benzene dianion. The DFT studies present a similar model to that developed by Cummins and co-workers (338) in that the two uranium centers are covalently bonded to the benzene via δ back-bonding with the benzene LUMOs. As mentioned previously (Section XIII.B.1) these complexes behave as latent divalent uranium species and engage in up to 4 e^- reductions per uranium center, vis-à-vis the synthesis of [(Cp*)$_2$U(NPh)$_2$] from [{(Cp*$_2$U}$_2$(μ^2-η^6:η^6-C_6H_6)] with 2 equiv of azobenzene and the elimination of benzene (346).

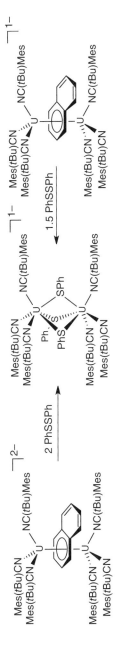

Scheme 48. Oxidation of $[\{(Mes(tBu)CN)_3U\}_2(\mu^2\text{-}\eta^6:\eta^6\text{-}napth)]^{2-/1-}$ by Ph_2S_2.

393

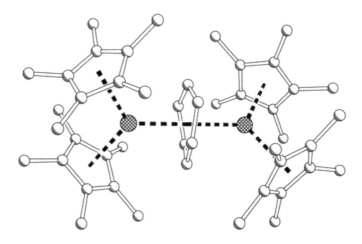

Figure 34.   Crystal structure of [{(Cp*$_2$U}$_2$(μ^2-η^6:η^6-C$_6$H$_6$)].

4. *Complex [{(NN^{fc})U}$_2$(μ^2-η^6:η^6-C$_7$H$_8$)] and [{(BIPM^{TMS}H)U(I)}$_2$(μ^2-η^6:η^6-C$_7$H$_8$)]*

Diaconescu and co-workers (347) built on these studies to construct [{(NN^{fc})U}$_2$(μ^2-η^6:η^6-C$_7$H$_8$)] from the reduction of [(NN^{fc})UI] in toluene with KC$_8$. This complex presents with similar metrical parameters for the bridging toluene as in [{(Ar(*t*Bu)N)$_2$U}$_2$(μ^2-η^6:η^6-C$_7$H$_8$)]. The complex eliminates toluene with the addition of pyrazine to form a tetracycle complex, [{(NN^{fc})U(μ^2-η^1:η^1-pyrazine}$_4$], in which each pyrazine has been reduced by 2 e$^-$. Liddle and co-workers (348) employed a related tridentate chelate to synthesize a dinuclear inverted sandwich complex. The reduction of [{(BIPM^{TMS}H)U(I)(μ-I}$_2$] [BIPM^{TMS}H = CH(PPh$_2$N(TMS))$_3$)$_2$] with 2 equiv of KC$_8$ in toluene gives [{(BIPM^{TMS}H)U(I)}$_2$(μ^2-η^6:η^6-C$_7$H$_8$)]. This complex also presents with solid-state bond metrics similar to [{(Ar(*t*Bu)N)$_2$U}$_2$(μ^2-η^6:η^6-C$_7$H$_8$)] with U—C bond lengths in the range of 2.553(7)–2.616(7) Å and a moderate lengthening of C—C bonds (0.02 Å) in the μ-toluene. Note that this complex possesses magnetic hysteresis at low temperature (1.8 K), which suggests that the complex may be a single-molecule magnet.

5. *Complex [{(OSi(OtBu)$_3$)$_3$U}$_2$(μ^2-η^6:η^6-C$_7$H$_8$)] and [{(Ts^{Xyl})U}$_2$(μ^2-η^6:η^6-C$_7$H$_8$)]*

Mazzanti and co-workers (224) employed the tris(siloxide) uranium(III) complex, [U(OSi(O(*t*Bu))$_3$)$_2$(μ-OSi(O(*t*Bu))$_3$)]$_2$, to synthesize a μ^2-η^6:η^6-C$_7$H$_8$ complex directly by crystallization of [U(OSi(O(*t*Bu))$_3$)$_2$(μ-OSi(O(*t*Bu))$_3$)]$_2$

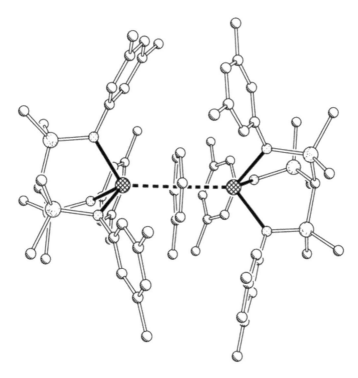

Figure 35.   Crystal structure of $[\{(Ts^{Xyl})U\}_2(\mu^2\text{-}\eta^6\text{:}\eta^6\text{-}C_7H_8)]$.

from toluene–hexane. This reaction yields the inverted sandwich complex $[\{(OSi(O(\textit{t}Bu))_3)_3U\}_2(\mu^2\text{-}\eta^6\text{:}\eta^6\text{-}C_7H_8)]$ in excellent yield. The authors give little analysis of the bonding in $[\{(OSi(O(\textit{t}Bu))_3)_3U\}_2(\mu^2\text{-}\eta^6\text{:}\eta^6\text{-}C_7H_8)]$. However, Liddle and co-workers (349) prepared another neutral analog of $[\{(Mes(\textit{t}Bu)CN)_3U\}_2$ $(\mu^2\text{-}\eta^6\text{:}\eta^6\text{-arene})]^{2-}$, $[\{(Ts^{Xyl})U\}_2(\mu^2\text{-}\eta^6\text{:}\eta^6\text{-}C_7H_8)]$ $(Ts^{Xyl} = HC(SiMe_2NAr)^{3-}$; Ar = 3,5-Me$_2C_6H_3$). This complex is prepared by the reduction of the uranium(IV) chloride, $[(Ts^{Xyl})UCl(thf)]$, in toluene with KC$_8$. The molecular structure of $[\{(Ts^{Xyl})U\}_2(\mu^2\text{-}\eta^6\text{:}\eta^6\text{-}C_7H_8)]$ as determined by XRD (Fig. 35) shows that the U$-$C bonds are long compared to those in $[\{(Ar(\textit{t}Bu)N)_2U\}_2(\mu^2\text{-}\eta^6\text{:}\eta^6\text{-}C_7H_8)]$ and span the range 2.651(4)–2.698(4) Å. However, these bond metrics are shorter than those observed in the uranium(III) mono-arene complexes discussed in Section XIII. A.1. The electronic absorption spectrum of $[\{(Ts^{Xyl})U\}_2(\mu^2\text{-}\eta^6\text{:}\eta^6\text{-}C_7H_8)]$ in toluene over the range 5000–25,000 cm^{-1} presents a clearly separated sharp peak at 6815 cm^{-1} ($\varepsilon = 120$ m^{-1} cm^{-1}) that is characteristic of U(V). Variable temperature magnetic moment measurements of $[\{(Ts^{Xyl})U\}_2(\mu^2\text{-}\eta^6\text{:}\eta^6\text{-}C_7H_8)]$ revealed a magnetic moment of 3.39 μ_B at 300 K, which decreases to 0.84 μ_B at 1.8 K, also typical for U(V). This data, in conjunction with DFT studies, lead the authors to propose the unusual formal electronic structure for the $[U_2(\mu^2\text{-}\eta^6\text{:}\eta^6\text{-}C_6H_6)]$ core with two pentavalent uranium centers bridged by a toluene tetraanion through two δ bonds, rather than tetravalent uranium centers bridged by a toluene dianion. Such an assignment is surprising based on consideration of the relevant redox potentials of a toluene tetraanion and U(V), but it does explain the observed spectroscopic, structural, and magnetic data. A more informed discussion of spectroscopic oxidation state in all these diuranium, inverted sandwich complexes awaits the synthesis of an isostructural series in all three charge states to be examined by uranium L_3-edge XANES (350–352). However, in support of the spectroscopic analysis presented by Liddle and co-workers (353), the group recently reported the characterization of the derivative, $[\{(Ts^{tol})U\}_2(\mu^2\text{-}\eta^6\text{:}\eta^6\text{-}C_7H_8)]$ $(Ts^{tol} = HC(SiMe_2NAr)^{3-}; Ar = 4\text{-}MeC_6H_4)$. This analysis of the complex extends that of $[\{(Ts^{Xyl})U\}_2(\mu^2\text{-}\eta^6\text{:}\eta^6\text{-}C_7H_8)]$ to include a 5 K EPR spectrum with an intense feature centered at $g \sim 2.7$. Best-fit analysis gives g-values of 3.3 and 2.7 (the third value is ill-defined). The authors argue that this feature can only be explained by a uranium(V), which is a Kramer's ion, as opposed to uranium(VI) (EPR silent), uranium(IV) (non-Kramer's ion, no signal expected under experimental conditions), and uranium(III) would not match the rest of the supporting characterization and might be expected to give a more anisotropic spectrum.

6.   Complex $[\{(X)_2U\}_2(\mu^2\text{-}\eta^6\text{:}\eta^6\text{-}Arene)]$ and Arene C–H Borylation

Arnold et al. (354) found that strong external reductants (e.g., KC_8 or Na or K mirrors) are not necessary to construct complexes with a $[U_2(\mu^2\text{-}\eta^6\text{:}\eta^6\text{-}C_6H_6)]$ core. Reaction of the protypical U(III) complexes $[(ODtbp)_3U]$ and $[((TMS)_2N)_3U]$ in benzene at $90\,^\circ C$ in a sealed tube for 6 days gives quantitative conversion to complexes of the type $[\{(X)_2U\}_2(\mu^2\text{-}\eta^6\text{:}\eta^6\text{-}C_6H_6)]$, where $X = N(TMS)_2$ or ODtbp. Bond metrics of the $[U_2(\mu^2\text{-}\eta^6\text{:}\eta^6\text{-}C_6H_6)]$ core of these complexes in the XRD determined molecular structure imply (by comparison to $[\{(Ar(Ad)N)_2U\}_2(\mu^2\text{-}\eta^6\text{:}\eta^6\text{-}C_7H_8)]$) the reduction of the benzene to the dianion. The coproducts of the disproportionation reactions are $[(ODtbp)_4U]$ and $[U(N(TMS)_2)_2(\kappa^2\text{-}(N,C)\text{-}CH_2SiMe_2N(SiMe_3))]$ (from $[((TMS)_2N)_4U]$ with production of $HN(TMS)_2$), respectively (Scheme 49). The authors propose that it is possible for one X ligand to bridge and subsequently transfer to another UX_3, concomitant with electron transfer from the nascent UX_2 fragment to the arene solvent. An experiment with added dihydroanthracene revealed no production of H_2, which is consistent with this proposed mechanism and the absence of radical species in solution. Additionally, borylation of the trapped arene is possible by addition of HBBN (9-bora-9-bicyclononane) and heating to $90\,^\circ C$ to produce H_2 and $[\{(ODtbp)_2U\}_2(\mu^2\text{-}\eta^6\text{:}\eta^6\text{-}PhBBN)]$ (Scheme 49). This reactivity pattern is incompatible with alkali metal reductants. It suggests that, in conjunction with the observation that benzene can exchange with the functionalized napthalene in $[\{(ODtbp)_2U\}_2(\mu^2\text{-}\eta^6\text{:}\eta^6\text{-}C_{10}H_7BBN)]$, this reaction may eventually be rendered catalytic.

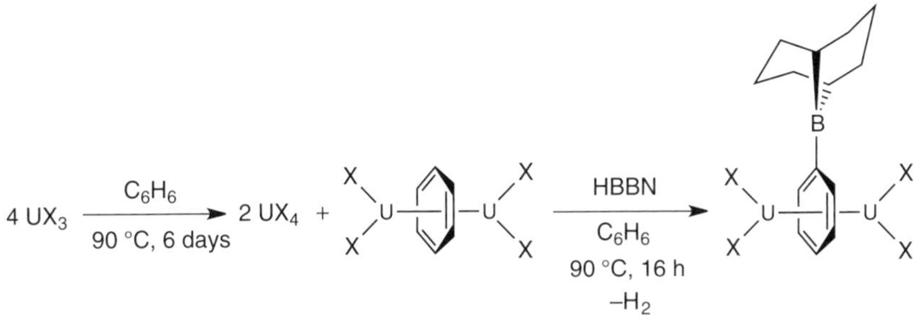

Scheme 49.   Synthesis and R_2BH reactivity of $[\{(X)_2U\}_2(\mu^2\text{-}\eta^6\text{:}\eta^6\text{-}C_6H_6)]$, where $X = N(TMS)_2$ or ODtbp.

C. Benzyne

Uranium(IV) benzyne complexes have been implicated in σ-bond metathesis reactions of uranium diphenyl complexes and the activation of sp^2 C–H bonds. Specifically, Marks and co-workers (162) observed that the phenyl ligands in $[(Cp^*)_2U(Ph)_2]$ become perdeuterated in C_6D_6 solvent, suggesting that ortho-hydrogen abstraction may occur to form a uranium benzyne complex, which in turn adds C_6D_6 via σ-bond metathesis. The authors report the spectroscopic characterization of the insertion of diphenylacetylene into the putative benzyne to give $[(Cp^*)_2U(C_6H_4(Ph)C=C(Ph))]$ (Scheme 50). Evans et al. (355) provided

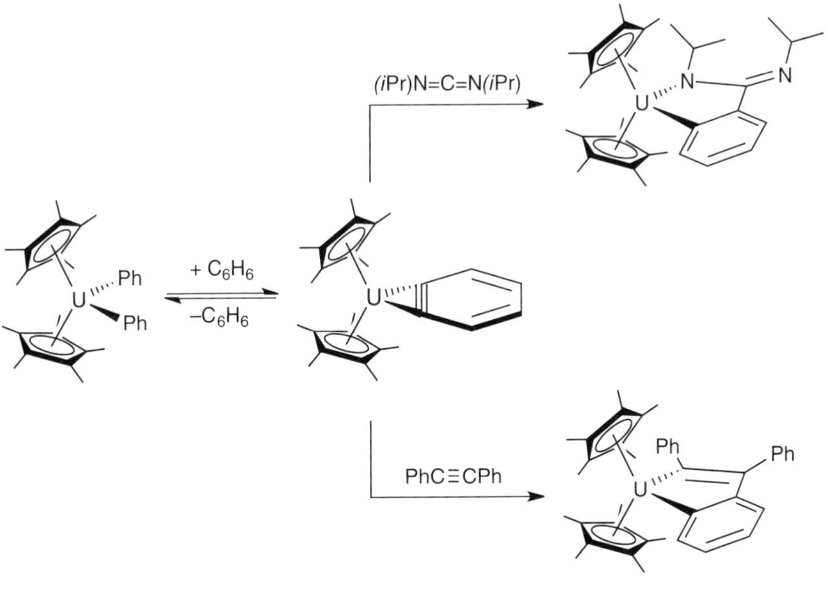

Scheme 50.   A σ-bond metathesis and 1,2-insertion chemistry of the putative benzyne complex $[(Cp^*)_2U(C_6H_4)]$.

further evidence for this hypothesis in the reaction of phenyl lithium with $[\{(Cp^*)_2U(Me)(OTf)\}_2]$ to give LiOTf, methane, and a U(IV) complex, whose ^1H NMR agrees with the formulation of $[(Cp^*)_2U(C_6H_4)]$ (Scheme 50). Trapping this intermediate with 1,3-diisopropylcarbodiimide gives the crystallographically characterized 1,2-insertion product $[(Cp^*)_2U(C_6H_4C(=N(iPr))N(iPr))]$.

XIV. CONCLUDING REMARKS

The field of uranium small molecule activation chemistry has grown rapidly in the last few years. However, as this chapter demonstrates, the rich and unique chemistry of uranium is just beginning to be explored. Many lines of inquiry have been started but remain unresolved. In particular, the development of systems with which to probe the electronic basis of structure and reactivity is necessary. These fundamental studies may provide the design criteria for valuable chemical processes and for the development of nuclear waste remediation technologies.

ACKNOWLEDGMENTS

Dr. Christina Hauser, Dr. Frank W. Heinemann, and Dr. Andreas Scheurer are thanked for their insightful discussions and help in preparing this chapter. HSL and KM would like to thank their mentors (John Arnold, Richard Andersen, Karl Wieghardt, and Christopher Cummins) for their inspiration to study uranium small-molecule reactivity. The work described herein by the Meyer group was accomplished through the diligent efforts of graduate students and postdoctoral scholars. In particular, we would like to thank Dr. Ingrid Castro-Rodriquez, Dr. Andreas Scheurer, Dr. Frank W. Heinemann, Dr. Christian Anthon, Prof. Suzanne Bart, Prof. Hidetaka Nakai, Prof. Hajime Kameo, Dr. Oanh P. Lam, Dr. Stefan Zuend, Dr. Boris Kosog, Sebastian Franke, and Corina Schmidt. Funding for the work described and the time to develop this chapter was provided by the Department of Energy (DOE), the Alfred P. Sloan Foundation, the Deutsche Forschungsgemeinschaft (DFG), the Bundesministerium für Bildung und Forschung (BMBF), the Bavarian–California Technology Center (BaCaTeC), the University of California, San Diego (UCSD), and the Friedrich-Alexander-University (FAU) Erlangen – Nuremberg.

ABBREVIATIONS

1,3-TMSC$_5$H$_3$	1,3-Trimethylsilylcyclopentadienyl anion
1D	One dimensional
2D	Two dimensional
15-C-5	1,4,7,10,13-Pentaoxacyclopentadecane

18-C-6	1,4,7,10,13,16-Hexaoxacyclooctadecane
((^{tBu}ArOH)$_3$mes)	1,3,5-Trimethyl-2,4,6-*tris*(2,4-di-*tert*-butylhydroxyben-zyl)-methylbenzene
((^{Ad}ArOH)$_3$N)	Tris(2-hydroxy-3-adamantyl-5-methylbenzyl)amine)
(^{Ad}ArOH)$_3$tacn	1,4,7-Tris-(3-adamantyl-5-*tert*-butyl-2-hydroxybenzylate)-1,4,7-triazacyclononane
(^{Neop,Me}ArOH)$_3$tacn	1,4,7-Tris(2-hydroxy-5-methyl-3-neopentylbenzyl)-1,4,7-triazacyclononane
(^{tBu}ArOH)$_3$tacn	1,4,7-Tris(3,5-di-*tert*-butyl-2-hydroxybenzylate)-1,4,7-triazacyclononane
ACN	Acetonitrile solvent
can	Acetonitrile ligand
BDE	Bond dissociaton energy
BIPM^{TMS}H	CH(PPh$_2$N(TMS)$_3$)$_2$ anion
tBuC$_5$H$_4$	*Tert*-butylcylclopentadienyl anion
^{13}C NMR	Carbon-13 nuclear magnetic resonance
CH$_4$	Methane
C$_5$HMe$_4$	Tetramethylcyclopentadienyl anion
C$_5$H$_4$Me	Methylcyclopentadienyl anion
C$_6$H$_3$Me$_3$	Mesitylene
CO	Carbon monoxide
CO$_2$	Carbon dioxide
COT	Cyclooctatetraenyl dianion
Cp	Cyclopentadienyl anion
Cp$'$	Substituted cyclopentadienyl anion
Cp$''$	1,3-(TMS)$_2$C$_5$H$_3$ anion
Cp'''	1,2,4-(tBu)$_3$C$_5$H$_2$ anion
Cp‡	1,3-(tBu)$_2$C$_5$H$_3$ anion
Cp*	Pentamethylcyclopentadienyl anion
DEPT	Distortionless enhancement by polarization transfer
DFT	Density functional theory
DME	1,2-Dimethoxyethane solvent
dme	1,2-Dimethoxyethane ligand
DMPE	1,2-Bis(dimethylphosphino)ethane
EPR	Electron paramagnetic resonance
Et$_2$O	Diethyl ether solvent
^{1}H NMR	Proton nuclear magnetic resonance
H$_2$	Dihydrogen
H$_2$O	Water
Ind	C$_7$H$_5$N$_2$, indazolate anion
IR	Infrared
KIE	Kinetic isotope effect

LUMO	Lowest unoccupied molecular orbital
Me$_3$NO	Trimethylamine N-oxide
mes	Mesityl
MeOH	Methanol solvent
MO	Molecular orbital
napth	Napthalene
NHC	N-Heterocyclic carbene
N$_2$	Dinitrogen
N$_3{}^{1-}$	Azide anion
NN^{fc}	Ferrocene(NSitBuMe$_2$)$_2$ dianion
NMR	Nuclear magnetic resonance
N$_2$O	Nitrous oxide
NO	Nitrogen monoxide
ODipp	O-2,6-iPr-C$_6$H$_3$ anion
ODtbp	O-2,6-tBuC$_6$H$_3$ anion
O$_2$	Dioxygen
OPPh$_3$	Triphenylphosphine oxide
OTf	Triflate
OTtbp	O-2,4,6-tBuC$_6$H$_2$ anion
O$_{yl}$	Terminal oxygen atom
PFTB	*Per*-fluoro-*tert*-butoxide
PhC(CPh	Diphenylacetylene
Pn*	Permethylpentalene
PPh$_3$	Triphenylphosphine
ppm	Parts per million
Py	Pyridine solvent
py	Pyridine ligand
rt	Room temperature
salan	Diamine bis(phenolate)
SQUID	Superconducting quantum interference device
terpy	Terpyridine
THF	Tetrahydrofuran solvent
thf	Tetrahydrofuran ligand
TMEDA	Tetramethylethylenediamine
TMS	Trimethylsilyl
TMSCHN$_2$	Trimethylsilyldiazomethane
TMSC$_5$H$_4$	Trimethylsilylcyclopentadienyl anion
tol	Toluene
Tp*	Hydro-tris(3,5-dimethylpyrazolyl)borate
tpa	Tris(2-pyridine)methylamine
TREN^{DMSB}	N(CH$_2$CH$_2$NSiMe$_2$$t$Bu)$_3$ trianion
TREN^{TIPSi}	N(CH$_2$CH$_2$NSi(iPr)$_3$)$_3$ trianion

trityl	Triphenylmethane
Ts^{tol}	HC(SiMe$_2$NAr)$^{3-}$; Ar = 4-MeC$_6$H$_4$
Ts^{Xyl}	HC(SiMe$_2$NAr)$^{3-}$; Ar = 3,5-Me$_2$C$_6$H$_3$
UV	Ultraviolet
vis	Visible
VT	Variable temperature
XANES	X-ray absorption near-edge spectroscopy
XRD	X-ray diffraction

REFERENCES

1. W. B. Tolman, Ed., *Activation of Small Molecules: Organometallic and Bioinorganic Perspectives*, Wiley–VCH, Weinheim, Germany, 2006.

2. D. L. Clark, A. P. Sattelberger, S. G. Bott, and R. N. Vrtis, *Inorg. Chem.*, *28*, 1771 (1989).

3. L. R. Avens, S. G. Bott, D. L. Clark, A. P. Sattelberger, J. G. Watkin, and B. D. Zwick, *Inorg. Chem.*, *33*, 2248 (1994).

4. I. A. Khan and H. S. Ahuja, *Inorg. Synth.*, *21*, 187 (1982).

5. J. L. Kiplinger, D. E. Morris, B. L. Scott, and C. J. Burns, *Organometallics*, *21*, 5978 (2002).

6. F. G. N. Cloke and P. B. Hitchcock, *J. Am. Chem. Soc.*, *124*, 9352 (2002).

7. W. J. Evans, S. A. Kozimor, J. W. Ziller, A. A. Fagin, and M. N. Bochkarev, *Inorg. Chem.*, *44*, 3993 (2005).

8. C. D. Carmichael, N. A. Jones, and P. L. Arnold, *Inorg. Chem.*, *47*, 8577 (2008).

9. D. D. Schnaars, G. Wu, and T. W. Hayton, *Dalton Trans.* 6121 (2008).

10. M. J. Monreal, R. K. Thomson, T. Cantat, N. E. Travia, B. L. Scott, and J. L. Kiplinger, *Organometallics*, *30*, 2031 (2011).

11. D. L. Clark, D. E. Hobart, and M. P. Neu, *Chem. Rev.*, *95*, 25 (1995).

12. L. R. Morss, N. M. Edelstein, J. Fuger, and J. J. Katz, Eds., *The Chemistry of the Actinide and Transactinide Elements*, Vol. *1*, 6th ed., Springer, Dordrecht, The Netherlands, 2010.

13. W. J. Evans and S. A. Kozimor, *Coord. Chem. Rev.*, *250*, 911 (2006).

14. S. Labouille, C. Clavaguera, and F. Nief, *Organometallics*, *Article ASAP*, DOI: 10.1021/om301018u.

15. M. Mazzanti, R. L. Wietzke, J. Pecaut, J. M. Latour, P. Maldivi, and M. Remy, *Inorg. Chem.*, *41*, 2389 (2002).

16. J. G. Brennan, S. D. Stults, R. A. Andersen, and A. Zalkin, *Organometallics*, *7*, 1329 (1988).

17. J. G. Brennan, S. D. Stults, R. A. Andersen, and A. Zalkin, *Inorg. Chimica Acta*, *139*, 201 (1987).

18. D. R. Brown and R. G. Denning, *Inorg. Chem.*, *35*, 6158 (1996).

19. R. G. Denning, in *Structure and Bonding*, D. M. P. Mingos, Ed., Vol. *79*, Springer–Verlag, Berlin, 1992, p. 215.

20. B. Kosog, H. S. La Pierre, F. W. Heinemann, S. T. Liddle, and K. Meyer, *J. Am. Chem. Soc.*, *134*, 5284 (2012).

21. H. S. La Pierre and K. Meyer, *Inorg. Chem.*, *52*, 529 (2013).

22. V. C. Williams, M. Muller, M. A. Leech, R. G. Denning, and M. L. H. Green, *Inorg. Chem.*, *39*, 2538 (2000).

23. D. R. Brown, R. G. Denning, and R. H. Jones, *J. Chem. Soc., Chem. Commun.*, 2601 (1994).

24. T. W. Hayton, J. M. Boncella, B. L. Scott, E. R. Batista, and P. J. Hay, *J. Am. Chem. Soc.*, *128*, 10549 (2006).

25. L. P. Spencer, R. L. Gdula, T. W. Hayton, B. L. Scott, and J. M. Boncella, *Chem. Commun.*, 4986 (2008).

26. L. P. Spencer, P. Yang, B. L. Scott, E. R. Batista, and J. M. Boncella, *J. Am. Chem. Soc.*, *130*, 2930 (2008).

27. T. W. Hayton, J. M. Boncella, B. L. Scott, P. D. Palmer, E. R. Batista, and P. J. Hay, *Science*, *310*, 1941 (2005).

28. T. W. Hayton, J. M. Boncella, B. L. Scott, and E. R. Batista, *J. Am. Chem. Soc.*, *128*, 12622 (2006).

29. L. P. Spencer, P. Yang, B. L. Scott, E. R. Batista, and J. M. Boncella, *Inorg. Chem.*, *48*, 2693 (2009).

30. R. E. Jilek, L. P. Spencer, R. A. Lewis, B. L. Scott, T. W. Hayton, and J. M. Boncella, *J. Am. Chem. Soc.*, *134*, 9876 (2012).

31. K. Tatsumi and R. Hoffmann, *Inorg. Chem.*, *19*, 2656 (1980).

32. W. R. Wadt, *J. Am. Chem. Soc.*, *103*, 6053 (1981).

33. P. Pyykko, L. J. Laakkonen, and K. Tatsumi, *Inorg. Chem.*, *28*, 1801 (1989).

34. N. Kaltsoyannis, *Inorg. Chem.*, *39*, 6009 (2000).

35. R. G. Denning, J. C. Green, T. E. Hutchings, C. Dallera, A. Tagliaferri, K. Giarda, N. B. Brookes, and L. Braicovich, *J. Chem. Phys.*, *117*, 8008 (2002).

36. S. Matsika and R. M. Pitzer, *J. Phys. Chem. A*, *105*, 637 (2001).

37. R. Thetford and M. Mignanelli, *J. Nucl. Mat.*, *320*, 44 (2003).

38. W. G. Van der Sluys, C. J. Burns, and D. C. Smith, US Patent, 5128112 (1992).

39. D. A. Petti, J. Buongiorno, J. T. Maki, R. R. Hobbins, and G. K. Miller, *Nucl. Eng. Des.*, *222*, 281 (2003).

40. M. Streit and F. Ingold, *J. Eur. Ceram. Soc.*, *25*, 2687 (2005).

41. C. B. Yeamans, G. W. C. Silva, G. S. Cerefice, K. R. Czerwinski, T. Hartmann, A. K. Burrell, and A. P. Sattelberger, *J. Nucl. Mater.*, *374*, 75 (2008).

42. A. C. Sather, O. B. Berryman, and J. Rebek, *J. Am. Chem. Soc.*, *132*, 13572 (2010).

43. S. Beer, O. B. Berryman, D. Ajami, and J. Rebek, *Chem. Sci.*, *1*, 43 (2010).

44. P. L. Arnold, D. Patel, C. Wilson, and J. B. Love, *Nature (London)*, *451*, 315 (2008).

45. P. L. Arnold, A. F. Pecharman, E. Hollis, A. Yahia, L. Maron, S. Parsons, and J. B. Love, *Nat. Chem.*, *2*, 1056 (2010).

46. P. L. Arnold, E. Hollis, F. J. White, N. Magnani, R. Caciuffo, and J. B. Love, *Angew. Chem., Int. Ed.*, *50*, 887 (2011).

47. D. R. Lovley, E. J. P. Phillips, Y. A. Gorby, and E. R. Landa, *Nature (London)*, *350*, 413 (1991).

48. G. Nocton, P. Horeglad, V. Vetere, J. Pecaut, L. Dubois, P. Maldivi, N. M. Edelstein, and M. Mazzanti, *J. Am. Chem. Soc.*, *132*, 495 (2010).

49. L. Natrajan, F. Burdet, J. Pecaut, and M. Mazzanti, *J. Am. Chem. Soc.*, *128*, 7152 (2006).

50. J. C. Berthet, P. Thuery, and M. Ephritikhine, *Chem. Commun.*, 3415 (2005).

51. M. B. Jones, A. J. Gaunt, J. C. Gordon, N. Kaltsoyannis, M. P. Neu, and B. L. Scott, *Chem. Sci.*, *4*, 1189 (2013).

52. Z. Kolarik, *Chem. Rev.*, *108*, 4208 (2008).

53. D. D. Schnaars, A. J. Gaunt, T. W. Hayton, M. B. Jones, I. Kirker, N. Katsoyannis, I. May, S. D. Reilly, B. L. Scott, and G. Wu, *Inorg. Chem.*, *51*, 8557 (2012).

54. S. G. Minasian, K. S. Boland, R. K. Feller, A. J. Gaunt, S. A. Kozimor, I. May, S. D. Reilly, B. L. Scott, and D. K. Shuh, *Inorg. Chem.*, *51*, 5728 (2012).

55. A. J. Gaunt and M. P. Neu, *C. R. Chim.*, *13*, 821 (2010).

56. R. A. Wigeland, T. H. Bauer, T. H. Fanning, and E. E. Morris, *Nucl. Technol.*, *154*, 95 (2006).

57. F. Haber, DE Patent, 229126 (1909).

58. M. S. Eisen, *Topics in Organometallic Chemistry*, *31* (C-X Bond Formation), 157 (2010).

59. M. Sharma and M. S. Eisen, in *Structure and Bonding (Organometallic and Coordination Chemistry of the Actinides)*, T. E. Albrecht-Schmidt, Ed., Vol. *127*, Springer, Berlin, 2008, p. 1.

60. T. Andrea and M. S. Eisen, *Chem. Soc. Rev.*, *37*, 550 (2008).

61. E. Barnea and M. S. Eisen, *Coord. Chem. Rev.*, *250*, 855 (2006).

62. M. S. Eisen and T. J. Marks, *J. Mol. Catal.*, *86*, 23 (1994).

63. C. J. Weiss and T. J. Marks, *Dalton Trans.*, *39*, 6576 (2010).

64. L. R. Morss, N. M. Edelstein, J. Fuger, and J. J. Katz, Eds., in *The Chemistry of the Actinide and Transactinide Elements*, Vol. 5, 6th Ed., Springer, Dordrecht, The Netherlands, 2010.

65. A. R. Fox, S. C. Bart, K. Meyer, and C. C. Cummins, *Nature (London)*, *455*, 341 (2008).

66. A. P. Amrute, F. Krumeich, C. Mondelli, and J. Perez-Ramirez, *Chem. Sci.*, *Accepted Manuscript*, DOI: 10.1039/C3SC22067B (2013).

67. S. J. Taylor, in *Metal Oxide Catalysis*, Wiley–VCH, Germany, 2009, p. 539.

68. M. Ephritikhine, *Dalton Trans.* 2501 (2006).

69. P. L. Arnold, *Chem. Commun.*, *47*, 9005 (2011).

70. J. L. Sessler, P. J. Melfi, and G. D. Pantos, *Coord. Chem. Rev.*, *250*, 816 (2006).

71. R. J. Baker, *Chem. -Eur. J.*, *18*, 16258 (2012).

72. L. R. Morss, N. M. Edelstein, J. Fuger, and J. J. Katz, Eds., in *The Chemistry of the Actinide and Transactinide Elements*, 6th Ed., Springer, Dordrecht, The Netherlands, 2010.

73. M. B. Jones and A. J. Gaunt, *Chem. Rev.*, *113*, 1137 (2013).

74. R. J. Baker, *Coord. Chem. Rev.*, *256*, 2843 (2012).

75. I. Korobkov and S. Gambarotta, *Prog. Inorg. Chem.*, *54*, 321 (2005).

76. P. L. Arnold, J. B. Love, and D. Patel, *Coord. Chem. Rev.*, *253*, 1973 (2009).

77. P. L. Arnold and I. J. Casely, *Chem. Rev.*, *109*, 3599 (2009).

78. S. Fortier and T. W. Hayton, *Coord. Chem. Rev.*, *254*, 197 (2010).

79. T. W. Hayton, *Dalton Trans.*, *39*, 1145 (2010).

80. T. W. Hayton, *Chem. Commun.*, *49*, 2956 (2013).

81. J. Qiu and P. C. Burns, *Chem. Rev.*, *113*, 1097 (2013).

82. M. Nyman and P. C. Burns, *Chem. Soc. Rev.*, *41*, 7354 (2012).

83. L. S. Natrajan, *Coord. Chem. Rev.*, *256*, 1583 (2012).

84. C. L. Cahill and M. B. Andrews, *Chem. Rev.*, *113*, 1121 (2013).

85. S. C. Bart and K. Meyer, in *Structure and Bonding*, (*Organometallic and Coordination Chemistry of the Actinides*), T. E. Albrecht-Schmidt, Ed., Vol. *127*, Springer, Berlin, 2008, p. 119.

86. K. Meyer and S. C. Bart, *Adv. Inorg. Chem.*, *60*, 1 (2008).

87. O. P. Lam, C. Anthon, and K. Meyer, *Dalton Trans.* 9677 (2009).

88. O. P. Lam and K. Meyer, *Polyhedron*, *32*, 1 (2012).

89. H. Gilman, R. G. Jones, E. Bindschadler, D. Blume, G. Karmas, G. A. Martin, Jr., J. F. Nobis, J. R. Thirtle, H. L. Yale, and F. A. Yeoman, *J. Am. Chem. Soc.*, *78*, 2790 (1956).

90. J. J. Katz and E. Rabinawitch, in *The Chemistry of Uranium*, McGraw-Hill, New York, 1951, pp. 356–364.

91. R. K. Sheline and J. L. Slater, *Angew. Chem., Int. Ed. Engl.*, *14*, 309 (1975).

92. J. L. Slater, R. K. Sheline, K. C. Lin, and W. Weltner, *J. Chem. Phys.*, *55*, 5129 (1971).

93. K. R. Kunze, R. H. Hauge, D. Hamill, and J. L. Margrave, *J. Phys. Chem.*, *81*, 1664 (1977).

94. D. C. Sonnenberger, E. A. Mintz, and T. J. Marks, *J. Am. Chem. Soc.*, *106*, 3484 (1984).

95. J. G. Brennan, R. A. Andersen, and J. L. Robbins, *J. Am. Chem. Soc.*, *108*, 335 (1986).

96. L. E. Schock, A. M. Seyam, M. Sabat, and T. J. Marks, *Polyhedron*, *7*, 1517 (1988).

97. L. Maron, O. Eisenstein, and R. A. Andersen, *Organometallics*, *28*, 3629 (2009).

98. L. E. Schock and T. J. Marks, *J. Am. Chem. Soc.*, *110*, 7701 (1988).

99. J. Parry, E. Carmona, S. Coles, and M. Hursthouse, *J. Am. Chem. Soc.*, *117*, 2649 (1995).

100. M. D. Conejo, J. S. Parry, E. Carmona, M. Schultz, J. G. Brennann, S. M. Beshouri, R. A. Andersen, R. D. Rogers, S. Coles, and M. Hursthouse, *Chem. -Eur. J.*, *5*, 3000 (1999).

101. W. J. Evans, S. A. Kozimor, G. W. Nyce, and J. W. Ziller, *J. Am. Chem. Soc.*, *125*, 13831 (2003).

102. W. J. Evans, K. J. Forrestal, and J. W. Ziller, *J. Am. Chem. Soc.*, *117*, 12635 (1995).

103. B. E. Bursten and R. J. Strittmatter, *J. Am. Chem. Soc.*, *109*, 6606 (1987).

104. B. E. Bursten, L. F. Rhodes, and R. J. Strittmatter, *J. Am. Chem. Soc.*, *111*, 2756 (1989).

105. B. E. Bursten, L. F. Rhodes, and R. J. Strittmatter, *J. Am. Chem. Soc.*, *111*, 2758 (1989).

106. K. Tatsumi and R. Hoffmann, *Inorg. Chem.*, *23*, 1633 (1984).

107. H. S. La Pierre, J. Arnold, R. G. Bergman, and F. D. Toste, *Inorg. Chem.*, *51*, 13334 (2012).

108. S. G. Minasian, J. L. Krinsky, V. A. Williams, and J. Arnold, *J. Am. Chem. Soc.*, *130*, 10086 (2008).

109. S. G. Minasian, J. L. Krinsky, J. D. Rinehart, R. Copping, T. Tyliszczak, M. Janousch, D. K. Shuh, and J. Arnold, *J. Am. Chem. Soc.*, *131*, 13767 (2009).

110. V. Vetere, P. Maldivi, B. O. Roos, and C. Adamo, *J. Phys. Chem. A*, *113*, 14760 (2009).

111. V. Vetere, P. Maldivi, and C. Adamo, *J. Comput. Chem.*, *24*, 850 (2003).

112. N. Kaltsoyannis, *Chem. Soc. Rev.*, *32*, 9 (2003).

113. I. Castro-Rodriguez and K. Meyer, *J. Am. Chem. Soc.*, *127*, 11242 (2005).

114. T. D. Tilley and R. A. Andersen, *J. Am. Chem. Soc.*, *104*, 1772 (1982).

115. O. T. Summerscales, F. G. N. Cloke, P. B. Hitchcock, J. C. Green, and N. Hazari, *Science*, *311*, 829 (2006).

116. W. Buchner, *Helv. Chim. Acta*, *46*, 2111 (1963).

117. L. Gmelin, *Ann. Phys. Chem.*, *4*, 31 (1825).

118. P. W. Lednor and P. C. Versloot, *J. Chem. Soc., Chem. Commun.*, 285 (1983).

119. G. Silvestri, S. Gambino, G. Filardo, G. Spadaro, and L. Palmisano, *Electrochim. Acta*, *23*, 413 (1978).

120. S. Coluccia, E. Garrone, E. Guglielminotti, and A. Zecchina, *J. Chem. Soc., Faraday Trans.*, *77*, 1063 (1981).

121. O. T. Summerscales, F. G. N. Cloke, P. B. Hitchcock, J. C. Green, and N. Hazari, *J. Am. Chem. Soc.*, *128*, 9602 (2006).

122. J. A. Pool, E. Lobkovsky, and P. J. Chirik, *Nature (London)*, *427*, 527 (2004).

123. P. J. Chirik, *Organometallics*, *29*, 1500 (2010).

124. N. Tsoureas, O. T. Summerscales, F. G. N. Cloke, and M. R. Roe, *Organometallics*, *Article ASAP*, DOI: 10.1021/om301045k.

125. A. S. Frey, F. G. N. Cloke, P. B. Hitchcock, I. J. Day, J. C. Green, and G. Aitken, *J. Am. Chem. Soc.*, *130*, 13816 (2008).

126. G. Aitken, N. Hazari, A. S. P. Frey, F. G. N. Cloke, O. Summerscales, and J. C. Green, *Dalton Trans.*, *40*, 11080 (2011).

127. D. McKay, A. S. P. Frey, J. C. Green, F. G. N. Cloke, and L. Maron, *Chem. Commun.*, *48*, 4118 (2012).

128. O. T. Summerscales and F. G. N. Cloke, in *Structure and Bonding*, (*Organometallic and Coordination Chemistry of the Actinides*), Vol. *127*, Springer, Berlin, 2008, p. 87.

129. R. A. Andersen, *Inorg. Chem.*, *18*, 1507 (1979).

130. P. L. Arnold, Z. R. Turner, R. M. Bellabarba, and R. P. Tooze, *Chem. Sci.*, *2*, 77 (2011).

131. S. M. Mansell, N. Kaltsoyannis, and P. L. Arnold, *J. Am. Chem. Soc.*, *133*, 9036 (2011).

132. B. M. Gardner, J. C. Stewart, A. L. Davis, J. McMaster, W. Lewis, A. J. Blake, and S. T. Liddle, *PNAS*, *109*, 9265 (2012).

133. P. Roussel, P. B. Hitchcock, N. Tinker, and P. Scott, *Chem. Commun.*, 2053 (1996).

134. P. Roussel, R. Boaretto, A. J. Kingsley, N. W. Alcock, and P. Scott, *J. Chem. Soc., Dalton Trans.*, 1423 (2002).

135. P. T. Wolczanski and J. E. Bercaw, *Acc. Chem. Res.*, *13*, 121 (1980).

136. J. M. Manriquez, P. J. Fagan, T. J. Marks, C. S. Day, and V. W. Day, *J. Am. Chem. Soc.*, *100*, 7112 (1978).

137. G. Paolucci, G. Rossetto, P. Zanella, K. Yunlu, and R. D. Fischer, *J. Organomet. Chem.*, *272*, 363 (1984).

138. C. Villiers, R. Adam, and M. Ephritikhine, *J. Chem. Soc., Chem. Commun.*, 1555 (1992).

139. M. Weydert, J. G. Brennan, R. A. Andersen, and R. G. Bergman, *Organometallics*, *14*, 3942 (1995).

140. S. J. Simpson and R. A. Andersen, *J. Am. Chem. Soc.*, *103*, 4063 (1981).

141. S. J. Simpson, H. W. Turner, and R. A. Andersen, *Inorg. Chem.*, *20*, 2991 (1981).

142. O. Benaud, J. C. Berthet, P. Thuery, and M. Ephritikhine, *Inorg. Chem.*, *50*, 12204 (2011).

143. O. Benaud, J. C. Berthet, P. Thuery, and M. Ephritikhine, *Inorg. Chem.*, *49*, 8117 (2010).

144. A. Dormond, A. Aaliti, A. Elbouadili, and C. Moise, *J. Organomet. Chem.*, *329*, 187 (1987).

145. P. L. Arnold, Z. R. Turner, A. I. Germeroth, I. J. Casely, G. S. Nichol, R. Bellabarba, and R. B. Tooze, *Dalton Trans.*, *42*, 1333 (2013).

146. W. J. Evans, N. A. Siladke, and J. W. Ziller, *Chem. -Eur. J.*, *16*, 796 (2010).

147. N. A. Siladke, J. W. Ziller, and W. J. Evans, *J. Am. Chem. Soc.*, *133*, 3507 (2011).

148. N. A. Siladke, J. LeDuc, J. W. Ziller, and W. J. Evans, *Chem. -Eur. J.*, *18*, 14820 (2012).

149. P. Zanella, G. Paolucci, G. Rossetto, F. Benetollo, A. Polo, R. D. Fischer, and G. Bombieri, *J. Chem. Soc., Chem. Commun.*, 96 (1985).

150. A. Dormond, A. A. Elbouadili, and C. Moise, *J. Chem. Soc., Chem. Commun.*, 749 (1984).

151. R. E. Cramer, R. B. Maynard, J. C. Paw, and J. W. Gilje, *Organometallics*, *1*, 869 (1982).

152. R. E. Cramer, K. T. Higa, S. L. Pruskin, and J. W. Gilje, *J. Am. Chem. Soc.*, *105*, 6749 (1983).

153. R. E. Cramer, K. T. Higa, and J. W. Gilje, *J. Am. Chem. Soc.*, *106*, 7245 (1984).

154. R. E. Cramer, K. T. Higa, and J. W. Gilje, *Organometallics*, *4*, 1140 (1985).

155. R. E. Cramer, J. H. Jeong, and J. W. Gilje, *Organometallics*, *5*, 2555 (1986).

156. J. W. Gilje and R. E. Cramer, *Inorg. Chim. Acta*, *139*, 177 (1987).

157. R. E. Cramer, J. H. Jeong, P. N. Richmann, and J. W. Gilje, *Organometallics*, *9*, 1141 (1990).

158. P. J. Fagan, J. M. Manriquez, S. H. Vollmer, C. S. Day, V. W. Day, and T. J. Marks, *J. Am. Chem. Soc.*, *103*, 2206 (1981).

159. C. Boisson, J. C. Berthet, M. Lance, M. Nierlich, and M. Ephritikhine, *J. Organomet. Chem.*, *548*, 9 (1997).

160. J. C. Berthet and M. Ephritikhine, *Coord. Chem. Rev.*, *178*, 83 (1998).

161. J. M. Manriquez, P. J. Fagan, and T. J. Marks, *J. Am. Chem. Soc.*, *100*, 3939 (1978).

162. P. J. Fagan, J. M. Manriquez, E. A. Maatta, A. M. Seyam, and T. J. Marks, *103*, 6650 (1981).

163. M. R. Duttera, P. J. Fagan, T. J. Marks, and V. W. Day, *J. Am. Chem. Soc.*, *104*, 865 (1982).

164. K. G. Moloy and T. J. Marks, *J. Am. Chem. Soc.*, *106*, 7051 (1984).

165. P. J. Fagan, K. G. Moloy, and T. J. Marks, *J. Am. Chem. Soc.*, *103*, 6959 (1981).

166. I. Castro-Rodriguez, H. Nakai, and K. Meyer, *Angew. Chem. Int. Ed.*, *45*, 2389 (2006).

167. T. E. Hanna, I. Keresztes, E. Lobkovsky, W. H. Bernskoetter, and P. J. Chirik, *Organometallics*, *23*, 3448 (2004).

168. L. R. Avens, D. M. Barnhart, C. J. Burns, S. D. Mckee, and W. H. Smith, *Inorg. Chem.*, *33*, 4245 (1994).

169. A. S. P. Frey, F. G. N. Cloke, M. P. Coles, and P. B. Hitchcock, *Chem. -Eur. J.*, *16*, 9446 (2010).

170. D. M. King, F. Tuna, E. J. L. McInnes, J. McMaster, W. Lewis, A. J. Blake, and S. T. Liddle, *Science*, *337*, 717 (2012).

171. N. A. Siladke, K. R. Meihaus, J. W. Ziller, M. Fang, F. Furche, J. R. Long, and W. J. Evans, *J. Am. Chem. Soc.*, *134*, 1243 (2012).

172. J. G. Brennan, Organoactinide Chemistry: Synthesis, Structure, and Solution Dynamics, Ph.D. Thesis, University of California, Berkeley, CA, 1985.

173. A. J. Lewis, P. J. Carroll, and E. J. Schelter, *J. Am. Chem. Soc.*, *135*, 511 (2013).

174. F. Dulong, J. Pouessel, P. Thuery, J. C. Berthet, M. Ephritikhine, and T. Cantat, *Chem. Commun.*, *Advance Article*, DOI: 10.1039/c3cc39163a (2013).

175. P. Roussel and P. Scott, *J. Am. Chem. Soc.*, *120*, 1070 (1998).

176. P. Roussel, N. D. Tinker, and P. Scott, *J. Alloys Compd.*, *271*, 150 (1998).

177. N. Kaltsoyannis and P. Scott, *Chem. Commun.*, 1665 (1998).

178. P. Roussel, W. Errington, N. Kaltsoyannis, and P. Scott, *J. Organomet. Chem.*, *635*, 69 (2001).

179. F. G. N. Cloke, J. C. Green, and N. Kaltsoyannis, *Organometallics*, *23*, 832 (2004).

180. F. M. Chadwick, A. Ashley, G. Wildgoose, J. M. Goicoechea, S. Randall, and D. O'Hare, *Dalton Trans.*, *39*, 6789 (2010).

181. W. J. Evans, S. A. Kozimor, and J. W. Ziller, *J. Am. Chem. Soc.*, *125*, 14264 (2003).

182. A. L. Odom, P. L. Arnold, and C. C. Cummins, *J. Am. Chem. Soc.*, *120*, 5836 (1998).

183. C. E. Laplaza and C. C. Cummins, *Science*, *268*, 861 (1995).

184. C. E. Laplaza, M. J. A. Johnson, J. C. Peters, A. L. Odom, E. Kim, C. C. Cummins, G. N. George, and I. J. Pickering, *J. Am. Chem. Soc.*, *118*, 8623 (1996).

185. W. G. Vandersluys and A. P. Sattelberger, *Inorg. Chem.*, *28*, 2496 (1989).

186. W. G. Van Der Sluys and A. P. Sattelberger, *Inorg. Chem.*, *28*, 2496 (1989).

187. I. Korobkov, S. Gambarotta, and G. P. A. Yap, *Angew. Chem. Int. Ed.*, *41*, 3433 (2002).

188. K. Meyer, E. Bill, B. Mienert, T. Weyhermuller, and K. Wieghardt, *J. Am. Chem. Soc.*, *121*, 4859 (1999).

189. J. J. Scepaniak, C. S. Vogel, M. M. Khusniyarov, F. W. Heinemann, K. Meyer, and J. M. Smith, *Science*, *331*, 1049 (2011).

190. C. Vogel, F. W. Heinemann, J. Sutter, C. Anthon, and K. Meyer, *Angew. Chem. Int. Ed.*, *47*, 2681 (2008).

191. J. F. Berry, E. Bill, E. Bothe, S. D. George, B. Mienert, F. Neese, and K. Wieghardt, *Science*, *312*, 1937 (2006).

192. J. L. Stewart, Tris[bis(trimethylsilyl)amido]uranium: Compounds with Tri-, Tetra-, and Penta-valent Uranium, Ph.D. Thesis, University of California, Berkeley, CA, 1988.

193. I. Castro-Rodriguez, H. Nakai, L. N. Zakharov, A. L. Rheingold, and K. Meyer, *Science*, *305*, 1757 (2004).

194. R. K. Thomson, C. R. Graves, B. L. Scott, and J. L. Kiplinger, *Eur. J. Inorg. Chem.*, 1451 (2009).

195. J. C. Berthet, M. Lance, M. Nierlich, J. Vigner, and M. Ephritikhine, *J. Organomet. Chem.*, *420*, C9 (1991).

196. W. J. Evans, K. A. Miller, J. W. Ziller, and J. Greaves, *Inorg. Chem.*, *46*, 8008 (2007).

197. S. Fortier, G. Wu, and T. W. Hayton, *Dalton Trans.*, *39*, 352 (2010).

198. A. J. Lewis, E. Nakamaru-Ogiso, J. M. Kikkawa, P. J. Carroll, and E. J. Schelter, *Chem. Commun.*, *48*, 4977 (2012).

199. O. Benaud, J. C. Berthet, P. Thuery, and M. Ephritikhine, *Chem. Commun.*, *47*, 9057 (2011).

200. M. J. Crawford, A. Ellern, and P. Mayer, *Angew. Chem. Int. Ed.*, *44*, 7874 (2005).

201. G. Nocton, J. Pecaut, and M. Mazzanti, *Angew. Chem. Int. Ed.*, *47*, 3040 (2008).

202. W. J. Evans, S. A. Kozimor, and J. W. Ziller, *Science*, *309*, 1835 (2005).

203. S. Fortier, G. Wu, and T. W. Hayton, *J. Am. Chem. Soc.*, *132*, 6888 (2010).

204. A. R. Fox, P. L. Arnold, and C. C. Cummins, *J. Am. Chem. Soc.*, *132*, 3250 (2010).

205. A. R. Fox and C. C. Cummins, *J. Am. Chem. Soc.*, *131*, 5716 (2009).

206. R. K. Thomson, T. Cantat, B. L. Scott, D. E. Morris, E. R. Batista, and J. L. Kiplinger, *Nat. Chem.*, *2*, 723 (2010).

207. D. M. King, F. Tuna, E. J. L. McInnes, J. McMaster, W. Lewis, A. J. Blake, and S. T. Liddle, *Nat. Chem.*, AOP DOI: 10.1038/nchem.1642 (2013).

208. M. R. Spirlet, J. Rebizant, C. Apostolidis, E. Dornberger, B. Kanellakopulos, and B. Powietzka, *Polyhedron*, *15*, 1503 (1996).

209. J. C. Berthet, J. F. Lemarechal, M. Nierlich, M. Lance, J. Vigner, and M. Ephritikhine, *J. Organomet. Chem.*, *408*, 335 (1991).

210. J. L. Brown, G. Wu, and T. W. Hayton, *Organometallics*, *32*, 1193 (2013).

211. S. Fortier, J. L. Brown, N. Kaltsoyannis, G. Wu, and T. W. Hayton, *Inorg. Chem.*, *51*, 1625 (2012).

212. L. Chatelain, V. Mougel, J. Pecaut, and M. Mazzanti, *Chem. Sci.*, *3*, 1075 (2012).

213. G. M. Jones, P. L. Arnold, and J. B. Love, *Angew. Chem. Int. Ed.*, *51*, 12584 (2012).

214. J. G. Brennan, R. A. Andersen, and A. Zalkin, *Inorg. Chem.*, *25*, 1756 (1986).

215. O. P. Lam, L. Castro, B. Kosog, F. W. Heinemann, L. Maron, and K. Meyer, *Inorg. Chem.*, *51*, 781 (2012).

216. S. C. Bart, F. W. Heinemann, C. Anthon, C. Hauser, and K. Meyer, *Inorg. Chem.*, *48*, 9419 (2009).

217. O. P. Lam, S. C. Bart, H. Kameo, F. W. Heinemann, and K. Meyer, *Chem. Commun.*, *46*, 3137 (2010).

218. N. W. Davies, A. S. P. Frey, M. G. Gardiner, and J. Wang, *Chem. Commun.*, 4853 (2006).

219. L. Castro, O. P. Lam, S. C. Bart, K. Meyer, and L. Maron, *Organometallics*, *29*, 5504 (2010).

220. O. P. Lam, S. M. Franke, F. W. Heinemann, and K. Meyer, *J. Am. Chem. Soc.*, *134*, 16877 (2012).

221. O. P. Lam, F. W. Heinemann, and K. Meyer, *Chem. Sci.*, *2*, 1538 (2011).

222. A. C. Schmidt, A. V. Nizovtsev, A. Scheurer, F. W. Heinemann, and K. Meyer, *Chem. Commun.*, *48*, 8634 (2012).

223. O. T. Summerscales, A. S. P. Frey, F. Geoffrey, N. Cloke, and P. B. Hitchcock, *Chem. Commun.*, 198 (2009).

224. V. Mougel, C. Camp, J. Pecaut, C. Coperet, L. Maron, C. E. Kefalidis, and M. Mazzanti, *Angew. Chem. Int. Ed.*, *51*, 12280 (2012).

225. O. P. Lam, F. W. Heinemann, and K. Meyer, *Angew. Chem. Int. Ed.*, *50*, 5965 (2011).

226. S. J. Zuend, O. P. Lam, F. W. Heinemann, and K. Meyer, *Angew. Chem. Int. Ed*, *50*, 10626 (2011).

227. E. M. Matson, W. P. Forrest, P. E. Fanwick, and S. C. Bart, *J. Am. Chem. Soc.*, *133*, 4948 (2011).

228. E. M. Matson, P. E. Fanwick, and S. C. Bart, *Organometallics*, *30*, 5753 (2011).

229. P. C. Leverd, M. Ephritikhine, M. Lance, J. Vigner, and M. Nierlich, *J. Organomet. Chem.*, *507*, 229 (1996).

230. K. W. Bagnall and E. Yanir, *J. Inorg. Nucl. Chem.*, *36*, 777 (1974).

231. F. Calderazzo, G. Dellamico, R. Netti, and M. Pasquali, *Inorg. Chem.*, *17*, 471 (1978).

232. S. C. Bart, C. Anthon, F. W. Heinemann, E. Bill, N. M. Edelstein, and K. Meyer, *J. Am. Chem. Soc.*, *130*, 12536 (2008).

233. K. G. Moloy and T. J. Marks, *Inorg. Chim. Acta*, *110*, 127 (1985).

234. W. J. Evans, J. R. Walensky, and J. W. Ziller, *Organometallics*, *29*, 945 (2010).

235. C. L. Webster, J. W. Ziller, and W. J. Evans, *Organometallics*, *31*, 7191 (2012).

236. W. J. Evans, J. R. Walensky, and J. W. Ziller, *Inorg. Chem.*, *49*, 1743 (2010).

237. W. J. Evans, N. A. Siladke, and J. W. Ziller, *C. R. Chimie*, *13*, 775 (2010).

238. E. Mora, L. Maria, B. Biswas, C. Camp, I. C. Santos, J. Pecaut, A. Cruz, J. M. Carretas, J. Marcalo, and M. Mazzanti, *Organometallics*, *Article ASAP*, DOI: 10.1021/om3010806.

239. C. Lescop, T. Arliguie, M. Lance, M. Nierlich, and M. Ephritikhine, *J. Organomet. Chem.*, *580*, 137 (1999).

240. N. A. Piro, M. F. Lichterman, W. H. Harman, and C. J. Chang, *J. Am. Chem. Soc.*, *133*, 2108 (2011).

241. S. Solomon, D. Qin, M. Manning, Z. Chen, M. Marquis, K. Averyt, M. M. B. Tignor, and H. L. Miller, Eds., in *Contribution of Working Group I to the Fourth Assessment Report of the Intergovernmental Panel on Climate Change, 2007*, Cambridge University Press, Cambridge, UK, 2007.

242. D. S. J. Arney and C. J. Burns, *J. Am. Chem. Soc.*, *117*, 9448 (1995).

243. D. S. J. Arney and C. J. Burns, *J. Am. Chem. Soc.*, *115*, 9840 (1993).

244. J. F.de Wet and J. G. H. du Preez, *J. Chem. Soc., Dalton Trans.*, 592 (1978).

245. S. Fortier, N. Kaltsoyannis, G. Wu, and T. W. Hayton, *J. Am. Chem. Soc.*, *133*, 14224 (2011).

246. S. P. McGlynn and J. K. Smith, *J. Mol. Spectrosc.*, *6*, 164 (1961).

247. W. W. Lukens, S. M. Beshouri, L. L. Blosch, and R. A. Andersen, *J. Am. Chem. Soc.*, *118*, 901 (1996).

248. Y. Suzuki, S. D. Kelly, K. M. Kemner, and J. F. Banfield, *Nature (London)*, *419*, 134 (2002).

249. G. Nocton, F. Burdet, J. Pecaut, and M. Mazzanti, *Angew. Chem. Int. Ed.*, *46*, 7574 (2007).

250. B. H. Gu, H. Yan, P. Zhou, D. B. Watson, M. Park, and J. Istok, *Environ. Sci. Technol.*, *39*, 5268 (2005).

251. B. Biswas, V. Mougel, J. Pecaut, and M. Mazzanti, *Angew. Chem. Int. Ed.*, *50*, 5744 (2011).

252. L. M. Mokry, N. S. Dean, and C. J. Carrano, *Angew. Chem. Int. Ed. Engl.*, *35*, 1497 (1996).

253. S. Takao, K. Takao, W. Kraus, F. Ernmerling, A. C. Scheinost, G. Bernhard, and C. Hennig, *Eur. J. Inorg. Chem.*, 4771 (2009).

254. V. Mougel, B. Biswas, J. Pecaut, and M. Mazzanti, *Chem. Commun.*, *46*, 8648 (2010).

255. J. C. Berthet, M. Nierlich, Y. Miquel, C. Madic, and M. Ephritikhine, *Dalton Trans.* 369 (2005).

256. L. Karmazin, M. Mazzanti, and J. Pecaut, *Inorg. Chem.*, *42*, 5900 (2003).

257. W. J. Evans, S. A. Kozimor, and J. W. Ziller, *Polyhedron*, *23*, 2689 (2004).

258. K. A. N. S. Ariyaratne, R. E. Cramer, G. B. Jameson, and J. W. Gilje, *J. Organomet. Chem.*, *689*, 2029 (2004).

259. J. C. Berthet, M. Ephritikhine, M. Lance, M. Nierlich, and J. Vigner, *J. Organomet. Chem.*, *460*, 47 (1993).

260. S. R. Daly, M. Ephritikhine, and G. S. Girolami, *Polyhedron*, *33*, 41 (2012).

261. S. Kannan, A. E. Vaughn, E. M. Weis, C. L. Barnes, and P. B. Duval, *J. Am. Chem. Soc.*, *128*, 14024 (2006).

262. H. W. Crandall, *J. Chem. Phys.*, *17*, 602 (1949).

263. H. Gordon and H. Taube, *J. Inorg. Nucl. Chem.*, *16*, 189 (1961).

264. D. Rios, M. D. Micheini, A. F. Lucena, J. Marcalo, and J. K. Gibson, *J. Am. Chem. Soc.*, *134*, 15488 (2012).

265. M. Ephritikhine, *Chem. Rev.*, *97*, 2193 (1997).

266. P. J. Fagan, J. M. Manriquez, T. J. Marks, C. S. Day, S. H. Vollmer, and V. W. Day, *Organometallics*, *1*, 170 (1982).

267. A. S. P. Frey, F. G. N. Cloke, M. P. Coles, L. Maron, and T. Davin, *Angew. Chem. Int. Ed.*, *50*, 6881 (2011).

268. J. M. Manriquez, D. R. Mcalister, R. D. Sanner, and J. E. Bercaw, *J. Am. Chem. Soc.*, *100*, 2716 (1978).

269. P. T. Wolczanski, R. S. Threlkel, and J. E. Bercaw, *J. Am. Chem. Soc.*, *101*, 218 (1979).

270. E. L. Werkema, L. Maron, O. Eisenstein, and R. A. Andersen, *J. Am. Chem. Soc.*, *129*, 6662 (2007).

271. E. A. Maatta and T. J. Marks, *J. Am. Chem. Soc.*, *103*, 3576 (1981).

272. R. W. Broach, A. J. Schultz, J. M. Williams, G. M. Brown, J. M. Manriquez, P. J. Fagan, and T. J. Marks, *Science*, *203*, 172 (1979).

273. Z. R. Lin and T. J. Marks, *J. Am. Chem. Soc.*, *109*, 7979 (1987).

274. W. J. Evans, K. A. Miller, S. A. Kozimor, J. W. Ziller, A. G. DiPasquale, and A. L. Rheingold, *Organometallics*, *26*, 3568 (2007).

275. D. Grant, T. J. Stewart, R. Bau, K. A. Miller, S. A. Mason, M. Gutmann, G. J. McIntyre, L. Gagliardi, and W. J. Evans, *Inorg. Chem.*, *51*, 3613 (2012).

276. H. W. Turner, S. J. Simpson, and R. A. Andersen, *J. Am. Chem. Soc.*, *101*, 2782 (1979).

277. S. J. Simpson, H. W. Turner, and R. A. Andersen, *J. Am. Chem. Soc.*, *101*, 7728 (1979).

278. M. Foyentin, G. Folcher, and M. Ephritikhine, *J. Chem. Soc., Chem. Commun.*, 494 (1987).

279. G. Just and Y. Kauko, *Z. Phys. Chem.*, *76*, 601 (1911).

280. J. P. Collman, L. M. Slaughter, T. A. Eberspacher, T. Strassner, and J. I. Brauman, *Inorg. Chem.*, *40*, 6272 (2001).

281. J. Halpern, *Adv. Catal.*, 9 (1957).

282. R. G. Peters, B. P. Warner, and C. J. Burns, *J. Am. Chem. Soc.*, *121*, 5585 (1999).

283. I. Castro-Rodriguez, H. Nakai, P. Gantzel, L. N. Zakharov, A. L. Rheingold, and K. Meyer, *J. Am. Chem. Soc.*, *125*, 15734 (2003).

284. P. L. Watson and G. W. Parshall, *Acc. Chem. Res.*, *18*, 51 (1985).

285. C. M. Fendrick and T. J. Marks, *J. Am. Chem. Soc.*, *106*, 2214 (1984).

286. W. J. Evans, K. A. Miller, A. G. DiPasquale, A. L. Rheingold, T. J. Stewart, and R. Bau, *Angew. Chem. Int. Ed.*, *47*, 5075 (2008).

287. W. J. Evans, E. Montalvo, S. A. Kozimor, and K. A. Miller, *J. Am. Chem. Soc.*, *130*, 12258 (2008).

288. W. J. Evans, K. A. Miller, J. W. Ziller, A. G. DiPasquale, K. J. Heroux, and A. L. Rheingold, *Organometallics*, *26*, 4287 (2007).

289. J. A. Pool, B. L. Scott, and J. L. Kiplinger, *J. Am. Chem. Soc.*, *127*, 1338 (2005).

290. J. A. Pool, B. L. Scott, and J. L. Kiplinger, *J. Alloys Compd.*, *418*, 178 (2006).

291. J. L. Kiplinger, B. L. Scott, E. J. Schelter, and J. A. P. D. Tournear, *J. Alloys Compd.*, *444*, 477 (2007).

292. M. J. Monreal, S. Khan, and P. L. Diaconescu, *Angew. Chem. Int. Ed.*, *48*, 8352 (2009).

293. S. Duhovic, M. J. Monreal, and P. L. Diaconescu, *J. Organomet. Chem.*, *695*, 2822 (2010).

294. M. J. Monreal and P. L. Diaconescu, *J. Am. Chem. Soc.*, *132*, 7676 (2010).

295. E. M. Broderick, N. P. Gutzwiller, and P. L. Diaconescu, *Organometallics*, *29*, 3242 (2010).

296. P. L. Diaconescu, *Acc. Chem. Res.*, *43*, 1352 (2010).

297. A. Shafir and J. Arnold, *J. Am. Chem. Soc.*, *123*, 9212 (2001).

298. A. Shafir, M. P. Power, G. D. Whitener, and J. Arnold, *Organometallics*, *19*, 3978 (2000).

299. O. P. Lam, P. L. Feng, F. W. Heinemann, J. M. O'Connor, and K. Meyer, *J. Am. Chem. Soc.*, *130*, 2806 (2008).

300. R. G. Peters, B. P. Warner, B. L. Scott, and C. J. Burns, *Organometallics*, *18*, 2587 (1999).

301. J. L. Kiplinger, D. E. Morris, B. L. Scott, and C. J. Burns, *Chem. Commun.*, 30 (2002).

302. W. J. Evans, S. A. Kozimor, and J. W. Ziller, *Chem. Commun.*, 4681 (2005).

303. B. Kosog, C. E. Kefalidis, F. W. Heinemann, L. Maron, and K. Meyer, *J. Am. Chem. Soc.*, *134*, 12792 (2012).

304. D. Weber and M. R. Gagne, *Org. Lett.*, *11*, 4962 (2009).

305. D. Weber, M. A. Tarselli, and M. R. Gagne, *Angew. Chem. Int. Ed.*, *48*, 5733 (2009).

306. D. J. Gorin and F. D. Toste, *Nature (London)*, *446*, 395 (2007).

307. M. A. Antunes, A. Domingos, I. C.dos Santos, N. Marques, and J. Takats, *Polyhedron*, *24*, 3038 (2005).

308. B. D. Stubbert, C. L. Stern, and T. J. Marks, *Organometallics*, *22*, 4836 (2003).

309. B. D. Stubbert and T. J. Marks, *J. Am. Chem. Soc.*, *129*, 4253 (2007).

310. D. Baudry, A. Dormond, and A. Hafid, *J. Organomet. Chem.*, *494*, C22 (1995).

311. B. S. Newell, A. K. Rappe, and M. P. Shores, *Inorg. Chem.*, *49*, 1595 (2010).

312. R. Boaretto, P. Roussel, N. W. Alcock, A. J. Kingsley, I. J. Munslow, C. J. Sanders, and P. Scott, *J. Organomet. Chem.*, *591*, 174 (1999).

313. A. R. Fox, S. E. Creutz, and C. C. Cummins, *Dalton Trans.*, *39*, 6632 (2010).

314. T. Straub, W. Frank, G. J. Reiss, and M. S. Eisen, *J. Chem. Soc., Dalton Trans.*, 2541 (1996).

315. C. R. Graves, B. L. Scott, D. E. Morris, and J. L. Kiplinger, *Organometallics*, *27*, 3335 (2008).

316. J. Q. Wang, A. K. Dash, J. C. Berthet, M. Ephritikhine, and M. S. Eisen, *Organometallics*, *18*, 2407 (1999).

317. G. F. Zi, L. Jia, E. L. Werkema, M. D. Walter, J. P. Gottfriedsen, and R. A. Andersen, *Organometallics*, *24*, 4251 (2005).

318. G. F. Zi, L. L. Blosch, L. Jia, and R. A. Andersen, *Organometallics*, *24*, 4602 (2005).

319. N. Barros, D. Maynau, L. Maron, O. Eisenstein, G. F. Zi, and R. A. Andersen, *Organometallics*, *26*, 5059 (2007).

320. E. M. Matson, P. E. Fanwick, and S. C. Bart, *Eur. J. Inorg. Chem.*, 5471 (2012).

321. *Modern Arene Chemistry*, D. Astruc, Ed., Wiley, Weinheim, 2002.

322. M. Cesari, U. Pedretti, A. Zazzatta, G. Lugli, and W. Marconi, *Inorg. Chim. Acta*, *5*, 439 (1971).

323. E. O. Fischer and J. Wawersik, *J. Organomet. Chem.*, *5*, 559 (1966).

324. F. A. Cotton and W. Schwotzer, *Organometallics*, *4*, 942 (1985).

325. G. C. Campbell, F. A. Cotton, J. F. Haw, and W. Schwotzer, *Organometallics*, *5*, 274 (1986).

326. F. A. Cotton, W. Schwotzer, and C. Q. Simpson, *Angew. Chem. Int. Ed. Engl.*, *25*, 637 (1986).

327. A. V. Garbar, M. R. Leonov, L. N. Zakharov, and Y. T. Struchkov, *Russ. Chem. Bull.*, *45*, 451 (1996).

328. D. Baudry, E. Bulot, and M. Ephritikhine, *J. Chem. Soc., Chem. Commun.*, 1369 (1988).

329. D. Baudry, E. Bulot, P. Charpin, M. Ephritikhine, M. Lance, M. Nierlich, and J. Vigner, *J. Organomet. Chem.*, *371*, 155 (1989).

330. W. G. Van Der Sluys, C. J. Burns, J. C. Huffman, and A. P. Sattelberger, *J. Am. Chem. Soc.*, *110*, 5924 (1988).

331. W. J. Evans, S. A. Kozimor, W. R. Hillman, and J. W. Ziller, *Organometallics*, *24*, 4676 (2005).

332. W. J. Evans, G. W. Nyce, K. J. Forrestal, and J. W. Ziller, *Organometallics*, *21*, 1050 (2002).

333. G. Y. Hong, M. Dolg, and L. M. Li, *Int. J. Quant. Chem.*, *80*, 201 (2000).

334. M. Dolg, *J. Chem. Inf. Comp. Sci.*, *41*, 18 (2001).

335. I. Infante, J. Raab, J. T. Lyon, B. Liang, L. Andrews, and L. Gagliardi, *J. Phys. Chem. A*, *111*, 11996 (2007).

336. A. W. Duff, K. Jonas, R. Goddard, H. J. Kraus, and C. Kruger, *J. Am. Chem. Soc.*, *105*, 5479 (1983).

337. T. L. Gianetti, G. Nocton, S. G. Minasian, N. C. Tomson, A. L. D. Kilcoyne, S. A. Kozimor, D. K. Shuh, T. Tyliszczak, R. G. Bergman, and J. Arnold, *J. Am. Chem. Soc., Article ASAP*, DOI: 10.1021/ja311966h (2013).

338. P. L. Diaconescu, P. L. Arnold, T. A. Baker, D. J. Mindiola, and C. C. Cummins, *J. Am. Chem. Soc.*, *122*, 6108 (2000).

339. B. Vlaisavljevich, P. L. Diaconescu, W. W. Lukens, Jr., L. Gagliardi, and C. C. Cummins, *Organometallics, Article ASAP*, DOI: 10.1021/om3010367 (2013).

340. J. Li and B. E. Bursten, *J. Am. Chem. Soc.*, *119*, 9021 (1997).

341. P. L. Diaconescu and C. C. Cummins, *J. Am. Chem. Soc.*, *124*, 7660 (2002).

342. W. L. Huang, S. I. Khan, and P. L. Diaconescu, *J. Am. Chem. Soc.*, *133*, 10410 (2011).

343. C. Elschenbroich and J. Heck, *J. Am. Chem. Soc.*, *101*, 6773 (1979).

344. P. L. Diaconescu and C. C. Cummins, *Inorg. Chem.*, *51*, 2902 (2012).

345. W. J. Evans, S. A. Kozimor, J. W. Ziller, and N. Kaltsoyannis, *J. Am. Chem. Soc.*, *126*, 14533 (2004).

346. W. J. Evans, C. A. Traina, and J. W. Ziller, *J. Am. Chem. Soc.*, *131*, 17473 (2009).

347. M. J. Monreal, S. I. Khan, J. L. Kiplinger, and P. L. Diaconescu, *Chem. Commun.*, *47*, 9119 (2011).

348. D. P. Mills, F. Moro, J. McMaster, J.van Slageren, W. Lewis, A. J. Blake, and S. T. Liddle, *Nat. Chem.*, *3*, 454 (2011).

349. D. Patel, F. Moro, J. McMaster, W. Lewis, A. J. Blake, and S. T. Liddle, *Angew. Chem. Int. Ed.*, *50*, 10388 (2011).

350. C. DenAuwer, J. Madic, J. C. Berthet, M. Ephritikhine, and J. Rehr, *Radiochim. Acta*, *76*, 211 (1997).

351. B. Kosog, H. S. La Pierre, M. A. Denecke, F. W. Heinemann, and K. Meyer, *Inorg. Chem.*, *51*, 7940 (2012).

352. W. W. Lukens, P. G. Allen, J. J. Bucher, N. N. Edelstein, E. A. Hudson, D. K. Shuh, T. Reich, and R. A. Andersen, *Organometallics*, *18*, 1253 (1999).

353. D. Patel, F. Tuna, E. J. L. McInnes, J. McMaster, W. Lewis, A. J. Blake, and J. Liddle, *Dalton Trans.*, *Advance Article*, DOI: 10.1039/c0xx00000x (2013).

354. P. L. Arnold, S. M. Mansell, L. Maron, and D. McKay, *Nat. Chem.*, *4*, 668 (2012).

355. W. J. Evans, J. R. Walensky, J. W. Ziller, and A. L. Rheingold, *Organometallics*, *28*, 3350 (2009).

Reactive Transition Metal Nitride Complexes

JEREMY M. SMITH

Department of Chemistry and Biochemistry, New Mexico State University, Las Cruces, NM 88003

Current Address: Department of Chemistry, Indiana University, Bloomington, IN 47405

CONTENTS

Progress in Inorganic Chemistry, Volume 58, First Edition. Edited by Kenneth D. Karlin.
© 2014 John Wiley & Sons, Inc. Published 2014 by John Wiley & Sons, Inc.

VI. REACTIVITY

    A. Three-Electron Reactions
        1. Intermetallic Nitrogen-Atom Transfer
        2. Reductive Coupling
    B. Two-Electron Reactions
        1. Intermetallic Nitrogen-Atom Transfer
        2. Reaction With Phosphines
        3. Reaction With CO, CNR, CN^-, and Carbenes
        4. Reaction With Carbanions
        5. Formation of Nitrosyl, Thionitrosyl, and Selenonitrosyl Ligands
        6. Reaction With Amines
        7. Reaction With Azide
        8. Reaction With Thiols
        9. Reaction With Nitriles
      10. Reaction With Alkenes
    C. One-Electron Reactions
    D. Reactions With No Change in Oxidation State
        1. Reaction With CO_2 and CS_2
        2. Reaction With Lewis Acids
        3. Reaction With Alkylating and Acylating Reagents
        4. Reaction With Other Electrophiles
        5. Other Reactions
    E. Formation of Ammonia, Ammine, and Amide Ligands: Reaction With H_2
    F. Activation of C$-$H Bonds
    G. Complete Nitrogen-Atom Transfer
        1. Electrosynthesis of Amino Acids
        2. Nitrile Synthesis by Complete Nitrogen-Atom Transfer
        3. Formation of Cyanate and Carbodiimide Anions

VII. NITRIDES AS CATALYST PRECURSORS AND INTERMEDIATES

    A. Catalytic Ammonia Formation
    B. Alkyne–Nitrile Metathesis

VIII. STRATEGIES FOR INCREASING NITRIDE REACTIVITY

    A. Changing the Oxidation State
    B. Changing the Ancillary Ligands
    C. Addition of *trans* Donors
    D. Addition of Lewis Acids and Electrophiles
    E. Photolysis

IX. CONCLUSIONS

ACKNOWLEDGMENTS

ABBREVIATIONS

REFERENCES

I. INTRODUCTION

Transition metal nitride complexes have been known for more than 160 years, following the synthesis of OsO_3N^- in 1847 (1). While these complexes have always been of interest for fundamental questions regarding structure and bonding, it is only in relatively recent times that serious efforts have been made into developing their reaction chemistry.

Given the isoelectronic nature of the $M{\equiv}N$ and $N{\equiv}N$ bonds, it should not be surprising that nitride complexes have been found to be important in schemes aimed at using N_2 as a reagent, including the catalytic formation of ammonia (2) and the synthesis of organic molecules (3). Nitride complexes are also important in the catalytic triple-bond metathesis of alkynes (4) and nitriles (5). Apart from their synthetic utility, transition metal nitrides are also important as potential models for catalytic intermediates of certain metalloenzymes, particularly nitrogenase (6).

II. SCOPE

The intended focus of this chapter is the chemistry of the terminal metal nitride fragment, $M{\equiv}N$ (i.e., the synthesis and reactivity of the nitride ligand in transition metal complexes). Bridging nitrides or nitrides that are part of clusters will typically not be covered. In addition to this focus on terminal metal nitride complexes, a further emphasis will be on nitrides that can be isolated or have been spectroscopically characterized. Transient species for which there is little direct characterization data will generally not be covered. For the most part, the focus will be on solution-phase complexes, although occasionally gas-phase or matrix-isolated species will be mentioned as appropriate.

While this chapter attempts to describe all the known types of reactions involving terminal nitrides, specific examples have been chosen for illustration and it is likely that not every single example of a particular reaction has been cataloged here. Thus, this chapter is not intended to be exhaustive in its coverage, but rather provide the reader with the state of play *circa* July 2012.

III. PREVIOUS REVIEWS

The last comprehensive review in this area dates back to the early 1990s (7, 8). Since then, a number of reviews have appeared that focus on aspects of transition metal nitride chemistry, typically focused on a small section of the periodic table (9–15). The more comprehensive reviews included material on the transition metal nitrides of groups 6–8 (16) and late transition metal nitrides (17). Moreover, the May 2003 issue of *Z. Anorg. Allg. Chem.* (9) was devoted to aspects of transition metal nitride chemistry.

IV. PROPERTIES OF THE NITRIDE LIGAND

A.   Simple Bonding Considerations

The properties of the nitride ligand have often been considered in terms of a simple, but incomplete bonding scheme that is included here for completeness. Since the nitride ligand is a very strong σ- and π-donor, it might be expected that it will have nucleophilic character in its reactions. However, simple bonding considerations show that the reactivity of the nitride ligand is in fact dependent on the relative energies of the metal and nitride fragment orbitals (16). In this model, two extreme cases involving the symmetry-allowed interaction of the metal and nitride orbitals can be considered (Fig. 1) (16, 18). These orbitals are typically considered as having π symmetry.

In one extreme, the metal d-orbitals are higher in energy than the nitride atomic orbitals, resulting in the formation of antibonding MOs that are mostly metal in character [Fig. 1(a)]. Therefore the nitride ligand is expected to be nucleophilic. In the other extreme, where the d-orbitals are lower in energy, the antibonding MOs are mostly ligand in character [Fig. 1(b)], and the nitride ligand is predicted to be electrophilic.

Based on these energetic considerations, early metal nitrides are expected to be nucleophilic, while late metal nitrides are expected to be electrophilic. Moreover, according to this simple MO picture, the nucleophilicity of the nitride is expected to increase with the metal oxidation state. Consistent with this picture, the partial charge at the nitride ligand, as determined from density functional theory by a

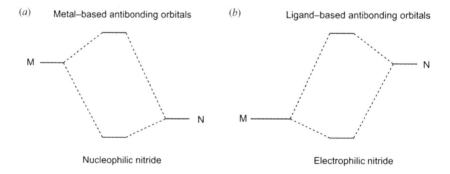

Figure 1.   Simplified molecular orbital (MO) diagrams for the M≡N interaction. [Adapted from (16).]

number of population analyses, has been shown to correlate with the expected reactivity of the nitride ligand in the complexes (salen)M(N) and $[M(N)Cl_4]^{2-}$ (M = V, Cr, Mn, Fe and salen = N,N'-bis(salicylidene)ethyldiamine dianion) (19). Thus, the partial negative charge at the nitride ligand decreases from left to right across the transition series, which follows the trend of increasing electrophilic reactivity across the series.

However, this simple bonding picture, while often useful, does not fully capture the factors that determine the reactivity of the nitride ligand, which may depend on more than the nature of the transition metal. As illustrated in subsequent sections, the ancillary ligands can play a key role in modifying the relative energies of the orbitals either by virtue of their donor strength or through orbital mixing.

B. Structural and Electronic Impact of the Nitride Ligand

Due to its strong donor properties, the nitride ligand has a very strong *trans* influence, which may even prevent ligands from binding in the *trans* position, helping to stabilize lower coordination numbers. Thus, square-pyramidal complexes with the nitride ligand in the axial position are often observed [e.g., Tc(VI) and Tc(V) complexes] (20). This *trans* influence may even modify the nature of bonding to *trans* ligands [e.g., causing cyclopentadienyl (Cp) ring slippage in CpV(N)Cl$_2$ (21) and Cp$_2$*Mo(N)N$_3$ (22)].

The nitride ligand often dominates the ligand field. Thus, for example, spectroscopic and computational data suggest that the electronic structure of square-pyramidal $[Cr^V{\equiv}N]^{2+}$ complexes with weak ancillary ligands can be more usefully viewed as being weakly perturbed pseudolinear complexes rather than strongly perturbed octahedral systems. The d-orbitals having $M-N$ σ^* and π^* character are most strongly destabilized, particularly with weak equatorial donor ligands (Fig. 2, right) (23–25). Indeed, the nitride ligand has a greater influence on the ligand field than the oxo ligand in isoelectronic VO^{2+} complexes (Fig. 2, left). The electron paramagnetic resonance (EPR) spectra of these complexes reveal very strong dπ–pπ interactions, as well as an enormous expansion of the 3d electron cloud due to the strong covalent Cr$\equiv$N bond (26, 27). Nonetheless, despite its overwhelming influence, the effect of the nitride ligand can be attenuated by the nature of the ancillary ligands. With stronger donors, such as the salen ligand in (salen)Cr(N), the splitting is similar to that of VO^{2+}.

C. Effects of Orbital Mixing

The nitride is not always dominant, and mixing of orbitals from the supporting ligand set can also diminish its effect on the ligand field. For example, in four-coordinate d^0 group 6 (VI B) (RO)$_3$M(N) complexes, significant mixing of the alkoxide ligand π-symmetry orbitals with the M$\equiv$N fragment orbitals is possible,

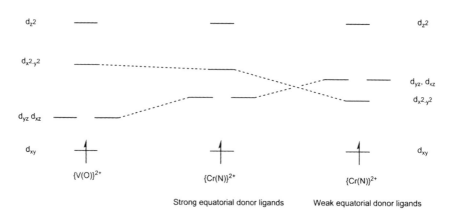

Figure 2.   Qualitative MO diagrams for d^1 complexes having metal–ligand multiple bonds showing the relative impact of the nitride ligand compared with that of the oxo ligand in isoelectronic complexes. [Adapted from (23).]

and the description of localized $M-N$ σ- and π-bonding orbitals is no longer valid (28). The extent of mixing depends largely on the nature of M. Thus for Mo and W, the highest energy $2a_1$ orbital in C_{3v} symmetry (Fig. 3, right) is largely $M\equiv N$ nonbonding and can donate electrons with little cost to the stability of the molecule. In the case of Cr, the a_2 orbital is highest in energy and the nitride is less nucleophilic (Fig. 3, left). As a result of these orbital contributions, the $W\equiv N$ bond is the most polarized and nucleophilic. The extent of orbital mixing is also determined by the nature of the ancillary ligands: when the alkoxide ligands in $(RO)_3Mo(N)$ are changed to thiolates in $(RS)_3Mo(N)$, the a_2 orbital is highest in energy (29).

In addition to mixing of ligand orbitals, the nitride may facilitate the mixing of metal-based orbitals (30). Such mixing has been proposed to occur in square-pyramidal $[M(N)(CN)_4]^{2-}$ complexes (31), where strong admixing of the metal-based d_{z^2} and p_z orbitals creates a new hybrid orbital with substantially reduced antibonding character. This mixing reduces the energy of this orbital sufficiently such that it becomes the lowest unoccupied molecular orbital (LUMO), in contrast to the orbital splitting observed in six-coordinate analogues, where this admixing is not as extensive (Fig. 4).

D.   Bonding Analogies

Analogies have been drawn between the $M\equiv N$ bond and more familiar triple bonds. For example, parallels between the $Os\equiv N$ bond in $TpOs(N)Cl_2$ and $C\equiv O$

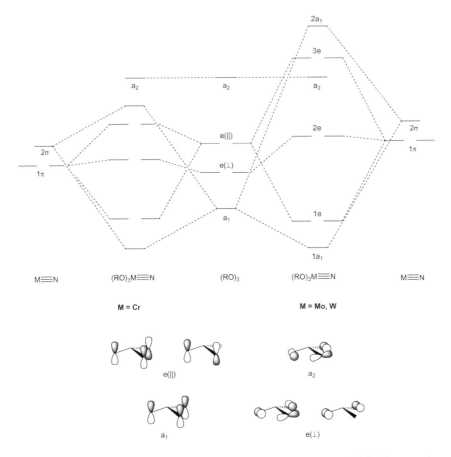

Figure 3. Qualitative diagram showing orbital mixing between the M≡N and (RO)$_3$ fragments in (N)M(OR)$_3$ complexes, M = Cr, Mo, W. The symmetry adapted linear combinations for the pπ orbitals of the (RO)$_3$ fragment are shown below; σ-bonding combinations are not shown. Symmetry labels are for the C_{3v} point group. [Adapted from (28).]

have been noted (32), an analogy that emphasizes the π-acidic nature and low σ-basicity of the nitride ligand in this complex. Following this analogy, both species act as π-acids to metals and both insert into B$-$C bonds of organoboranes by similar mechanisms, namely, via formation of an adduct with the borane, followed by alkyl group migration. Similarly, an analogy has been drawn between the C≡N bond in organic nitriles and the M≡N bond in four-coordinate (RO)$_3$M(N) complexes (28), in this case emphasizing the σ-basicity and low π-acidity of these nitride ligands.

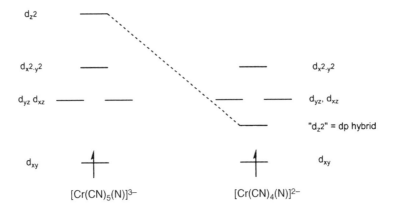

Figure 4.   Effect of orbital mixing in $Cr(CN)_4N^{2-}$. [Adapted from (31).]

E.   Noninnocence of the Nitride Ligand

In some complexes, the nitride ligand may be noninnocent, making simple bonding pictures even less appropriate (33). Indirect evidence for nitride non-innocence was first obtained from EPR studies of the Cr(V) nitride complexes (salen)Cr(N) and $(dbm)_2Cr(N)$ (dbm = dibenzoylmethanolate) and their Lewis acid adducts (34). Specifically, the ^{14}N hyperfine coupling constants in these complexes increase with increasing Cr−N bond length, an unexpected trend that has been attributed to the contribution of the Cr^{3+}–triplet N^- resonance structure to the ground state of the complex, in addition to the expected $Cr^{5+}-N^{3-}$ resonance structure.

Stronger evidence for the nitride ligand noninnocence has been obtained in the four-coordinate Ir(IV) complex (PNP)Ir(N), where PNP = [N(CHCHP*t*-$Bu_2)_2]^-$, (35). Electronic structure calculations, in conjunction with EPR and ENDOR (where ENDOR = electron–nuclear double resonance) spectroscopic measurements, provide strong evidence for unpaired electron density on the nitride ligand, that is, partial contribution of a nitridyl resonance structure (Ir^{3+}–doublet N^{2-}) to the electronic structure of the complex. This resonance structure has been proposed to be important in facilitating the dimerization of nitride complexes to yield M−N≡N−M dimers.

V.   SYNTHESIS OF TRANSITION METAL NITRIDES

Synthetic routes to transition metal nitrides have been extensively covered in previous reviews (7, 36). In this section, the most commonly used methods of

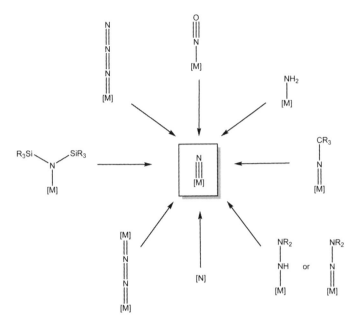

Scheme 1.   Some of the more common routes to transition metal nitride complexes.

nitride synthesis will be summarized (Scheme 1). New mechanistic insights and new reagents will be described in more detail. Unusual transformations that lead to metal nitrides will also be briefly mentioned.

The strength of the M≡N bond that is formed likely provides a key role in determining the applicability of the synthetic method. Weaker bonds, which are more likely to be reactive, require additional driving forces, for example, the formation of very stable species (e.g., N_2). More synthetic routes will be available when the formation of a strong M≡N bond is the likely driving force. These stronger M≡N bonds are often formed by reactions that may be viewed as the reverse of nitrogen atom transfer reactions (see Section VI).

A.   Azide Decomposition

The thermal or photochemical decomposition of precursor azide complexes remains a very common synthetic method for preparing terminal nitrides. While this reaction is typically conducted in solution, thermolysis in the solid state is a synthetic alternative that may avoid intermolecular decomposition reactions (37–39).

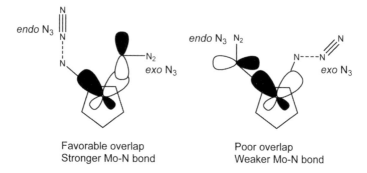

Favorable overlap   Poor overlap
Stronger Mo-N bond   Weaker Mo-N bond

Figure 5.   Enhanced π interactions stabilize the *exo* N_3^- ligand in $Cp_2^*Mo(N_3)_2$. Thermolysis results in N_2 dissociation from the *endo* N_3^- ligand. [Adapted from (22).]

Despite the importance of this method for the synthesis of nitride complexes, there is little mechanistic understanding of the thermally driven azide decomposition reaction. One such study investigated the thermolysis of $Cp^*_2Mo(N_3)_2$ to yield $Cp^*_2Mo(N_3)(N)$. Based on a combination of experimental and computational methods, nitride formation is proposed to occur by N_2 dissociation from the *endo* azide ligand in $Cp^*_2Mo(N_3)_2$ (Fig. 5). The *exo*-Mo-N_3 bond in the transition state is stronger due to better orbital overlap between the well-oriented azide and the b_1 orbital of the metallocene fragment. Thus N_2 dissociation from the *endo* azide is favored (22). Following N_2 loss, rotation of the *exo* azide to the *endo* position gives the final product.

More recently, the mechanism of azide thermolysis in a β-diketiminate V(III) azide complex to yield the corresponding V(V) nitride complex has been investigated. Mechanistic investigations suggest that azide decomposition likely occurs via Lewis acid assisted activation of the azide ligand. (Scheme 2) (40).

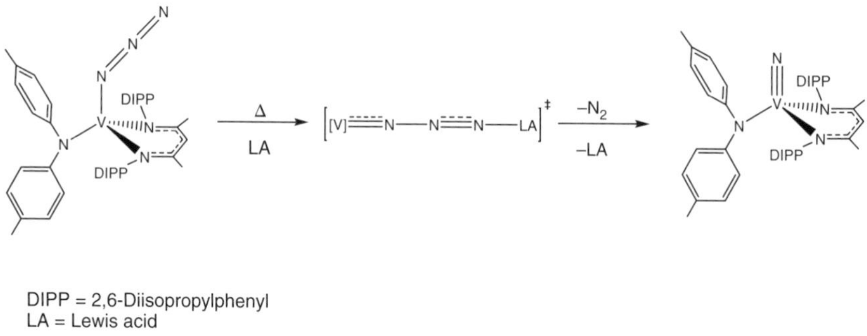

DIPP = 2,6-Diisopropylphenyl
LA = Lewis acid

Scheme 2.   Lewis acid assisted thermolysis of a V(III) azido complex to the corresponding V(V) nitride. [Adapted from (41).]

There have also been few studies on the photochemical decomposition of azide complexes. In principle, photolysis could cleave either the $M-N_3$ bond to eject N_3 (photoreduction) or the $N-N$ bond to eject N_2 (photooxidation). In the case of six-coordinate, low-spin Fe(III) azido complexes supported by functionalized amine macrocycles, photooxidation occurs to provide the corresponding Fe(V) nitrides (42). The photooxidation pathway is proposed as being favored over the competing photoreduction reaction due to the low reorganizational energy associated with formation of the iron(V) nitride. In the case of related high-spin Fe(III) complexes, there is a larger reorganizational energy associated with formation of the nitride complex, and photoreduction occurs instead (43).

Nitride formation may also be critically dependent on the wavelength of irradiation (44, 45). In the case of a Mn containing polyoxometallate, it has been proposed that irradiating at a frequency that leads to an excited state with azide $\pi\pi^*$ character results in photooxidation to the nitride, whereas irradiation at a frequency that contains $N_3 \rightarrow Mn$ charge-transfer character results in M_n-N bond homolysis and photoreduction.

In some cases, photoreduction of the azide complex in liquid solution may be prevented by conducting the photolysis reaction in frozen solution (43, 45–47). It is proposed that the reduced mobility of the azide radical in frozen solution makes the photoreduction pathway partially reversible, increasing the yield of the photooxidation product (i.e., the metal nitride). Similarly, encapsulation of the azide complex (e.g., in zeolites) may also increase the yield of the nitride product (48).

B. Ammonia Oxidation

The synthesis of nitride complexes from ammonia was noted in previous nitride reviews (7, 36) and continues to be a method of nitride synthesis, particularly for metals in high oxidation states. In aqueous solution, this oxidation is pH depen-dent, as may be expected (49, 50). It is notable that a number of Mn^VN complexes supported by salen and salen-like ligands can be easily prepared from ammonia using inexpensive reagents, namely, ammonium hydroxide as the nitrogen source and bleach as the oxidant (51–54).

C. Nitrogen-Atom Transfer

A relatively recent development for the synthesis of nitride complexes has been the introduction of compounds that transfer single nitrogen atoms. Two types of these compounds are known: transition metal complexes that can undergo three- or two-electron nitrogen-atom transfer, and organic molecules that facilitate two-electron nitrogen-atom transfer (Fig. 6).

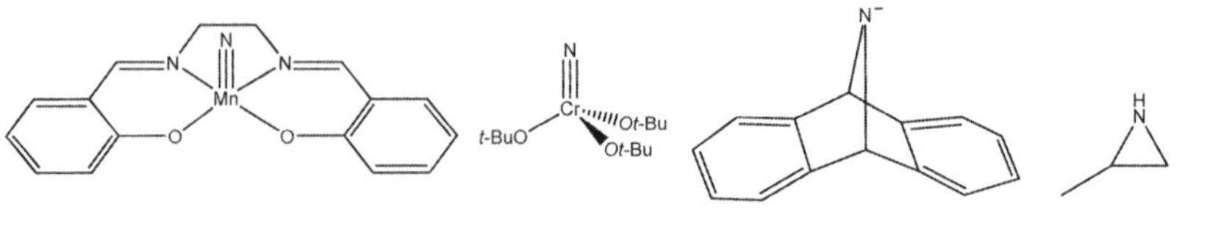

| (salen)MnN | Cr(O*t*-Bu)$_3$CrN | dbabh | 2-Methylaziridine |

Figure 6.   Nitrogen-atom transfer reagents for the synthesis of nitride complexes in solution.

The insights obtained from earlier work in (54) has led to the development of a number of easily prepared transition metal nitrides for the synthesis of less accessible nitrides by nitrogen-atom transfer. The most commonly used complex for this purpose has been (salen)Mn(N) and its derivatives (25, 52, 56), which typically react in two- or three-electron nitrogen-atom transfer reactions. These are air-stable complexes that can be prepared on a multigram scale from (salen)MnCl, ammonia, and bleach. Their solubility and atom-donor abilities can in principle be tuned by changes to the substituents on the salen ligand.

Thus, for example, reaction of $CrCl_3(thf)_3$, thf = tetrahydrofuran (ligand), with (salen)Mn(N) results in the synthesis of the simple Cr(V) complex, $[Cr(N)Cl_4]^{2-}$ (23, 25). This complex serves as an excellent starting material for the synthesis of other Cr(V) nitride complexes through substitution of the chloro ligands (23). Another transition metal complex that has been used to deliver nitrogen atoms is (t-BuO)$_3$Cr(N) (57–60), which reacts, for example, with a masked V(II) complex to yield the corresponding V(V) nitride (59).

An organic reagent for the synthesis of nitride complexes is the anion of 2,3:5,6-dibenzo-7-azabicyclo[2.2.1]hepta-2,5-diene (61). Reaction of this anion with a suitable precursor complex leads to an amido complex that loses anthracene in a two-electron oxidation step to ultimately provide the nitride complex (8). 2-Methylaziridine has also been found to be a source of nitrogen atoms, as shown by the reaction of (silox)$_3$WCl, where silox = t-BuSiO$^-$, with 2 equiv of 2-methylaziridine, affording the nitride complex (silox)$_3$W(N), propylene, and 2-methylaziridinium chloride (62).

D.   Cleavage of N—Si Bonds

The use of N(SiMe$_3$)$_3$ to prepare metal nitrides was described in previous reviews (7), and has continued to find application (63). Typically, a high-valent metal complex having multiple chloride ligands is treated with N(SiMe$_3$)$_3$; formation of the nitride ligand is accompanied by the production of 3 equiv of Me$_3$SiCl. This type of reaction is observed for the heavier mid-transition metals, where the formation of strong M≡N and Si—Cl bonds likely provides the driving force for this reaction. The same driving forces are likely at play when HN(SiMe$_3$)$_2$

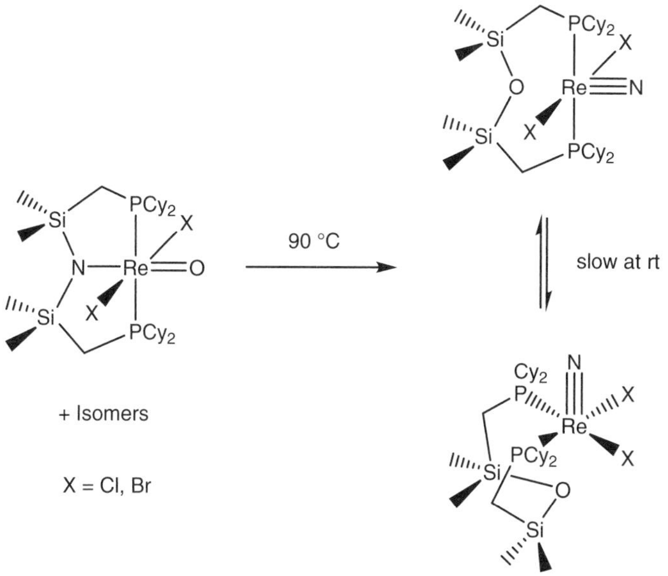

Scheme 3. Double Si migration to yield a nitride ligand. [Adapted from (68).]

is used to prepare metal nitride complexes (64, 65), although a base is sometimes required to neutralize the HCl byproduct (66).

Intramolecular N–Si bond cleavage leading to nitride ligands has also been observed. Thus, for example, the addition of nitrogen bases to $V(NSiMe_3)Cl_3$ induces elimination of Me_3SiCl to afford V(V) complexes $V(N)Cl_2L_2$ [L = py (pyridine), 4-t-Bupy(4-$tert$-butylpyridine); L_2 = TMEDA (TMEDA = tetramethylethylenediamine)] (66, 67). A more remarkable transformation occurs upon thermolysis of the pincer complexes $(PNP^{Cy})Re(O)X_2$ (X = Cl, Br and Cy = cyclohexyl): double Si migration from N to O yields $(POP^{Cy})Re(N)X_2$ (Scheme 3, POP = [O(CHCHPCy_2)]) (68). A similar transformation occurs when an Os(II) complex supported by a related pincer is treated with oxygen-atom transfer reagents (69).

E. Reductive Cleavage of N_2

While atmospheric nitrogen is the most inexpensive and abundant nitrogen source available for the synthesis of metal nitrides, due to the difficulty of N_2 reduction, only a limited number of complexes have been made this way.

The three-coordinate Mo(III) tris(amido) complex, $Mo(t\text{-}BuNAr)_3$, cleaves N_2 to afford the nitride $NMo(t\text{-}BuNAr)_3$ (Fig. 7) (70, 71). The reaction mechanism

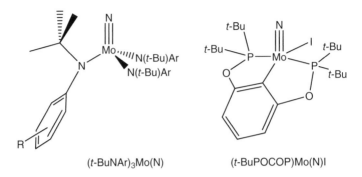

$(t\text{-BuNAr})_3\text{Mo(N)}$ $(t\text{-BuPOCOP})\text{Mo(N)I}$

Figure 7.   Nitride complexes prepared by N_2 reductive cleavage $(t\text{-BuBOCOP})=C_6H_3\text{-}1,3\text{-CO(P-}t\text{-}$ $Bu)_2]_2$.

involves initial binding of N_2 to the three-coordinate center, which reacts with another equivalent of the Mo complex to give a bridged Mo–N=N–Mo interme- diate $(S = 1)$ that undergoes N–N bond cleavage via a zigzag transition state (70). Cleavage of the N–N bond can also be effected photochemically (72). The reaction is catalyzed by reductants (73) and bases (74). Bimetallic dinitrogen cleavage as well as nitrogen-atom transfer to other metal centers leads to complexes such as $(t\text{-BuO})_3\text{Mo(N)}$ (70) and $[(i\text{-PrNAr})_3\text{Nb(N)}]^-$ (75).

Interestingly, the related three-coordinate fragment Mo(Mes)$_3$ (Mes = 2,4,6- trimethylphenyl) can only cleave N_2 photochemically, likely a consequence of the lower basicity of Mo when coordinated by aryl ligands (76). While similar reductive cleavage of N_2 by other three-coordinate metal complexes has yet to be realized, computational investigations suggest that three-coordinate tris(anilido) W(III), Ta(II), and Nb(II) complexes would be more favorable for N_2 cleavage (77, 78), although this would likely come at the expense of subsequent nitride reactivity.

The only other confirmed example of a complex that can reductively cleave N_2 to a terminal nitride ligand is also based on molybdenum, in this case, supported by a pincer ligand. Reduction of $(t\text{-BuPOCOP})\text{MoI}_2$ complex under N_2 results in formation of the Mo(V) nitride complex $(t\text{-BuPOCOP})\text{Mo(N)I}$ (Fig. 7) (79). It is proposed that the key N–N cleavage step occurs by a Mo(II)(μ-N$_2$)Mo(II) dimer.

F. Cleavage of N–N Bonds

A number of nitride complexes have been prepared by N–N bond cleavage in hydrazido(1–) and hydrazido(2–) complexes. This cleavage reaction is most commonly observed for the heavier mid-transition elements in higher oxidation states, where particularly strong metal–nitride bonds are formed. Thus, for example, Tc(V) (80–82), Tc(VII) (83), Re(V) (82, 84–88), Re(VII) (84),

Mo(IV) (89), Mo(VI) (90, 91), W(IV) (92, 93), Ru(VI) (94), and Os(VI) (95) nitride complexes have been reported via N−N cleavage reactions.

Mechanistic investigations into N−N bond cleavage upon protonation of the hydrazido(2−) complexes $(dppe)_2M(NNC_5H_{10})$ (M = Mo, W and dppe = diphenylphosphinoethane) have been reported (93, 96). It is proposed that initial protonation of the hydrazido(2−) ligand by the acid allows for solvent coordination at the metal center and subsequent N−N bond cleavage. When M = W, the rate of reaction is an order of magnitude faster than for M = Mo, a result of the higher basicity of the tungsten complex.

G.   Cleavage of C−N Bonds

The cleavage of C−N bonds to give metal nitrides remains an unusual way of synthesizing metal nitrides, and is likely driven by the formation of strong M≡N bonds. There are a number of examples in which the C−N bonds in metal imido complexes are cleaved to yield nitrides, particularly with *tert*-butylimides (97–100). Thus, for example, the reaction of $RuO_2Cl_3{}^-$ with t-BuNCO is proposed to involve a [2 + 2] cycloaddition to form the imido species $[Ru(Nt\text{-}Bu)(t\text{-}BuNC(O)Nt\text{-}Bu)Cl_3]^-$, which in turn undergoes a cycloelimination reaction to provide isobutylene, HCl, and the four-coordinate Ru(VI) nitride complex, $[Ru(N)(t\text{-}BuNC(O)Nt\text{-}Bu)Cl_2]^-$ (Scheme 4) (99).

In a potentially related reaction, the dioxoruthenium(VI) porphyrins $Ru(por)(O)_2$ (por = 2,4,6-Me_3TPP, 3,4,5-MeO_3TPP, where TPP = tetraphenylporphyrin) react with $HN=Ct\text{-}Bu_2$ to afford the corresponding nitride complexes $Ru(por)(N)(OH)$ (96). The reaction pathway is very sensitive to the nature of the porphyrin ligand: Electron-rich porphyrins provide μ-oxo dimers, while less electron-rich porphyrins provide nitrosyl complexes. Moreover, nitride formation is dependent on the nature of the ketimine since reaction of $Ru(3,4,5\text{-}MeO_3TPP)(O)_2$ with $HN=CPh_2$ provides the bis(methyleneamido) complex $Ru(3,4,5\text{-}MeO_3TPP)(N=CPh_2)_2$, without further reaction to afford a nitride complex (101).

Osmium complexes can also facilitate C−N cleavage to provide nitrides. In the presence of water, the nitrogen atom of acetonitrile is abstracted by the Os(IV) complex, $TpOsCl_2(OTf)$. A proposed mechanism involves hydrolysis

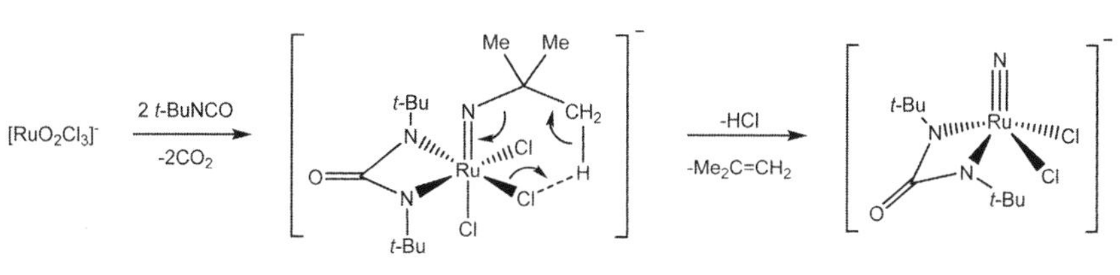

Scheme 4.   Proposed mechanism for the synthesis of $[Ru(N)(t\text{-}BuNC(O)Nt\text{-}Bu)Cl_2]^-$ from $[RuO_2Cl_3]^-$ and t-BuNCO. [Adapted from (99).]

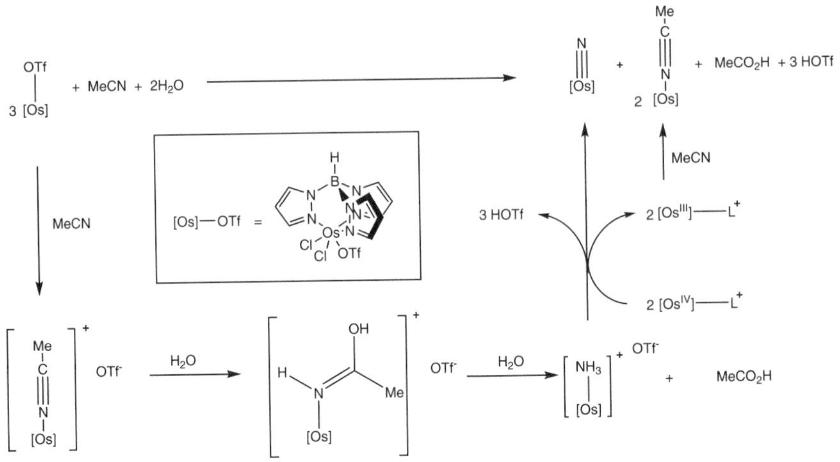

Scheme 5.   Proposed mechanism of MeCN hydrolysis and oxidation to a nitride ligand. [Adapted from (101).]

of coordinated acetonitrile to provide the ammine complex, which is then oxidized by Os(IV) to the corresponding nitride complex (Scheme 5) (102). Similar CN bond cleavage of β-ketonitriles by low-valent $Mo(N_2)(dppe)_2$ is proposed to involve a mechanism in which the cyano carbon of coordinated nitrile is protonated by a second molecule of the β-ketonitrile. Slow proton shift from the α-position to the amido carbon is followed by rapid C−N cleavage to yield the nitride complex and a ketone (Scheme 6) (103).

H.   Nitrosyl Deoxygenation

Deoxygenation of coordinated NO ligands has been reported for low-coordinate group 6 (VI B) metal complexes. Deoxygenation of an NO ligand was first reported for the diamagnetic trisamido complexes $(RNR')_3Cr(NO)$ ($R = R' = i$-Pr; $R = C(CD_3)_2Me$, $R' = 2,5$-C_6H_3FMe; $R = C(CD_3)_2Me$, $R' = 3,5$-$C_6H_3Me_2$). Thermolysis of these complexes in the presence of the masked three-coordinate complex $V(Mes)_3(thf)$ yields the Cr(VI) nitride complexes $(RNR')_3Cr(N)$, along with the bridging oxo complex $O[V(Mes)_3]_2$ (104).

Similar deoxygenation of $(silox)_3W(NO)$ to afford $(silox)_3W(N)$ can be effected by $(silox)_3M$ ($M = V$, Ta, Nb) fragments (62, 105). The rate of deoxygenation is found to be metal dependent, and detailed mechanistic investigations involving experimental and computational studies reveal that binding of the $(silox)_3M$ fragment to $(silox)_3W(NO)$ requires access to a triplet state. For $M = Ta$, the

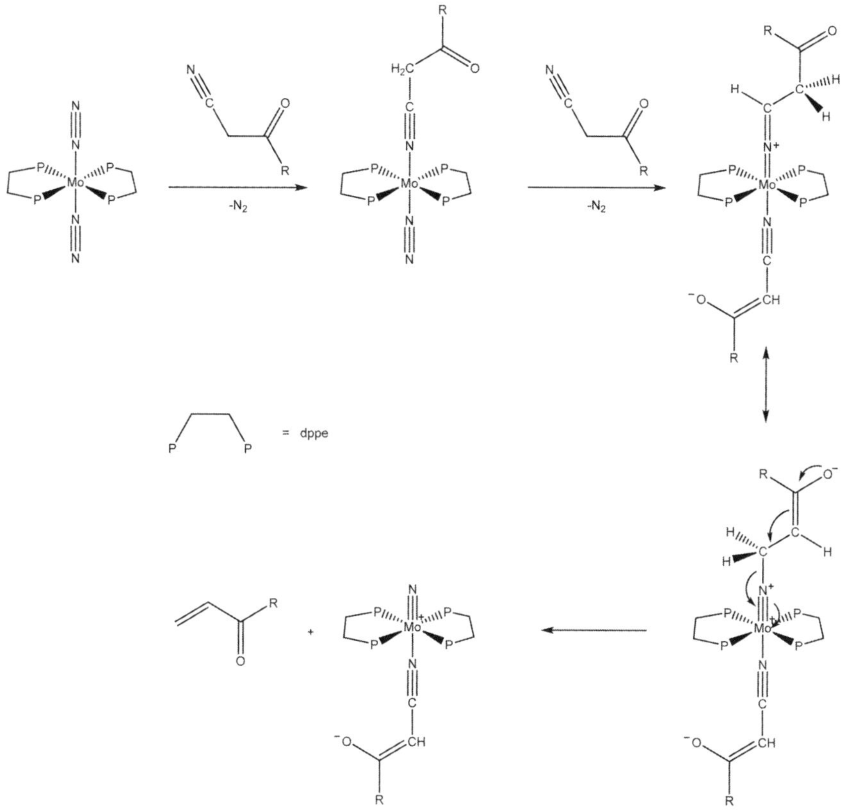

Scheme 6.   Proposed mechanism of C≡N cleavage leading to nitride ligand formation. [Adapted from (103).]

large singlet–triplet gap hinders the deoxygenation reaction. The required inter-system crossing can be facilitated by a reduction in symmetry. Specifically, if the N−O−M angle (θ) deviates from 180°, the symmetry is lowered from C_{3v} to C_s, allowing for mixing between orbitals that formerly had pure σ and π character. Since this mixing lowers the barrier for intersystem crossing, the deoxygenation reaction is also subject to steric constraints (Fig. 8).

I.   Cleavage of N_2O N−N Bond

A unique example of nitride formation involves the reaction of three-coordinate Mo(III) trisanilido complexes $(RNAr)_3Mo$ [R = t-Bu, $C(CD_3)_2Me$] with N_2O, giving two products, namely, $(RNAr)_3Mo(N)$ [R = t-Bu, $C(CD_3)_2Me$] and

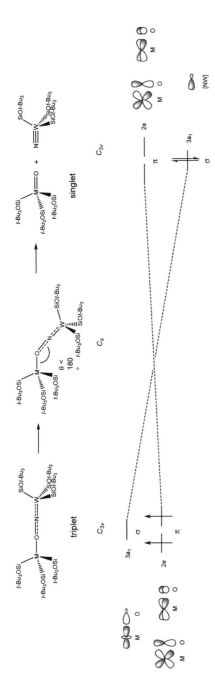

Figure 8. Key orbital interactions in the deoxygenation of (silox)₃W(NO) by (silox)₃M (M = V, Ta, Nb and silox = t-Bu₃SiO) fragments. Deviation of θ from 180° lowers the symmetry, allowing σ- and π-orbitals to be mixed, which facilitates intersystem crossing.

(RNAr)$_3$Mo(NO) (71). Solution calorimetry measurements reveal that N−N bond cleavage to give the nitride, rather than the anticipated N−O bond cleavage to the oxo, is under kinetic control. Mechanistic investigations excluded the intermediacy of free NO and are most consistent with rate-determining binding of N$_2$O to Mo, with bimetallic N−N cleavage occurring after this step (106, 107).

J. Other N−E Cleavage Reactions

Arylazides have been reported to react with some metal complexes to yield metal nitride complexes with loss of the NNAr radical (71, 108). Thus, for example, reaction of the tungsten dimer [W(TPP)]$_2$ with p-tolylazide affords (TPP)W(N) in moderate yield (108).

Reaction of [TcOCl$_4$]$^-$ with 2 equiv of Me$_3$SiNPPh$_3$ yields the Tc(V) complex Tc(N)Cl$_2$(Ph$_2$NH)$_2$ by a unique N−P bond cleavage reaction (109). This result is the reverse of the commonly observed reaction between nitrides and phosphines, and suggests the Tc≡N bond has a large thermodynamic stability.

Related two-electron reductions that yield nitrides from isocyanate ligands also have been observed. Thus, photolysis of Re(NCO)Cl$_2$(PMe$_2$Ph)$_3$ leads to irreversible CO loss, analogously to N$_2$ loss from azide ligands, yielding Re(N)Cl$_2$(PMe$_2$Ph)$_3$ (110), while one-electron reduction of (t-BuNAr)$_3$Nb(OCN) induces CO loss to give[(t-BuNAr)$_3$Nb(N)]$^-$ (111).

VI. REACTIVITY

The reactivity of nitrides is often described according to the perceived electrophilic–nucleophilic character of this ligand. While the simple MO considerations discussed above suggest the reactivity of the nitride ligand can be predicted based on the location of the transition metal in the periodic table, reality is more complicated. The ancillary ligands can modify the reactivity of the nitride ligand and completely change its reactivity. Thus, for example, the nitride in Cp*Os(CH$_2$SiMe$_3$)$_2$(N) is nucleophilic, reacting with MeOTf to provide the imido complex [Cp*Os(CH$_2$SiMe$_3$)$_2$NMe]OTf (112). By contrast, TpOsCl$_2$(N), which features the same metal in the same oxidation state, is electrophilic, reacting with PhMgCl to provide the amido complex TpOsCl$_2$(N(H)Ph) after aerobic workup (32). Moreover, some nitrides are ambiphilic, for example, (TPP)Mo(N), which reacts with PBu$_3$ to yield (TPP)Mo(NPBu$_3$) (113) and with MeI to alkylate the nitride ligand, yielding [(TPP)Mo(NMe)(H$_2$O)]$^+$. Finally, while some reactions appear to be electrophilic (e.g., reactions with phosphines), nucleophilic interactions between the nitride highest occupied molecular orbital (HOMO) and substrate LUMO may also be important (114).

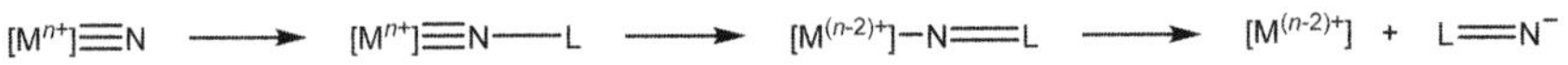

$$[M^{n+}]{\equiv}N \quad\longrightarrow\quad [M^{n+}]{\equiv}N{-\!\!-}L \quad\longrightarrow\quad [M^{(n-2)+}]{-}N{=\!\!=}L \quad\longrightarrow\quad [M^{(n-2)+}] \;+\; L{\equiv}N^-$$

Scheme 7.   Two-electron nitrogen-atom transfer considered as an electron-transfer reaction.

In this chapter, the reactivity of terminal nitride complexes has been organized according to the change in the formal oxidation state of the metal center, which may be between zero and three, with most common oxidation state changes being zero and two. Thus, this scheme considers only the nature of the reactant and product and not the mechanistic details of the transformation. Note that nitrides that are considered nucleophilic often react with no change in the oxidation state of the metal center, whereas nitrides that are considered electrophilic commonly participate in two-electron reactions.

Since this scheme is based on redox state changes, reactions can be considered in terms of the factors that affect electron-transfer reactions, for example, a two-electron nitrogen-atom transfer reaction can be considered as involving the formation of a precursor complex, followed by inner-sphere electron transfer (Scheme 7). Dissociation of the successor complex to new products may or may not occur. This scheme also allows for comparison with other atom- and group-transfer reactions, (e.g., oxo transfer, particularly the thermodynamic considerations that drive atom-transfer reactions) (115–117).

Due to their potential importance in organic synthesis and N_2 fixation, reactions related to ammonia formation and C$-$H activation are described in Sections VI.E and VI.F. However, note that these reactions could also be considered in terms of the overall change in oxidation state: Ammonia formation typically involves three electrons, while C$-$H activation is typically a two-electron process.

A.   Three-Electron Reactions

1.   Intermetallic Nitrogen-Atom Transfer

Three-electron intermetallic nitrogen-atom transfer was reviewed in 1993 (55) and will not be extensively described here. The synthetic utility of this reaction for the preparation of new nitride complexes that are unavailable by other routes, particularly with (salen)Mn(N) as the nitrogen atom source, has been described in Section V.C.

Mechanistic studies on intermetallic nitrogen-atom transfer reactions have suggested that the reaction proceeds via a bridged nitride intermediate, $M(\mu\text{-}N)M'$. This intermediate was first isolated from the reaction of the Mo(VI) complex $Mo(N)(NMe_2)_3$ with the three-coordinate Mo(III) complex

$Mo(NRAr_F)_3$ [R = $C(CD_3)_2Me$; Ar_F = 4-C_6H_4F] (118). When the reaction is conducted at low temperature, the bridged nitride complex $(NMe_2)_3Mo-N-Mo(NRAr_F)_3$ can be isolated. At higher temperatures, this species decomposes to provide $(NRAr_F)_3Mo(N)$, the product resulting from complete nitrogen-atom transfer, establishing the bridged nitride as a competent intermediate in intermetallic nitrogen-atom transfer.

2. Reductive Coupling

Certain nitride complexes are not stable and decompose by reductive coupling to yield bridged N_2 complexes. This reaction was first proposed in the oxidation of $[Os(NH_3)_5CO]^{2+}$, which yields $[(Os(NH_3)_4CO)_2(N_2)]^{2+}$ (119), and has been directly observed in a number of instances, including polypyridyl Os complexes (120, 121) and $(BP_3i-Pr)Fe(N)$ (8). Interestingly, N−N bond formation in (salen)M(L)(N) [M = Ru (122), Os (123), L = trans ligand] by reductive coupling is accelerated by changing the solvent or the trans ligand, activating the nitride for N−N bond formation. In addition to these homocoupling reactions, heterocoupling between $TpOs(N)Cl_2$ and $(Et_2NCS_2)_3Mo(N)$ also has been observed (124).

B. Two-Electron Reactions

A wide range of two-electron nitrogen-atom transfer reactions have been observed (Scheme 8). In addition to simple atom-transfer reactions, the nitride may also be inserted into covalent bonds. In this section, nitrogen-atom transfer to a range of substrates is described.

1. Intermetallic Nitrogen-Atom Transfer

In addition to the three-electron nitrogen-atom transfer reaction described above, two-electron intermetallic nitrogen-atom transfer is also known. This reaction typically involves exchange of the nitride ligand on one complex for a halide on the second.

A detailed mechanistic study of two-electron nitrogen-atom transfer between Mn(V) (porphyrin)Mn(N) and Cr(III) (porphyrin)Cr(X) (X = F^-, Cl^-, $CF_3CO_2^-$, N_3^-, SCN^-, ClO_4^-) complexes established a mechanism in which initial ionization of the Cr−X bond is followed by nucleophilic attack of the nitride ligand on Cr, followed by nitrogen-atom transfer and coordination of the halide to Mn (125, 126). Consistent with this mechanism, the rate of atom transfer increases with solvent dielectric constant, except when coordinating solvents are

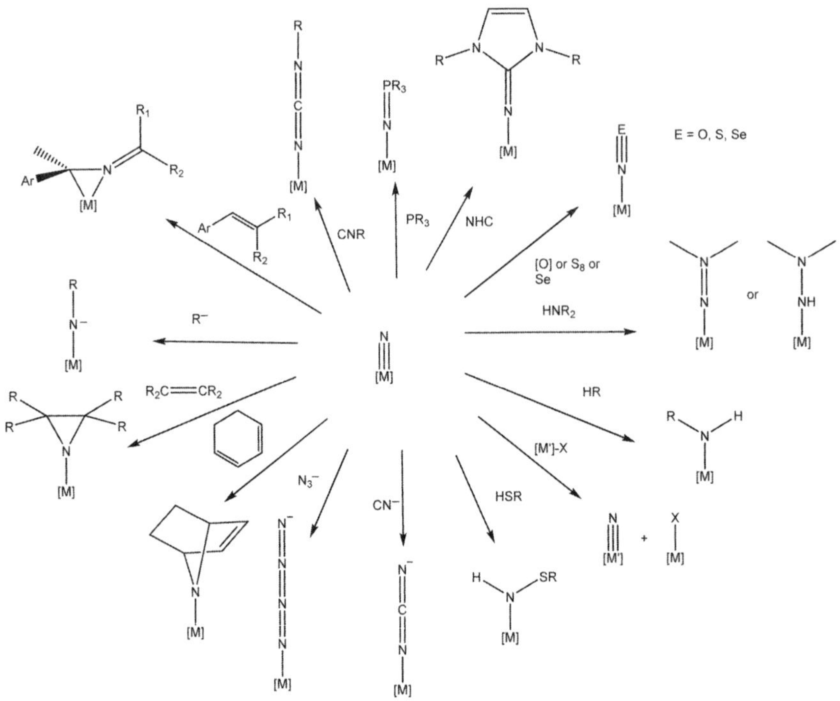

Scheme 8.   Summary of known two-electron nitrogen-atom transfer reactions. Not all nitrides react with all substrates and in some cases, further transformations occur after the nitrogen-atom transfer step.

used, in which case the reaction slows. The rate constants are also inversely related to the ligand field strength of X. Electron-withdrawing substituents on (porphyrin)Cr(X) facilitate the reaction and bulky porphyrin substituents slow the reaction.

Additional mechanistic insight into intermetallic nitrogen-atom transfer has come from studies into nitrogen-atom transfer between Mn(III) and Mn(V) salen, porphyrin, and corrole complexes. These studies reveal that nitrogen-atom transfer is principally driven by the ability of the supporting ligand to stabilize higher oxidation states (i.e., the thermodynamic stability of the higher valent nitride complex). The ability of the solvent to stabilize the products is also important in the rate of reaction (127).

A key observation is that readily prepared (salen)Mn(N) also reacts by two-electron intermetallic nitrogen-atom transfer (128), which has led to this complex being used as a reagent for the preparation of new nitride complexes by nitrogen-atom transfer.

2. Reaction With Phosphines

The two-electron reaction of nitrides with phosphines, leading to phosphoraniminato ligands, $^-N=PR_3$, is one of the most commonly reported reactions of nitride complexes, and has been discussed in previous reviews (7, 36). Due to the prevalence of this reaction, not all examples will be described. Suffice it to say that phosphoraniminato ligand formation by reaction with a suitable phosphine has been observed for early [e.g., $V(N)Cl_2(PMe_2Ph)_2$ (65)], middle [e.g., $(TPP)Mo(N)$] (113), and late [e.g., $Os(terpy)Cl_2(N)$, terpy = 2,2′,6′,2″-terpyridine] (129, 130)] transition metal nitrides.

Phosphoraniminato ligand formation reaction is often used as a synecdoche for electrophilic reactivity. A kinetic investigation of the reaction between PPh_3 and $[(salen)Os(N)(MeOH)]^+$ complexes supports this contention (131). Specifically, a Hammett study revealed that the rate of reaction is increased by electron-withdrawing substituents on the salen ligand [$\rho = 1.9(1)$]. Similarly, the reaction of $TpOs(N)X_2$ with PPh_3 is found to be accelerated by more electron-withdrawing X ligands (132). This finding is consistent with nucleophilic attack at the nitride ligand to give $TpOs(NPPh_3)X_2$, although the effect is very small and the rates vary by only 10^2. The relative rates correlate with the peak reduction potentials of $TpOs(N)X_2$, although the latter vary over a much greater range ($> = 10^{10}$ in K_{eq} for electron transfer).

In contrast to these two studies, a Hammett study for the reaction of $PhB(MesIm)_3Fe(N)$ (MesIm = 1-(2,4,6-trimethylphenyl)imidazole-2-ylidene) with *para*-substituted triarylphosphines revealed that the rate of reaction increased with electron-withdrawing substituents (114). While the effect is relatively modest, the trend is opposite to that expected for an electrophilic nitride ligand. A combined experimental and theoretical investigation has attributed this unexpected selectivity to a dual-nature transition state involving a σ-symmetry interaction involving the nitride LUMO and a π-symmetry interaction involving the nitride HOMO. While it is not clear at this point how general this observation is, it does serve as a cautionary tale to viewing phosphoraniminato ligand formation as being indicative of an electrophilic nitride ligand.

3. Reaction With CO, CNR, CN⁻, and Carbenes

The reaction of nitrides with CO and CNR has been observed across the transition series. Both early transition metal nitride complexes, such as $(\beta\text{-diketiminate})V(N)(NAr_2)$ (40), and late transition metal nitrides, such as $(PNPt\text{-}Bu)Ru(N)$ (133), react with CO to yield isocyanate ligands by two-electron nitrogen-atom transfer. When a coordinatively unsaturated isocyanate complex is formed, additional CO ligands can bind to increase the coordination number of the metal center (133, 134). In a so far unique example, upon reduction of a V(IV) tris(anilido) complex, the isocyanate ligand spontaneously dissociates to form a three-coordinate V(III) complex and the cyanate anion (135).

The reaction of nitride ligands with isocyanides (CNR) also proceeds in a two-electron reduction reaction to provide carbodiimido ligands. While there are few examples of this reaction, note that both early V (40) and late Fe (134) and Ru (136) metal nitrides have been observed to react with isocyanides. As with CO, coordinatively unsaturated products are able to bind additional isocyanide (134, 136). In one instance, complete loss of the carbodiimido ligand as an anion has been observed (134).

Given the analogies between phosphines and N-heterocyclic carbenes (NHC), it is perhaps not surprising that NHCs have been reported to react with metal nitrides. Thus, the Os(VI) nitride complex $(Q')Os(N)Cl$ reacts with bis(1,3-dialkylimidazolin-2-ylidenes) to generate the corresponding Os(IV) azavinylidenes, $(Q')Os(N=NHC)Cl$ (137). A ruthenium(VI) nitride supported by a polyoxometalate ligand, $[PW_{11}O_{39}Ru(N)]^{4-}$, has also been reported to react with the NHC 2,5-bis(2,4,6-trimethylphenyl)imidazol-2-ylidene (IMes). Protonation of the product releases a guanidine that is derived from the NHC reagent (138).

There are few examples of nitrides reacting with the cyanide anion. The highly oxidizing Os(VI) complexes, $[(terpy)OsCl_2(N)]^+$ and $(bpy)OsCl_3(N)$, where $bpy = 2,2'$-bipyridine react with CN^- to yield the Os(IV) cyanamido complexes $(terpy)OsCl_2(NCN)$ and $[(bpy)OsCl_3(NCN)]^-$, respectively (139).

4. Reaction With Carbanions

Perhaps surprisingly, there are few documented examples in which a nitride ligand reacts with a carbanion. The electrophilic Os(VI) complex $TpOs(N)Cl_2$ reacts with PhMgCl to yield the Os(IV) anilido complex $TpOs(NHPh)Cl_2$ following aqueous workup (32, 140). Mechanistic studies suggest that the reaction proceeds with initial weak coordination of Mg to the nitride ligand, followed by migration of the carbanion to nitrogen, which emphasizes that the nucleophilic properties of the nitride ligand are important for facilitating an apparently electrophilic reaction. There is no evidence for initial aryl coordination at the metal.

5. Formation of Nitrosyl, Thionitrosyl, and Selenonitrosyl Ligands

There are multiple examples where oxygen-atom transfer to the nitride ligand gives a nitrosyl. The oxygen-atom donor Me_3NO has been shown to provide nitrosyl ligands in reaction with the nitrides $TpOsCl_2(N)$ (141), [(terpy)OsCl$_2$(N)]$^+$ (142), L$_{OEt}$Ru(N)Cl$_2$ (143), and (PNP)Ir(N) (39). Other oxygen atom donors, such as N$_2$O (144), also have been used to prepare nitrosyl ligands. Other reagents that react with nitrides to yield nitrosyl ligands include hydroxylamine (145, 146) and hydroxide (147). The mechanistic details of these reactions remain obscure.

There are multiple reports on the formation of thionitrosyl ligands from nitrides. A number of reagents have been used to effect this transformation, including S$_8$ (113, 141), propylene sulfide (141), SPPh$_3$ (148), Li$_2$S (149), NaS$_2$O$_3$ (143), S$_2$Cl$_2$ (150), and SOBr$_2$ (20). The synthesis of a selenonitrosyl ligand from elemental Se and an Os(VI) nitride has been reported (141).

6. Reaction With Amines

Only a limited number of nitride complexes are sufficiently oxidizing to react with amines. Certain high-valent osmium nitrides react with secondary amines to yield hydrazido complexes. For example, [(terpy)OsCl$_2$(N)]$^+$ reacts with morpho-line, piperidine, and diethylamine to yield the Os(V) complexes [(terpy) OsCl$_2$(NNR$_2$)]$^+$, along with one-half of an equivalent of the Os(II) dimer (terpy)Cl$_2$Os(NN)Os(terpy)Cl$_2$ (151, 152). The proposed mechanism involves initial nucleophilic attack of the secondary amine on the Os(VI) nitride to give a protonated Os(IV)–hydrazido intermediate. This species is deprotonated by unreacted amine and oxidized by unreacted Os(VI) nitride to yield the final Os(V) hydrazido product (152). The Os(V) nitride undergoes N$-$N coupling to yield the Os N$_2$ dimer as a byproduct.

The analogous reaction between [(terpy)OsCl$_2$(N)]$^+$ and NH$_3$ follows a slightly different course. Initial nucleophilic attack of the amine and deprotonation yields the Os(IV) hydrazido intermediate [(terpy)OsCl$_2$(NNH$_2$)]$^+$, as with the secondary amines. However, with NH$_3$, reaction with another equivalent of the Os(VI) nitride results in N$-$N coupling to yield the bridged azido complex (terpy)Cl$_2$Os(μ-1,3-N$_3$)Os(terpy)Cl$_2$ (153).

A number of Os(VI) nitride complexes react with hydroxylamines, giving a range of products depending on the nature of the complex and the hydroxyl-amine (146). The complexes *trans*- and *cis*-[(terpy)OsCl$_2$(N)]$^+$ react with H$_2$NOH to yield the corresponding Os(II) nitrosyl complexes. With substituted hydroxyl-amines, Me(H)NOH, H$_2$NOMe, and Me(H)NOMe, a range of products are formed, including nitrosyl, dinitrogen, and hydrazido(2$-$) ligands. The product distribution depends on the nature of the Os complex and the hydroxylamine. Interestingly, for *fac*-[(tpm)OsCl$_2$(N)]$^+$, where tpm = tris(pyrazoly)methane, all reagents provide the same final product, *fac*-[(tpm)OsCl$_2$(N$_2$)]$^+$.

The only other example of a nitride reacting with an amine involves the highly electrophilic nitride [(salchda)Ru(N)(MeOH)]$^+$, where salchda = N,N$'$-bis(salicylidene)-*o*-cyclohexyldiamine dianion. In nonpolar solvents, the complex reacts with morpholine and pyrrolidine to yield the Ru(IV) hydrazido complex [(salchda)Ru(N(H)NR$_2$)(HNR$_2$)]$^+$. Kinetic investigations reveal the rate constant for this reaction to be four orders of magnitude faster than for the Os complexes described above and suggest this particular complex to be among the most electrophilic nitrides known (122).

7. Reaction With Azide

A limited number of Os(VI) nitride complexes have been reported to react with the azide anion. The reaction of N$_3$$^-$ with [(terpy)OsCl$_2$N]$^+$ provides the Os(II) dinitrogen complex (terpy)OsCl$_2$(N$_2$) along with an equivalent of N$_2$ (154, 155). Isotopic labeling studies have established that these reactions involve initial attack of azide at the nitride ligand to give an N$_4$$^-$ ligand, (terpy)OsCl$_2$(NNNN). In the case of the related nitride complex (bpy)Os(Cl$_3$)N, this species could be isolated from the reaction with N$_3$$^-$ (156).

8. Reaction With Thiols

Highly electrophilic Os and Ru nitride complexes react with thiols to provide sulfilamido complexes. Reaction of [(terpy)OsCl$_2$(N)]$^+$ with aromatic thiols ArSH yields [(terpy)OsCl$_2$(NS(H)Ar)]$^+$ (Ar = Ph, 4-MeC$_6$H$_4$, 3,5-Me$_2$C$_6$H$_3$) (157, 158). Similarly, [(salen)Ru(N)(MeOH)]$^+$ reacts with RSH to yield [(salen)Ru(N(H)SR)]$^+$, R = t-Bu, Ph (159).

9. Reaction With Nitriles

The Ru(VI) complexes (Q)$_2$Ru(N)Cl react with NCCH$_2$CN–piperidine to afford ruthenium(II) dicyanoimine products (Scheme 9). The reaction with NaTCNE, where TCNE = tetracyanoethylene radical anion, is solvent dependent, affording a diimine complex in MeOH and an imino–oxazolone complex in mixed MeOH–acetone solvent, respectively (160).

10. Reaction With Alkenes

There are also few examples of nitrides that react directly with unsaturated hydrocarbons. All reported cases so far feature metals in high oxidation states, and the nitride complexes are often positively charged.

The cationic Os(VI) complex cis-[(terpy)Os(N)Cl$_2$]$^+$ inserts the nitride ligand into carbon–carbon double bonds of stilbenes and dienes to form azaallenium products under mild conditions (161, 162). More electron-rich alkenes are more reactive, consistent with an electrophilic nitride ligand. The related complexes

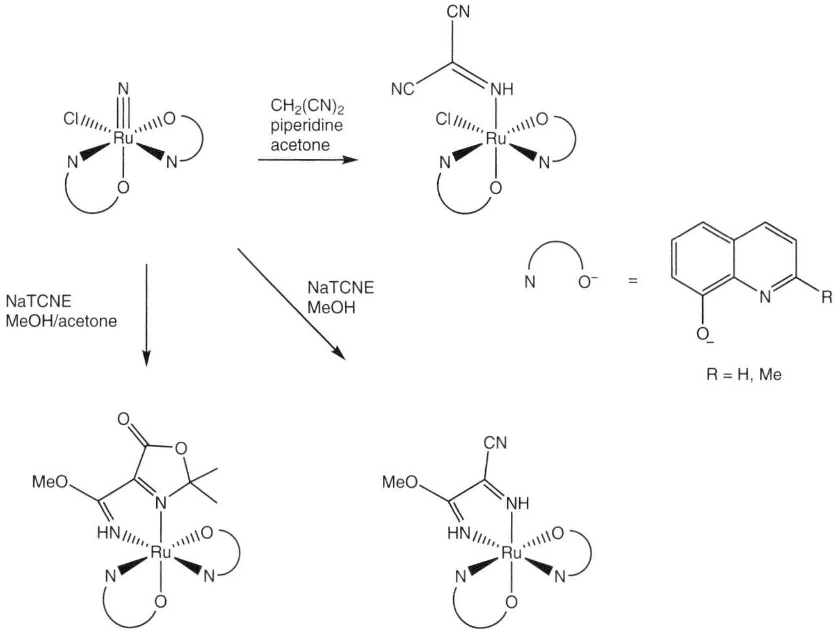

Scheme 9.   Reaction of a Ru(VI) nitride with TCNE and CH₂(CN)₂. [Adapted from (160).]

trans-[(terpy)Os(N)Cl₂]⁺ and [(Tpm)Os(N)Cl₂]⁺ react similarly, however, the insertion reaction is only observed with conjugated dienes. Alkenes (e.g., styrene) do not give tractable products.

An interesting cycloaddition reaction has been observed for an osmium(VI) nitride. The complex TpOs(N)Cl₂ undergoes a [4 + 1] cycloaddition reaction with 1,3-cyclohexadiene to provide the bicyclic osmium amido complex TpOs(NC₆H₈)Cl₂ (163). The more electron-rich 1-methoxy-1,3-cyclo-hexadiene reacts more rapidly, however, the amido product is not stable and an equilibrium between the nitride and cycloaddition complexes is established. The cationic analogues, *cis/trans*-[(terpy)Os(N)Cl₂]⁺ and [(Tpm)Os(N)Cl₂]⁺ react similarly, albeit at faster rates. An analogy between this reaction and the Diels–Alder reaction has been proposed. As noted above, acyclic dienes do not undergo the cycloaddition reaction; instead C=C bond cleavage occurs.

So far, the lone example of aziridination by a nitride complex involves the ruthenium(VI) nitride [(salchda)Ru(MeOH)]⁺ (Scheme 10) (164). Upon activation with bases (e.g., py), the nitride reacts rapidly with 2,3-dimethyl-2-butene to yield the Ru(IV) aziridino complex [(salen)Ru(NC₆H₁₂)(py)]⁺, followed by slower conversion to the Ru(III) complex [(salchda)Ru(N(H)C₆H₁₂)(py)]⁺. The second

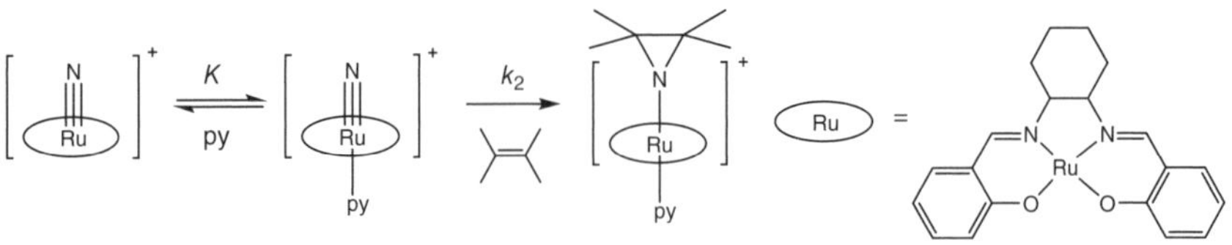

Scheme 10.   Formation of an aziridino ligand from a Ru(VI) salen nitride complex. Further reaction to yield an aziridine ligand occurs. [Adapted from (164).]

step involves ring opening of the aziridino ligand to provide a species that can transfer hydrogen atoms to other aziridino complexes. Styrenes react similarly, although the aziridino intermediate is not observed. There is no stereochemical change, consistent with a concerted reaction.

C.   One-Electron Reactions

Examples of one-electron reactions are extremely rare, although it is possible that some of the multielectron reactions described above occur by mechanisms involving multiple one-electron steps.

One-electron electrochemical reduction of the Os(VI) nitride $(Tp)Os(Cl_2)(N)$, where $Tp = tris(pyrazolyl)borate$, in the presence of HPF_6 results in formation of the Os(V) imide complex $(Tp)Os(Cl_2)(NH)$. This synthetic strategy also allows for the synthesis of $(bpy)Os(Cl)_3(NH)$ (165).

There are a few examples of nitride complexes that react with substrates having weak $X-H$ bonds. It is likely that these reactions occur by hydrogen-atom abstraction, although other mechanisms cannot be discounted. Thus, for example, as described in Section VI.F, $[(salen)Ru(N)(MeOH)]^+$ abstracts hydrogen atoms from dihydroanthracene and xanthene (166), while the iron(V) complex $[PhB(t-BuIm)_3Fe(N)]^+$ reacts with dihydroanthracene to provide anthracene (167). The only instance in which there is mechanistic evidence for hydrogen-atom abstraction is the reaction of $PhB(MesIm)_3Fe(N)$ with 3 equiv of TEMPO-H (TEMPO-H = 2,2,6,6-tetramethylpiperidine-1-ol), which results in the formation of ammonia and the iron(II) complex $PhB(MesIm)_3Fe(TEMPO)$. An investigation involving thermodynamic and kinetic studies reveals that formation of the first $N-H$ bond involves hydrogen-atom transfer from the TEMPO-H to the nitride ligand (168).

The only known example of $C-N$ bond formation by a one-electron reaction also involves the iron(IV) nitride $PhB(MesIm)_3Fe(N)$. This complex reacts with the triphenylmethyl radical (Gomberg's dimer) to yield the iron(III) imido complex $PhB(MesIm)_3Fe(NCPh_3)$ (168).

D. Reactions With No Change in Oxidation State

These reactions are common for nitrides that are considered to be nucleophilic (Scheme 11). Reactions observed include adduct formation with Lewis acids, alkylation, acylation, and heterocumulene functionalization. Metathesis reactions involving nitriles and alkynes, which also do not change the oxidation state of the metal, will be covered in Section VII.B.

1. Reaction With CO_2 and CS_2

The heterocumulenes (CO_2 and CS_2) react with very nucleophilic nitride ligands. The four-coordinate V(V) trisanilido complexes $(RNAr)_3V(N)$ (R = t-Bu, Ad), react with CO_2 and CS_2 to yield the N-bound dicarbamate $[(RNAr)_3V(NCO_2)]^-$ and dithiocarbamate complexes, $[(RNAr)_3V(NCS_2)]^-$, respectively. Interestingly, the

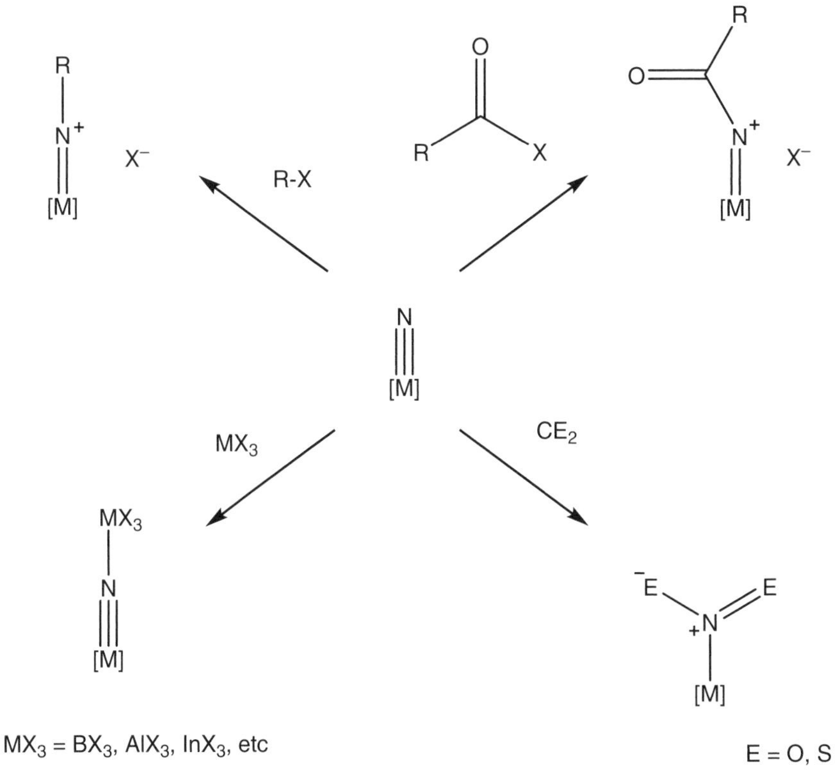

$MX_3 = BX_3, AlX_3, InX_3, etc$

$E = O, S$

Scheme 11.   Summary of reactions without an oxidation state change at the metal.

dithiocarbamate complex [(t-BuNAr)$_3$V(NCS$_2$)]$^-$ extrudes the thiocyanate anion to yield the neutral vanadium(V) sulfide complex (t-BuNAr)$_3$V(S) via a first-order process, possibly through the intermediacy of a V−N−C−S metallacycle (169). The analogous niobium nitride [(t-BuNAr]$_3$Nb(N)]$^-$ reacts similarly and when treated with CO$_2$ provides the N-bound carbamate complex [(t-BuNAr]$_3$Nb(NCO$_2$)]$^-$ (170). The carbamate can be converted to an isocyanate ligand upon reaction with acyl halides or acetic anhydride.

2. Reaction With Lewis Acids

There are numerous examples of nitrides that react with boranes, invariably forming adducts between the nitride ligand and the borane. Most of these cases are for the mid-transition elements, particularly Mo (113, 171, 172), Tc (173, 174), and Re (172, 175–181). Thus, for example, the complex (Pc)Re(N) reacts with a range of boranes BX$_3$ (X = Cl, Br, Et, Ph, and C$_6$F$_5$) to yield (Pc)Re(NBX$_3$) (180, 182). Formation of the adduct attenuates the *trans* influence of the nitride ligand and may allow new ligands to coordinate *trans* to the newly functionalized nitride. Adduct formation is sometimes reversible (172, 173, 180). Thus, for example, in the case of (Pc)Re(NBX$_3$), the stability of the adduct generally increases as the electron-withdrawing ability of the groups on boron increase (180).

While in most cases, formation of the adduct is the final product, reaction of TpOs(N)X$_2$ (X = trifluoroacetate, trichloroacetate, tribromoacetate, bromide; X$_2$ = oxalate, nitrate) with BPh$_3$ results in B−C cleavage to provide the borylamido products, TpOs[N(Ph)(BPh$_2$)]X$_2$ (132, 183). Mechanistic investigations show that this reaction proceeds via initial formation of the acid–base adduct, followed by migration of the phenyl group from boron to nitrogen.

The formation of similar adducts involving other Lewis acids (e.g., AlX$_3$, GaX$_3$, InX$_3$) is also observed for nitrides from the mid-transition series (180, 184–186). For example, (Pc)Re(N) reacts with GaCl$_3$, GaBr$_3$, AlCl$_3$, and InCl$_3$, to give the corresponding adducts, with the Ga adducts being more stable than those of Al and In (180).

3. Reaction With Alkylating and Acylating Reagents

A number of nitrides have been reported to react with alkylating agents, providing the corresponding imido ligands without a change in the oxidation state of the metal. This reaction is taken as an indicator of a nucleophilic nitride and is most commonly observed for nitrides of the mid-transition series. Metal nitrides that have been observed to react with alkyl halides include complexes of Mo (187–190), W (191), Re (192, 193), Ru (194), and Os (112, 195). Thus, for example, $L_{OEt}Re(N)(PPh_3)Cl$ reacts with MeOTf, $PhCH_2Br$, and Ph_3CBF_4, to provide the imido complexes $[L_{OEt}Re(NMe)(PPh_3)Cl]OTf$, $[L_{OEt}Re(NCH_2Ph)(PPh_3)Cl]Br$, and $[L_{OEt}Re(NCPh_3)(PPh_3)Cl]BF_4$, respectively (193).

Similarly to reactions with alkylating reagents, many nitrides react with acyl halides or acid anhydrides to yield acylimido ligands (196–199). As with alkylating agents, these nitrides are typically found in the middle of the transition series. The reaction with acid anhydrides in particular has been developed as a method for activating nitrides toward nitrogen-atom transfer (e.g., to alkenes), albeit as the acylimido group (51, 52).

4. Reaction With Other Electrophiles

The nucleophilic reactivity of some anionic nitride complexes has provided a platform for the construction of exotic ligands. Thus, for example, the four-coordinate complexes $[(RNAr)_3M(N)]^-$ ($M = V$, Nb; $R = (CD_3)_2Me$; $Ar = 3,5\text{-}Me_2C_6H_3$), react with PCl_3 to yield dichlorophosphine complexes $(RNAr)_3M(NPCl_2)$ (200). Subsequent transformations provide the iminophosphinimide complexes $(RNAr)_3M(N\text{-}P=Nt\text{-}Bu)$.

A modified supporting amido ligand has allowed this methodology to be extended to the synthesis of new main group compounds. Two equivalents of $[(NpNAr)_3V(N)]^-$ (Np = neopentyl) react with PCl_3 to provide the bridged dimer $[(NpNAr)_3V(N)]_2PCl$ along with $(NpNAr)_3V(Cl)$ (201). The former compound serves as a precursor to a stable phosphorus radical.

5. Other Reactions

The four-coordinate W(VI) nitride $(i\text{-}PrNAr)_3W(N)$ undergoes unusual reactions with electrophiles. Reaction with oxalyl chloride $(C(O)Cl)_2$ results in formation of the six-coordinate complex $(i\text{-}PrNAr)_3W(OCN)Cl_2$ along with concomitant formation of CO. Interestingly, the cyanato ligand is believed to be bound to tungsten through the oxygen atom. This complex also reacts with PCl_5 to provide the trichlorophosphinimide complex $(i\text{-}PrNAr)_3W(NPCl_3)Cl_2$ (202).

The Re(V) complex $(PPhMe_2)_3Re(N)Cl_2$ reacts with 2 equiv of dimercaption-succinic acid dimethyl ester $(DMSMe_2)$ in acetone to yield the ketimide complex $(DMSMe_2)_2Re((NC(Me_2)_2(PPhMe_2))$. It is proposed that the reaction involves substitution of two phosphines ligand by $DMSMe_2$. In a step that resembles the reactivity of frustrated Lewis pairs (203), the nitride ligand attacks the solvent molecule in the presence of liberated phosphine, providing a ligand-based carbocation that reacts with a second displaced phosphine ligand (Scheme 12) (204).

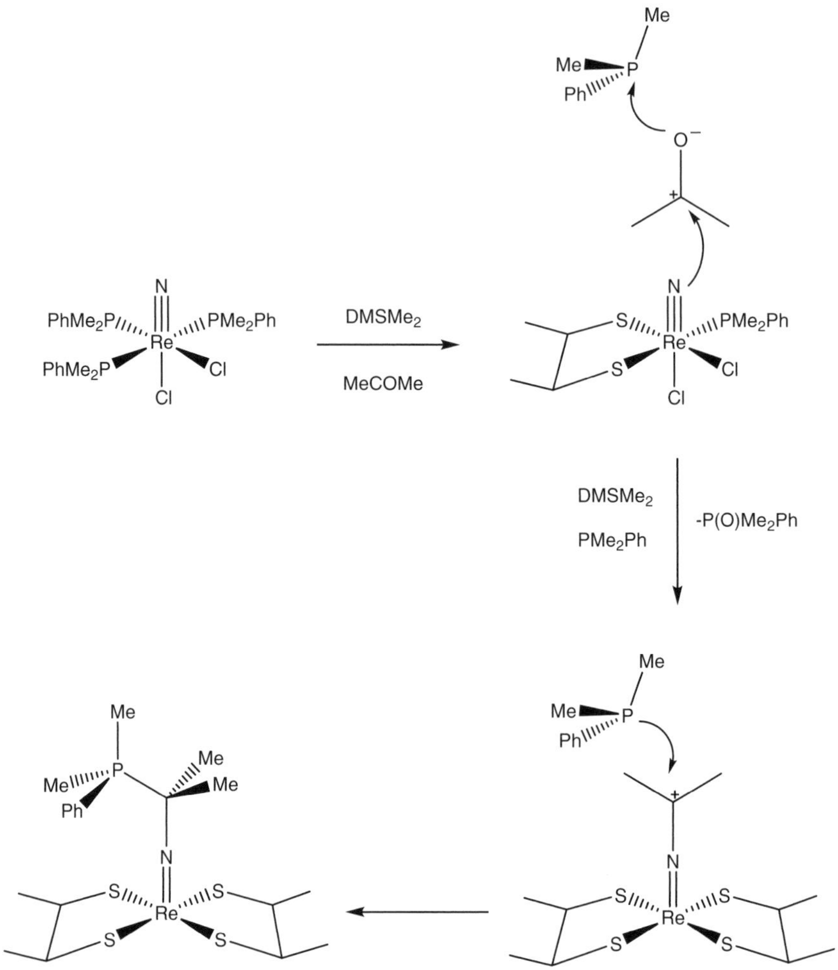

Scheme 12. Proposed mechanism for the reaction of Re(N)(PMe₂Ph)₃Cl₂ with DMSMe₂ in acetone. [Adapted from (204).]

E. Formation of Ammonia, Ammine, and Amide Ligands: Reaction With H₂

The formation of ammonia by protonolysis has been described in earlier reviews (7, 36) and will not be further described here. These reactions typically maintain the same oxidation state at the metal center and have most commonly been observed with nucleophilic nitrides in the mid-transition series.

A more recent development has been to conduct these protonation reactions in the presence of a suitable electron source (reductant), resulting in three-electron reduction of the metal center concomitant with ammonia formation. For example, the Mo(VI)N complex supported by a very bulky tris(amido)amine ligand, (HIPTN$_3$N)Mo(N), where HIPTN$_3$N=[(RNCH$_2$CH$_2$)$_3$N]$^{3-}$, R = 3,5-(2,4,6-i-Pr$_3$C$_6$H$_2$)$_2$C$_6$H$_3$), reacts with acid to provide the corresponding imido complex [(HIPTN$_3$N)MoNH]$^+$. When treated with acid in the presence of Cp$_2$Co as reductant, the ammine complex (HIPTN$_3$N)Mo(NH$_3$) is formed in good yield (205, 206). This strategy has been successful in forming ammonia from other metal nitrides, for example, the iron(IV) nitride complex (BP$_3$$i$-Pr)Fe(N) reacts with acid and reductant to provide ammonia in moderate yield (8).

The reaction of nitrides with H$_2$ to provide ammonia has so far only been observed with low-coordinate late metal nitride complexes. The square-planar iridium complex (PDI)Ir(N) reacts with H$_2$ under mild heating to provide the amido complex (PDI)Ir (NH$_2$) in a reaction that can be catalyzed by acid (207). Theoretical calculations suggest that the reaction involves direct attack of H$_2$ at the nitride ligand of the complex. The proton affinity of the nitride ligand is calculated to be larger than that of trimethylamine.

In contrast to the reaction above, the square-planar ruthenium complex (PNP)Ru(N) reacts with H$_2$ to produce ammonia and the ruthenium tetrahydrido complex (PNHP)RuH$_4$, in which an N−H bond is formed at the supporting pincer ligand (133). A computational investigation of the mechanism implicates cooperativity between the metal center and the amido group of the pincer ligand: H$_2$ addition occurs across this Ru−N bond, followed by migration of the hydride ligands to form new N−H bonds at the nitride ligand.

There are also few examples of N−H bond formation by hydrogen-atom transfer. The low-coordinate Fe(IV) nitride complex PhB(MesIm)$_3$Fe(N) reacts with TEMPO-H to provide ammonia and an Fe(II) TEMPO complex (168). Thermodynamic and kinetic mechanistic studies suggest that the first N−H bond is formed by hydrogen-atom transfer from TEMPO-H to the nitride ligand.

The ruthenium nitride complex L$_{OEt}$Ru(N)Cl$_2$ reacts with excess Et$_3$SiH to yield the ammine complex L$_{OEt}$Ru(NH$_3$)Cl$_2$. No reaction occurs with H$_2$. Reaction with the ruthenium hydride complex L$_{OEt}$Ru(PPh$_3$)(CO)H yields instead the bridged amido complex L$_{OEt}$RuCl$_2$(μ-NH)L$_{OEt}$Ru(CO)PPh$_3$ (208).

F. Activation of C−H Bonds

Thermolysis of diruthenium azido complexes results in amination of the C−H bonds of the supporting N,N'-diphenylformamidinate ligands (Scheme 13; dPhf = N,N'-diphenylformamidinate) (38, 209). Photolysis of the azido complex in frozen solution allows the nitride ligand to be spectroscopically characterized. Upon thawing, this species aminates the C−H bonds, establishing that amination involves the intermediacy of a nitride ligand. Temperature-dependent thermodynamic and kinetic investigations have validated a computationally determined reaction pathway in which there is electrophilic attack of the nitride on the aromatic ring. The dimetal fragment is found to be critical to the observed reactivity (210).

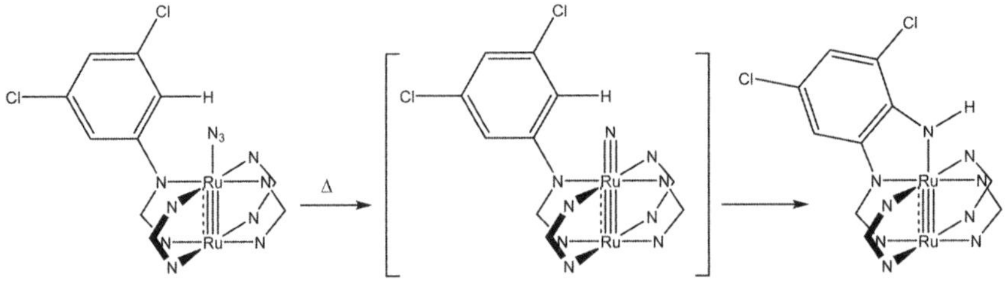

Scheme 13.   Intramolecular arene amination by a transient RuRuN species supported by dPhf. [Adapted from (245).]

The four-coordinate iridium complex (PDI)Ir(N) and the putative species (PDI)Rh(N) and (PDI)Co(N) are able to intramolecularly aminate benzylic C−H bonds (211, 212). The proposed mechanism involves hydrogen-atom abstraction from the C−H bond to yield the imido (PDI)M(NH), followed by a rebound step to insert the nitrogen atom into the C−H bond (Scheme 14).

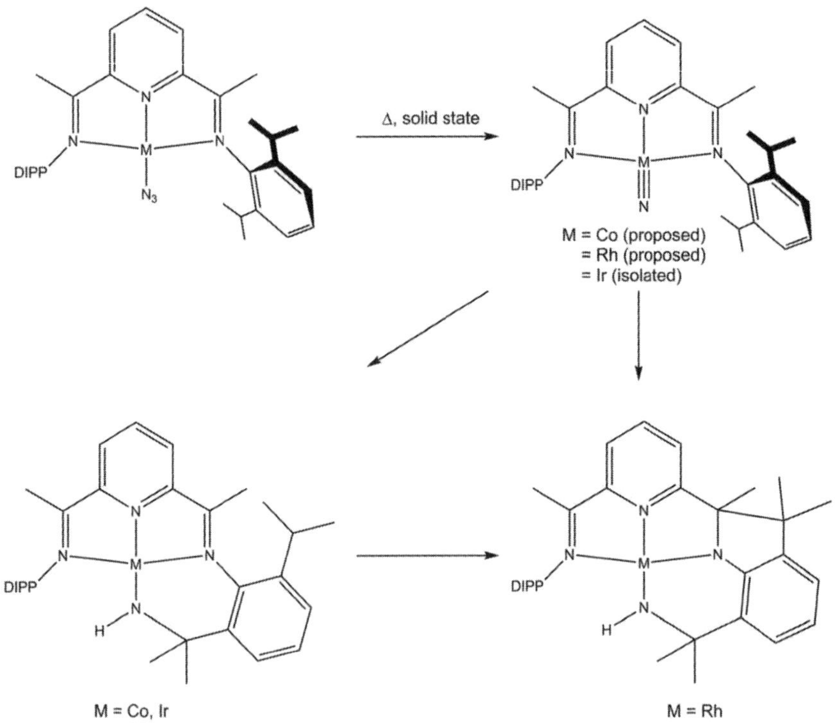

Scheme 14.   Range of products resulting from thermolysis of group 9 (VIII B) azide complexes.

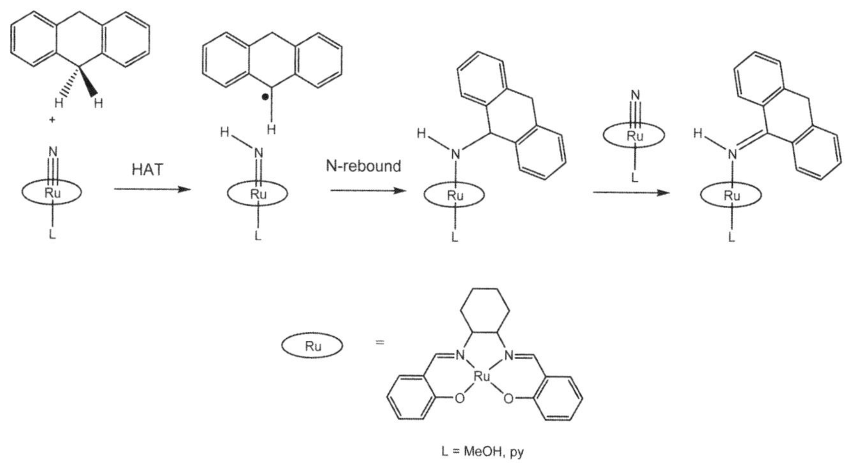

Scheme 15.   Rebound mechanism for C−H functionalization by a Ru(VI) nitride complex (HAT = hydrogen-atom transfer). [Adapted from (166).]

Interestingly, for Rh the species that is isolated is the product of two C−H activation steps, suggesting a slower rebound step than for Ir or Co. In the case of Rh, replacing the isopropyl groups with chloro groups does not prevent ligand degradation; in this case, products consistent with insertion of the nitride ligand into the C−Cl bond are obtained (213).

In acetonitrile, the isolable ruthenium(VI) nitride [(salchda)Ru(N)(MeOH)]$^+$ reacts with xanthene and dihydroanthrancene to yield the imine complexes, [(salchda)Ru(HN=Ar)(NCMe)]$^+$ (Ar = xanthene(−2H), dihydroanthracene(−2H)) (214). These reactions are proposed to occur by a hydrogen-atom abstraction, radical-rebound mechanism to yield amine products that are further dehydrogenated to the imine (Scheme 15). No reaction occurs with substrates with C−H bonds having bond dissociation enthalpy (BDE) > 78 kcal mol^{-1}. However, addition of pyridine to the nitride complex increases its reactivity, and hydrogen-atom abstraction from stronger bonds, including those in cyclohexane (C−H BDE = 95 kcal mol^{-1}) becomes possible although the formation of rebound products is not observed with these substrates. Clearly, the very challenging task of controlling the relative rates of hydrogen-atom abstraction and rebound will be critical to further developing this methodology for hydrocarbon functionalization by nitride ligands.

G.   Complete Nitrogen-Atom Transfer

In these reactions the nitride ligand is completely transferred to give a new molecule, generally organic. These are typically multistep syntheses that start from the nitride complex, although there are more recent examples of single-step reactions. Most notably, organic molecules have been prepared using nitrides derived from atmospheric N_2.

1. Electrosynthesis of Amino Acids

The Mo(IV) complex trans-MoCl(N)(dppe)$_2$ has been used to prepare amino acids (188, 189). Alkylation with MeOC(O)CH$_2$I provides trans-[MoCl(NCH$_2$C(O)OMe)(dppe)$_2$]I. Deprotonation by NEt$_3$ leads to the neutral alkenylamide trans-[MoCl(NCHC(O)OMe)(dppe)$_2$], which can be alkylated at the α-carbon by MeI to yield trans-[MoCl(NCH(Me)C(O)OMe)(dppe)$_2$]I. Constant potential electrolysis of both these complexes in the presence of acetic acid results in release of the amino acid methyl esters of glycine and DL-alanine (Scheme 16) (188).

A different organic product can be released under basic conditions. Deprotonation of trans-[MoCl(NCH$_2$C(O)OEt)(dppe)$_2$] by excess KOt-Bu in the presence of CO (to trap the Mo product) provides the cyanoformate ester N≡C–CO$_2$Et and cis–trans-Mo(CO)$_2$(dppe)$_2$. In the absence of CO, the cyanoformate ester and trans-Mo(N$_2$)(dppe)$_2$ are formed (189).

2. Nitrile Synthesis by Complete Nitrogen-Atom Transfer

Nucleophilic Nb, Mo, and W nitrides have been shown to react with acyl halides to provide organic nitriles in a single step. A common mechanism involving the formation of a metallacyclobutene has been proposed to account for nitrile formation (Scheme 17). These reactions are typically driven by the formation of strong M=O bonds.

The four-coordinate tungsten(VI) nitride (i-PrNAr)$_3$W(N) reacts with organic acyl chlorides ROCl (R = t-Bu, Ad) to yield metastable acylimido chloride

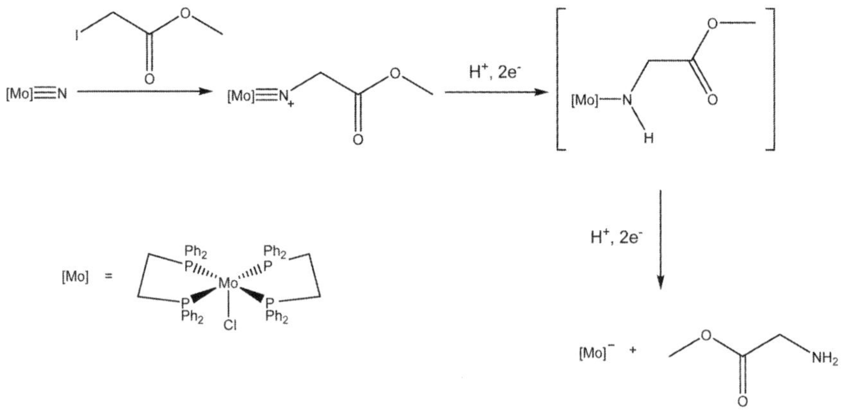

Scheme 16.   Electrosynthesis of an amino acid ester from a molybdenum(IV) nitride. The liberated complex undergoes further reactions with an added ligand trap. [Adapted from (188).]

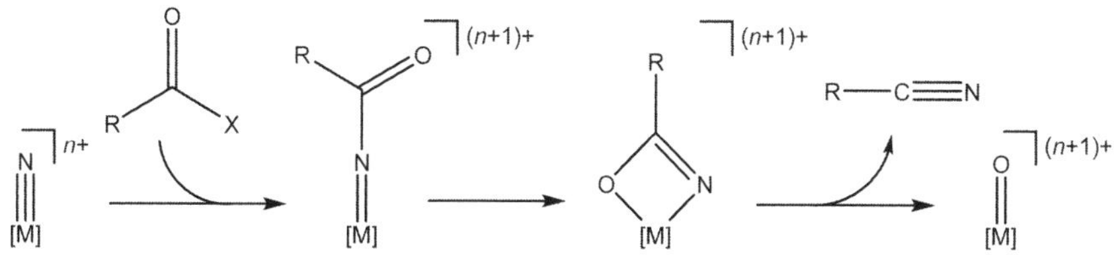

Scheme 17.   General mechanism for the synthesis of organic nitriles from nucleophilic nitrides. The intermediacy of all these species has not been established in all cases.

complexes (i-PrNAr)$_3$W(RC(O)N)(Cl), which decay to give an oxo-chloride product (i-PrNAr)$_3$W(O)Cl with concomitant formation of the corresponding organic nitrile RCN (215). In this case, the formation of a strong W$=$O bond, as well as expansion of the W coordination shell, provides the driving force for the reaction.

Subsequent work has shown that nitriles can be prepared from N$_2$ using the nucleophilic four-coordinate niobium(V) nitride complex [(NpNAr)$_3$Nb(N)]$^-$, which can be prepared from N$_2$ in a heterobimetallic N$_2$ cleavage reaction (215). The negative charge on this complex imparts high nucleophilicity, and the complex reacts with acyl chlorides to yield organic nitriles and the oxo complex [(NpNAr)$_3$Nb (O)]. Interestingly, the niobium reagent can be recycled, leading to a noncatalytic cycle for the synthesis of organic nitriles from N$_2$. Since N$_2$ cleavage involves two metal complexes, this means that only one-half of the N$_2$ ends up in the nitrile product.

To more efficiently use the nitrogen atoms, a route to nitriles directly from (RNAr)$_3$Mo(N) is required. However, in contrast to the case for tungsten, the corresponding Mo complexes, (RNAr)$_3$Mo(N), do not react with acyl chlorides. Synthesis of the corresponding acyl imido complexes, [(RNAr)$_3$Mo(RC(O)N)]$^+$ requires activation of the acyl halide (e.g., by in $situ$ generation of the acyl triflate). Furthermore, the nitrile elimination reaction is also less facile than for W and Nb, and additional reagents are required to effect this reaction. This step can nonetheless be coupled to regeneration of the Mo complex, allowing for a synthetic cycle for the synthesis of nitriles from N$_2$ to be developed (3). These results highlight an inherent difficulty with using N$_2$ to prepare organic molecules, particularly using nitride intermediates (viz., that the thermodynamic driving force required for cleavage of N$_2$ can overstabilize the nitride products and make further elaboration very difficult).

Additional evidence for the proposed mechanism of nitride formation has been obtained from the nucleophilic Mo pincer complex [(t-BuOCO)Mo(N)(NMe$_2$)]$^-$, which reacts with acyl halides in dimethylformamide (DMF) solvent to yield the corresponding acylimido complexes (t-BuOCO)Mo(NC(O)R)(dmf)(NMe$_2$) (R $=$ SiMe$_3$, Ph, t-Bu) (216). This acylimido complex also extrudes the corresponding organic nitrile with formation of the oxo complex (t-BuOCO)Mo(O)(NMe$_2$). Kinetic analysis is consistent with initial loss of the coordinated DMF solvent, which suggests that an open coordination site is required, consistent with the idea of metallacycle formation prior to nitrile loss (Scheme 18).

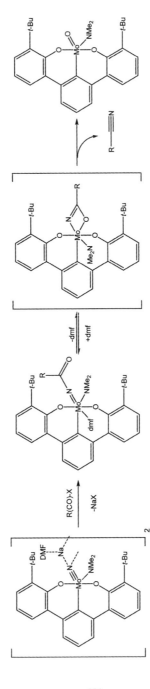

Scheme 18. Nitrile synthesis from a Mo nitride complex (dmf = dimethylformamide ligand). [Adapted from (216).]

3. Formation of Cyanate and Carbodiimide Anions

A number of nitride complexes have been reported to react with unsaturated substrates, releasing anions in an overall two-electron reaction. The highly electrophilic vanadium(V) nitride complex, $[(t\text{-BuNAr})_3V(N)]^-$ reacts with CO to release the cyanate anion, NCO^-, and form the coordinatively unsaturated V(III) complex $(t\text{-BuNAr})_3V$ (217). The coordinatively unsaturated V(III) complex does not bind CO. It is surprising that the apparently nucleophilic nitride complex reacts with CO, although the mechanism of reaction is unknown.

While the vanadium nitride does not react with isonitriles, the iron(IV) nitride $PhB(MesIm)_3Fe(N)$ reacts with $t\text{-BuNC}$ to release the carbodiimide anion $t\text{-BuNCN}^-$ (134). The other product of this reaction is the six-coordinate iron(II) complex $[PhB(MesIm)_3Fe(CN t\text{-Bu})_3]^+$, whose formation likely provides the driving force for the reaction. In contrast to the vanadium nitride, when the iron nitride is treated with CO, the cyanate ligand remains bound to the metal center in the final product, $PhB(MesIm)_3Fe(CO)_2(NCO)$.

VII. NITRIDES AS CATALYST PRECURSORS AND INTERMEDIATES

A. Catalytic Ammonia Formation

A number of Mo(VI) nitrides are precatalysts for the formation of ammonia from N_2. The majority of these complexes are based on very bulky tris(amido) amine ligands (Scheme 18) (2, 218). Extensive reviews on this topic have been published and so only some key points will be highlighted here (2, 219).

Although this has not been definitively established, the catalytic cycle for ammonia formation is presumed to involve a Mo(VI) nitride intermediate. Studies relevant to the catalytic cycle reveal complexity with many intermediate species that could themselves serve as proton and/or electron sources. It is, however, unlikely that the reaction mechanism involves a linear sequence through all the intermediates in Scheme 19 (220, 221). Subtle variations in the triamidoamine ligand typically lead to the loss of catalytic activity (222–227), while attempts to develop similar nitride precatalysts for the synthesis of NEt_3 from N_2 have so far been unsuccessful (228). Related tungsten (229) and chromium (230) complexes are also catalytically inactive.

B. Alkyne–Nitrile Metathesis

Certain four-coordinate group 6 (VI) nitrides have been found to be catalysts and precatalysts for triple-bond metathesis. The tungsten(VI) complex $(t\text{-BuO})_3W(N)$

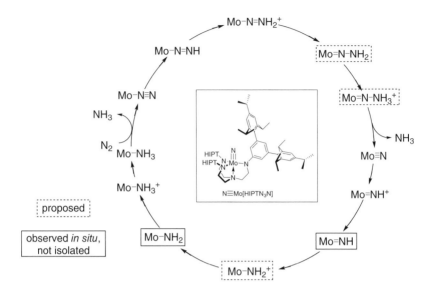

Scheme 19.   Proposed catalytic cycle for the synthesis of ammonia using (HIPTN$_3$N)Mo(N).
Complexes not shown in boxes can be isolated. [Adapted from (2).]

catalyzes room temperature nitrogen-atom exchange between organic nitriles (5, 60).
A key step in the proposed reaction mechanism is a 2 + 2 cycloaddition reaction
between the nitride complex and the nitrile to give a metalladiazacyclobutadiene
intermediate (Scheme 20). While nitrile exchange for the related complex
(t-BuO)$_3$Mo(N) only occurs at elevated temperatures, replacing the $tert$-butoxide
ligands with the fluorinated OC(CF$_3$)$_2$Me$^-$ and OC(CF$_3$)$_3^-$ ligands allows for
metathesis at room temperature (231).

   Due to their relative ease of synthesis, group 6 (VI) nitrides have been investigated
as precursors for alkyne metathesis. While the nitride ligand is generally more stable
than the alkylidyne ligand for these types of complexes, increasing the electro-
negativity of the metal center helps stabilize the alkylidyne ligand over the nitride (4).
Thus, reaction of (OC(CF$_3$)$_2$Me)$_3$Mo(N) and (OC(CF$_3$)$_3$)$_3$Mo(N) with butyne affords

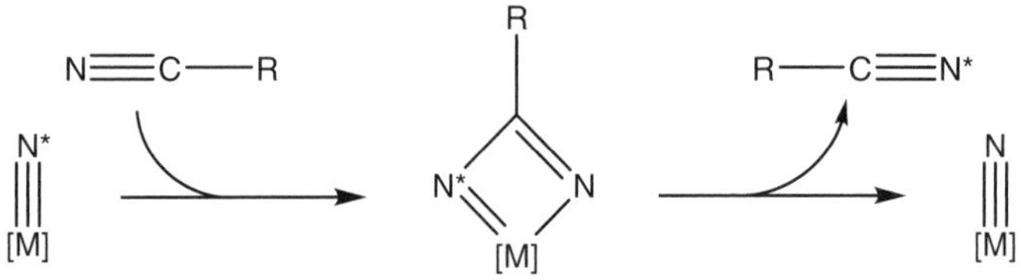

Scheme 20.   Proposed mechanism of nitrile metathesis.

the alkyne metathesis catalysts $(OC(CF_3)_2Me)_3Mo(CEt)$ and $(OC(CF_3)_3)_3Mo(CEt)$, respectively. Likewise, the readily accessible complex $(Ph_3SiO)_3(py)Mo(N)$ also serves as a precatalyst for alkyne metathesis (232). The air- and moisture-stable complex $(Ph_3SiO)_3Mo(N)(bpy)$, which is activated by metal salts (e.g., $MnCl_2$) is an even more practical catalyst precursor (233).

When bound by electronegative alkoxide ligands, electronegative tungsten(VI) nitrides have been found to catalyze the nitrile–alkyne cross-metathesis reaction. Specifically, $(OC(CF_3)_2Me)_3W(N)$ and $(OC(CF_3)_3)_3W(N)$ catalyze the exchange of R and R′ groups of a nitrile (RCN) and an alkyne (R′CCR′) (234, 235).

VIII. STRATEGIES FOR INCREASING NITRIDE REACTIVITY

The foregoing discussion may imbue the reader with the impression that nitride complexes are highly reactive. In fact, the vast majority of nitride complexes are completely unreactive at the nitride ligand. Instead reactions often occur at other locations on the complex (e.g., ancillary ligand substitution). For example, when the Mo(VI) complex $(t\text{-}BuO)_3Mo(N)$ is treated with the Grignard reagent $t\text{-}BuCH_2MgBr$, no reaction occurs at the nitride ligand. Rather, metathesis of the alkoxide ligands to give the tris(alkyl) complex $(t\text{-}BuCH_2)_3Mo(N)$ occurs (236).

However, the accumulated knowledge regarding the reactivity of metal nitride complexes described above allows some possible strategies for increasing the reactivity of the nitride ligand to be postulated. A number of these strategies are outlined in the following discussion, with specific examples. Some of these reactions may have been mentioned in earlier sections.

A. Changing the Oxidation State

As would be expected from the simple MO picture, changing the oxidation state of the metal complex is likely to have an impact on the reactivity of the nitride ligand. However, there are few examples of nitride complexes that are stable in more than one oxidation state. One such case is $(C_5H_{11})_8PcW(N)$, which is unreactive toward $t\text{-}BuMe_2SiCl$, however, upon one-electron reduction with KC_8, the complex reacts with $t\text{-}BuMe_2SiCl$ to provide $(C_5H_{11})_8PcW(NSiMe_2t\text{-}Bu)$ (180). This increase in nucleophilicity is what would be expected from the simple MO considerations described earlier (Fig. 1).

B. Changing the Ancillary Ligands

It should not be surprising that the ancillary ligand set can have a profound effect on the reactivity of the nitride ligand. Thus, while $(Me_2N)_3Mo(N)$ reacts with phenols to replace the amido with alkoxide ligands, all the while maintaining the nitride functionality (237), the nitride ligand in $(t\text{-BuCH}_2)_3\text{Mo(N)}$ is protonated by phenol to yield the five-coordinate complex $(t\text{-BuCH}_2)_3\text{Mo(NH)}$ (OPh) (236).

Changing the ancillary ligands can therefore beget new reactivity. For example, $\text{Re(N)(PMe}_2\text{Ph})_3\text{Cl}_2$ is stable in acidic methanol (179). When treated with 1,3-diethylimidazol-2-ylidene (IEt), most of the ancillary ligands are substituted to yield $[(\text{IEt})_4\text{Re(N)Cl}]^+$. The nitride ligand in this complex is more reactive than in the starting material and the complex decomposes in hot methanol to yield $[(\text{IEt})_4\text{ReO}_2]^-$ (238). Similar results are observed with other NHC ligands. An obvious drawback of this approach is that changing the supporting ligands completely changes the nature of the metal complex.

C. Addition of *trans* Donors

In certain complexes, the nitride ligand can be activated by changes to the *trans* ligand. As described in Section VI.10, the complex $[(\text{salchda})\text{Ru(N)}]^+$ does not react with alkenes. However, on addition of pyridine or 1-methylimidazole, the complex reacts with alkenes to provide aziridines (164). Kinetic measurements of the pyridine-accelerated reaction establish that the first step of the reaction involves binding of the nitrogenous base to give *trans*-$[(\text{salchda})\text{Ru(N)(py)}]^+$. It is speculated that binding of the base to ruthenium both weakens the $\text{Ru}\equiv\text{N}$ bond and pulls the metal closer to the plane of the salchda ligand, reducing the reorganization energy for atom transfer.

This strategy obviously requires the complex be in a geometry that provides an accessible *trans* coordination site. Even if this is the case, the large *trans* effect of the nitride ligand may prevent axial ligands from binding. One solution is to use ligands with larger *trans* effects, which has reached its likely apotheosis in the design of complexes having a $\text{M}-\text{M}-\text{N}$ motif (i.e., a metal–metal multiple bond trans to the nitride ligand) (37, 38, 209, 210). Thus, for example, the $\text{Ru}\equiv\text{N}$ bond in $[\text{Ru}_2(\text{dPhf})_4(\text{N})]$, where $\text{dPhf} = N,N'$-diphenylformamidinate, is substantially longer than in other Ru complexes (37) and is activated to the intramolecular amination of aryl groups (38, 209).

D. Addition of Lewis Acids and Electrophiles

The addition of Lewis acids to nitride ligands with sufficiently nucleophilic character may weaken the $\text{M}\equiv\text{N}$ bond sufficiently to increase its reactivity. For example, although $(t\text{-BuO})_3\text{Mo(N)}$ is not active as a precatalyst for alkyne metathesis, in combination with MgBr_2, catalytic alkyne metathesis of 1-phenyl-1-propyne occurs (239). It is possible that the Lewis acid activates the complex by formation of an adduct with the nitride ligand, as observed with Lewis acids such as $B(C_6F_5)_3$ (240), however, other modes of activation cannot be excluded at present.

More generally, electrophiles have been shown to activate certain nitride ligands for nitrogen-atom transfer. The most widely investigated are five-coordinate (salen)Mn(N), its derivatives and analogues, particularly for alkene functionalization. Addition of trifluoroacetic anhydride to these complexes activates the nitride ligand for transfer, most likely as the $NCOCF_3$ group, to suitable substrates, such as silyl enol ethers (51, 241) and styrene (51, 54). A drawback of this strategy is that the group transferred is not the nitrogen atom, but rather the $NCOCF_3$ group. Trifluoroacetic acid and BF_3 have been shown to have a similar influence on the reactivity of these nitride complexes, however, for these activators net NH group transfer occurs to convert styrenes to the corresponding aziridines (242). Moreover a competitive imide coupling reaction decreases the extent of styrene aziridination (243).

E. Photolysis

There have been few reports on the photochemical reactivity of metal nitride complexes. In one example, photolysis of $[Os(N)(NH_3)_4]^{3+}$ in the presence of an electron donor, such as 1,4-dimethoxybenzene or C_6Me_6 in acetonitrile, results in N–N coupling to form the μ-N_2 complex $[(Os(NH_3)_4(MeCN))_2N_2]^{5+}$ (244). There are apparently no examples where photolysis results in nitrogen-atom transfer to substrates and this would appear to be an area ripe for exploration.

IX. CONCLUSIONS

As described at the outset, the goal of this chapter is to provide an authoritative, if not exhaustive, summary of transition metal nitride chemistry since the early 1990s. While many of the transition metal nitrides do not react at the nitride ligand, researchers have uncovered a range of complexes with varied reactivity, ranging from complexes that facilitate three-electron atom-transfer reactions to those that participate in C–H activation processes. The most striking use of transition metal nitrides has been in catalytic processes, as well as stoichiometric reactions for functionalizing N_2. The increasing variety of nitride complexes (both the nature of the transition metal and the geometry of the complex), coupled with an understanding of their electronic structures suggests that it is likely that more diverse reactivity patterns, both stoichiometric and catalytic, will be uncovered.

ACKNOWLEDGMENTS

Support from the Camille and Henry Dreyfus Foundation, as well as the DOE-BES (DE-FG02-08ER15996) is gratefully acknowledged. Wei-Tsung Lee, Mei Ding, Salvador Muñoz, and the reviewers are thanked for their critical feedback on this manuscript.

ABBREVIATIONS

BDE	Bond dissociation enthalpy
bipy	2,2$'$-Bipyridine
t-BuOCO	$[2,6\text{-}C_6H_3(6\text{-}t\text{-}BuC_6H_3O)_2]^{3-}$
t-BuPOCOP	$[C_6H_3\text{-}1,3\text{-}(OP(t\text{-}Bu)_2)]^-$
4-t-BuPy	4-t-Butylpyridine
$(C_5H_{11})_8Pc$	Octapentylphthalocyanine dianion
Cp*	Pentamethylcyclopentadienyl
Cy	Cyclohexyl
dbabh	2,3:5,6-Dibenzo-7-aza bicyclo[2.2.1]hepta-2,5-diene
dbm	Dibenzoylmethanolate
DIPP	2,6-Diisopropylphenyl
dmf	Dimethylformamide (ligand)
DMF	Dimethylformamide (solvent)
DMSMe$_2$	Dimercaptionsuccinic acid dimethyl ester
dppe	Diphenylphosphinoethane
dPhf	N,N'-Diphenylformamidinate
ENDOR	Electron–nuclear double resonance
EPR	Electron paramagnetic resonance
HAT	Hydrogen-atom transfer
HIPTN$_3$N	$[(RNCH_2CH_2)_3N]^{3-}$, $R = 3,5\text{-}(2,4,6\text{-}i\text{-}Pr_3C_6H_2)_2C_6H_3)$
HOMO	Highest occupied molecular orbital
IEt	2,5-Diethylimidazol-2-ylidene
Im	Imidazol-2-ylidene
IMes	2,5-Bis(2,4,6-trimethylphenyl)imidazol-2-ylidene
L$_{OEt}$	Kläui's tripodal ligand, $[CpCoP(O)(OEt)_3)]^-$
LUMO	Lowest unoccupied molecular orbital
Mes	2,4,6-Trimethylphenyl
MesIm	1-(2,4,6-Trimethylphenyl)imidazol-2-ylidene
MO	Molecular orbital
NHC	N-Heterocyclic carbene
Np	neo-Pentyl, CH_2CMe_3
Pc	tetra-$tert$-Butylphthalocyanine

PDI	Aryl-substituted bis(imino)pyridine
PhB(t-BuIm)$_3^-$	Phenyl(tris(3-*tert*-butylimidazol-2-ylidene))borate
PhB(MesIm)$_3^-$	Phenyl(tris(3-(2,4,6-trimethylphenyl)imidazol-2-ylidene))borate
BP$_3i$-Pr	[PhB(CH$_2$Pi-Pr$_2$)$_3$]$^-$
PNP	[N(CHCHPt-Bu$_2$)$_2$]$^-$
PNPt-Bu	[N(CH$_2$CH$_2$Pt-Bu$_2$)$_2$]$^-$
por	Porphyrin
py	Pyridine
POP^{Cy}	[O(CHCHPCy$_2$)$_2$]
Q	8-Hydroxyquinoline or 2-methyl-8-hydroxyquinoline
Q$'$	5-Nitro-8-quinolato
salchda	N,N'-Bis(salicylidene)-O-cyclohexyldiamine dianion
salen	N,N'-Bis(salicylidene)-O-cyclohexyldiamine dianion
silox	t-Bu$_3$SiO$^-$
TCNE$^-$	Tetracyanoethylene radical anion
TEMPO	2,2,6,6-Tetramethylpiperidine 1-oxyl
TEMPO-H	2,2,6,6-Tetramethylpiperidine 1-ol
TMEDA	Tetramethylethylenediamine
thf	Tetrahydrofuran (ligand)
Tp	Tris(pyrazolyl)borate
Tpm	Tris(pyrazolyl)methane
TPP	Tetraphenylporphyrin
terpy	2,2$'$;6$'$,2$''$-Terpyridine

REFERENCES

1. J. Fritzsche and H. Struve, *J. Prakt. Chem. 41*, 103 (1847).

2. R. R. Schrock, *Acc. Chem. Res. 38*, 955–962 (2005).

3. J. J. Curley, E. L. Sceats, and C. C. Cummins, *J. Am. Chem. Soc. 128*, 14036 (2006).

4. R. L. Gdula and M. J. A. Johnson, *J. Am. Chem. Soc. 128*, 9614 (2006).

5. M. H. Chisholm, E. E. Delbridge, A. R. Kidwell, and K. B. Quinlan, *Chem. Commun.* 126 (2003).

6. B. M. Hoffmann, D. R. Dean, and L. C. Seefeldt, *Acc. Chem. Res. 42*, 609–619 (2009).

7. K. Dehnicke and J. Strähle, *Angew. Chem. Int. Ed. Engl. 31*, 955 (1992).

8. T. A. Betley and J. C. Peters, *J. Am. Chem. Soc. 126*, 6252 (2004).

9. U. Abram, B. Schmidt-Brücken, A. Hagenbach, M. Hecht, R. Kirmse, and A. Voigt, *Z. Anorg. Allg. Chem. 629*, 838 (2003).

10. C.-M. Che, *Pure Appl. Chem. 67*, 225 (1995).

11. M. P. Mehn and J. C. Peters, *J. Inorg. Biochem. 100*, 634 (2006).

12. J. M. Smith and D. Subedi, *Dalton Trans.* *41*, 1423 (2012).

13. J. Strähle, *Z. Anorg. Allg. Chem.* *633*, 1757 (2007).

14. K. Nakamoto, *J. Mol. Struct.* *408–409*, 11 (1997).

15. J. Straehle, *Z. Anorg. Allg. Chem.* *629*, 828 (2003).

16. R. A. Eikey and M. M. Abu-Omar, *Coord. Chem. Rev.* *243*, 83 (2003).

17. J. Berry, *Comments Inorg. Chem.* *30*, 28 (2009).

18. W. A. Nugent and J. M. Mayer, *Metal–Ligand Multiple Bonds*, John Wiley & Sons, Inc., New York, 1988.

19. E. D. Hedegård, J. Bendix, and S. P. A. Sauer, *J. Mol. Struct.: THEOCHEM 913*, 1 (2009).

20. J. Baldas, S. F. Colmanet, and G. A. Williams, *Inorg. Chim. Acta 179*, 189 (1991).

21. C. E. Johnson, E. A. Kysor, M. Findlater, J. P. Jasinski, A. S. Metell, J. W. Queen, and C. D. Abernethy, *Dalton. Trans.* *39*, 3482 (2010).

22. J. H. Shin, B. M. Bridgewater, D. G. Churchill, M.-H. Baik, R. A. Friesner, and G. Parkin, *J. Am. Chem. Soc.* *123*, 10111 (2001).

23. T. Birk and J. Bendix, *Inorg. Chem.* *42*, 7608 (2003).

24. J. Bendix, T. Birk, and T. Weyhermueller, *Dalton Trans.* 2737 (2005).

25. J. Bendix, *J. Am. Chem. Soc.* *125*, 13348 (2003).

26. A. Hori, T. Ozawa, H. Yoshida, Y. Imori, Y. Kuribayashi, E. Nakano, and N. Azuma, *Inorg. Chim. Acta 281*, 207 (1998).

27. N. Azuma, Y. Imori, H. Yoshida, K. Tajima, Y. Li, and J. Yamauchi, *Inorg. Chim. Acta 266*, 29 (1997).

28. S. Chen, M. H. Chisholm, E. R. Davidson, J. B. English, and D. L. Lichtenberger, *Inorg. Chem.* *48*, 828 (2009).

29. M. H. Chisholm, E. R. Davidson, M. Pink, and K. B. Quinlan, *Inorg. Chem.* *41*, 3437 (2002).

30. J. J. Scepaniak, M. D. Fulton, R. P. Bontchev, E. N. Duesler, M. L. Kirk, and J. M. Smith, *J. Am. Chem. Soc.* *130*, 10515 (2008).

31. J. Bendix, R. J. Deeth, T. Weyhermueller, E. Bill, and K. Wieghardt, *Inorg. Chem.* *39*, 930 (2000).

32. T. J. Crevier, B. K. Bennett, J. D. Soper, J. A. Bowman, A. Dehestani, D. A. Hrovat, S. Lovell, W. Kaminsky, and J. M. Mayer, *J. Am. Chem. Soc.* *123*, 1059 (2001).

33. P. J. Chirik, *Inorg. Chem.* *50*, 9737 (2011).

34. J. Bendix, C. Anthon, M. Schau-Magnussen, T. Brock-Nannestad, J. Vibenholt, M. Rehman, and S. P. Sauer, *Angew. Chem. Int. Ed.* *50*, 4480 (2011).

35. M. G. Scheibel, B. Askevold, F. W. Heinemann, E. J. Reijerse, B. de Bruin, and S. Schneider, *Nat. Chem.* *4*, 552 (2012).

36. K. Dehnicke and J. Strähle, *Angew. Chem. Int. Ed. Engl.* *20*, 413 (1981).

37. J. S. Pap, S. DeBeer George, and J. F. Berry, *Angew. Chem. Int. Ed.* *47*, 10102 (2008).

38. A. K. Long, G. H. Timmer, J. S. Pap, J. L. Snyder, R. P. Yu, and J. F. Berry, *J. Am. Chem. Soc.* *133*, 13138 (2011).

39. J. Schoffel, A. Y. Rogachev, S. DeBeer George, and P. Burger, *Angew. Chem. Int. Ed.* *48*, 4734 (2009).

40. B. L. Tran, M. Pink, X. Gao, H. Park, and D. J. Mindiola, *J. Am. Chem. Soc.* *132*, 1458 (2010).

41. B. L. Tran, J. Krzystek, A. Ozarowski, C.-H. Chen, M. Pink, J. A. Karty, J. Telser, K. Meyer, and D. J. Mindiola, *Eur. J. Inorg. Chem.* 3916 (2013).

42. C. A. Grapperhaus, B. Mienert, E. Bill, T. Weyermuller, and K. Wieghardt, *Inorg. Chem. 39*, 5306 (2000).

43. Y.-F. Song and J. F. Berry, *Inorg. Chem. 46*, 2208 (2007).

44. G. Izzet, E. Ishow, J. Delaire, C. Afonso, J. C. Tabet, and A. Proust, *Inorg. Chem. 48* (24), 11865 (2009).

45. K. Meyer, J. Bendix, N. Metzler-Nolte, T. Weyhermueller, and K. Wieghardt, *J. Am. Chem. Soc. 120*, 7260 (1998).

46. C. A. Grapperhaus, E. Bill, T. Weyhermueller, F. Neese, and K. Wieghardt, *Inorg. Chem. 40*, 4191 (2001).

47. K. Meyer, E. Bill, B. Mienert, T. Weyermuller, and K. Wieghardt, *J. Am. Chem. Soc. 121*, 4859 (1999).

48. P. Formentin, J. V. Folgado, V. Fornes, H. Garcia, F. Marquez, and M. Sabater, *J. Phys. Chem. 104*, 8361 (2000).

49. G. M. Coia, K. D. Demadis, and T. J. Meyer, *Inorg. Chem. 39*, 2212 (2000).

50. E.-S. El-Samanody, K. D. Demadis, T. J. Meyer, and P. S. White, *Inorg. Chem. 40*, 3677 (2001).

51. J. Du Bois, C. S. Tomooka, J. Hong, E. M. Carreira, and M. W. Day, *Angew. Chem., Int. Ed. Engl. 36*, 1645 (1997).

52. J. Du Bois, J. Hong, E. M. Carreira, and M. W. Day, *J. Am. Chem. Soc. 118*, 915 (1996).

53. C. S. Tomooka and E. M. Carreira, *Helv. Chim. Acta 85*, 3773 (2002).

54. F. R. Perez, J. Belmar, Y. Moreno, R. Baggio, and O. Penna, *New J. Chem. 29*, 283 (2005).

55. L. K. Woo, *Chem. Rev. 93*, 1125 (1993).

56. G. Golubkov and Z. Gross, *Angew. Chem. Int. Ed. 42*, 4507 (2003).

57. C.-Y. Tsai, M. J. A. Johnson, D. J. Mindiola, C. C. Cummins, W. T. Klooster, and T. F. Koetzle, *J. Am. Chem. Soc. 121*, 10426 (1999).

58. H.-T. Chiu, Y.-P. Chen, S.-H. Chuang, J.-S. Jen, G.-H. Lee, and S.-M. Peng, *Chem. Commun.* 139 (1996).

59. B. L. Tran, M. Singhal, H. Park, O. P. Lam, M. Pink, J. Krzystek, A. Ozarowski, J. Telser, K. Meyer, and D. J. Mindiola, *Angew. Chem. Int. Ed. 49*, 9871 (2010).

60. B. A. Burroughs, B. E. Bursten, S. Chen, M. H. Chisholm, and A. R. Kidwell, *Inorg. Chem. 47*, 5377 (2008).

61. D. J. Mindiola and C. C. Cummins, *Angew. Chem., Int. Ed. 37*, 945 (1998).

62. A. S. Veige, L. M. Slaughter, E. B. Lobkovsky, P. T. Wolczanski, N. Matsunaga, S. A. Decker, and T. R. Cundari, *Inorg. Chem. 42*, 6204 (2003).

63. C. E. Pohl-Ferry, J. W. Ziller, and N. M. Doherty, *Chem. Commun.* 1815 (1999).

64. G.-S. Kim and C. W. DeKock, *Polyhedron 19*, 1363 (2000).

65. A. Hills, D. L. Hughes, G. J. Leigh, and R. Prieto-Alcón, *J. Chem. Soc., Dalton Trans.* 3609 (1993).

66. C. M. Jones, M. E. Lerchen, C. J. Church, B. M. Schomber, and N. M. Doherty, *Inorg. Chem. 29*, 1679 (1990).

67. K. L. Sorensen, M. E. Lerchen, J. W. Ziller, and N. M. Doherty, *Inorg. Chem. 31*, 2678 (1992).

68. O. V. Ozerov, H. F. Gerard, L. A. Watson, J. C. Huffman, and K. G. Caulton, *Inorg. Chem. 41*, 5615 (2002).

69. N. Tsvetkov, M. Pink, H. Fan, J.-H. Lee, and K. G. Caulton, *Eur. J. Inorg. Chem.* 4790 (2010).

70. C. E. Laplaza, M. J. A. Johnson, J. C. Peters, A. L. Odom, E. Kim, C. C. Cummins, G. N. George, and I. J. Pickering, *J. Am. Chem. Soc. 118*, 8623 (1996).

71. C. E. Laplaza and C. C. Cummins, *Science 268*, 861 (1995).

72. J. J. Curley, T. R. Cook, S. Y. Reece, P. Mueller, and C. C. Cummins, *J. Am. Chem. Soc. 130*, 9394 (2008).

73. J. C. Peters, J.-P. F. Cherry, C. J. Thomas, L. Baraldo, D. J. Mindiola, W. M. Davis, and C. C. Cummins, *J. Am. Chem. Soc. 121*, 10053 (1999).

74. Y.-C. Tsai and C. C. Cummins, *Inorg. Chim. Acta 345*, 63 (2003).

75. D. J. Mindiola, K. Meyer, J.-P. F. Cherry, T. A. Baker, and C. C. Cummins, *Organometallics 19*, 1622 (2000).

76. E. Solari, C. Da Silva, B. Iacono, J. Hesschenbrouck, C. Rizzoli, R. Scopelliti, and C. Floriani, *Angew. Chem. Int. Ed. 40*, 3907 (2001).

77. Q. Cui, D. G. Musaev, M. Svensson, S. Sieber, and K. Morokuma, *J. Am. Chem. Soc. 117*, 12366 (1995).

78. G. J. Christian, R. Stranger, and B. F. Yates, *Inorg. Chem. 45*, 6851 (2006).

79. T. J. Hebden, R. R. Schrock, M. K. Takase, and P. Muller, *Chem. Commun. 48*, 1851 (2012).

80. A. Duatti, A. Marchi, and R. Pasqualini, *J. Chem. Soc., Dalton Trans.* 3729 (1990).

81. R. Pasqualini and A. Duatti, *J. Chem. Soc., Chem. Commun.* 1354 (1992).

82. F. Mevellec, F. Tisato, F. Refosco, A. Roucoux, N. Noiret, H. Patin, and G. Bandoli, *Inorg. Chem. 41*, 598 (2002).

83. M. J. Abrams, Q. Chen, S. N. Shaikh, and J. Zubieta, *Inorg. Chim. Acta 176*, 11 (1990).

84. J. R. Dilworth, P. Jobanputra, J. R. Miller, S. J. Parrott, Q. Chen, and J. Zubieta, *Polyhedron 12*, 513 (1993).

85. J. R. Dilworth, J. S. Lewis, J. R. Miller, and Y. Zheng, *J. Chem. Soc., Dalton Trans.* 1357 (1995).

86. J. W. Buchler, A. D. Cian, J. Fischer, S. B. Kruppa, and R. Weiss, *Chem. Ber. 123*, 2247 (1990).

87. A.-M. Lebuis and A. L. Beauchamp, *Can. J. Chem. 71*, 441 (1993).

88. F. Demaimay, A. Roucoux, N. Noiret, and H. Patin, *J. Organomet. Chem. 575*, 145 (1999).

89. D. Watanabe, S. Gondo, H. Seino, and Y. Mizobe, *Organometallics 26*, 4909 (2007).

90. J.-P. F. Cherry, P. L. Diaconescu, and C. C. Cummins, *Can. J. Chem. 83*, 302 (2005).

91. M. B. O'Donoghue, W. M. Davis, and R. R. Schrock, *Inorg. Chem. 37*, 5149 (1998).

92. K. H. Horn, N. Bores, N. Lehnert, K. Mersmann, C. Nather, G. Peters, and F. Tuczek, *Inorg. Chem. 44*, 3016 (2005).

93. K. Mersmann, K. H. Horn, N. Bores, N. Lehnert, F. Studt, F. Paulat, G. Peters, I. Ivanovic-Burmazovic, R. van Eldik, and F. Tuczek, *Inorg. Chem. 44*, 3031 (2005).

94. L. Bonomo, E. Solari, R. Scopelliti, and C. Floriani, *Angew. Chem., Int. Ed. 40*, 2529 (2001).

95. S. K.-Y. Leung, J.-S. Huang, N. Zhu, and C.-M. Che, *Inorg. Chem. 42*, 7266 (2003).

96. A. Dreher, K. Mersmann, C. Nather, I. Ivanovic-Burmazovic, R. van Eldik, and F. Tuczek, *Inorg. Chem. 48*, 2078 (2009).

97. S. K.-Y. Leung, J.-S. Huang, J.-L. Liang, C.-M. Che, and Z.-Y. Zhou, *Angew. Chem., Int. Ed. 42*, 340 (2003).

98. A. A. Danopoulos, G. Wilkinson, T. K. N. Sweet, and M. B. Hursthouse, *J. Chem. Soc. Dalton Trans.* 205 (1995).

99. W.-H. Leung, G. Wilkinson, B. Hussain-Bates, and M. B. Hursthouse, *J. Chem. Soc., Dalton Trans.* 2791 (1991).

100. M. Brown and C. Jablonski, *Can. J. Chem. 79*, 463 (1991).

101. J.-S. Huang, S. K.-Y. Leung, K.-K. Cheung, and C.-M. Che, *Chem. Eur. J. 6*, 2971 (2000).

102. B. K. Bennett, S. Lovell, and J. M. Mayer, *J. Am. Chem. Soc. 123*, 4336 (2001).

103. Y. Tanabe, H. Seino, Y. Ishii, and M. Hidai, *J. Am. Chem. Soc. 122*, 1690 (2000).

104. A. L. Odom, C. C. Cummins, and J. D. Protasiewicz, *J. Am. Chem. Soc. 117*, 6613 (1995).

105. A. S. Veige, L. M. Slaughter, P. T. Wolczanski, N. Matsunaga, S. A. Decker, and T. R. Cundari, *J. Am. Chem. Soc. 123*, 6419 (2001).

106. J.-P. F. Cherry, A. R. Johnson, L. M. Baraldo, Y.-C. Tsai, C. C. Cummins, S. V. Kryatov, E. V. Rybak-Akimova, K. B. Capps, C. D. Hoff, C. M. Haar, and S. P. Nolan, *J. Am. Chem. Soc. 123*, 7271 (2001).

107. A. R. Johnson, W. M. Davis, C. C. Cummins, S. Serron, S. P. Nolan, D. G. Musaev, and K. Morokuma, *J. Am. Chem. Soc. 120*, 2071 (1998).

108. J. C. Kim, V. L. Goedken, and B. M. Lee, *Polyhedron 15*, 57 (1996).

109. U. Abram and A. Hagenbach, *Z. Anorg. Allg. Chem. 628*, 1719 (2002).

110. E. Bonfada, U. Abram, and J. Strähle, *Z. Anorg. Allg. Chem. 624*, 757 (1998).

111. M. G. Fickes, A. L. Odom, and C. C. Cummins, *Chem. Commun.* 1993 (1997).

112. R. W. Marshman, J. M. Shusta, S. R. Wilson, and P. A. Shapley, *Organometallics 10*, 1671 (1991).

113. J. C. Kim, B. M. Lee, and J. I. Shin, *Polyhedron 14*, 2145 (1995).

114. J. J. Scepaniak, C. G. Margarit, J. N. Harvey, and J. M. Smith, *Inorg. Chem.* 50, 9508 (2011).

115. R. H. Holm, *Chem. Rev.* 87, 1401 (1987).

116. R. H. Holm and J. P. Donahue, *Polyhedron 12*, 571 (1993).

117. J. P. Donahue, *Chem. Rev.* 106, 4747 (2006).

118. M. J. A. Johnson, P. M. Lee, A. L. Odom, W. M. Davis, and C. C. Cummins, *Angew. Chem., Int. Ed. Engl.* 36, 87 (1997).

119. J. D. Buhr and H. Taube, *Inorg. Chem.* 18, 2208 (1979).

120. D. C. Ware and H. Taube, *Inorg. Chem.* 30, 4598 (1991).

121. K. D. Demadis, T. J. Meyer, and P. S. White, *Inorg. Chem.* 36, 5678 (1997).

122. W.-L. Man, T.-M. Tang, T.-W. Wong, T.-C. Lau, S.-M. Peng, and W.-T. Wong, *J. Am. Chem. Soc. 126*, 478 (2004).

123. W. L. Man, G. Chen, S. M. Yiu, L. Shek, W. Y. Wong, W. T. Wong, and T. C. Lau, *Dalton. Trans. 39*, 11163 (2010).

124. S. B. Seymore and S. N. Brown, *Inorg. Chem. 41*, 462 (2002).

125. L. A. Bottomley and F. L. Neely, *Inorg. Chem. 36*, 5435 (1997).

126. F. L. Neely and L. A. Bottomley, *Inorg. Chem. 36*, 5432 (1997).

127. G. Golubkov and Z. Gross, *J. Am. Chem. Soc. 127*, 3258 (2005).

128. C. J. Chang, D. W. Low, and H. B. Gray, *Inorg. Chem. 36*, 270 (1997).

129. M. Bakir, P. S. White, A. Dovletoglou, and T. J. Meyer, *Inorg. Chem. 30*, 2835 (1991).

130. K. D. Demadis, M. Bakir, B. G. Klesczewski, D. S. Williams, P. S. White, and T. J. Meyer, *Inorg. Chim. Acta 270*, 511 (1998).

131. T.-W. Wong, T.-C. Lau, and W.-T. Wong, *Inorg. Chem. 38*, 6181 (1999).

132. A. Dehestani, W. Kaminsky, and J. M. Mayer, *Inorg. Chem. 42*, 605 (2003).

133. B. Askevold, J. T. Nieto, S. Tussupbayev, M. Diefenbach, E. Herdtweck, M. C. Holthausen, and S. Schneider, *Nat. Chem. 3*, 532 (2011).

134. J. J. Scepaniak, R. P. Bontchev, D. L. Johnson, and J. M. Smith, *Angew. Chem. Int. Ed. 50* (29), 6630 (2011).

135. J. S. Silva and C. C. Cummins, *J. Am. Chem. Soc. 131*, 446 (2009).

136. H.-K. Kwong, W.-L. Man, J. Xiang, W.-T. Wong, and T.-C. Lau, *Inorg. Chem. 48*, 3080 (2009).

137. C.-F. Leung, T.-W. Wong, T.-C. Lau, and W.-T. Wong, *Eur. J. Inorg. Chem. 2005*, 773 (2005).

138. C. Besson, J. H. Mirebeau, S. Renaudineau, S. Roland, S. Blanchard, H. Vezin, C. Courillon, and A. Proust, *Inorg. Chem. 50*, 2501 (2011).

139. M. H. V. Huynh, P. S. White, C. A. Carter, and T. J. Meyer, *Angew. Chem., Int. Ed. 40*, 3037 (2001).

140. T. J. Crevier and J. M. Mayer, *J. Am. Chem. Soc. 120*, 5595 (1998).

141. T. J. Crevier, S. Lovell, J. M. Mayer, A. L. Rheingold, and I. A. Guzei, *J. Am. Chem. Soc. 120*, 6607 (1998).

142. D. S. Williams, T. J. Meyer, and P. S. White, *J. Am. Chem. Soc. 117*, 823 (1995).

143. X.-Y. Yi, T. C. H. Lam, Y.-K. Sau, Q.-F. Zhang, I. D. Williams, and W.-H. Leung, *Inorg. Chem. 46*, 7193 (2007).

144. A. Walstrom, M. Pink, H. Fan, J. Tomaszewski, and K. G. Caulton, *Inorg. Chem. 46*, 7704 (2007).

145. J. Lu and M. J. Clarke, *J. Chem. Soc., Dalton Trans.* 1243 (1992).

146. M. H. V. Huynh, T. J. Meyer, M. A. Hiskey, and D. L. Jameson, *J. Am. Chem. Soc. 126*, 3608 (2004).

147. A. Wu, A. Dehestani, E. Saganic, T. J. Crevier, W. Kaminsky, D. E. Cohen, and J. M. Mayer, *Inorg. Chim. Acta 359*, 2842 (2006).

148. M. H. V. Huynh, P. S. White, and T. J. Meyer, *Inorg. Chem. 39*, 2825 (2000).

149. C.-F. Leung, D. T. Y. Yiu, W.-T. Wong, S.-M. Peng, and T.-C. Lau, *Inorg. Chim. Acta 362*, 3576 (2009).

150. M. Reinel, T. Hoecher, U. Abram, and R. Kirmse, *Z. Anorg. Allg. Chem. 629*, 853 (2003).

151. M. H. V. Huynh, E.-S. El-Samanody, K. D. Demadis, T. J. Meyer, and P. S. White, *J. Am. Chem. Soc. 121*, 1403 (1999).

152. M. H. V. Huynh, E.-S. El-Samanody, K. D. Demadis, P. S. White, and T. J. Meyer, *Inorg. Chem. 39*, 3075 (2000).

153. M. H. V. Huynh, T. J. Meyer, A. Labouriau, D. E. Morris, and P. S. White, *J. Am. Chem. Soc. 125*, 2828 (2003).

154. K. Demadis, E.-S. El-Samanody, T. J. Meyer, and P. S. White, *Inorg. Chem. 37*, 838 (1998).

155. K. D. Demadis, T. J. Meyer, and P. S. White, *Inorg. Chem. 37*, 3610 (1998).

156. M. H. V. Huynh, R. T. Baker, D. L. Jameson, A. Labouriau, and T. J. Meyer, *J. Am. Chem. Soc. 124*, 4580 (2002).

157. M.-H. V. Huynh, P. S. White, and T. J. Meyer, *Angew. Chem., Int. Ed. 39*, 4101 (2000).

158. M. H. V. Huynh, R. T. Baker, D. E. Morris, P. S. White, and T. J. Meyer, *Angew. Chem., Int. Ed. 41*, 3870 (2002).

159. W. L. Man, W. W. Lam, H. K. Kwong, S. M. Peng, W. T. Wong, and T. C. Lau, *Inorg. Chem. 49* (1), 73 (2010).

160. C.-F. Leung, S.-M. Yiu, J. Xiang, and T.-C. Lau, *Chem. Commun. 46*, 7575 (2010).

161. S. N. Brown, *J. Am. Chem. Soc. 121*, 9752 (1999).

162. A. G. Maestri, S. D. Taylor, S. M. Schuck, and S. N. Brown, *Organometallics 23*, 1932 (2004).

163. A. G. Maestri, K. S. Cherry, J. J. Toboni, and S. N. Brown, *J. Am. Chem. Soc. 123*, 7459 (2001).

164. W.-L. Man, W. W. Y. Lam, S.-M. Yiu, T.-C. Lau, and S.-M. Peng, *J. Am. Chem. Soc. 126*, 15336 (2004).

165. M. H. V. Huynh, P. S. White, K. D. John, and T. J. Meyer, *Angew. Chem. Int. Ed. 40*, 4049 (2001).

166. W. L. Man, W. W. Lam, H. K. Kwong, S. M. Yiu, and T. C. Lau, *Angew. Chem. Int. Ed. 51*, 9101 (2012).

167. J. J. Scepaniak, C. S. Vogel, M. M. Khusniyarov, F. W. Heinemann, K. Meyer, and J. M. Smith, *Science 331*, 1049 (2011).

168. J. J. Scepaniak, J. A. Young, R. P. Bontchev, and J. M. Smith, *Angew. Chem. Int. Ed. 48*, 3158 (2009).

169. J. K. Brask, V. Durà-Vilà, P. L. Diaconescu, and C. C. Cummins, *Chem. Commun.* 902 (2002).

170. J. S. Silvia and C. C. Cummins, *J. Am. Chem. Soc. 132*, 2169 (2010).

171. J. Cugny, H. W. Schmalle, T. Fox, O. Blacque, M. Alfonso, and H. Berke, *Eur. J. Inorg. Chem.* 540 (2006).

172. L. H. Doerrer, A. J. Graham, and M. L. H. Green, *J. Chem. Soc., Dalton Trans.* 3941 (1998).

173. A. Hagenbach and U. Abram, *Z. Anorg. Allg. Chem. 631*, 2303 (2005).

174. A. Hagenbach and U. Abram, *Z. Anorg. Allg. Chem. 628*, 31–33 (2002).

175. U. Abram, A. Hagenbach, A. Voigt, and R. Kirmse, *Z. Anorg. Allg. Chem. 627*, 955 (2001).

176. U. Abram, E. Schulz Lang, S. Abram, J. Wegmann, J. R. Dilworth, R. Kirmse, and J. D. Woollins, *J. Chem. Soc., Dalton Trans.* 623 (1997).

177. U. Abram, F. J. Kohl, K. Oefele, W. A. Herrrmann, A. Voigt, and R. Kirmse, *Z. Anorg. Allg. Chem. 624*, 934 (1998).

178. U. Abram, *Z. Anorg. Allg. Chem. 625*, 839 (1999).

179. B. Schmidt-Bruecken and U. Abram, *Z. Anorg. Allg. Chem. 627*, 1714 (2001).

180. S. Verma and M. Hanack, *Z. Anorg. Allg. Chem. 629*, 880 (2003).

181. S. Ritter and U. Abram, *Inorg. Chim. Acta 231*, 245 (1995).

182. K. Frick, U. Ziener, and M. Hanack, *Eur. J. Inorg. Chem.* 1309 (1999).

183. T. J. Crevier and J. M. Mayer, *Angew. Chem., Int. Ed. 37*, 1891 (1998).

184. U. Abram, B. Schmidt-Bruecken, and S. Ritter, *Polyhedron 18*, 831 (1999).

185. J. Zeller, S. Büschel, B. K. H. Reiser, F. Begum, and U. Radius, *Eur. J. Inorg. Chem.* 2037 (2005).

186. S. Ritter, R. Hübener, and U. Abram, *J. Chem. Soc., Chem. Commun.* 2047 (1995).

187. E. L. Sceats, J. S. Figueroa, C. C. Cummins, N. M. Loening, P. Van der Wel, and R. G. Griffin, *Polyhedron 23*, 2751 (2004).

188. D. L. Hughes, S. K. Ibrahim, C. J. Macdonald, H. M. Ali, and C. J. Pickett, *J. Chem. Soc., Chem. Commun.* 1762 (1992).

189. S. A. Fairhurst, D. L. Hughes, S. K. Ibrahim, M.-L. Abasq, J. Talarmin, M. A. Queiros, A. Fonseca, and C. J. Pickett, *J. Chem. Soc., Dalton Trans.* 1973 (1995).

190. J. C. Kim, W. S. Jr. Rees, and V. L. Goedken, *Inorg. Chem. 34*, 2483 (1995).

191. K. R. Powell, P. J. Perez, L. Luan, S. G. Feng, P. S. White, M. Brookhart, and J. L. Templeton, *Organometallics 13*, 1851 (1994).

192. U. Abram, A. Voigt, and R. Kirmse, *Polyhedron 19*, 1741 (2000).

193. W.-H. Leung, E. Y. Y. Chan, T. C. Y. Lai, and W.-T. Wong, *J. Chem. Soc., Dalton Trans.* 51 (2000).

194. A. Walstrom, H. Fan, M. Pink, and K. G. Caulton, *Inorg. Chim. Acta 363*, 633 (2010).

195. R. W. Marshman and P. A. Shapley, *J. Am. Chem. Soc. 112*, 8369 (1990).

196. W.-H. Leung, J. L. C. Chim, I. D. Williams, and W.-T. Wong, *Inorg. Chem. 38*, 3000 (1999).

197. V. Lahootun, J. Karcher, C. Courillon, F. Launay, K. Mijares, E. Maatta, and A. Proust, *Eur. J. Inorg. Chem.* 4899 (2008).

198. L. A. Bottomley and F. L. Neely, *Inorg. Chem. 29*, 1860 (1990).

199. Z.-Y. Li, W.-Y. Yu, C.-M. Che, C.-K. Poon, R.-J. Wang, and T. C. W. Mak, *J. Chem. Soc., Dalton Trans.* 1657 (1992).

200. J. K. Brask, M. G. Fickes, P. Sangtrirutnugul, V. Durà-Vilà, A. L. Odom, and C. C. Cummins, *Chem. Commun.* 1676 (2001).

201. P. Agarwal, N. A. Piro, K. Meyer, P. Muller, and C. C. Cummins, *Angew. Chem. Int. Ed. 46*, 3111 (2007).

202. C. R. Clough, P. Muller, and C. C. Cummins, *Dalton. Trans.* 4458 (2008).

203. D. W. Stephan and G. Erker, *Angew. Chem. Int. Ed. 46*, 46 (2010).

204. S. Seifert, P. Leibnitz, and H. Spies, *Z. Anorg. Allg. Chem. 625*, 1037 (1999).

205. D. V. Yandulov and R. R. Schrock, *J. Am. Chem. Soc. 124*, 6252 (2002).

206. D. V. Yandulov, R. R. Schrock, A. L. Rheingold, C. Ceccarelli, and W. M. Davis, *Inorg. Chem. 42*, 796 (2003).

207. J. Schoeffel, A. Y. Rogachev, S. DeBeer George, and P. Burger, *Angew. Chem., Int. Ed. 48*, 4734 (2009).

208. X. Y. Yi, H. Y. Ng, I. D. Williams, and W. H. Leung, *Inorg. Chem. 50*, 1161 (2011).

209. A. K. M. Long, R. P. Yu, G. H. Timmer, and J. F. Berry, *J. Am. Chem. Soc. 132*, 12228 (2010).

210. G. H. Timmer and J. F. Berry, *Chem. Sci. 3*, 3038 (2012).

211. J. Schoeffel, N. Susnjar, S. Nueckel, D. Sieh, and P. Burger, *Eur. J. Inorg. Chem.* 4911 (2010).

212. C. C. H. Atienza, A. C. Bowman, E. Lobkovsky, and P. J. Chirik, *J. Am. Chem. Soc. 132*, 16343 (2010).

213. D. Sieh, J. Schoffel, and P. Burger, *Dalton. Trans. 40*, 9512 (2011).

214. W. L. Man, W. W. Lam, H. K. Kwong, S. M. Yiu, and T. C. Lau, *Angew. Chem. Int. Ed. 51*, 9101 (2012).

215. C. R. Clough, J. B. Greco, J. S. Figueroa, P. L. Diaconescu, W. M. Davis, and C. C. Cummins, *J. Am. Chem. Soc. 126*, 7742 (2004).

216. S. Sarkar, K. A. Abboud, and A. S. Veige, *J. Am. Chem. Soc. 130*, 16128 (2008).

217. J. S. Silvia and C. C. Cummins, *J. Am. Chem. Soc. 131*, 446 (2009).

218. D. V. Yandulov and R. R. Schrock, *Science 301*, 76 (2003).

219. R. R. Schrock, *Chem. Commun.* 2389 (2003).

220. D. V. Yandulov and R. R. Schrock, *Inorg. Chem.* *44*, 1103 (2005).

221. T. Munisamy and R. R. Schrock, *Dalton. Trans.* *41*, 130 (2012).

222. V. Ritleng, D. V. Yandulov, W. W. Weare, R. R. Schrock, A. S. Hock, and W. M. Davis, *J. Am. Chem. Soc.* *126*, 6150 (2004).

223. M. R. Reithofer, R. R. Schrock, and P. Mueller, *J. Am. Chem. Soc.* *132*, 8349 (2010).

224. W. W. Weare, X. Dai, M. J. Byrnes, J. M. Chin, R. R. Schrock, and P. Muller, *Proc. Natl. Acad. Sci. USA* *103*, 17099 (2006).

225. J. M. Chin, R. R. Schrock, and P. Muller, *Inorg. Chem.* *49*, 7904 (2010).

226. W. W. Weare, R. R. Schrock, A. S. Hock, and P. Mueller, *Inorg. Chem.* *45*, 9185 (2006).

227. R. R. Schrock, *Angew. Chem. Int. Ed.* *47*, 5512 (2008).

228. T. Kupfer and R. R. Schrock, *J. Am. Chem. Soc.* *131*, 12829 (2009).

229. D. V. Yandulov and R. R. Schrock, *Can. J. Chem.* *83*, 341 (2005).

230. N. C. Smythe, R. R. Schrock, P. Mueller, and W. W. Weare, *Inorg. Chem.* *45*, 7111 (2006).

231. R. L. Gdula, M. J. A. Johnson, and N. W. Ockwig, *Inorg. Chem.* *44*, 9140 (2005).

232. M. Bindl, R. Stade, E. K. Heilmann, A. Picot, R. Goddard, and A. Fuerstner, *J. Am. Chem. Soc.* *131*, 9465 (2009).

233. J. Heppekausen, R. Stade, R. Goddard, and A. Fuerstner, *J. Am. Chem. Soc.* *132*, 11045 (2010).

234. A. M. Geyer, R. L. Gdula, E. S. Wiedner, and M. J. A. Johnson, *J. Am. Chem. Soc.* *129*, 3800 (2007).

235. A. M. Geyer, E. S. Wiedner, J. B. Gary, R. L. Gdula, N. C. Kuhlmann, M. J. A. Johnson, B. D. Dunietz, and J. W. Kampf, *J. Am. Chem. Soc.* *130*, 8984 (2008).

236. W. A. Herrmann, S. Bogdanovic, R. Poli, and T. Priermeier, *J. Am. Chem. Soc.* *116*, 4989 (1994).

237. E. S. Wiedner, K. J. Gallagher, M. J. Johnson, and J. W. Kampf, *Inorg. Chem.* *50*, 5936 (2011).

238. H. Braband, E. Oehlke, and U. Abram, *Z. Anorg. Allg. Chem.* *632*, 1051 (2006).

239. A. M. Geyer, M. J. Holland, R. L. Gdula, J. E. Goodman, M. J. A. Johnson, and J. W. Kampf, *J. Organomet. Chem.* *708–709*, 1 (2012).

240. A. D. Finke and J. S. Moore, *Chem. Commun.* *46*, 7939 (2010).

241. N. Svenstrup, A. Bøgevig, R. G. Hazell, and K. A. Jørgensen, *J. Chem. Soc., Perkin 1* 1559 (1999).

242. C.-M. Ho, T.-C. Lau, H.-L. Kwong, and W.-T. Wong, *J. Chem. Soc., Dalton Trans.* 2411 (1999).

243. S.-M. Yiu, W. W. Y. Lam, C.-M. Ho, and T.-C. Lau, *J. Am. Chem. Soc.* *134*, 803 (2007).

244. H.-W. Lam, C.-M. Che, and K.-Y. Wong, *J. Chem. Soc., Dalton Trans.* 1411 (1992).

Subject Index

Absorption spectroscopy, metal-metal
　　bonding, 238–240
Acylating reagents, nitride oxidation,
　　446–447
Adamantyl derivatives, uranium small
　　molecule activation, uranium
　　(III)-carbon dioxide bond
　　cleavage, 345–346
Adduct formation:
　alkyne-uranium(III) small molecule
　　activation, 383
　alkyne-uranium(IV) small molecule
　　activation, 386
Aerobic oxidation, dipalladium
　　catalysts, 283–284
Alkenes:
　dithiolate ligands, tris(dithiolene)
　　chemistry, 8–15
　　carbon disulfide, 10–13
　　elemental sulfur, 9–10
　　miscellaneous sulfur sources, 14–15
　　phosphorus pentasulfide, 13–14
　nitride two-electron reaction, 442–444
　uranium(IV) small molecule activation,
　　insertion chemistry, 385–386
Alkoxides, water-uranium(IV) complexes,
　　hydrolysis and
　　oxidation, 369–370
Alkylating reagents, nitride
　　oxidation, 446–447
Alkynes:
　nitride metathesis, 455–457
　uranium(III) small molecule
　　activation, 380–383
　　adduct coupling, 383
　　reductive coupling, 380–383
　　terminal protonolysis, 383

uranium(IV) small molecule
　　activation, 384–387
　adduct formation, 386
　cycloaddition and metallacycle
　　formation, 386–387
　insertion chemistry, 385–386
　terminal protonolysis, 384–385
Allylic oxidations, dirhodium
　　complexes, 264
Amide ligands, nitride synthesis,
　　448–449
Amines, nitride two-electron
　　reaction, 441–442
Amino acid ligands:
　carbonyl complexes, 209–210
　nitride electrosynthesis, 452
Ammines, nitride synthesis, 448–449
Ammonia:
　nitride catalyst formation, 455
　nitride synthesis and oxidation of, 427,
　　448–449
Ancillary ligands, nitride reactivity,
　　457–458
Antitrigonal prismatic geometry, tris
　　(dithiolene) chemistry,
　　reduced complexes,
　　29–35
Arene dithiolate ligands, tris(dithiolene)
　　chemistry, 5–7
Arenes, uranium small molecule
　　activation, 387–398
　benzyne, 397–398
　dinuclear, inverted sandwich
　　complexes, 390–397
　half-sandwich complexes, 387–390
Arylazides, nitride synthesis, cleavage
　　reactions, 435

Progress in Inorganic Chemistry, Volume 58, First Edition. Edited by Kenneth D. Karlin.
© 2014 John Wiley & Sons, Inc. Published 2014 by John Wiley & Sons, Inc.

Cumulative Index, Volumes 1–58

Progress in Inorganic Chemistry, Volume 58, First Edition. Edited by Kenneth D. Karlin.
© 2014 John Wiley & Sons, Inc. Published 2014 by John Wiley & Sons, Inc.